Energetics Science and Technology

An integrated approach

Online at: https://doi.org/10.1088/978-0-7503-3943-8

Energetics Science and Technology

An integrated approach

Edited by
Adam S Cumming
School of Chemistry, University of Edinburgh, Edinburgh, UK

IOP Publishing, Bristol, UK

ISBN 978-0-7503-3943-8 (ebook)
ISBN 978-0-7503-3941-4 (print)
ISBN 978-0-7503-3944-5 (myPrint)
ISBN 978-0-7503-3942-1 (mobi)

DOI 10.1088/978-0-7503-3943-8

Version: 20221201

IOP ebooks

British Library Cataloguing-in-Publication Data: A catalogue record for this book is available from the British Library.

Published by IOP Publishing, wholly owned by The Institute of Physics, London

IOP Publishing, No.2 The Distillery, Glassfields, Avon Street, Bristol, BS2 0GR, UK

US Office: IOP Publishing, Inc., 190 North Independence Mall West, Suite 601, Philadelphia, PA 19106, USA

Contents

Foreword

This is a book about energetic materials technology. That means explosives, gun and rocket propellants and pyrotechnics. It addresses many aspects of what is a complex scientific area.

Energetics, as they are often called, are sometimes dismissed as a mature technology far less interesting than the other technological aspects that accompany them. Yet without them the desired effects—space launch etc—would not be possible, and their properties can be critical. They are energetic which means that they release energy and indeed are designed to do so, but in a controlled way. This behaviour poses risks, and those risks must be understood. It is important to remember that the safety and performance of energetic materials are inextricably linked.

This field covers a wide range of technological areas. Some of this is comparable to materials technology, but with the added complication of the chemical energy delivery. The energy and the reactivity of the materials makes the design a challenge. To this can be added the complication of time as that reactivity can and does change the material with time so that the risks can change.

This volume attempts to address some of these areas in a way that shows how they are interconnected. It aims to illustrate how those connections can be used to produce better understanding and therefore more effective and safer materials.

Modelling plays an increasingly important part in the development, not just in predicting performance but in also predicting properties such as toxicity, and this is covered. Further novel approaches such as machine learning and the application of artificial intelligence as tools to assist in looking to the future are also discussed.

The technology therefore covers all aspects of ingredients and their characteristics. Understanding and controlling the properties of these building blocks can produce more highly specified materials more efficiently and more cost effectively so that the benefits of the improved properties can be applied. Understanding is extended to the behaviour over time and in contact with other materials as such contact may produce risk. Knowing the risks means understanding both the risk of explosive reaction as well as the possible toxic risk to both biosystems and the environment. Energetic materials are reactive chemicals and so can also be bioactive. This is an area that will continue to develop and needs further attention.

The individual components then need to be assembled into a composite material—one that is well characterised. This introduces processing and here too new methods are changing the way things are done. With the introduction of additive processing this will continue to change.

All these changes affect the choices available for the applications and this is discussed with reference to propellants.

This approach applies to all the forms of energetics as is shown by the chapter on smokes. Equally, disposal is a common requirement with tools needed for both the routine disposal and also the emergency disposal of munitions.

As is mentioned, there are both opportunities and constraints to be both understood and considered. It bears repeating that this is a multidisciplinary topic, from synthesis and basic physics to processing, measurement, modelling and environmental science. These factors produce the need for multidisciplinary teams and that these are forming.

These teams will become more common and as the members learn from each other the science and understanding will improve and the risks reduce. While there is a need for the science there is a need to protect all who come into contact with the products and effects.

This book is a snapshot from world experts of the present state as well as where the technology is likely to go. All the contributors hope that you find it a useful and stimulating compilation.

I am grateful to all the authors who have taken part and contributed to this volume. Many countries are represented—the UK, the USA and Europe as well as India and China—and this reflects the international nature of the research.

All have willingly spent time and energy in preparing their contribution and without them this book would not have been possible. I must also thank the publisher for both agreeing to accept and commission this volume and for support during its production.

Adam Cumming
March 2022

Editor biography

Adam S Cumming

Professor Adam Cumming has over 40 years of experience in energetic materials research. He is an established world expert in the field with current interests in environmental issues and the design of new materials. From 1976–2014, he provided technical advice within the UK Ministry of Defence, where he managed several international research activities. Since 2013, he has been an honorary professor in the School of Chemistry at the University of Edinburgh.

List of contributors

C Alliod
ECMO Team—Laboratoire de Biologie Tissulaire et d'Ingénierie (LBTI)—UMR 5305 CNRS—Université Lyon 1. Institut de Biologie et Chimie des Protéines (IBCP), 7 passage du Vercors, 69367 Lyon, France

Mark Ashcroft
QinetiQ

Zois Boukouvalas
Department of Mathematics and Statistics, American University, Washington, DC 20016, USA

Son Chen
Xi'an Modern Chemistry Research Institute, 710065 Xi'an, China

Peter W Chung
Department of Mechanical Engineering, University of Maryland, College Park, MD 20742, USA

Adam Cumming
School of Chemistry, University of Edinburgh, Edinburgh, UK

Alain Dejeaifve
R and D department, Eurenco Clermont, Engis, Belgium

R Denis
Institut Camille Jordan, UMR 5208 CNRS, Université Lyon 1, 43 boulevard du 11 Novembre 1918, 69622 Villeurbanne Cedex, France

Rowan Dobson
R and D department, Eurenco Clermont, Engis, Belgium

Ruth Doherty
Energetics Technology Center, Indian Head, MD, USA

Eleftheria (Licia) Dossi
Centre for Defence Chemistry Cranfield University Defence Academy of UK Shrivenham, Oxfordshire, SN6 8LA, UK

William Eck
Defense Centers for Public Health—Aberdeen, Aberdeen Proving Ground, MD 21010 USA

Carlos Ferreira
ADAI, Department of Mechanical Engineering, University of Coimbra, Rua Luís Reis Santos, Pólo II, 3030-788 Coimbra, Portugal

G Jacob
Ariane Group, SAS Centre de Recherches du Bouchet, 9 Rue Lavoisier, 91710 Vert le Petit, France

Rajiv Kumar
Department of Space Engineering and Rocketry, Birla Institute of Technology, Mesra 835215 Ranchi, India

Melissa Ladyman
Cranfield Defence Chemistry, Cranfield University at Shrivenham, Shrivenham, UK

R Mendes
Univ. Coimbra, ADAI, LEDAP, Department of Mechanical Engineering, Rua Luís Reis Santos, Pólo II, 3030-788 Coimbra, Portugal

Ning Ma
Xi'an Modern Chemistry Research Institute, 710065 Xi'an, China
and
Shaanxi Normal University, 710119 Xi'an, China

Jared Moretti
U.S. Army, Combat Capabilities Development Command, Armaments Center, Picatinny Arsenal, NJ 07806, USA

Elan Moritz
Energetics Technology Center, Indian Head, MD, USA

Weiqiang Pang
Xi'an Modern Chemistry Research Institute, 710065 Xi'an, China

Ruth Pettifer
QinetiQ

J Pimenta
Univ. Coimbra, ADAI, LEDAP, Department of Mechanical Engineering, Rua Luís Reis Santos, Pólo II, 3030-788 Coimbra, Portugal

William G Proud
Institute of Shock Physics, Imperial College London, Prince Consort Road, London, SW7 2AZ, UK

J Quaresma
Univ. Coimbra, ADAI, LEDAP, Department of Mechanical Engineering, Rua Luís Reis Santos, Pólo II, 3030-788 Coimbra, Portugal

J Ribeiro
Univ. Coimbra, ADAI, LEDAP, Department of Mechanical Engineering, Rua Luís Reis Santos, Pólo II, 3030-788 Coimbra, Portugal

L Rodrigues
Department of Mechanical Engineering, University of Aveiro, Campus Universitário de Santiago, 3810-193 Aveiro, Portugal

Anthony Shaw
U.S. Army, Combat Capabilities Development Command, Armaments Center, Picatinny Arsenal, NJ 07806, USA

Xiaopeng Sun
Xi'an Modern Chemistry Research Institute, 710065 Xi'an, China

Tracey Temple
Cranfield Defence Chemistry, Cranfield University at Shrivenham, Shrivenham, UK

R Terreux
ECMO Team—Laboratoire de Biologie Tissulaire et d'Ingénierie (LBTI)—UMR 5305 CNRS—Université Lyon 1. Institut de Biologie et Chimie des Protéines (IBCP), 7 passage du Vercors, 69367 Lyon, France
and
Institut des Sciences Pharmaceutiques et Biologiques (ISPB), Université Lyon 1, 69373 Lyon, France

Mohan Varma
Department of Space Engineering and Rocketry, Birla Institute of Technology, Mesra 835215, Ranchi, India

William H Wilson
Energetics Technology Center, Indian Head, MD, USA

Zhongyuan Xie
Xi'an Modern Chemistry Research Institute, 710065 Xi'an, China

Rebecca Yates
Frazer-Nash Consultancy

Guangbin Zhang
Shaanxi Normal University, 710119 Xi'an, China

Zhe Zhang
Xi'an Modern Chemistry Research Institute, 710065 Xi'an, China

Chapter 1

Introduction: formulation design—an integrated approach

Adam S Cumming

1.1 Introduction

It has become harder to produce new formulations and compositions cost effectively. Not only are there new constraints to negotiate, but there are also increasing demands on the performance required. This has influenced the approach taken by several groups, but there is a need to take a wider and more systematic view. This has been discussed recently [1] and is being partially applied in many research groups.

The existing constraints of vulnerability and life management have already affected choices and have limited the materials available for application. When environmental impact and toxicology are added to the constraints it is hard to see how any material can negotiate the hurdles placed in its path. The only option is to look at the problem differently. The aim of this book is to illustrate some of these issues and the approaches being taken in research and development.

The growing integration approach will be discussed, and will take into account the performance requirements as well as the constraints provided by vulnerability, life management, and increasingly disposal and other environmental aspects. The level of understanding will be discussed, as will the tools for managing these as far as is possible. This will include discussion of new materials and variants of existing materials; novel concepts for predicting all aspects of behaviour; and the iterative approach to improving these aspects, including machine learning. Finally, the applications of energetics will be discussed.

The approach will cover the advantages and limitations of the approach with discussion of where investment is needed.

Later chapters will explore many of the aspects in much greater detail. Together they present a complex and interconnected field of science. There is a mix of overview and detailed discussion.

doi:10.1088/978-0-7503-3943-8ch1

1.2 Designing formulations for applications

In addition to the constraints mentioned above, there are also new materials and processes to consider. These need careful evaluation and assessment, in part to provide the data to support decision making and predictions.

It would be possible therefore to regard formulation and production as the last step in an integrated process, the step in which the end product is created to meet the requirement in the most cost effective and effective way possible.

Any such approach must build on understanding and be supported by critically assessed data.

1.3 Approach

There is therefore a need to take a systematic view. However, the requirements remain largely the same for modern munitions. These are as follows.

Improved system performance while providing *reduced vulnerability*. This in turn means meeting *insensitive munitions requirements* (STANAG 4439).

Naturally, there is a requirement for *reduced cost of ownership*, which can now be defined as minimising the costs for the whole life cycle from 'cradle to grave' and which also includes longer service life.

There is an increased desire for *reduced environmental impact*, which acts as a further constraint on design options, and which is coupled with a need for *improved ease of disposal*—the so-called 'Design for Disposal' as specified in NATO STANAG 4518. This must all be done at *minimal cost*—this requirement might be better stated as minimal acceptable cost.

Previous reports have attempted to address this issue. One [2] very important one was presented in 2007, entitled 'A new Paradigm for R&D to Implement New Energetic Materials in Munitions', presented by R Behrens and R Swanson.

Their basic conclusions include the two below; both still have relevance.

- *To develop national initiatives to provide new experimental tools.* Develop national initiatives to provide the materials scientist/engineer with the experimental tools needed to design new formulations required to meet system requirements of the future.
- *To explore the concept of a virtual laboratory.* We should explore the concept of a national virtual laboratory enterprise or consortium to bring the scientific and engineering communities together to understand, measure, model, develop, and most importantly predict the actions and interrelationships of complex energetic molecular formulations so that application developments can progress at a faster pace with greater assurances of safety, reliability, and performance.

Since 2007 advances have been made, and a partially integrated approach is now generally used though driven by costs and constraints. It is similar to that being introduced in other technical areas [3].

This is shown in the table below, first published by the Swedish Defence Research Agency (FOI) in Sweden. It captures the developed linear process now often used.

It is important to note that this is not always the approach used. It is also important to note that there is a need for excellence and depth in the various topic areas as well as interconnection and the awareness of mutual dependence.

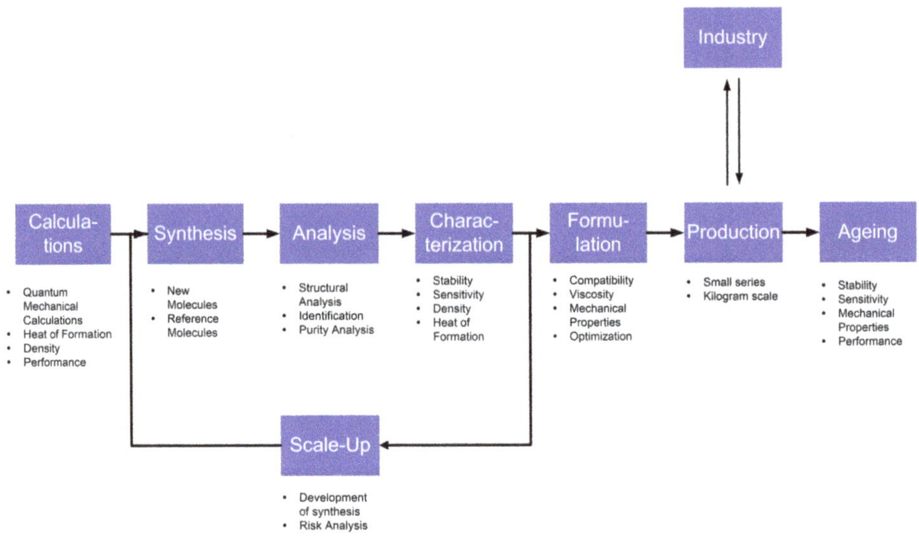

1.4 Summary

The balancing act is driven by the competing demands of required properties and the limitations produced by other constraints.

- Demands
  - Performance
  - Availability
  - Cost
- Constraints
  - Vulnerability [4]
  - Life management [5]
  - Environmental impact [5, 6]
  - Cost

1.5 Methods of approach

This can be expressed as an increasing awareness of the need to be efficient with increasing awareness of the interconnectedness of the science. It produces several challenges as it introduces areas not normally considered and can be seen as a threat to the traditional approach and not an opportunity. It needs to deliver new options rapidly and must include the possibility of responding to changing technology that needs rapid evaluation. However, safety must still drive the approach, and this must now include toxicity in manufacturing, use, and disposal. Prediction of basic properties is becoming feasible [4]. All of this must respect cost and the available resources.

These requirements need a range of approaches, and new materials are only one of these. Thus there is a continuing need to examine options and technology for reengineered materials, that is, traditional materials whose properties can be enhanced; cocrystals form part of this, as do high-quality crystals of existing materials—RS-RDX [7] is the best known of these. It must include optimising of particle quality and both shape and size.

Processing remains key since everything must be capable of manufacture, and new options may provide better solutions than traditional methods. And, finally, the whole life of the material needs consideration at the concept phase to cover its *lifespan*, *disposal*, and *environmental management* [5, 6].

This approach includes modelling in its various forms; it requires fully characterised *ingredients*, that is, solids, binders, and plasticisers, together with the characterisation of trace-level additives for burn rate modification, for example. This characterisation provides confidence in handling and of the quality of the end product. It also eases the approach to *processing* and *manufacturing*, including scale up. Finally, the understanding generated in this combination of characterisation supports reproducible performance effects including *detonation*, *blast*, *metal-driving*; *Platonisation*; *thrust vectoring*, etc, and also in use giving greater confidence in *life management* and *ageing* and assisting in developing and managing *disposal plans*. It should also prove to be cost efficient.

This approach would produce a different diagram indicating the interconnectedness of the process:

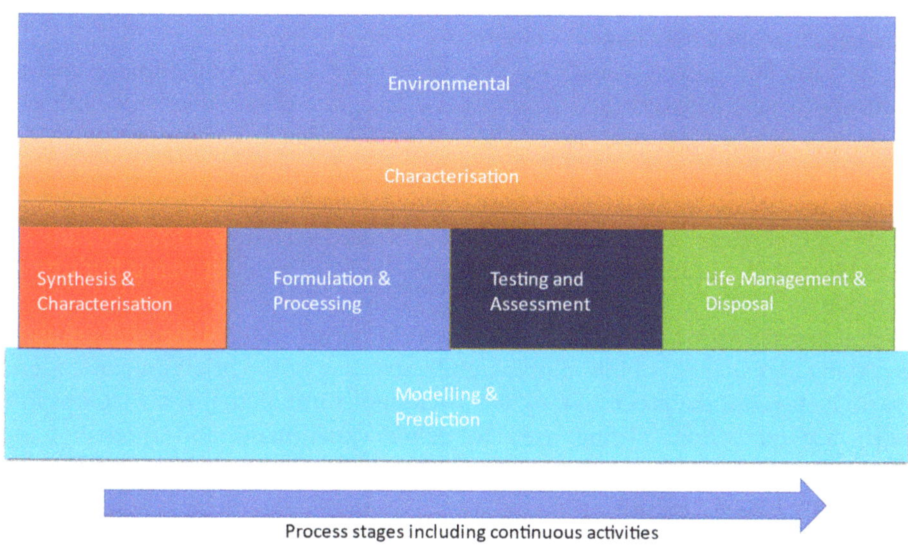

This is a schematic for the approach to designed systems where the links are known and operate iteratively. It is a step further than 'Design for Life'. It is important to emphasise that the interconnected nature of the process means that the process must not only operate in one direction. In other words, this needs to be an iterative approach.

The impact of developing technology and new approaches should assist and promote the following aspects to be designed into systems:

- High and accurately understood performance
- Flexibility to be able to meet future needs
- Low vulnerability
- Managed and long service life
- Reduced environmental impact and ease of disposal
- However, it must include manufacturing and processing.

This necessitates a design toolbox drawing from across the whole technical area.

1.6 Basic properties and management

Can we, and should we be able to:

Design new molecules and formulations? Have enough information and component control to manufacture inexpensively and reproducibly? Specify the component requirements, energetic or otherwise, to meet the requirement in a cost-effective way? And, finally, can we be sufficiently flexible in approach to be able to see new opportunities?

Is this an opportunity for *machine learning*? Can we use AI to simplify the approach and the work needed? [3]

Similarly, can we:

Predict the behaviour of a material or a component, now and after ageing, including environmental impact? Operate a virtual design and testing laboratory with at least 80% effectiveness? Be flexible enough to respond rapidly and cost effectively to new demands and new threats? Integrate them effectively with other components making the best use of the properties of each to produce an optimal system or subsystem with the materials available?

These are goals to be attained and there is significant progress being made worldwide that suggests that some aspects can be used even now. However, any such approach must be tested and tested repeatedly so as to avoid it being limiting and restrictive.

The approach to modelling provides a useful example. Modelling and simulation are playing a greater part in developing understanding of the behaviour of materials in various stages of the process [4, 6], from synthesis through to disposal. Gaps in the models are increasingly being used to define experiments and modelling now includes the prediction of performance, vulnerability, mechanical properties, and, increasingly, toxicity [4, 6].

In addition, the use of integrated chemical models is being used in synthesis to seek candidate molecules with some success. However, it is important to note that models and predictions will need updating with routine validation to maintain their adequate utility.

Similar integrated approaches are developing in other areas, as noted above and in the references.

The actual penultimate stage is the production of a material which fits all the requirements and uses the best available materials and technology to achieve that. Novel methods are being and have always been examined as they may change the approach and allow better materials to be provided more rapidly. There are several being developed at present such as resonance acoustic mixing [8] and additive processing [9]. Others have been looked into in the past but often failed to realise their potential. However, that is not a reason not to examine them as even in failure much is learned that can still be applied, and success will generate dividends and opportunities. To date the two methods mentioned above show very real cost-effective promise.

The general approach can be summarised below.

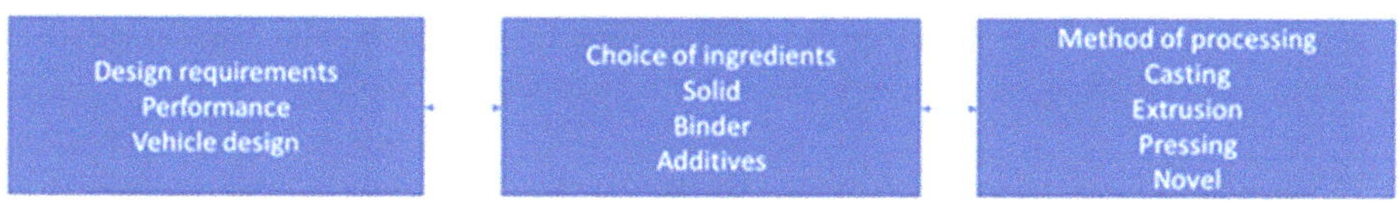

1.7 Conclusions

Formulation is not a single area but is tied to the whole science and technology of energetic materials. This integrated scheme is an approach wherein new concepts can be introduced easily and developed productively.

This integrated approach requires coordination between area specialists and generalists together with awareness of current developments in the field.

New options are presented—will they follow others as being found inadequate? However, as mentioned, they might add to existing capabilities and so must not be dismissed.

Details of the various components are discussed in the chapters that follow. In addition, some novel approaches to applications are discussed. Work is in progress across the world, and this produces new perspectives, and new approaches and solutions. Therefore, some of the following chapters are from groups spread across the globe, including India and China. It remains important to understand that no one group or nation has a monopoly of the truth.

References

[1] Cumming A S 2007 Modern approaches to formulation design and production *Innovative Energetic Materials: Properties, Combustion Performance and Application* ed W Pang, L T DeLuca, A A Gromov and A S Cumming (Berlin: Springer) ch 13 2020

[2] Behrens R and Swanson R L 2007 A new paradigm for R&D to implement new energetic materials in munitions *Insensitive Munitions and Energetic Materials Technology Symp.*

[3] Panchal J H, Choi H-J, Allen J K and McDowell D L 2007 A systems-based approach for integrated design of materials, products and design process chains *J. Comput.-Aided Mater. Des.* **14** 265–93

[4] Michalchuk A A L, Trestman M, Rudić S, Portius P, Fincham P T, Pulham C R and Morrison C A 2019 Predicting the reactivity of energetic materials: an *ab initio* multiphonon approach *J. Mater. Chem. A* **7** 19539–553

[5] Ferreira C, Friere F and Riberio J 2019 Overview of REACH regulation and its implications for the military sector *Energetic Materials and Munitions: Life Cycle Management, Environmental Impact and Demilitarization* (New York: Wiley-VCH) ch 9

[6] Alliod C, Chemelle J-A, Jacob G and Terreux R 2017 Ames test prediction on high energy molecules by On-the-Fly QSAR; *Propellants Explosives and Pyrotechnics* **42** 24–35

[7] Johansen Ø H, Kristiansen J D, Gjersø R, Berg A, Halvorsen T, Smith K -T and Nevstad G O 2008 RDX and HMX with reduced sensitivity towards shock initiation–RS-RDX and RS-HMX *Propellants Explos. Pyrotech.* **33** 20–4

[8] Alexander B 2019 Resonant acoustic mixing (RAM) technology for coated explosive material manufacturing *NDIA IMEMTS (Seville)*

[9] Woods H, Boddorff A, Ewaldz E, Adams Z, Ketcham M, Jang D J, Sinner E, Thadhani N and Brettmann B 2019 Rheological considerations for binder development in direct ink writing of energetic materials *Propellants Explos. Pyrotech.* **45** 26–35

IOP Publishing

Energetics Science and Technology
An integrated approach
Adam S Cumming

Chapter 2

From raw ingredients to energetic materials

Eleftheria (known as Licia) Dossi

2.1 Introduction

The discovery of energetic materials has its roots in ninth century China. While searching ambitiously for the 'elixir of life', Chinese alchemists stumbled upon an unexpectedly explosive mixture of raw materials. It is said that they mixed saltpeter, sulphur, and charcoal, a combination we would later describe as *black powder*, or more commonly *gunpowder*. The recipe was refined and variations of it spread around the world as enemies and allies alike were awed by its power.

From these primitive roots, a multidisciplinary scientific community has emerged, and now works in pristine laboratories to finely tweak the properties of novel energetic materials. The scientists use their understanding of materials to perfect each energetic substance to their chosen specifications by keeping in mind factors such as raw ingredient availability, safety, performance, and costs.

Energetic materials contribute to both military and civilian applications by rapidly releasing large amounts of stored energy [1–4]. Most of their energetic output reflects three mechanisms: (i) the oxidation of their chemical elements, (ii) the high positive heat of formation (ΔH_f) [5], and (iii) high levels of cage or ring strain. High-molecular-weight energetic materials combine the general properties of polymers (i.e. a variety of molecular structures, thermal and chemical stability, and processability) with many energetic groups, and produce large volumes of hot gases by combustion, making them excellent binders for composite explosive formulations [6, 7]. The synthesis and manufacturing of explosives have become well-established processes. Important energetic molecules, such as 2,4,6-trinitrotoluene (TNT), nitroglycerine (NG), cyclotrimethylenetrinitramine (RDX), and pentaerythritol tetranitrate (PETN) [8], are produced worldwide in large amounts and achieve excellent performance. The energetics community continually monitors other manufacturing fields, (e.g. pharmaceuticals, food, and agrochemicals) in order to improve the manufacturing of energetic materials with new and advanced technologies.

New energetic compounds have been synthesised recently to improve the performance of existing explosives, propellants and pyrotechnics, and to meet challenging new requirements such as tailored performance, reduced sensitivity, low vulnerability, high stability, and environmental safety. Future generations of energetic materials must meet these specifications if they are to be accepted and adopted widely for applications such as space propulsion, insensitive munitions (IMs) [8], and gas generator systems.

New protocols for the manufacturing of explosives adopt safety improvements, Registration, Evaluation, Authorisation and Restriction of Chemicals (REACH) [9], and focus on the availability of raw materials, International Traffic in Arms Regulations (ITAR, obsolescence) [10] and process economy. The key objectives include the replacement of toxic raw materials with ingredients from green, sustainable and renewable resources, the reduction of qualification processes and material obsolescence, and the compatibility of manufacturing processes with safety at work, economic feasibility, and a clean environment. It is important to note that the major energetic molecules in today's market are difficult to replace given their excellent performance and widespread availability.

This chapter will summarise the logistics of the manufacturing phase as part of the concept, assessment, demonstration, manufacture, in-service and disposal (CADMID) cycle [11]. Current manufacturing methods for energetic materials and options for the future will be presented in the context of safety (REACH), the availability of raw materials (ITAR, obsolescence) and the economy of explosives manufacturing.

2.2 Manufacturing of energetic materials: a short overview

Manufacturing methods for the preparation of the most common energetics have been in use for 80 years or more. More recently, there has been a significant effort to produce materials to higher specifications, as well as focusing on novel energetic molecules and advanced synthesis and processing methods [12]. Information from the aerospace, agricultural, and pharmaceutical industries has been used to develop novel materials, methods, and processes, especially for large-scale production.

We recognise various motivations for the introduction of new manufacturing processes for energetics, but we must also pay close attention to the extremely important logistics behind the explosives manufacturing process because the hazardous energetic properties of such materials are added to their conventional chemical properties.

In the past 25 years, materials research has identified new explosive molecules that have already been commercialised and exploited in several energetic formulations (FOX-7) and many others developed at the laboratory scale (tetrazoles) or pilot scale (CL-20) [1, 4, 13]. Aerospace companies, space missions, and IM applications seek low-risk, high-performance options, allowing the evaluation of new energetic polymers/binders, offering a variety of structural architectures [14–18].

2.2.1 Manufacturing logistics

Manufacturing logistics involves the management and optimisation of all stages of the process from the selection and purchase of raw materials to the release of a product. We start with the assumption that the manufacturing site has a license for its manufacturing activities. This is dictated from agencies for health and safety at work—the Health and Safety Executive (HSE) in the UK—and applies to all industries, including agriculture, food, textiles, construction, and several other sectors [19].

According to the HSE, '*A licence is a document that allows the holder to store and/or manufacture explosives. The holder of the licence is the person who controls the activities covered by that licence i.e. manufacturing and/or storing the explosives, and complies with the conditions of that licence.*'

Of course, an explosives manufacturing site is subject to health and safety at work rules. The manufacture of energetic substances is a comprehensive subject based on the chemistry and physics of explosives, and the engineering and manufacturing processes. The energetic materials must possess definite and/or innovative qualities to meet the needs of current and new munitions as they emerge, each with their own specific functions and purposes.

The manufacturing process includes several stages of accurate assessments to cover all activities from different perspectives in order to fulfil health and safety standards. All manufacturing processes require an *in-depth* knowledge of the hazardous nature of the process and the materials, including the starting materials, intermediates, desired products, and any side products that might be found as impurities in the reaction mixtures or crude products. Most production processes require the use of highly concentrated acids, which possess corrosive, toxic, and oxidising properties [20]. For a process to be economically feasible and environmentally safe, spent nitrating acids must be recovered and recycled. Nitric and sulphuric acids must be reconcentrated. All organic waste must be destroyed, and the water used for washing must be decontaminated before discharge. Any solvents used in the purification of the explosives must be collected and recycled.

Pictograms describing the hazardous properties of materials used in the explosives manufacturing process are summarised in figure 2.1.

Figure 2.1. Pictograms describing some of the hazardous properties of materials used during the manufacture of explosives. From [21]. Contains public sector information published by the Health and Safety Executive and licensed under the Open Government Licence.

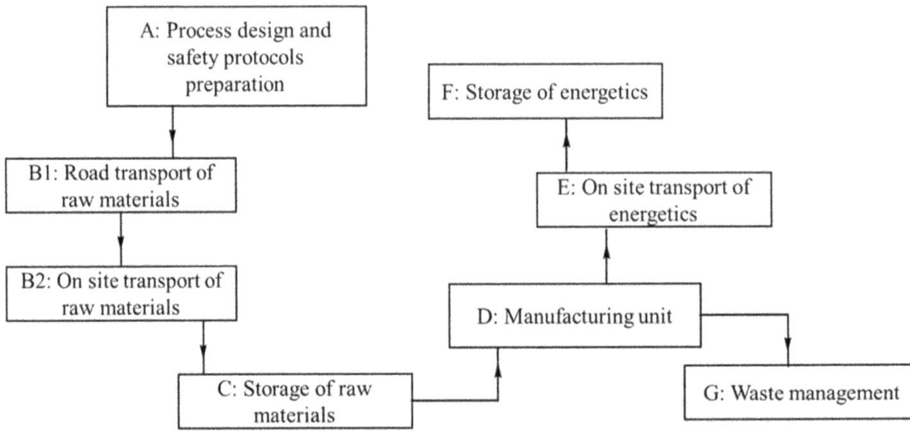

Figure 2.2. Manufacturing logistics for the preparation of energetic materials.

Manufacturing activities are designed and planned before the process begins (figure 2.2). The interdisciplinary nature of the process includes the combined activities of modellers, (industrial) chemists, physicists, plant specialists, and chemical engineers, focusing on the following aspects:

- The relationship between the structure and physicochemical properties of molecular (high) explosive materials and how they are made.
- The stability and hazardous characteristics of the energetic materials and the principles of their ordnance formulations.
- The cost, restricted substance risk, and ecological impact.
- Early-stage multi-scale modelling of the product, to gain confidence and develop new ideas through the systematic analysis and validation of results, thus ensuring consistency across engineering teams.

Once the activities are planned, risk assessments are prepared according to the Health and Safety at Work Act 1974, leading to the authorisation of the manufacturing process [22]. The risk assessments are focused on the safe design, construction and use of infrastructure, equipment, and substances without risks to health when properly used, and on the general duties of employers to their employees, thus ensuring so far as is reasonably practicable the health, safety, and welfare at work of all employees [22].

All manufacturing, storage, transportation, and waste collection procedures must be clearly documented and authorised. Process assessments (i.e. general risk assessment, COSHH, and DSEAR) must be in place, and REACH regulations must be considered before starting the nitration process. Permission to work and the continuous training of personnel offer the safest environment for the manufacture of explosives [23, 24].

2.2.2 Nitration reaction

The main process in the manufacturing of explosives is the nitration reaction, which means the introduction of energetic nitro groups (–NO_2) into inert, commercially available substrates to create energetic compounds (figure 2.3) [8, 20].

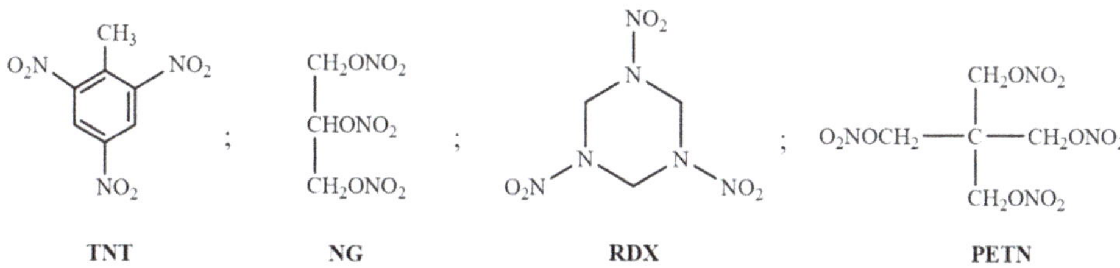

Figure 2.3. Chemical structures of commonly used secondary explosives. TNT = 2,4,6-trinitrotoluene, NG = nitroglycerine, RDX = cyclotrimethylenetrinitramine, PETN = pentaerythritol tetranitrate.

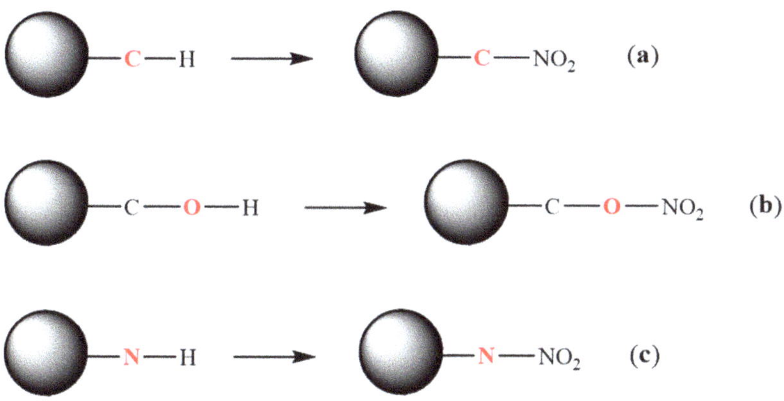

Figure 2.4. Types of nitration reaction producing (a) C-nitro compounds, (b) O-nitro compounds, and (c) N-nitro compounds. The general nitration reaction is *fuel molecule + nitric acid → energetic molecule + water.*

Concentrated nitric acid (HNO3) is the main ingredient for the nitration of fuel molecules such as toluene, glycerol, hexamethylenetetramine, and penthaerythritol to produce TNT, NG, RDX, and PETN, respectively. The $-NO_2$ group can be added to carbon, nitrogen, or oxygen atoms in the fuel molecule, generating C-nitro compounds such as TNT, N-nitro compounds such as RDX, and O-nitro compounds such as NG, respectively (figure 2.4) [8]. The different nitration systems are summarised in table 2.1.

RDX can be manufactured with HNO_3 as the only nitrating agent. Nitration occurs via the Woolwich process, and involves a complex reaction sequence between $(CH_2)_6N_4$ (hexamethylenetetramine, known as hexamine) and HNO_3.

The most common nitration system is a mixture of HNO_3 and sulphuric acid (H_2SO_4). The ratio of the mixture depends on how difficult it is to nitrate the inert substrate, and the degree of nitration required. For the manufacture of N-nitro materials such as RDX and 1,3,5,7-tetranitro-1,3,5,7-tetrazocane (HMX) via the Bachman process, the presence of sulphuric acid would cause product degradation, so acetic acid (CH_3COOH) is used instead, and the anhydride of the acid removes water liberated during the process. The inclusion of ammonium nitrate with nitric acid (the German process) improves the efficiency of hexamine nitration, reduces the amount of HNO_3 required, and achieves a higher yield of RDX. This converts six moles of $(CH_2)_6N_4$, four moles of HNO_3, and two moles of NH_4NO_3 into two moles

Table 2.1. Common nitrating systems used to manufacture energetic compounds.

Nitrating agent	Chemical formula	Byproducts	Features
Nitric acid	HNO_3	Dilute HNO_3	Low nitrating strength Limited applicability Recycling problems
Mixture of nitric acid and sulphuric acid	$HNO_3 + H_2SO_4$ (oleum[1])	Dilute mixed acids	High nitrating strength Wide substrate choice Recycling problems
Potassium nitrate and nitric acid	$KNO_3 + H_2SO_4$ (oleum[1])	Dilute mixed acids	Moderate nitrating strength Limited substrate choice Recycling issues Less expensive
Nitric acid and acetic anhydride	$HNO_3 + (CH_3COO)_2O$	Acetyl nitrate Acetic acid	Moderate nitrating strength Limited substrate choice Hazardous (detonable)
Dinitrogen pentoxide and nitric acid	$N_2O_5 + HNO_3$	Concentrated HNO_3	High nitrating strength Wide substrate choice Recyclable
Dinitrogen pentoxide and dichloromethane	$N_2O_5 + CH_2Cl_2$	Potentially none	Moderate nitrating strength Wide substrate choice Clean reactions

[1] Oleum is a solution of sulphur trioxide (SO_3) in concentrated sulphuric acid (H_2SO_4).

of RDX and six moles of water. Finally, the manufacturing of RDX and HMX via the Bachman process retains the nitric acid and ammonium nitrate but replaces sulphuric acid with acetic acid or mixtures of CH_3COOH and its anhydride.

An alternative and less expensive nitration method uses a mixture of nitrate salt in sulphuric acid (table 2.1). The amount of water is reduced by the replacement of nitric acid with the solid nitrate salt. This advantage has been overtaken by the ready availability of 'anhydrous' nitric acid in modern plants, and the process is now rarely used [20].

Dinitrogen pentoxide (N_2O_5), an environmentally friendly compound, has been investigated as an alternative nitrating agent for organic substrates (table 2.1). N_2O_5 is the anhydride of nitric acid, and it was first used for the nitration of cellulose. The oxide may undergo heterolysis to form ionic materials in solution (figure 2.5). N_2O_5 can be produced by electrogeneration in a polar solvent. Recent advances in the

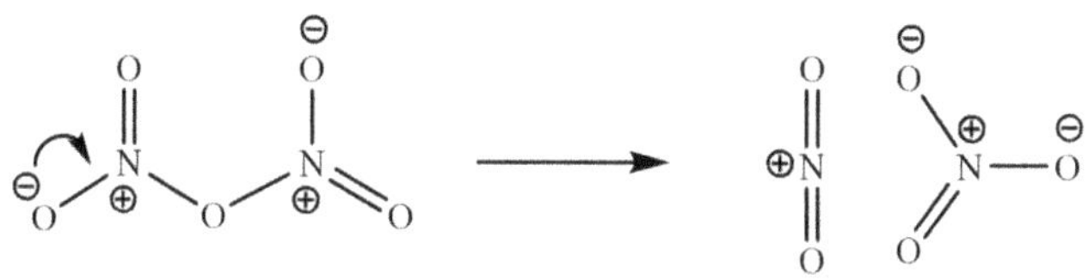

Figure 2.5. The heterolysis of dinitrogen pentoxide.

ozonolysis of N_2O_4 to prepare N_2O_5 suggest that it may soon be readily available for the production of explosives and other nitrated materials.

N_2O_5 is thermally unstable and must be stored at temperatures below −20 °C or used *in situ*. Its decomposition produces highly toxic nitrogen dioxide gas (NO_2) and oxygen gas (O_2). N_2O_5 in HNO_3 is used for the synthesis of nitramines, whereas N_2O_5 in dichloromethane (or chloroform) is used as a two-phase system for the manufacture of poly(NIMMO) and poly(GLYN), two common energetic binders (table 2.1). This nitrating medium allows acid-sensitive substrates to be nitrated selectively. The resulting mono-nitrate esters, NIMMO and GLYN, are then polymerised to produce the desired energetic binder.

Many other methods can introduce nitro groups into organic molecules, but they are often of little practical value due to the high relative cost of the materials. In practice, the utilisation of mixed acids is the predominant nitration method due to the relatively low cost of the readily available starting materials. As the use of HMX and RDX increases relative to that of TNT, the mixed acid nitration method is being replaced by more environmentally friendly systems based on N_2O_5.

All nitration reactions are exothermic, so the temperature must be controlled to avoid *thermal runaway* or *fume off* during the manufacturing process. Heat release must be monitored even for small (<5 g) laboratory-scale operations. In larger-scale processes, more efficient cooling in the reactor is required and this is achieved by using internal heat exchange coils/pipes. For industrial-scale processes, efficient stirring is necessary for effective heat dissipation to avoid thermal runaway. Dump valves are still used in medium-scale and large-scale batch reactions for the synthesis of nitrate esters, where they collect and desensitise the reaction mixture if there is a system fault [20].

2.2.3 Batch process

Two major processing modes can be used in a chemical manufacturing plant: batch processing and continuous processing. A batch process is a *discrete* process that consists of a sequence of steps performed in a defined order and requires starting materials that are introduced before processing begins ($t = 0$; figure 2.6) and a product that is discharged at the end of the process (batch product; figure 2.6). The manufacturing plant can then be cleaned and readied for the next batch of materials. The *unit explosives risk* is defined as the quantity of explosives present in the reaction vessel at any time during the process. In a batch operation, this is the whole materials charge.

Batch processes are typically used to manufacture higher-value products, such as specialty chemicals, pharmaceuticals, and structured products. The advantage of a

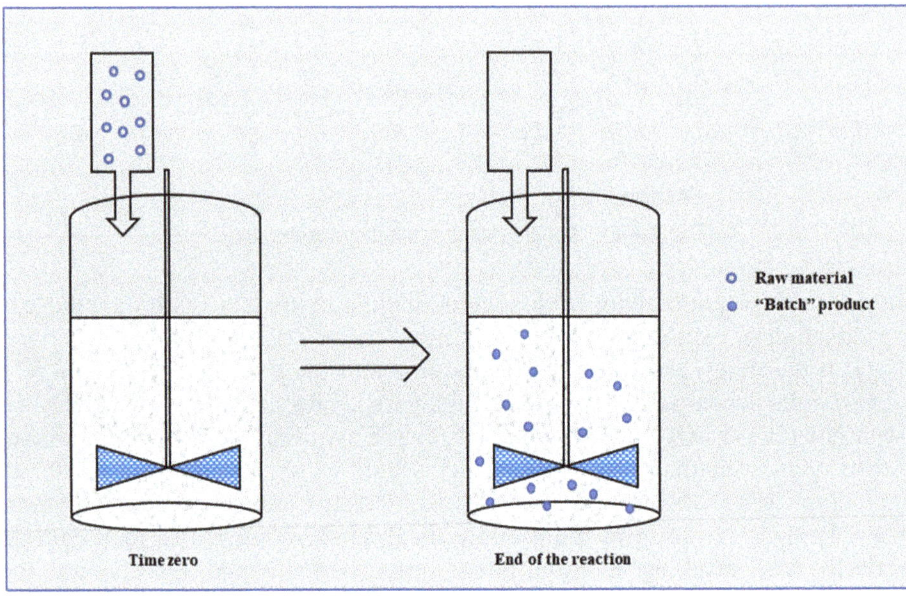

Figure 2.6. Schematic representation of a batch process.

batch process is its versatility and the control of the thermal regime, especially for exothermic reactions with safety implications [25].

There are several economic and technical factors that make batch processes favourable over continuous processes [26]:

- Batch processes often consist of simple processing units such as mixers and stirrers.
- A batch processing unit can be used in several process phases and can also be used to support multi-product manufacturing within the facility.
- Batch manufacturing plants are comparatively more robust than continuous plants.

2.2.4 Continuous process

In a continuous process, the starting materials are fed into the system and product is removed simultaneously throughout the process (figure 2.7) [27]. The daily throughput of a continuous plant tends to be larger than that of a batch plant of the same size. The main difference in the two operations is the *unit explosives risk*. As stated above, this is the whole materials charge in a batch process, but the quantity of explosive products in the reactor at any instant is relatively small in a continuous process (because the product is removed as soon as it is formed), and the unit risk is considerably lower. Continuous processing is ideal for large work orders because the chemical is produced constantly until the order is filled. Other advantages include savings in the use and storage of acids, less waste, quality improvements because it is easier to identify and correct errors, increased productivity, and easier adaptation to the needs of customers.

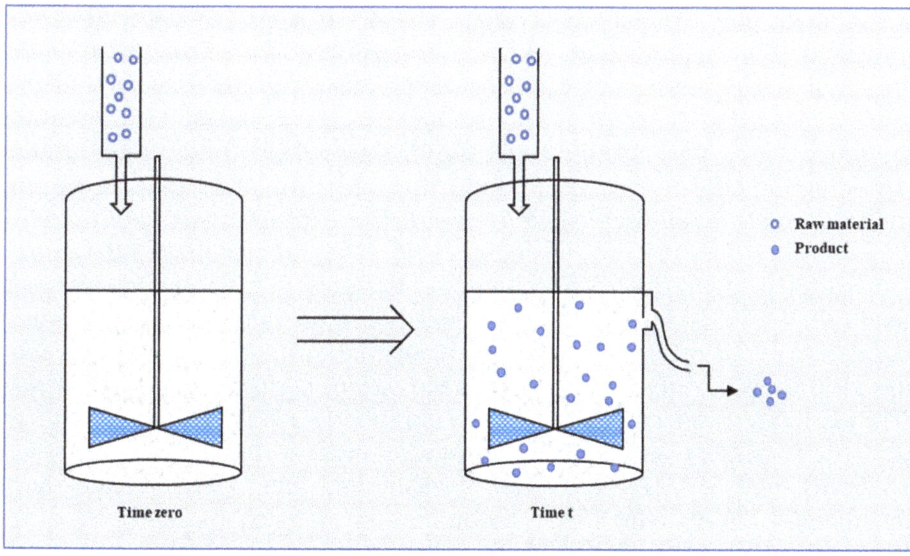

Figure 2.7. Schematic representation of a continuous process.

Several factors can influence the choice between batch and continuous manufacturing, but the main parameters are safety (exothermic reactions, unit explosives risk) and process economy. Small batches (gram scale) of explosives can be made in the laboratory using batch or flow reactors (see below). At the pilot-plant scale, batch or flow processes are used for multi-kilogram production runs or for the demonstration of new technologies or processes [28]. Continuous production can address some of the safety issues raised by exothermic reactions because the unit explosives risk is lower and heat transfer is more efficient. Although the up-front investment costs are likely to be higher for a continuous process compared to a batch process [29], a continuous process usually achieves greater productivity than a batch process [30].

2.2.5 Flow chemistry

Flow chemistry is a series of chemical processes that occur in a continuous flowing stream. This strategy is widely used in the chemical industry because it improves overall process selectivity, efficiency. and safety. More recently, this technology has also been adopted by the pharmaceutical and fine chemical industries. Heat transfer is more efficient during a flow process, resulting in rapid reaction mixing with precise control over the reaction parameters, which is beneficial for exothermic reactions. Further advantages of flow chemistry for nitration reactions include reproducibility, scalability, and compatibility with automation [31]. Flow nitration is therefore superior for the synthesis of explosives because the instability of the reaction mixtures means that the scaling-up of batch reactors can be hazardous [32, 33].

Flow reaction equipment includes two or more piston, peristaltic, syringe or gear centrifugal pumps that are used to deliver predefined and reproducible amounts of solvents and reagents [34]. The reagents are introduced in small volumes, the streams are mixed at a T-piece, and the mixture is directed to a coil reactor that provides the residence time for the reaction. Alternatively, a column reactor can be packed with solid reagents and catalysts. The temperature of the reactor is controlled for the duration of the flow process while a back pressure regulator controls the pressure of the system. The reaction mixture is collected at the outlet in a solvent and the high-quality product is recovered.

As stated above, flow chemistry increases the accessibility of nitration reactions by avoiding the need to store and handle toxic, reactive, and explosive reagents. Various attempts have been made to synthesise explosives and plasticizers in laboratory-scale flow reactors, including TNT [35], 2,6-diamino-3,5-dinitropyrazine-1-oxide (LLM-105) [36], 5-substituted 1H-tetrazoles [37, 38], and nitroethylnitramines (NENAs) [38]. However, flow chemistry is most suitable for highly soluble explosives to avoid precipitation and accumulation in the column reactor, and it seems ideal for the synthesis of low-molecular-weight plasticizers such as NENAs.

The feasibility of flow chemistry for the nitration of 2,4-DNT to 2,4,6-TNT was recently demonstrated, and the main advantages included the use of safer reagents (98% H_2SO_4 and 65% HNO_3 instead of oleum and fuming HNO_3) and shorter reaction times (20–30 min) [35]. The authors investigated the influence of key parameters such as the HNO_3/DNT molar ratio, residence time, and process temperature, achieving a >99% conversion rate from 2,4-DNT to TNT in only 20 min, compared to <58% for a comparable batch reaction.

LLM-105 was synthesised using a commercially available flow micro-reactor system [36]. The optimised flow nitration conditions were reported, along with multi-gram scale-up, the analysis of sample purity, and quenching conditions for purity and crystal morphology. However, the process yield was no higher than the typical 50%–60% achieved in a batch process.

In contrast, excellent yields of 5-substituted 1H-tetrazoles (68%–98%) were achieved in a reaction time of 10–15 min using a flow process in which HN_3 was generated *in situ* from NaN_3 and acetic acid, and then reacted with nitrile at 220 °C–260 °C and 36 bar overpressure [37]. The reaction stream was cooled in a heat exchanger and residual HN_3 decomposed by quenching with aqueous $NaNO_2$. Recently, another efficient flow process for the synthesis of 5-substituted 1H-tetrazoles was described, involving a reaction between organic nitriles and a polymer-supported triorganotin azide [38].

2.3 Classes of energetics

2.3.1 Nitro compounds

This class of energetic compounds includes 2,4,6-trinitrotoluene (TNT) and the related molecules hexanitrostilbene (HNS), 1,3,5-triamino-2,4,6-trinitrobenzene (TATB), and 2,4-dinitroanisole (DNAN) (figure 2.8).

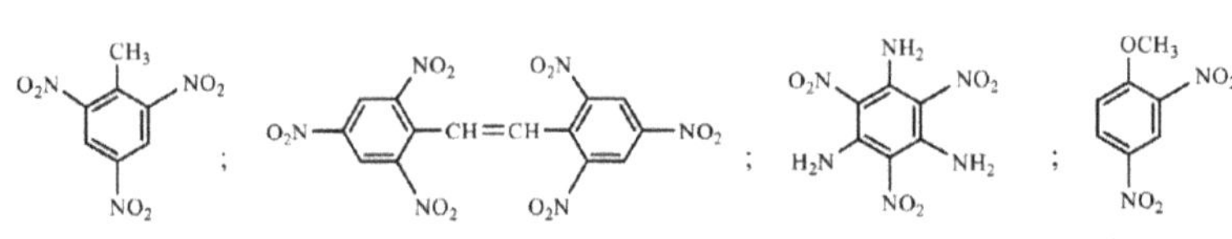

Figure 2.8. The chemical structures of TNT, HNS, TATB, and DNAN.

TNT is generally manufactured in continuous production plants using a mixture of concentrated nitric and sulphuric acids and toluene [39, 40]. The introduction of nitro groups into the benzene ring of toluene requires increasingly more vigorous conditions (high temperature, absence of water) as the number of nitro groups increases from mononitrotoluene (MNT) to dinitrotoluene (DNT) and finally TNT (figure 2.9). This is achieved in modern continuous process plants by having up to 21 reactors arranged in series, with the organic feed flowing in one direction and the mixed acid flowing in the opposite direction.

The fresh nitrating mixture is introduced at one end of the plant and becomes weaker as it flows through the reactors due to the formation of H2O, while crude TNT is collected from the last reactor. In this way, the acid concentration is lowest where the toluene is introduced, reducing the heat liberation and thus the risk of fire. The formation of MNT is the most exothermic step and sufficient heat can be liberated by this process to ignite the toluene. The conversion of toluene to TNT therefore occurs in two steps: the conversion of toluene to MNT and the subsequent conversion of MNT to TNT. The acids diluted by the production of water during the nitration reaction are separated from the organic materials, purified, and reused (figure 2.10). The process also generates a small proportion (4%–5% w/w) of unwanted nitrated products that act as impurities, lowering the melting point of

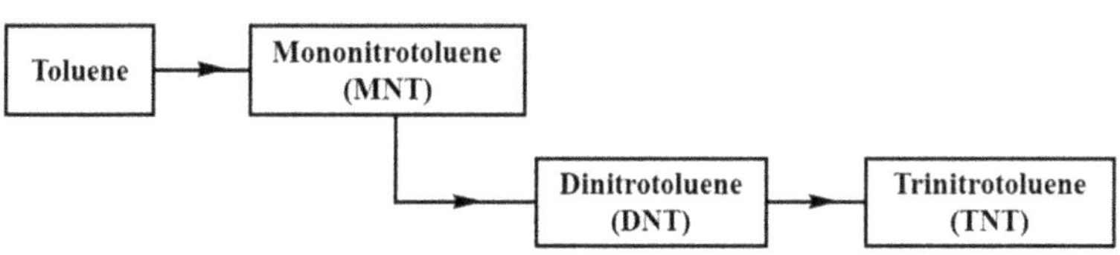

Figure 2.9. Conversion of toluene to TNT.

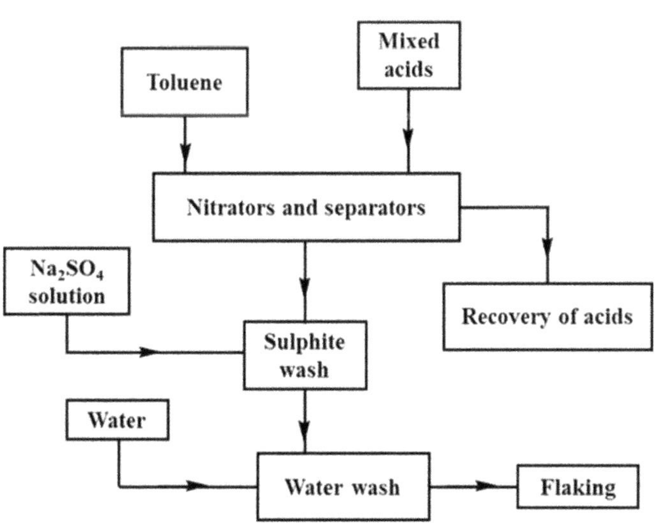

Figure 2.10. Flow diagram for the manufacture of TNT.

the TNT. These are removed by sulphite washing. Further washing with water removes any remaining salts or acids, and the TNT can then be cooled until it solidifies, allowing its collection often in the form of TNT flakes.

The TNT production wastewater (*red water*) is an important environmental pollution and public health concern because many of the components are toxic and mutagenic even at low concentrations. Ground contamination is still present in very old plants. Common treatment methods for TNT wastewater currently include activated carbon adsorption and evaporation to dryness (the toxic components are not volatile), followed by incineration, biodegradation, and other disposal methods [41].

Toluene is classified as highly toxic and teratogenic [42]. TNT produces typical nitro symptoms but also others (pale face, grey lips, stomach ache, constipation, and flatulence). Inhalation of TNT dust, ingestion, or absorption through the skin can cause cyanosis, liver and kidney damage, and even death. Skin absorption is enhanced if the skin is oily or sweaty. TNT is toxic to aquatic life, with long-lasting effects [43].

HNS is a heat-resistant explosive with a melting point of 316 °C–320 °C. It is relatively insensitive to impact and friction. HNS is manufactured by the oxidative coupling of TNT using the Schipp process [44] (figure 2.11) and a solution of sodium hypochlorite as the oxidising agent. The process produces impure HNS containing up to 6% (w/w) hexanitrobibenzyl (HNBiB) and unreacted TNT. All raw materials, byproducts, solvents, and the product are toxic and/or corrosive (table 2.1).

Impurities are removed by multiple washes, resulting in a recrystallised material (HNS-II) with a new geometry and a larger particle size, which is undesirable for many applications. The recrystallisation of HNS-I to produce HNS-II improves the bulk density of the material and can be achieved using a single solvent or a binary system. Current manufacturers produce all four types of HNS:

- HNS-I—for initiation (WS 5003F specifications): more stable than stipulated in the military specifications thanks to a special purification process.
- HNS-II—for oil and gas drilling: a special quality of HNS, coated with a high-temperature resistant polymer, is produced according to WS 5003F specifications.
- HNS-III—for the nucleation of TNT, produced under German and USA specifications.
- HNS-IV—for exploding foil initiator detonators: a form of HNS characterised by a higher sensitivity and thus a more reliable initiation [45].

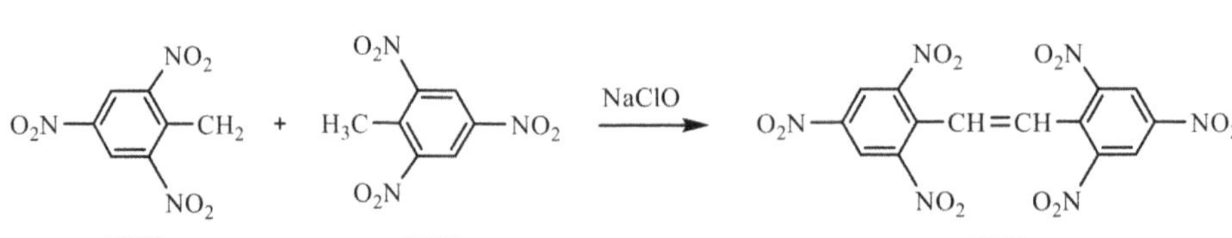

Figure 2.11. The synthesis of HNS from TNT.

TATB is a heat-stable explosive with low sensitivity and very low solubility in common solvents, which affects its processability. It is manufactured using the two-step Benziger process, in which 1,3,5-trichlorobenzene (TCB) is first nitrated using a mixture 30% (v/v) oleum and fuming nitric acid to produce the intermediate compound 1,3,5-trichloro-2,4,6-trinitrobenzene (TCTNB). The latter is subsequently aminated to produce the final product (figure 2.12).

The trinitration reaction is kinetically slow and the whole nitration process takes ~4 h to complete, and is accompanied by the formation of two byproducts (products 1 and 2; figure 2.13). At the end of the first step, the reaction mixture is cooled and diluted and the intermediate product filtered off. The spent acid is reconcentrated. The intermediate TCTNB and byproducts 1 and 2 are insensitive when wet, and even in the dry state their figure of insensitiveness (FoI) is 169.

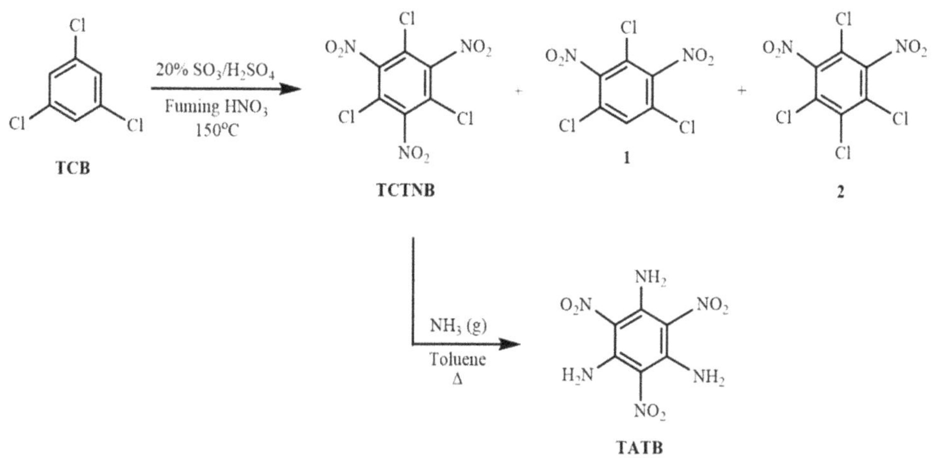

Figure 2.12. Synthesis of 1,3,5-triamino-2,4,6-trinitrobenzene (TATB) from trichloro-trinitrobenzene (TBC).

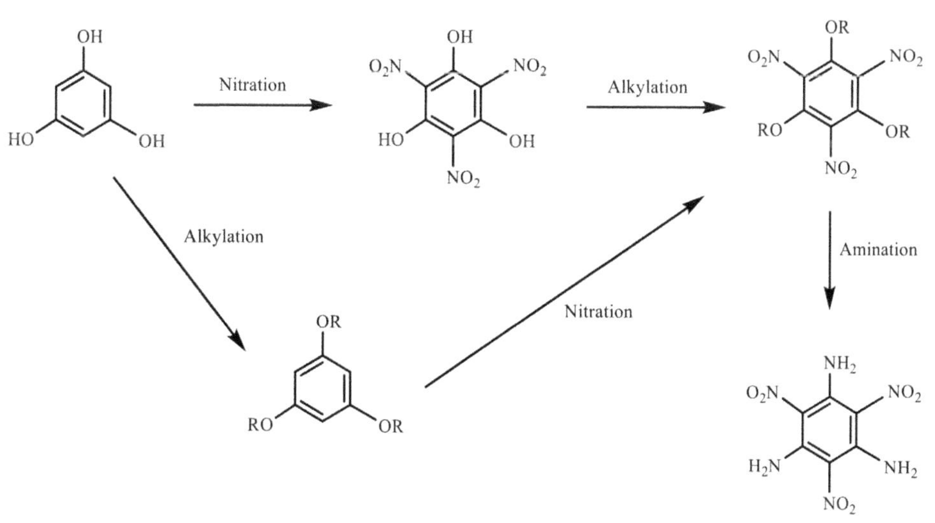

Figure 2.13. Halogen-free synthetic route to 1,3,5-triamino-2,4,6-trinitrobenzene (TATB).

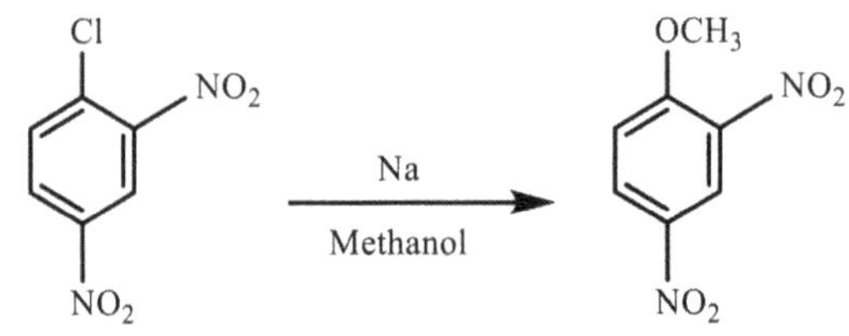

Figure 2.14. Synthesis of DNAN.

The washed TCTNB is dissolved in toluene and transferred to the reaction vessel. Further toluene and water are added, and the vessel is sealed and heated. Gaseous ammonia is added, and the amination reaction proceeds at 152 °C and 5 bar pressure for 4–6 h. The TATB is recovered, washed with water, steamed in a press, washed again, partly dried with compressed air, and finally blended and packaged. The solvent is recovered by distillation.

New routes for TATB synthesis have been investigated due to the toxicity of TBC and related compounds, as well as high costs and ITAR issues. An alternative, halogen-free synthetic route to TATB has been reported with three steps (figure 2.13) [46, 47]. First, $NaNO_2$ in HNO_3 or with N_2O_5 in H_2SO_4 mixtures are used to nitrate phloroglucinol, a natural glycoside derivative. The alkylation step is induced by K_2CO_3 dimethyl sulphate or a trialkylorthoformate (typically trimethyl orthoformate or triethyl orthoformate). Finally, amination is achieved by mixing with NH_3 in a solvent. On a small scale, it is convenient to pass ammonia gas into a solution in an organic solvent at low temperature, but on a larger scale the use of liquid ammonia is preferred, either at −33 °C (the boiling point of liquid ammonia) or under a pressure of 0.8–0.9 MPa at ambient temperature. The solid substrate can be added directly to the liquid ammonia, or as a solution in an organic solvent. The type of TATB produced depends on the conditions during the amination step [48] and whether or not water and/or emulsifier are present during the reaction.

DNAN is synthesised with yields of up to 97% by the methoxylation (etherification) of 1-chloro-2,4-dinitrobenzene (DNCB) (figure 2.14), an important chemical intermediate used in the production of *sulphur black* and many other dyes. In animal studies, DNAN is a mild ocular and skin irritant with a significant potential for dermal absorption. In humans, DNAN was a component of louse powder (prior to the development of DDT) with no reported safety concerns. However, its handling and use as a munition component presents a potential occupational hazard via inhalation and dermal exposure routes. The recommended workplace environmental exposure limit for DNAN is 0.1 mg m^{-2} 8 h time-weighted average [49, 50].

2.3.2 Nitroesters

This class of energetic compounds is produced by the nitration of one or more oxygen atoms. Common examples include pentaerythritol tetranitrate (PETN), nitroglycerine (NG), and nitrocellulose (NC) (figure 2.15). Nitroesters are

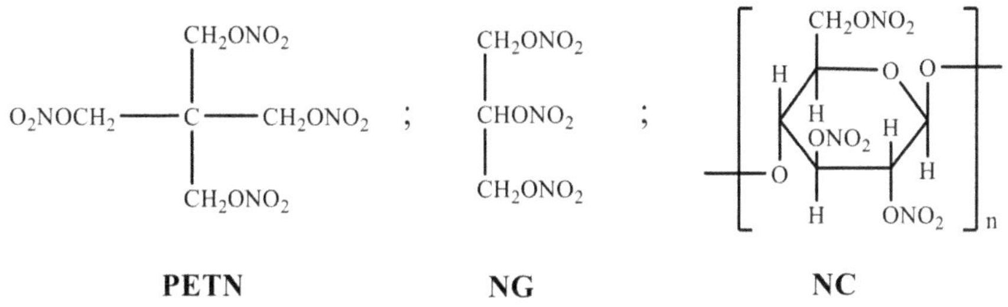

Figure 2.15. Chemical structures of PETN, NG, and NC.

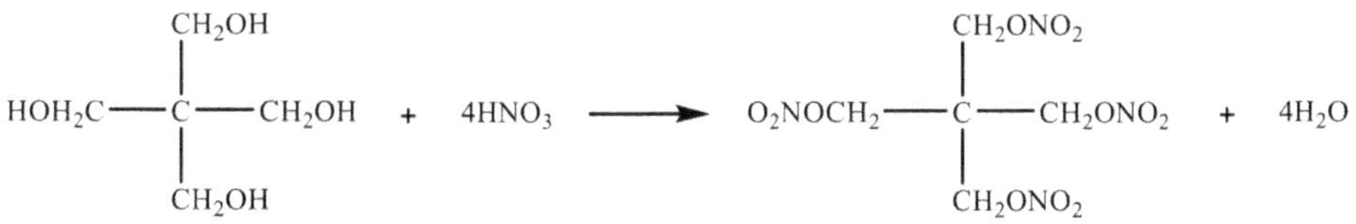

Figure 2.16. Synthesis of PETN.

characterised by a higher oxygen balance, heat of explosion, and sensitivity than nitro compounds and nitramines, but a lower stability [8]. The manufacture of NG is probably the most dangerous large-scale process for the preparation of explosive materials.

PETN is prepared by the nitration of pentaerythritol using 98%–99% (v/v) HNO_3 and is the only nitrate ester currently produced using nitric acid alone (figure 2.16). The exothermic reaction is typically carried out as a continuous process maintained at a temperature below 16 °C. Crude PETN containing less-nitrated analogues is separated from the spent acid by adding 30%–80% (v/v) water, reducing the solubility of the desired product and stabilising the mixture.

The crude product is washed with water and then with aqueous acetone-ammonia to remove acids and neutralise the residues. Specific particle sizes are achieved by recrystallisation from acetone/water mixtures at controlled stirring rates and temperatures. The PETN crystals are phlegmatized with 10% wax in water at 85 °C to reduce sensitivity and facilitate pressing.

PETN is classified as flammable and explosive and is used in small quantities with other nitroesters in pharmaceutical formulations [51]. The nitroester (nitrate) drugs are converted to nitrite in the body; nitrate therapy has proven beneficial for the treatment of coronary artery diseases [52].

NG has a low FoI and is highly susceptible to accidental initiation. It is manufactured by the continuous nitration of glycerol, and manufacturing plants were formerly constructed on natural or man-made hills to avoid the need for any form of pumping to move the reactants and intermediates around. In modern plants, the same is often achieved by pressurising the starting materials. The two main manufacturing processes are the Biazzi process and the Nilssen and Brunnberg (NAB) injector process. These differ mainly in the strategy used to mix the acids with glycerol (figure 2.17), and the size and geometry of the nitrator [20].

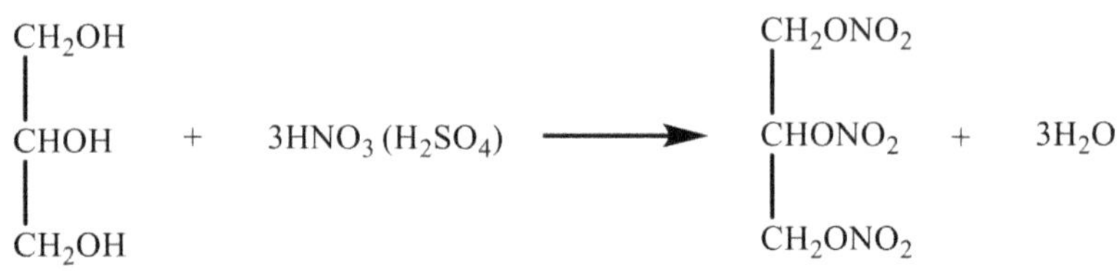

Figure 2.17. Nitration of glycerol to produce NG.

Table 2.2. Reaction conditions for glycerol nitration.

	Reaction conditions[1]				Product		
Process	HNO_3[2]	H_2SO_4[2]	H_2O	Acids/ glycerol	Unit explosion risk	Production (kg h^{-1})	Yield (%)
Biazzi	51	49	0	4.85	80–120 kg h[–3]	2000–3000	96
NAB	27	63	10	12.1	3.5 kg h[–3]	500–650	94

[1] At 10 °C–15 °C.
[2] Plus 98% HNO_3 and 98% H_2SO_4.
[3] NG amount 'at risk' at any one time (amount in separator).

The acid composition and substrate/acid ratio therefore differ between the two processes, but they both achieve similar high yields (table 2.2). To ensure safety, the Biazzi process features an automatic shutdown of the feed and dump facilities, with the materials highly diluted in water in the event of excess temperature in the nitrator or separator. The NAB injector process instead has a failure-safe shutdown operated by the closing off (or pressure reduction) of the mixed acids at the injector, which greatly reduces the quantity of reagents entering the nitrator (figure 2.18).

In the Biazzi process, the reaction vessel is cooled to <15 °C and the nitrated product overflows into a separator (figure 2.18), away from the spent acids. The NAB injector process allows the reaction mixture to reach ~45 °C for half a second followed by rapid cooling. In both production plants, the product is washed with sodium carbonate solution to neutralise residual acids in the NG emulsion. NG is commonly used *in situ* to manufacture propellants or commercial explosives without storage.

The two processes differ considerably in terms of production capacity and unit explosion risk (table 2.1). A large Biazzi plant can produce 2000–3000 kg h^{-1} but only 80–120 kg is at risk at any given time point. In the NAB injector process, both the production capacity and the unit explosion risk are lower (500–650 kg h^{-1} and 3.5 kg of NG, respectively). NG is classified as toxic if swallowed, inhaled, or in contact with the skin [53] and is used as a vasodilator in pharmaceutical formulations [54, 55].

NC is manufactured from the natural polymer cellulose, a linear polysaccharide containing β-1,4 glucosidic bonds, using a batch nitration process (figure 2.19). The characteristics and applications of NC vary with the percentage nitration, which

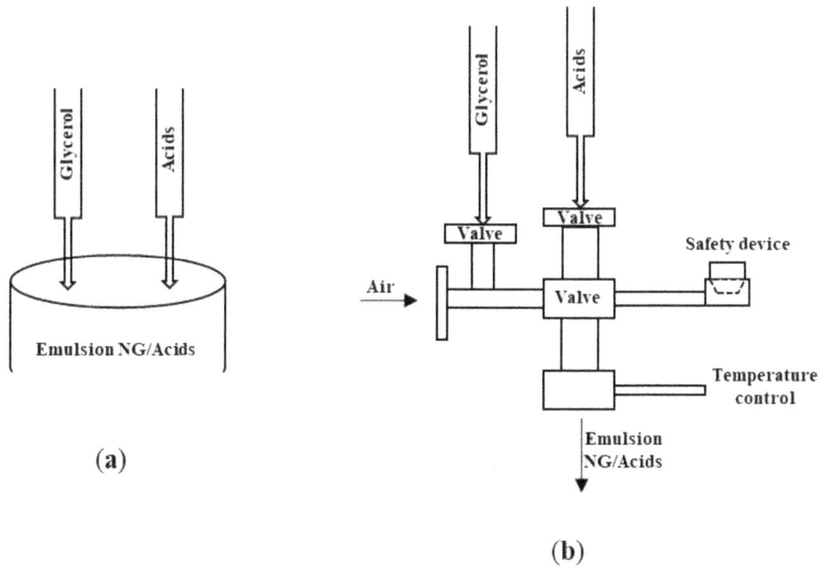

Figure 2.18. Schematic representation of (a) the Biazzi process and (b) the NAB injector process. (a) The Biazzi process includes an automatic shutdown of the feed. (b) The NAB injector process includes a safety device that comes into contact with the nitrating acid if the flow of acid is adequate.

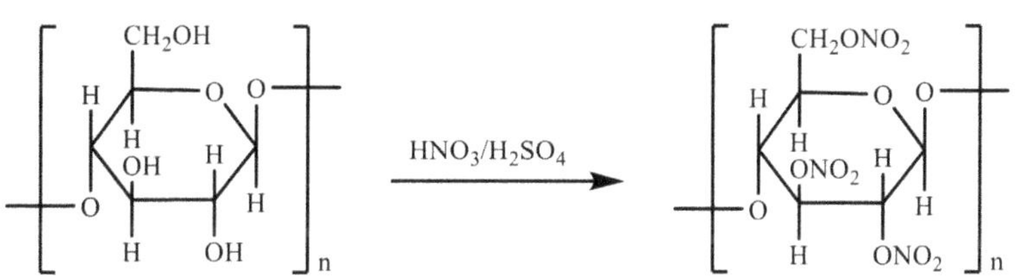

Figure 2.19. The synthesis of NC.

influences properties such as solubility, hazardous behaviour, and stability [20]. Complete nitration of all hydroxyl groups (three per glucose unit) gives a nitrogen content of 14.15% by weight. However, the properties of NC also depend on the quality of the starting material, which differs by geographic origin and due to seasonal variations. Such variations result in lengthy and expensive authorisation processes. Alternatives to NC are discussed in section 2.4.

The richest source of cellulose is cotton fibre, although it can also be isolated from wood and paper. All NC production methods require the pretreatment of raw cellulose before nitration with a mixture of HNO_3/H_2SO_4. The concentrations of the acids, and the temperature and duration of nitration all influence the final nitrogen content and the quality of the product. The crude NC is stabilised by a process of washing and boiling. Dry NC is susceptible to initiation by friction and impact, and is generally handled with a controlled water content to reduce such risks. The overall process is summarised in figure 2.20.

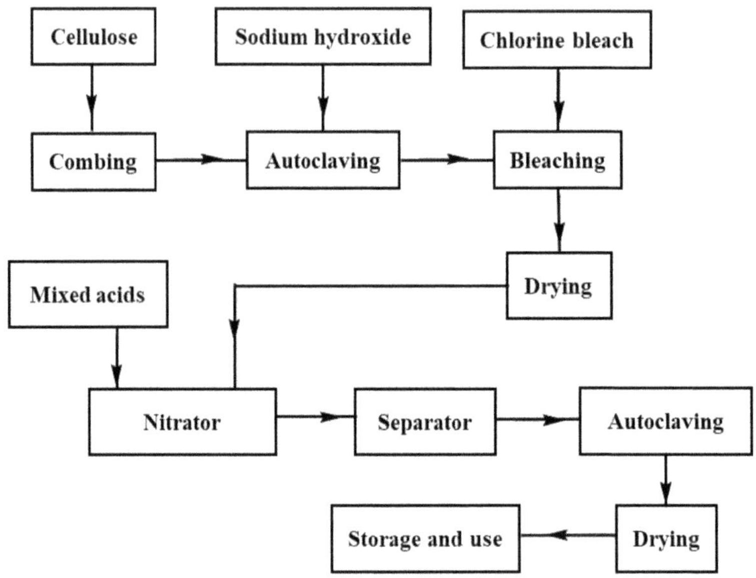

Figure 2.20. A flow diagram showing the manufacturing process for NC.

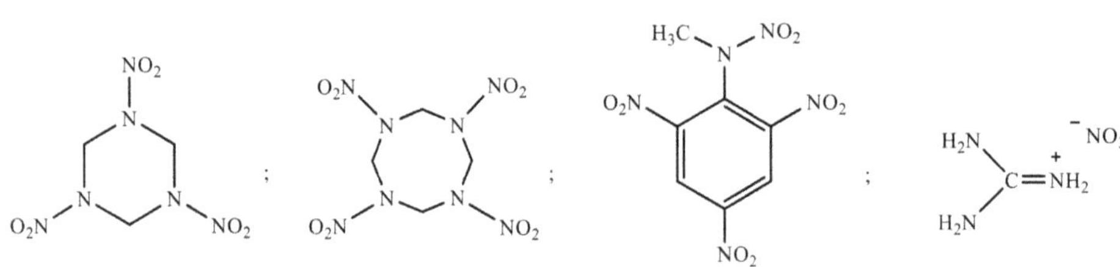

Figure 2.21. Chemical structures of RDX, HMX, tetryl, and NQ.

2.3.3 Nitramines

Nitramines, also known as nitroamines, are organic compounds containing a nitro group bonded to an amine group. They include hexahydro-1,3,5-trinitro-1,3,5-triazine (RDX), octahydro-1,3,5,7-tetranitro-1,3,5,7-tetrazine (HMX), tetryl, and nitroguanidine (NQ) (figure 2.21). Nitration on a nitrogen atom generally increases the sensitivity of the compound but is a more expensive process than C-nitration.

RDX, as stated in Section 1.2, is produced on a large scale for military and civilian applications using several different processes. The Woolwich process converts hexamine to pure RDX using 98%–99% HNO_3 (figure 2.22).

A continuous production plant typically feeds hexamine and HNO_3 into a series of nitrators by cocurrent flow. The complex nitration reaction produces ammonia and formaldehyde as byproducts, and these toxic gases must be removed and recycled to comply with health and safety mandates. A urea solution can be used as

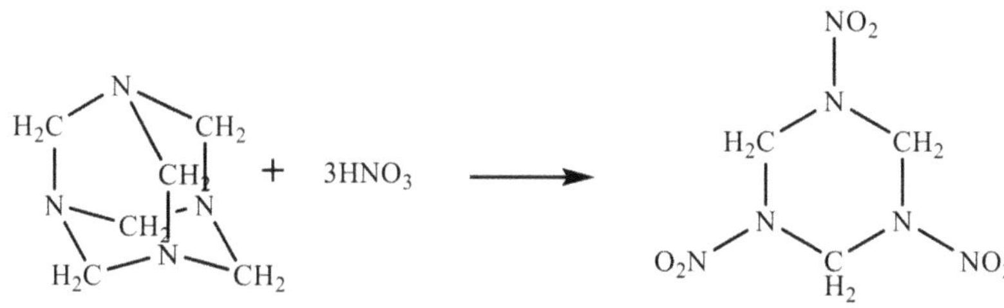

Figure 2.22. Synthesis of RDX via the Woolwich process.

$$(CH_2)_6N_4 \cdot 2HNO_3 + 2(NH_4NO_3 \cdot HNO_3) + 6(CH_3CO)_2O \longrightarrow 2(CH_2)_3(NNO_2)_3 + 12CH_3COOH$$

Hexamine nitrate — Ammonium nitrate-nitric acid — Acetic anhydride — RDX — Acetic acid

Figure 2.23. Synthesis of RDX via the Bachmann process.

Table 2.3. Types of RDX arising from different manufacturing processes.

Manufacture method	Purity[1] (w/w)	Type of RDX
Woolwich	100%	A
Bachmann	⩽12%	B
	<5% HMX content	I
	4%–17% HMX content	II

[1] Using acetone, methyl ethyl ketone (MEK), or cyclohexanone as solvents.

a dumping facility in case of thermal runaway or the evolution of excess nitrogen oxides. The crude product is recovered from the reaction, filtered, washed with water, and recrystallised from cyclohexanone to yield pure RDX.

The Bachmann process produces RDX using a mixture of HNO_3 and ammonium nitrate, glacial acetic acid, and acetic anhydride [20] (figure 2.23). The RDX obtained via the Bachmann process is always impure, containing varying amounts of HMX. Several types of RDX can be produced by recrystallisation in different solvents (table 2.3). The Bachmann process is the most widely used manufacturing process today, although a few manufacturers still use the Woolwich process [57, 58]. RDX is toxic [59] and a moderate irritant. The ingestion or inhalation of RDX dust may cause acute or chronic poisoning.

HMX is a byproduct of the Bachmann process for the manufacture of RDX. By controlling the reaction conditions, it is possible to favour the production of HMX over RDX, although the yields are typically no higher than 65%. The nitrolysis of hexamine using a HNO_3/ammonium nitrate system containing paraformaldehyde with acetic acid/acetic anhydride as a solvent produces an HMX-rich product. RDX can be removed from the product by rinsing in hot water.

HMX is the best-performing explosive produced in industrial quantities. There are several types of HMX depending on the manufacturing process, purity

Table 2.4. Crystal forms of HMX and their relative stability.

Form	α	β	γ	δ
Stability	Metastable	Stable	Metastable	Labile

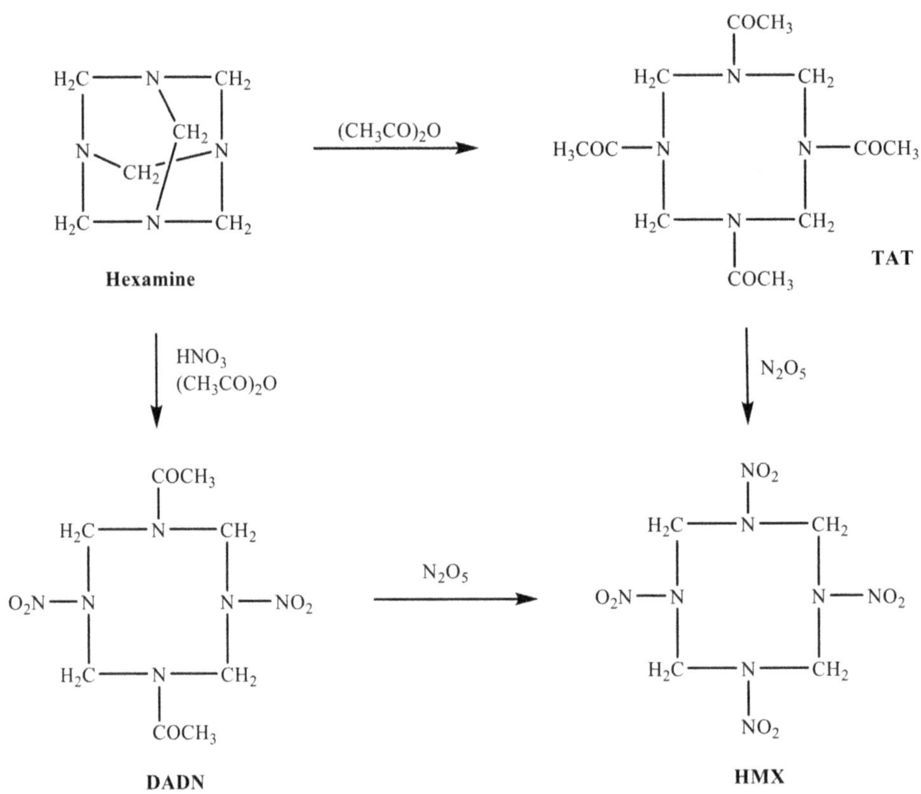

Figure 2.24. Synthesis of HMX using N_2O_5.

(recrystallisation from cyclohexanone), and crystal size (table 2.4). HMX costs significantly more to produce than RDX, reflecting the larger amounts of acetic anhydride required and the extra purification steps.

An alternative method to produce HMX using N_2O_5 is being investigated as a cleaner, more efficient, and possibly more economical process, and has been demonstrated at laboratory scale [20]. Hexamine is first converted to the non-energetic intermediate 1,3,5,7-tetraacetyl-1,3,5,7-tetraazacyclooctane (TAT), which is then cleanly nitrolysed to HMX by N_2O_5 with yields exceeding 80% (w/w) and high purity (figure 2.24). HMX is classified as toxic [60] if swallowed or in contact with the skin.

Tetryl is manufactured by the nitration of N,N-dimethylaniline using conventional nitrating mixtures [61]. The overall oxidation and nitration reaction is shown in figure 2.25.

A typical continuous nitrating plant would consist of two heated nitrating vessels and a cooled diluter for product precipitation. The crude product is washed with water to remove acids, then recrystallised from acetone to remove the byproduct 2,4-dinitrodimethylaniline, thus yielding tetryl in a highly pure form. Tetryl is classified as a skin irritant [62]. In one processing plant employing more than 5000 people, 23% of the workers suffered from the symptoms of tetryl illness and 75% of the affected group developed tetryl dermatitis. Interestingly, 20% of the affected group had no direct contact with tetryl but were nevertheless sensitised, and were thus prone to develop a rash at any time no matter where in the plant they worked. Tetryl has largely been replaced with less toxic explosives.

NQ is mainly used in the preparation of triple-base propellants. The material is prepared by reacting guanidine nitrate with H_2SO_4 (figure 2.26) [63]. Guanidine nitrate is readily produced by heating calcium carbide in a stream of nitrogen, acidifying the product, and reacting it with ammonium nitrate. The nitration reaction is rapid, and NQ is recovered by precipitation in water with yields exceeding 90%. The crude product is washed to remove acids and recrystallised from water.

CL-20 (2,4,6,8,10,12-hexanitro-2,4,6,8,10,12-hexaaza-tetracyclo[5,5,0,03.11,05.9] dodecane), also known as hexanitrohexaazaisowurtzitane (HNIW), was first synthesised in 1997 [64] and is the most energetic high-explosive compound. It is used in several polymer-bonded explosive (PBX) compositions (e.g. PAX-11) as well as rocket and gun propellants [65]. The preparation of CL-20 is a multistep process involving several substituted isowurtzitane precursors [20, 66]. First, 2,4,6,8,10,12-hexabenzyl-2,4,6,8,10,12-hexaazaisowurtzitane (HBIW) is synthesised by reacting

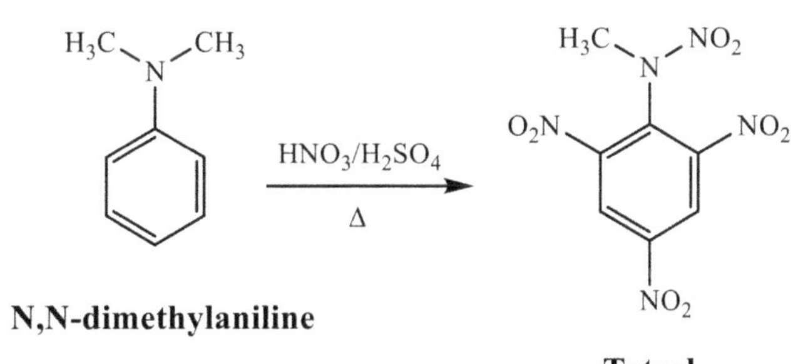

Figure 2.25. The synthesis of tetryl.

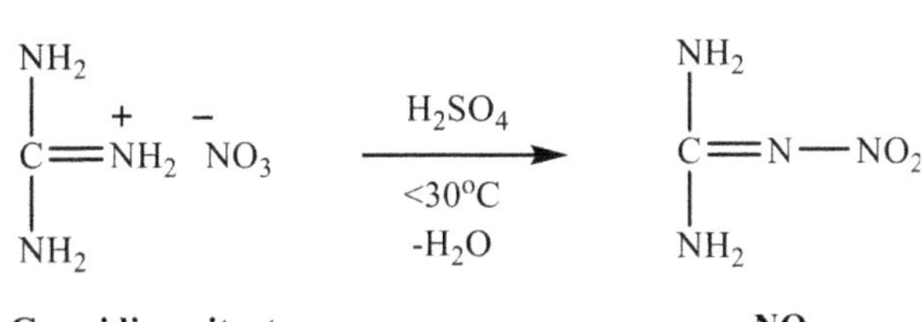

Figure 2.26. The synthesis of NQ.

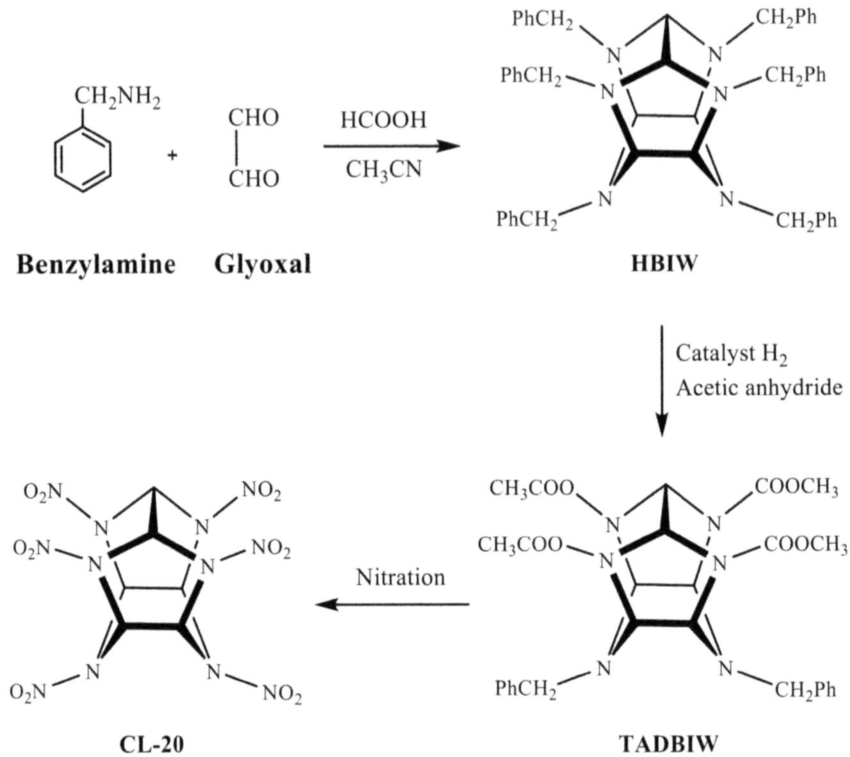

Figure 2.27. The synthesis of CL-20.

benzylamine with aqueous glyoxal (40% v/v) in aqueous acetonitrile containing a catalytic amount of formic acid (figure 2.27). The reductive acetylation of HBIW with palladium hydroxide on carbon in acetic anhydride containing a catalytic amount of bromobenzene yields the tetra-acetate 2,6,8,12-tetraacetyl-4,10-dibenzyl-2,4,6,8,10,12-hexaazaisowurtzitane (TADBIW) (figure 2.27).

The last step in the preparation of CL-20 involves the nitrolysis of TADBIW with N_2O_5 or nitrosonium tetrafluoroborate ($NOBF_4$) in sulfolane, followed by further treatment with nitronium tetrafluoroborate (NO_2BF_4) in the same vessel. The yield of CL-20 in this reaction is 90% [66]. $NOBF_4$ and NO_2BF_4 are expensive reagents, so alternative routes using the precursor 4,10-diamino-2,6,8,12-tetraacetyl-2,4,6,8,10,12-hexaazaisowurtzitae and nitrating agents such as 95%–99% HNO_3 have been attempted. The reaction parameters were optimised, resulting in a CL-20 yield of 85%. CL-20 exists in four crystalline forms, stable at different temperatures. A CL-20 manufacturing facility has been established by Thiokol (USA).

2.3.4 High-nitrogen-content explosives

NTO (3-nitro-1,2,4-triazole-5-one) is synthesised in a two-step reaction, the first involving semicarbazide hydrochloride and formic acid followed by nitration with

99% (v/v) HNO_3. Formic acid is added to solid semicarbazide hydrochloride at ambient temperature, and the resulting solution is heated to 65 °C and stirred to produce the intermediate 1,2,4-triazol-5-one (TO) (figure 2.28) [20, 66]. The reaction mixture is cooled to ambient temperature before adding HNO_3. After 3 h, the mixture is filtered and the resulting solid is washed with cold water. Recrystallisation from hot water yields NTO as a white crystalline solid. The method is simple and economical. NTO is an irritant to the eyes, respiratory system, and skin [68].

FOX-7 is synthesised in a single step by nitrating 2-methyl-4,6-dihydroxypyrimidine using a mixture of concentrated HNO_3 and H_2SO_4 [69]. This produces the tetranitrated intermediate 2-dinitromethylene-5,5-dinitropyrimidine-4,6(1H,5H)-dione (figure 2.29), which is hydrolysed in water at ambient temperature to produce FOX-7, dinitromethane, and CO_2 after several hours (figure 2.30).

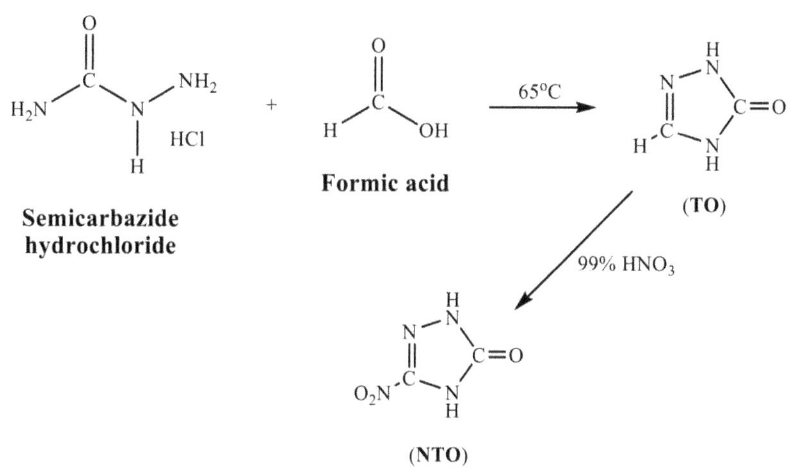

Figure 2.28. The synthesis of NTO.

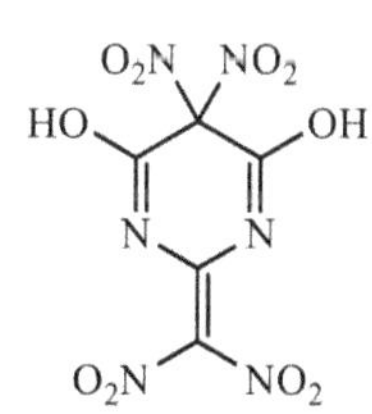

Figure 2.29. The chemical structure of 2-dinitromethylene-5,5-dinitropyrimidine-4,6(1H,5H)-dione.

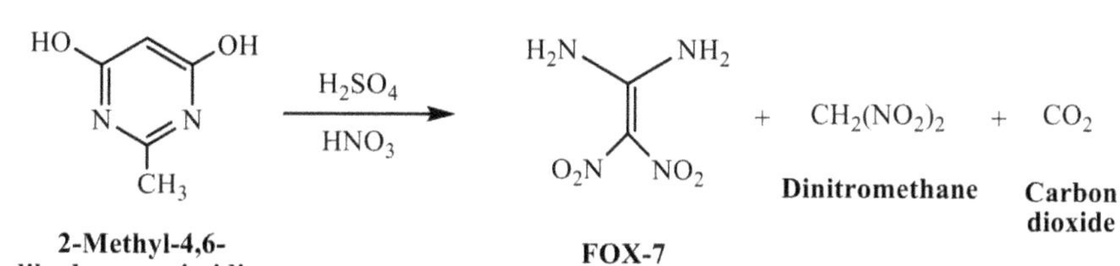

Figure 2.30. The synthesis of FOX-7.

The product separates out as it is formed, and is recovered by filtration before washing in water. Dimethyl sulfoxide (DMSO)/water, N-methylpyrrolidin-2-one/water and dilute HCl are often used for FOX-7 recrystallisation.

FOX-7 is classified as harmful if swallowed and irritating to the eyes, respiratory system, and skin, with sensitisation potentially caused by skin contact [70]. The manufacturing process for FOX-7 is summarised in figure 2.31.

GUDN (guanylurea dinitramide) also known as FOX-12, is a low-sensitivity, heat-stable explosive potentially suitable for the filling of IMs and as a precursor for the synthesis of the solid oxidiser ammonium dinitramide (ADN) via potassium dinitramide. GUDN is synthesised in a two-step process (figure 2.32) involving (1) the production of guanylurea sulphate from dicyandiamide, followed by (2) the nitration of ammonium sulphamate, which reacts with guanylurea sulphate to give GUDN as a white crystalline solid [71]. GUDN is only sparingly soluble in water, which reflects its layered structure based on the presence of intermolecular hydrogen bonds within the layers and van der Waals interactions between the layers [72].

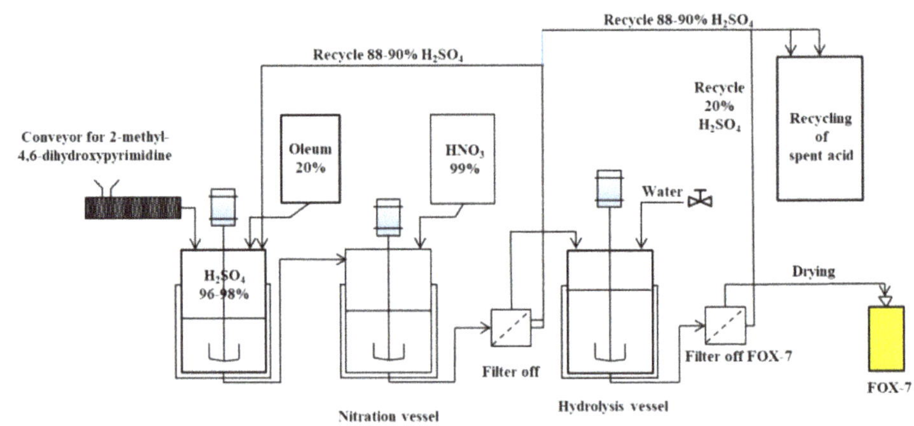

Figure 2.31. Plant schematic for the manufacture of FOX-7.

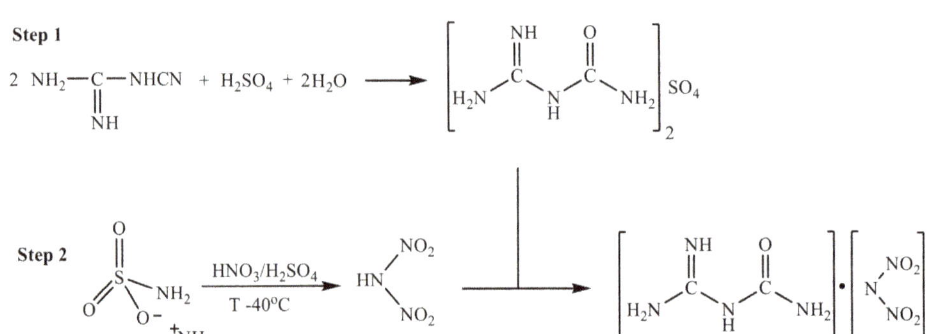

Figure 2.32. The synthesis of GUDN.

ADN is an environmentally benign and hygroscopic material that can be synthesised by the direct nitration of sulphamic acid salts using mixed nitric and sulphuric acids (figure 2.33) followed by neutralisation and separation of the product. ADN was first synthesised at the Zelinsky Institute of Russia in the 1970s. EURENCO maintains a pilot-scale production facility based on work carried out by Swedish Defence Research Agency (FOI).

LLM-105 (2,6-diamino3,5-dinitropyrazine-1 oxide) is a dense, low-sensitivity explosive material that was first synthesised at the Lawrence Livermore National Laboratory (USA) [66]. The batch synthesis of LLM-105 has been studied for nearly 20 years, and several methods have been developed [66]. Two promising routes are shown in figure 2.34. The 2,6-diaminopyrazine-1-oxide (DAPO) process is favoured because only one energetic material is handled (LLM-105). In contrast, the 2,6-diamino-3,5-dinitropyrazine (ANPZ) process involves the partial oxidation of ANPZ to LLM-105, and three energetic materials are involved: 2,6-dimethoxy-3,5-dinitropyrazine (DMDNP), ANPZ, and LLM-105. The overall yield from the commercially available substrates 2,6-dimethoxypyrazine and DAPO are comparable, at 50% and 50%–70%, respectively [66, 73–75]. LLM-105 has also been synthesised using a commercially available flow micro-reactor system [36].

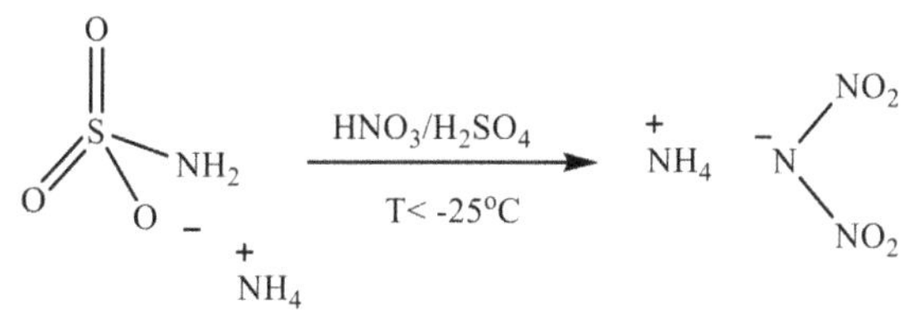

Figure 2.33. The synthesis of ADN.

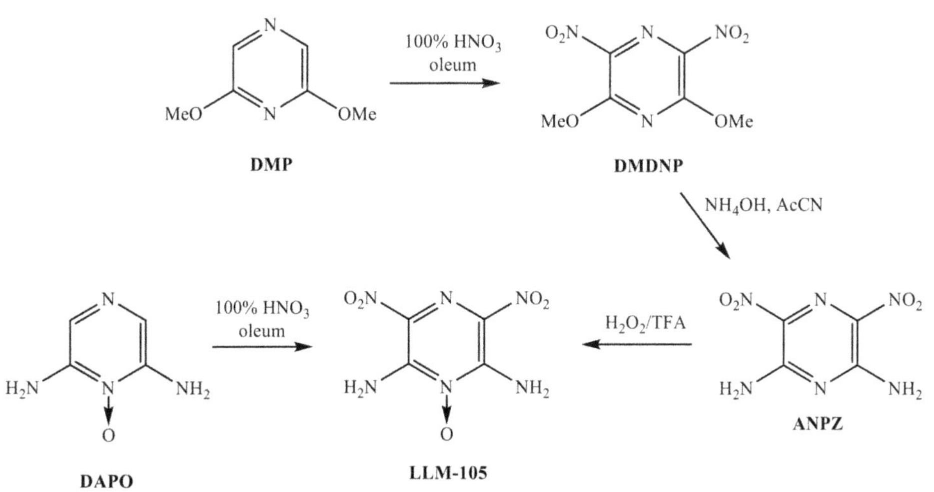

Figure 2.34. The synthesis of LLM-105.

TEX (4,10-dinitro-2,6,8,12-tetraoxa-4,10-diazaisowurtzitane) is a CL-20 analogue manufactured by Thiokol [76–78]. TEX is superior to LLM-105 in terms of density and insensitivity to impact stimuli, but has a comparable velocity of detonation. It is synthesised from 1,4-diformyl-2,3,5,6-tetrahydroxypiperazine and glyoxal trimers using a mixture of HNO_3/H_2SO_4 (figure 2.35), but the product yield and purity achieved in this process are not fully established [76].

DAAF (diaminoazoxy furazan) was first synthesised in the Soviet Union in the 1980s and was then investigated in the USA as part of the Joint Insensitive Munitions Technology Program for the replacement of vulnerable booster formulations. The production method (figure 2.36) was developed at Los Alamos National Laboratory in the early 2000s, achieving good yields from inexpensive starting materials and producing a small amount of nonhazardous waste [79, 80].

TNAZ (N-3,3-trinitroazetidine) is a high-performance explosive due to the strain from its square ring and the high density of its geometric structure, which increase its stored energy [81] (figure 2.37). TNAZ can be melt-casted, and has therefore been proposed as a replacement for TNT. Like TNT, it contracts by 10%–12% as it solidifies, but in addition it also sublimes due to its low molecular weight. Several multistep processes have been described for the production of TNAZ, but the number of steps reduces the yield to ~30% and scalability is also a challenge [82].

ONC (octanitrocubane) is another high-performance explosive material reflecting the energy conferred by its ring strain and cage strain (figure 2.38). The potential of this material for future applications in explosives and propellants has drawn the attention of researchers, particularly in the USA, but its applications are limited by

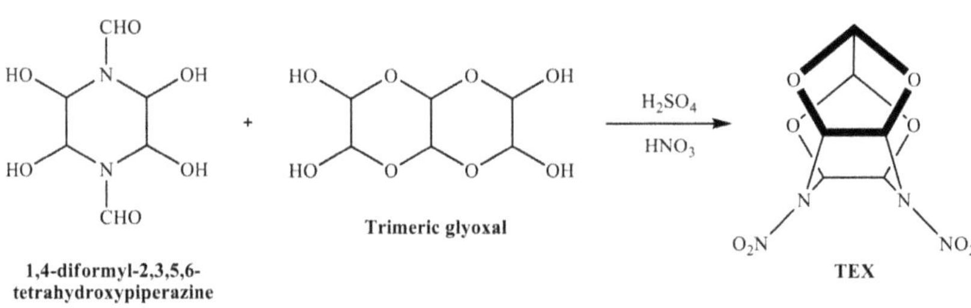

Figure 2.35. The synthesis of TEX.

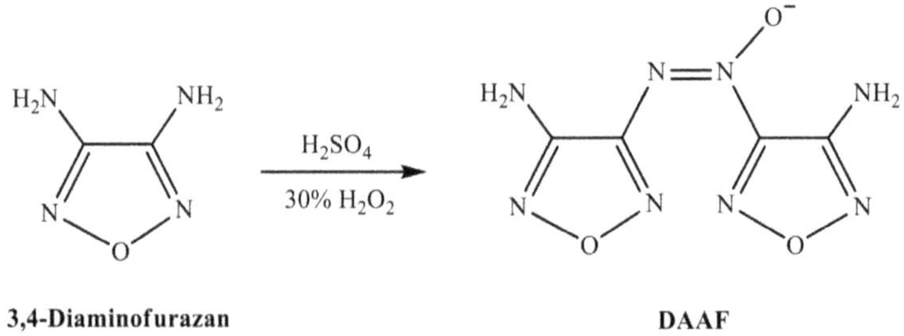

Figure 2.36. The synthesis of DAAF.

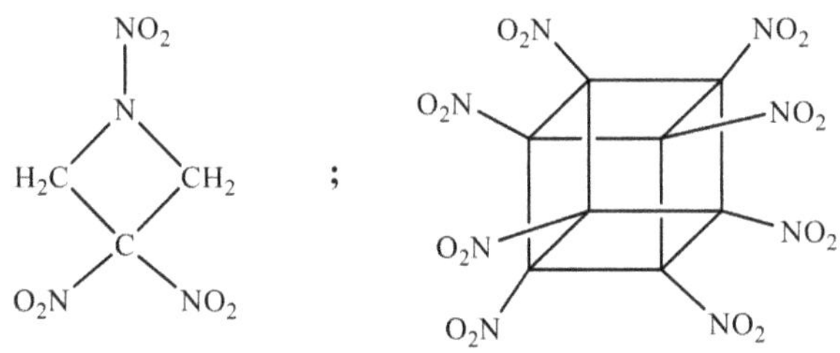

TNAZ **ONC**

Figure 2.37. The chemical structures of TNAZ and ONC.

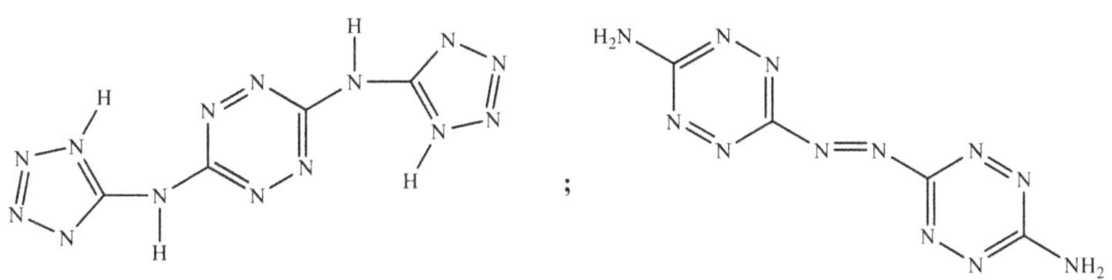

BTATz **DAAT**

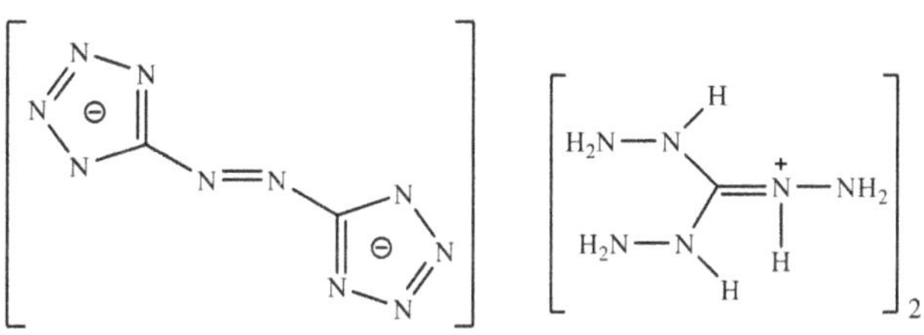

TAG-AT

Figure 2.38. The chemical structures of BTATz, DAAT, and TAGAT.

its high sensitivity [83, 84]. It is produced using a two-phase nitration reaction and through an extremely lengthy and expensive process.

Polynitrogen compounds containing multiple N≡N bonds release large amounts of energy (954 kJ mol^{-1}) and are therefore considered as potential energetic molecules based on theoretical quantum mechanical calculations. Such compounds are generally unstable and detonate violently [66]. The polynitrogen compound 3,6-bis(3-azido-1,2,4-triazol-1-yl)-1,2,4,5-tetrazine (BTATz) has a very fast burn rate and may be suitable as a sensitive gas generator (figure 2.30) [83]. Other gas-generating compounds in this class include 3,3-azobis-6-amino-1,2,4,5-tetrazine (DAAT) and triamino guanidinium azotetrazolate (TAGAT) (figure 2.38) [83].

TKX-50 (dihydroxylammonium 5,50-bistetrazole-1,10-diolate) [85–89] is a nitrogen-rich salt and a promising candidate to replace secondary explosives such as HMX

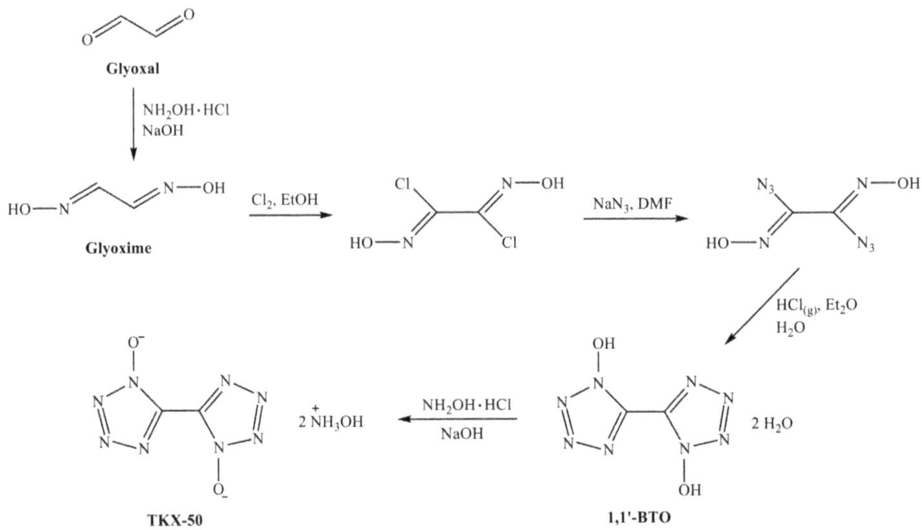

Figure 2.39. The synthesis of TKX-50.

and RDX in IMs. TKX-50 is currently being evaluated as a shock-insensitive melt-cast formulation. It is synthesised from inexpensive or gaseous reactants and achieves yields of up to 85%. The five-step procedure involves the oximation of glyoxal followed by chlorination, azidation, and cyclization reactions to form the intermediate 5,5′-bistetrazole-1,1′-diol dihydrate (1,1′-BTO) before nitration [87] (figure 2.39).

2.3.5 Polymers in energetics

Oligomers and polymers containing energetic nitrato, azido, and tetrazole groups are used as binders in energetic formulations as components of explosives, propellants, and pyrotechnics, and contribute to the overall energy of the formulated explosive system (table 2.5) [90].

The binder holds together the solid granular ingredients in the energetic formulation and primarily affects motor reliability, mechanical properties, propellant processing complexity, storability, aging, and costs. Generally, binders should exhibit the following properties:

- Chemical and thermal stability.
- Low toxicity.
- Compatibility with energetic fillers and additives.
- Commercial availability.
- Ease of processing (i.e. small number of reagents, low viscosity, room temperature mixing, and long pot life).
- Appropriate cure characteristics (i.e. not violently exothermic and reasonably short curing period).
- Low glass transition temperature.

Table 2.5. Polymers in energetic formulations. Adapted from [66, 90] with permission.

	Polymeric binder	
Precursor	Name	Chemical structure
1,3-Butadiene	Hydroxyl-terminated polybutadiene (HTPB) [18]	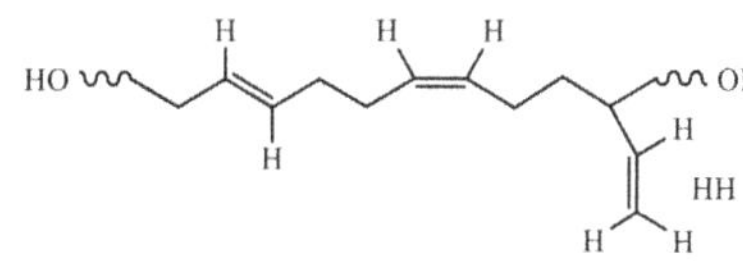
Cellulose	Nitrocellulose (NC) [91]	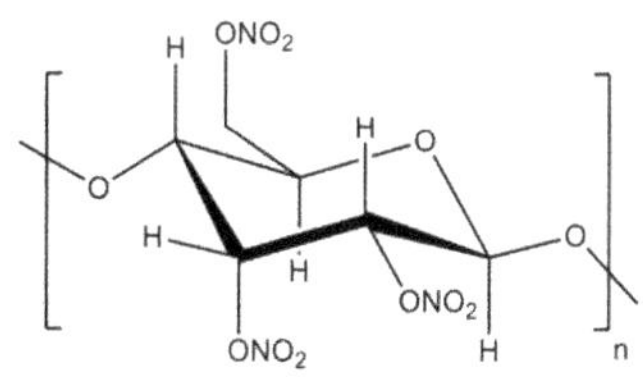
Glycidyl nitrate	Polyglycidyl nitrato (PGN) [92–94]	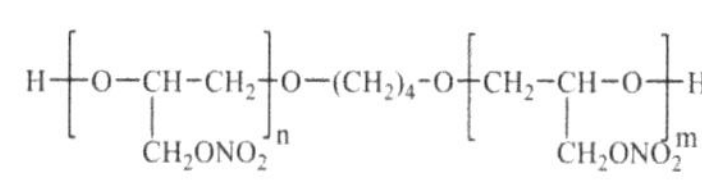
3-Nitratomethyl-3-methyl oxetane	Poly(NIMMO) [95]	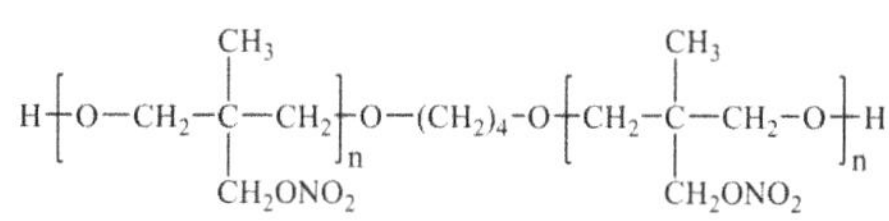
3-Azidomethyl-3-methyl oxetane	Poly 3-azidomethyl-3-methyl oxetane Poly(AMMO) [96]	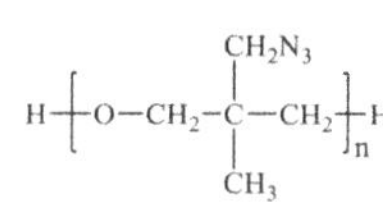
Epichlorohydrin	Glycidyl azide polymer (GAP) [97–99]	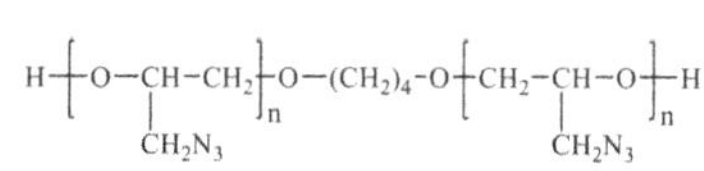
3,3-Bis-azidomethyl oxetane	Poly 3,3-bis-azidomethyl oxetane Poly(BAMO) [90]	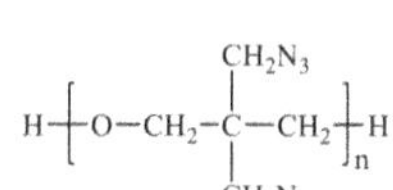
Bis-Trifluoroethoxy 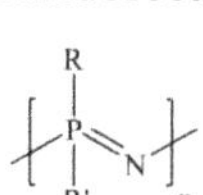	polyphosphazenes	Energetic polyphosphazenes [15]

(*Continued*)

Table 2.5. (*Continued*)

	Polymeric binder	
Precursor	Name	Chemical structure
HTPB	Nitrated hydroxyl-terminated polybutadiene (NHTPB) [100]	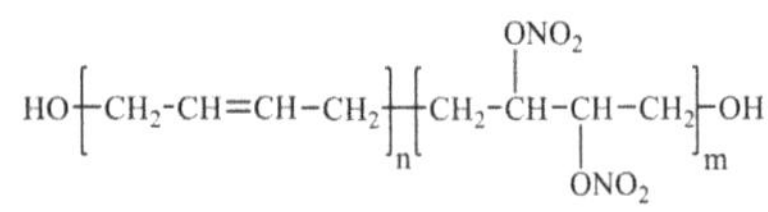
Cyclodextrins	Nitrated polycyclodextrins (PolyNCD) [101, 102]	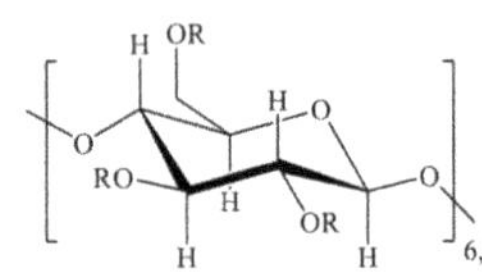 R = ONO_2 and/or EPY or isocyanate cross-linker
Cyclodextrins	Nitrated polycyclodextrins (PolyNCXCD) [16, 17]	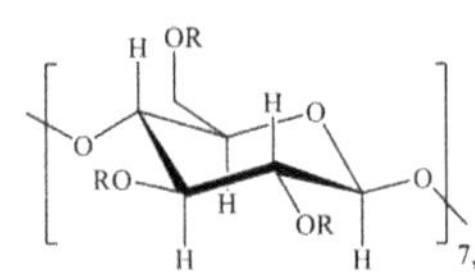 R = H, ONO_2 and/or polyethylene glycol cross-linker
Vinyl tetrazole	Poly(vinyl tetrazole) [90]	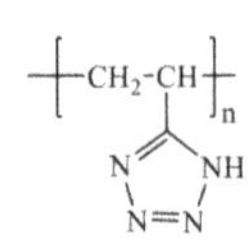

Additional functional groups at the ends of the polymer chain can react and contribute to the formation of flexible matrixes that facilitate binding. The binder undergoes cross-linking and branch chaining during the curing of the formulation.

Polyurethanes produced by the cross-linking of hydroxyl-terminated polybutadiene (HTPB), polyglycidyl nitrate (PGN), and other hydroxyl-terminated oligomers with bifunctional or multifunctional isocyanates [6, 18] are favoured as binders in energetic formulations because they increase the solid fraction and improve the physical properties of the formulation at its temperature limits. Isocyanates react readily with hydroxyl groups in HTPB under mild conditions [6] but their high reactivity and volatility raise safety issues under REACH regulations [7, 9]. The replacement of isocyanate cross-linkers is an important research field, as discussed more extensively in section 2.3.

Nitrato polymers contain pendant nitrato (ONO_2) groups on the polymer backbone (table 2.4) and release energy by oxidation of the macrochain and the scission of the nitrato groups [90]. NC (table 2.4) is manufactured by the nitration of cellulose, as discussed above [91]. High-nitrogen-content NC formulations

(11.5%–14%) are the major ingredients in solid propellants, and the typical nitrogen content is 12.6%. PGN (table 2.4) is prepared by cationic ring-opening polymerisation of the glycidyl nitrate monomer in the presence of an acid catalyst and polyfunctional alcohol initiator. The nitrated monomer is prepared by the nitration of glycidol with N_2O5 in dichloromethane (two-phase nitration). PGN has an atactic stereochemistry, and the highly viscous liquid cured with multifunctional isocyanates is used as a binder for solid propellants. A different process was developed for the synthesis of crystalline isotactic PGN, which is used as a hard block for PGN thermoplastic elastomeric copolymers [92, 94].

Poly(NIMMO) and poly(AMMO) (table 2.4) are hydroxyl-terminated polymers prepared from 3-nitratomethyl-3-methyl oxetane and 3-azidomethyl-3-methyl oxetane, respectively. They are cured with isocyanates to yield energetic binders [95, 96]. Glycidyl azide polymer (GAP) pre-polymer containing terminal hydroxyl groups (table 2.4) is synthesised by the cationic polymerisation of epichlorohydrin, followed by azidation with sodium azide [90]. GAP was the first energetic elastomeric prepolymer made during the early 1990s [97]. It is available in the USA [98] and via EURENCO in Europe [99]. A pilot plant to manufacture 5 kg/batch of branched GAP is operational at the Defence Research Establishment [90]. The ΔH_f of GAP is ~1 kJ g^{-1} compared to 0.02 kJ g^{-1} for HTPB [99].

Poly(BAMO) (table 2.4) is synthesised from 3,3-bis(chloromethyl) oxetane monomers by reacting them with sodium azide, followed by cationic polymerisation [90]. Crystalline poly(BAMO) has been copolymerised with other monomers such as oxetane, tetrahydrofuran, and caprolactone to minimise the chain regularity and reduce the glass transition temperature.

Copolymerisation with other energetic monomers provides the energetic binders with thermoplastic properties. Polyphosphazenes are made by grafting bis-trifluoroethoxy polyphosphazenes using lengthy hydroxyl-terminated side groups that are easily nitrated via conventional nitration processes [15].

Nitrated hydroxyl-terminated polybutadiene (NHTPB) (table 2.4) is prepared by the partial epoxidation of the main-chain double bonds followed by a nitration reaction. The properties of NHTPB depend on the percentage of double bonds converted into nitrate ester groups. NHTPB has a low viscosity for processing purposes, and can be cured with aliphatic and aromatic isocyanates [100].

Nitrated polycyclodextrins (PolyNCDs) (table 2.4) are prepared by cross-linking the cyclic 1,4-oligosaccharides β cyclodextrin (βCD) and γ cyclodextrin (γCD) with epichlorohydrin or isocyanate, and are subsequently nitrated using traditional nitration methods. They have been used to develop energetic formulations that are less vulnerable to accidental stimuli [101, 102]. The lipophilic cavity of cyclodextrins can encapsulate a wide variety of nitrate esters and nitramines, forming molecular energetic complexes and acting as a protective shield for energetic materials that are sensitive to impact, shock, friction, and electric discharge. The level of nitration affects the energetic performance and physicochemical properties of the βNCD. Nitrated βCD and γCD polymers are also prepared by using nontoxic polyethylene glycol diglycidyl ethers (table 2.4) [16, 17], and are further discussed in section 2.4 of this chapter.

Tetrazole polymers and their salts are energetic polymers known for their high nitrogen content and high ΔHf values. The tetrazole group is more stable and versatile than the azide group but releases a comparable amount of energy during combustion [90]. Poly(vinyl tetrazole) is prepared by the free-radical polymerisation of the vinyl tetrazole monomer. Low-molecular-weight multifunctional hydroxyethyl tetrazoles are cross-linked with isocyanates in the presence of polyethylene glycol to produce fully amorphous networks that act as desensitising energetic binders in PBX formulations.

Energetic thermoplastic elastomers (TPEs) are synthesised by the copolymerisation of energetic pre-polymers such as BAMO/AMMO, BAMO/GAP, BAMMO/NIMMO, and segmented copolyurethane elastomers based on GAP or poly (NIMMO) [90]. The BAMO block provides the hard segment in the TPE whereas AMMO, GAP, and NIMMO chains contribute to the soft phase due to their lower glass transition temperatures. The ratio of the blocks and their molecular weights influence the properties of the TPE.

2.3.6 Primaries

The preparation of lead azide from sodium azide and lead acetate is straightforward, but control of the reaction is critical because the secondary growth of lead azide crystals can generate dangerous products. Lead azide is sensitive to shock, friction, and low-energy electrical discharge, so stringent precautions are required to avoid dangers from electrostatic charges during the drying, sieving, weighing, and pressing of the material. All apparatus must remain at earth electrical potential and operatives must not build up an electrostatic charge while working with the material. All sensitive primary explosives are produced in small batches due to their inherent hazards. Recent efforts to reduce the toxicity of primary explosives that contain lead has led to the development of 'green explosives' such as 2-diazo-4,6-dinitrophenol (DDNP), potassium 4,6-dinitrobenzofuroxan (KDNBF), copper (I) 5-nitrotetrazole (DBX-1), and potassium 7-hydroxy-4,6-dinitrobenzofuroxan (KDNP) [103–106].

2.4 Replacement of isocyanates

As discussed in section 2.2, polymers are used as binders in energetic formulations; that is, they hold together the solid ingredients in the energetic formulation, making it more resilient to the effects of external stimuli. The binder is typically cross-linked with bifunctional or multifunctional isocyanates (figure 2.40) to form polyurethanes [107–109].

Polyurethanes are organic polymers used in a wide range of '*foams, rubbers and plastics of varying densities and hardness for products such as varnish, glue, insulation, packaging and car parts, including tyres*' [110, 111]. HTPB and other hydroxyl-terminated pre-polymers are generally cross-linked with isocyanates to produce thermoplastic polyurethane elastomers (figure 2.41) [112].

Isocyanates are versatile organic compounds that undergo a variety of chemical reactions. Bifunctional and multifunctional isocyanates achieve different degrees of cross-linking that affect the mechanical properties of the resulting polyurethanes [108].

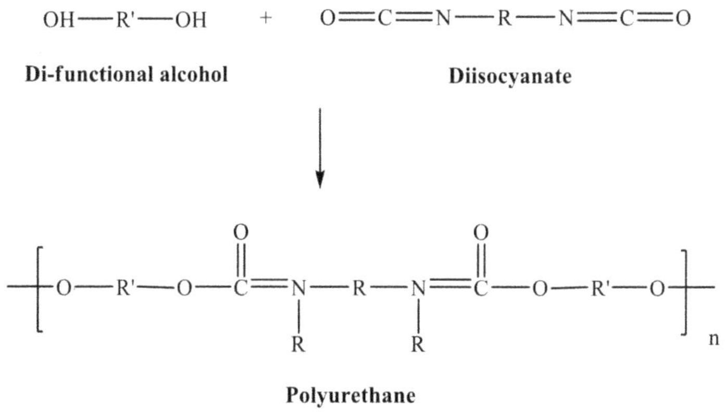

Figure 2.40. The synthesis of polyurethanes.

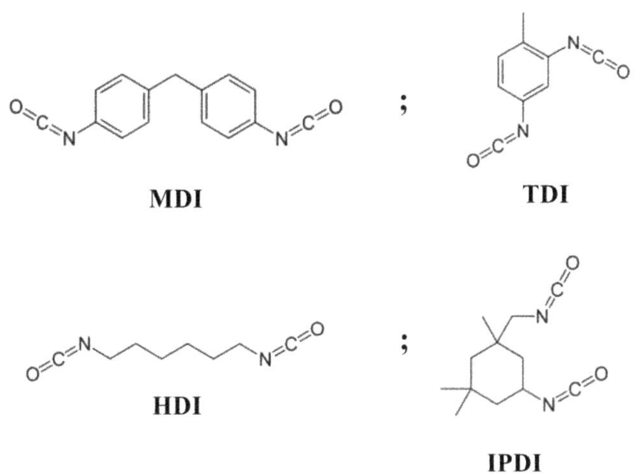

Figure 2.41. The chemical structures of methylene diphenyl diisocyanate (MDI), toluene diisocyanate (TDI), hexamethylene diisocyanate (HDI), and isophorone diisocyanate (IPDI).

For example, toluene diisocyanate (TDI) is mainly used to make flexible polyurethane whereas methylene diphenyl diisocyanate (MDI) is primarily used to make rigid polyurethane (figure 2.40). The curing time of polyurethanes can range from minutes to several days, and is dependent on the temperature, reagent (type of isocyanate and polyol), and in some cases the presence of moisture [113], although curing in the context of PBX is usually complete in 3–7 d.

Isocyanates are highly reactive, volatile, toxic, and also carcinogenic, thus raising significant health and safety issues [113–118]. REACH regulations in the European Union, adopted to protect humans and the environment from the risks posed by chemicals, have introduced restrictions on the manufacture, marketing, and use of certain diisocyanates [119] (table 2.6).

Table 2.6. Toxicological properties of isocyanates from safety data sheets[1].

Diisocyanate	H phrases
TDI	H315 Causes skin irritation
	H317 May cause an allergic skin reaction
	H319 Causes serious eye irritation
	H330 Fatal if inhaled
	H334 May cause allergy or asthma symptoms or breathing difficulties if inhaled
	H335 May cause respiratory irritation
	H351 Suspected of causing cancer
	H412 Harmful to aquatic life with long-lasting effects
MDI	H315 Causes skin irritation
	H317 May cause an allergic skin reaction
	H319 Causes serious eye irritation
	H332 Harmful if inhaled
	H334 May cause allergy or asthma symptoms or breathing difficulties if inhaled
	H335 May cause respiratory irritation
	H351 Suspected of causing cancer
	H373 May cause damage to organs (respiratory system) through prolonged or repeated exposure if inhaled
HMDI	H302 Harmful if swallowed
	H314 Causes severe skin burns and eye damage
	H317 May cause an allergic skin reaction
	H330 Fatal if inhaled
	H334 May cause allergy or asthma symptoms or breathing difficulties if inhaled
	H335 May cause respiratory irritation.
IPDI	H315 Causes skin irritation
	H317 May cause an allergic skin reaction
	H319 Causes serious eye irritation
	H330 Fatal if inhaled
	H334 May cause allergy or asthma symptoms or breathing difficulties if inhaled
	H335 May cause respiratory irritation
	H411 Toxic to aquatic life with long-lasting effects

[1]Merck | United Kingdom (sigmaaldrich.com).

The safety data sheet of TDI reveals several areas of concern: TDI is highly toxic by inhalation (vapour pressure = 1.33 Pa at 25 °C), mildly toxic by ingestion, and dermal (skin) contact can result in inflammation, eczema, and dermatitis [115]. Inhalation and dermal contact with TDI can also cause immune system sensitisation and asthma, even at very low exposure levels. Long-term exposure to TDI can result in significantly reduced lung function, whereas acute exposure to high concentrations can result in coma, pulmonary oedema, or severe respiratory distress, any of which can result in death.

Isocyanates are manufactured by reacting phosgene with a primary amine (for TDI, this is toluene diamine) [116]. Phosgene is a highly toxic pulmonary irritant and as such its use should also be limited where possible [120]. TDI is also considered a potential carcinogen in humans by the World Health Organization [115], although a consensus on its carcinogenetic properties has not been reached. When ingested by rats, TDI hydrolyses on aqueous tissue surfaces creating toluene diamine [121], a mutagen and rodent carcinogen [117], although it is unclear whether it is also a carcinogen in humans [122]. Many of these toxicological properties are characteristic of the N–C=O functional group, and therefore probably apply to most isocyanates [118].

Non-isocyanate polyurethanes (NIPUs) can be formed when a cyclocarbonate reacts with an amine (figure 2.42) [123]. These polymers show greater chemical resistance than isocyanate polyurethanes but lower permeability and porosity [124]. NIPUs with mechanical properties comparable to isocyanate polyurethanes are marketed as 'Green Polyurethane' by NanoTech Industries [125] Current NIPU research focuses on hard coatings and foams, but given the versatility of polyurethanes it is likely that NIPUs will also be suitable for PBX formulations.

An alternative method describes the formation of liquid rubber from para-aminobenzoic acid and HTPB (figure 2.43) by introducing a cyclocarbonate-terminated cross-linker [126]. The liquid rubber was evaluated as a toughening agent for epoxy resin based on the impact strength of the modified networks. The epoxy-ATPB formulation cured at ambient temperature may offer an alternative

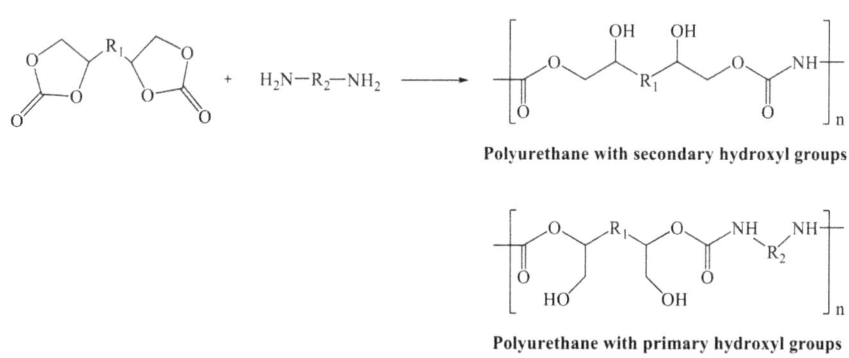

Figure 2.42. Isocyanate-free polyurethane formation. Reprinted from [123].

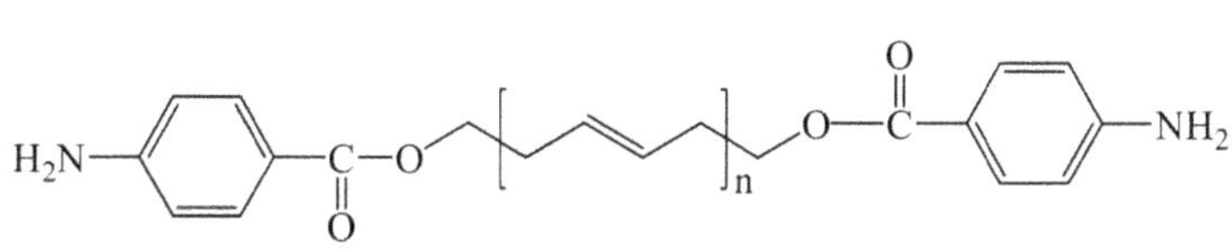

(ATPB)

Figure 2.43. The chemical structure of ATPB. Reprinted from [126].

toughened composite. One potential drawback of both these options is the incompatibility of amines with many explosives, including RDX and HMX.

A series of studies on the formation of new urethanes and phosgene-free isocyanates, as well as information about the distribution of cyclocarbonate-containing triglycerides from vegetable oils (i.e. renewable raw materials for polyurethanes), can serve as a quantitative basis for the development of relevant green chemistry processes [121].

The replacement of isocyanates with epoxy-terminated pre-polymers has also been investigated [127]. The epoxy-terminated pre-polymers react with hydroxyl-terminated elastomers in the presence of a catalyst to form cross-linked polymers. Several combinations were tested and were found to be compatible and suitable for use with energetic materials such as poly(GLYN), GAP, and poly(NIMMO).

Triethylene glycol diglycidyl ether (TEGDGE) is a nontoxic molecule that has been used to cross-link several samples of commercially available hydroxyl-terminated pre-polymers in an attempt to overcome REACH restrictions [112]. Most of the curing mixtures produced soluble polymers due to chain extension, but HTPB (Poly BD R45 HTLO) and polyacrylic ester (HyTemp 4454) formed thermally stable networks. Pre-aging the polybutadiene samples encouraged the formation of ethylene bridges between the pre-polymer molecules, which facilitated cross-linking. Both pre-polymers were formulated with the energetic fillers RDX and PETN and were successfully cross-linked with TEGDGE. The sensitivity of energetic fillers formulated with polybutadiene and polyacrylic ester pre-polymers and TEGDGE was generally lower than that of pure RDX and PETN. Energetic formulations of the polyacrylic ester pre-polymer with TEGDGE and PETN may also be suitable for mining applications.

Recently, β-cyclodextrin polymers have been developed by using polyethylene glycol diglycidyl ethers as alternative cross-linkers [16, 17, 128]. The polymeric samples, in their inert or nitrato forms, have interesting thermal properties, and have the potential to replace NC in several energetic compositions, as discussed in section 2.5.

2.5 Replacement of nitrocellulose

Qualification is the assessment of an energetic material by the national authority according to STANAG 4170 and AOP-7. It aims to determine whether the properties of that material make it safe and appropriate for its intended role. The qualification of NC, as stated above, depends on the raw material (cellulose, typically from cotton linters) because it is a natural polymer and its properties can vary from batch to batch. Accordingly, the source and properties of the cotton linters influence the properties and performance of the final NC product [91].

The specifications for military-grade NC require cotton linters to have specific chemical and physical properties. The manufacturer must also meet certain manufacturing controls. The linter grade selected by the manufacturer depends on the required processability and product characteristics. The qualification of energetic materials is a lengthy and expensive process in general, but NC requires repeated qualification due to the variability of the raw materials.

Cyclodextrins have been proposed as alternative polysaccharides to overcome the limitations of cellulose because they have reproducible chemical structures and predicable properties [16, 17]. They are cyclic organic molecules produced from starch by enzymatic conversion and have the ability to envelop smaller molecules in their inner cavity. The best-known types are α-cyclodextrin, β-cyclodextrin, and γ-cyclodextrin (figure 2.44), which are toroidal macrocycles comprising six, seven, and eight glucopyranose units, respectively [129]. Cyclodextrins with larger rings have been isolated but they are not used for the preparation of energetic materials [130].

Most publications dealing with cyclodextrins focus on their pharmaceutical applications, and many functionalized variants are commercially available [130–133]. Polymeric cyclodextrins form primarily due to a reaction between the primary and/or secondary hydroxyl groups with functional groups of other molecules, or due to guest–host interactions. Various cross-linkers can be used to form three-dimensional structures that are typically insoluble in common solvents [134]. The first cross-linked cyclodextrins were formed by reacting monomers with epichlorohydrin under basic conditions (table 2.4) [135].

A family of new β and γ cyclodextrin polymers has been developed by using polyethylene glycol diglycidyl ether cross-linkers as an alternative to isocyanates [16, 17, 128] (table 2.4). The raw materials are widely available from commercial suppliers, and both the products and the process are REACH compliant. Varying the type and length of the cross-linker influences the solubility and thermal/mechanical properties of the polymer, with shorter links producing brittle materials characterised by high glass transition temperatures ($T_g > 0$ °C) but longer links yielding softer materials with glass transition temperatures below −10 °C.

The inert polymers have been converted into nitrato analogues with good thermomechanical properties and glass transitions temperatures as low as −14 °C. The thermal stability of the polymers is similar to that of NC (~200 °C). They are

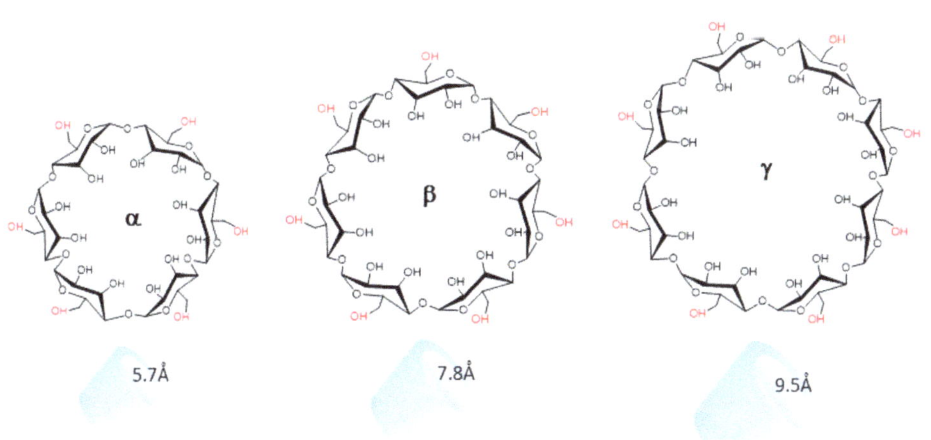

Figure 2.44. Chemical (top) and toroidal (bottom) structures of α, β, and γ cyclodextrins. Reprinted from [17] with permission.

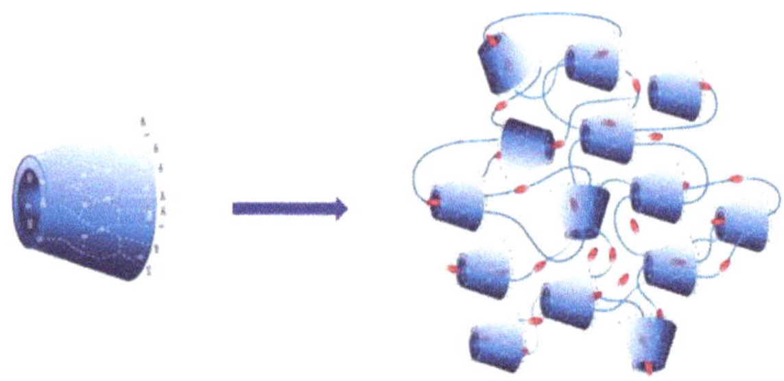

Figure 2.45. Schematic representation of the conversion of βCD (left) to polymeric βCD (right) with the ability to accommodate guest molecules (red ovals).

compatible with a variety of common energetic fillers, and small hazard tests confirmed that the nitrated materials are safe to handle. The new polymers (figure 2.45) offer solutions for the development of inert and energetic binders that could replace NC in propellants and other energetic compositions while offering improvements in processability and many other properties.

2.6 Conclusions

This chapter provides a brief overview of the manufacturing processes and logistics for common energetic materials, and discusses some more recent developments focusing on chemical structures, raw materials, and synthetic routes that can be used to produce low-molecular-weight and high-molecular-weight energetic products.

The production of energetic materials can involve single-step or multistep processes that must be evaluated carefully to ensure compliance with health and safety protocols. We need to recognise that the major energetic compounds on the market today are difficult to replace given their excellent performance and widespread availability. We also need to acknowledge that the development of new molecules and synthesis protocols that improve the performance, safety, and sustainability of energetic materials will continue to be an important aspect of international research.

References

[1] Agrawal J P 2010 *High Energy Materials.* (Weinheim: Wiley-VCH)

[2] Chen D, Jing D, Zhang Q, Xue X, Gou S and Li H *et al* 2017 Study of six green insensitive high energetic coordination polymers based on alkali/alkali-earth metals and 4,5-bis (tetrazol-5-yl)-2 H-1,2,3-triazole *Chem. Asian J.* **12** 3141–149

[3] Millar D 2012 *Energetic Materials at Extreme Conditions.* (Berlin: Springer)

[4] Amendola F 2020 *MSc Thesis* Cranfield University

[5] Hammerl A, Klapötke T M, Mayer P, Weigand J J and Holl G 2005 Synthesis, structure, molecular orbital calculations and decomposition mechanism for tetrazolylazide CHN7, its phenyl derivative PhCN7 and tetrazolylpentazole CHN9 *Propellants Explos. Pyrotech.* **30** 17–26

[6] Badgujar D M, Talawar M B, Zarko V E and Mahulikar P P 2017 New directions in the area of modern energetic polymers: an overview *Combust Explos Shock Waves* **53** 371–87
[7] Colclough M E, Desai H, Millar R W, Paul N C, Stewart M J and Golding P 1994 Energetic polymers as binders in composite propellants and explosives *Polym. Adv. Technol.* **5** 554–60
[8] Akhavan J 2011 *The Chemistry of Explosives* 3rd edn (Cambridge: Royal Society of Chemistry)
[9] How to comply with REACH chemical regulations GOV.UK (www.gov.uk)
[10] Understand The ITAR—DDTC Public Portal (state.gov)
[11] 20161002-LEGACY_JSP886-V7P01-ILSPol-FINAL-O.pdf (publishing.service.gov.uk)
[12] Cumming A 2020 Advanced processing—the changing face of manufacturing *Propellants Explos. Pyrotech.* **45** 7–8
[13] Klapötke T M 2005 New nitrogen-rich high explosives ed T M Klapötke ed *High Energy Density Materials* pp 85–121 (Berlin: Springer)
[14] Cheng T 2019 Review of novel energetic polymers and binders—high energy propellant ingredients for the new space race *Des. Monomers Polym.* **22** 54–65
[15] Golding P, Duncan Mearns D, Bellamy A J, Contini A E and Dossi E 2017 *Polyphosphazenes* US 9556314 https://patents.google.com/patent/US9556314
[16] Luppi F, Cavaye H and Dossi E 2018 Nitrated cross-linked β-cyclodextrin binders exhibiting low glass transition temperatures *Propellants Explos. Pyrotech.* **43** 1023–031
[17] Dossi E, Bolton M, Kister G and Afsar A 2021 Cross-linking of γ-cyclodextin using non-toxic polyethylene glycol spacer units *ChemistrySelect* **6** 7727–731
[18] Dossi E, Earnshaw J, Ellison L, Rabello dos Santos G, Cavaye H and Cleaver D 2021 Understanding and controlling the glass transition of HTPB oligomers *Polym. Chem.* **12** 2606–17
[19] HSE explosives—licensing
[20] Dossi E and Manton G The manufacture of explosives substances *JS333* MOD, ch 10
[21] Hazard symbols and hazard pictograms—chemical classification (https://www.hse.gov.uk/)
[22] Health and safety at work etc. Act 1974 (https://www.legislation.gov.uk/)
[23] The manufacture and storage of explosives regulations 2005 (https://www.legislation.gov.uk/)
[24] The Explosives Regulations 2014 (https://www.legislation.gov.uk/)
[25] Dimian A, Bildea C and Kiss A 2014 *Integrated Design and Simulation of Chemical Processes* 2nd edn (Amsterdam: Elsevier)
[26] Barker M, Rawtani J and Mackay S 2005 *Practical Batch Process Management* (Amsterdam: Elsevier) pp 1–16
[27] *Types of Reactors: Basic and Pharma Classification* Pharma GxP
[28] *Chemical Manufacturing: Batch Processing Vs. Continuous* (ecolink.com)
[29] *Batch vs. Continuous* | AIChE
[30] *Difference between Continuous and Batch Process | Continuous vs Batch Process | MindsMapped*
[31] *Flow Chemistry* (https://www.organic-chemistry.org/)
[32] Movsisyan M, Delbeke E I P, Berton J K E T, Battilocchio C, Ley S V and Stevens C V 2016 Taming hazardous chemistry by continuous flow technology *Chem. Soc. Rev.* **45** 4892–928
[33] Fleming J 2019 Advanced manufacturing of energetics: previous experience and new opportunities, Explosives Engineering, Institute of Explosives Engineers *J. Inst. Explos. Eng.*
[34] Darvas F, Hessel V and Dorman G 2014 *Fundamentals: Fundamentals and Applications* (Berlin: De Gruyter)

[35] Kyprianou D, Berglund M, Emma G, Rarata G, Anderson D, Diaconu G and Exarcho V 2020 Synthesis of 2,4,6-trinitrotoluene (TNT) using flow chemistry *Molecules* **25** 3586–601
[36] Zuckerman N B, Shusteff M, Pagoria P F and Gash A E 2015 Microreactor flow synthesis of the secondary high explosive 2,6-diamino-3,5-dinitropyrazine-1-oxide (LLM-105 *J. Flow Chem.* **5** 178–82
[37] Hessel V, Kralisch D, Kockmann N, Nol T and Wang Q 2013 Novel process windows for enabling, accelerating, and uplifting flow chemistry *ChemSusChem* **6** 746–89
[38] Carpentier F, Felpin F-X, Zammattio F and Le Grognec E 2020 Synthesis of 5-substituted 1H-tetrazoles from nitriles by continuous flow: application to the synthesis of valsartan *Organ. Process Res. Dev.* **24** 752–61
[39] O'Neil M J (ed) 2001 *The Merck Index: An Encyclopedia of Chemicals, Drugs, and Biologicals* 13th edn (Whitehouse Station, NJ: Merck and Co., Inc.) p 1733
[40] Urbanski T 1983 *Chemistry and Technology of Explosives* vol 4 (Amsterdam: Elsevier) pp 138–95
[41] Technical Fact Sheet—2,4,6-Trinitrotoluene (TNT) (https://www.epa.gov/)
[42] Safety Data Sheet (https://www.sigmaaldrich.com/IE/en)
[43] Karta Charakterystyki 2 4 6 trinitrotoluen ENG v 12.2 (https://www.nitrochem.com.pl/)
[44] Golding P and Hayes G F 1983 A parametric study of the synthesis of 2,2',4,4',6,6'-hexanitrostilbene from trinitrotoluene and sodium hypochlorite *Propellants Explos. Pyrotech.* **8** 35–9
[45] Kang D I, Hwang J S, Lee M H, Lee S M, Jeong W B, Kim M J, Bang G B and Kim S H 2003 Crystallization and characterization of HSN IV for modern slapper detonators *34th Int. Annual Conf. of ICT* 134/1–34/7
[46] Bellamy A J, Golding P and Ward S 2001 *Patent* GB 2 355 713 A
[47] Bellamy A J, Ward S and Golding P 2002 A new synthetic route to 1,3,5-triamino-2,4,6-trinitrobenzene (TATB) *Propellants Explos. Pyrotech.* **27** 49–58
[48] Boddu V M, Viswanath D S, Ghosh T K and Damavarapu R 2010 2,4,6-Triamino-1,3,5-trinitrobenzene (TATB) and TATB-based formulations—a review *J. Haz. Mat.* **181** 1–8
[49] No authors, 2014 2,4-Dinitroanisole (DNAN) *Toxicol Ind Health* **34** 2–7
[50] Davies P and Provatas A 2006 *Characterisation of 2,4-Dinitroanisole: An Ingredient for Use in Low Sensitivity Melt Cast Formulations.* (Edinburgh: Weapons Systems Division)
[51] Münzel T, Meinertz T, Tebbe U, Schneider H T, Stalleicken D, Wargenau M, Gori T and Klingmann I; CLEOPATRA Study Investigators 2014 Efficacy of the long-acting nitro vasodilator pentaerithrityl tetranitrate in patients with chronic stable angina pectoris receiving anti-anginal background therapy with beta-blockers: a 12-week, randomized, double-blind, placebo-controlled trial *Eur. Heart J.* **35** 895–903
[52] Fung H L 1991 Nitrate therapy: is there an optimal substance and formulation? *Eur. Heart J.* **12** 9–12 PMID: 1874273
[53] Y:\wercs.ps (https://www.pfizer.com/)
[54] Divakaran S and Loscalzo J 2017 The role of nitroglycerin and other nitrogen oxides in cardiovascular therapeutics *J. Am. Coll. Cardiol.* **70** 2393–410
[55] Dupuis J, Lalonde G, Lemieux R and Rouleau J L 1990 Tolerance to intravenous nitroglycerin in patients with congestive heart failure: role of increased intravascular volume, neurohumoral activation and lack of prevention with N-acetylcysteine *J. Am. Coll. Cardiol.* **16** 923–31

[56] Production, import/export, use, and disposal—Toxicological profile for RDX—NCBI Bookshelf (https://www.nih.gov/)
[57] Spyckerelle C 2008 Reduced sensitivity RDX obtained from Bachmann RDX *Propellants Explos. Pyrotech.* **33** 14–9
[58] Johansen Ø H *et al* 2008 RDX and HMX with reduced sensitivity towards shock initiation —RS RDX and RS HMX *Propellants Explos. Pyrotech.* **33** 20–4
[59] Safety-Data-Sheet-RDX-eng.pdf (https://multihost.no/)
[60] Safety-Data-Sheet-HMX-eng.pdf (https://multihost.no/)
[61] Batchelder G W and Mantua N J 1946 *Manufacture of tetryl U.S. Patent 2,405,340*
[62] CDC—NIOSH Pocket Guide to Chemical Hazards—Tetryl
[63] Koch E-C 2019 Insensitive high explosives: III. Nitroguanidine—synthesis—structure—spectroscopy—sensitiveness *Propellants Explos. Pyrotech.* **44** 267–92
[64] Nielsen A T 1997 *Caged polynitramine compound* U.S. Patent #5693794
[65] Nair U R, Sivabalan R, Gore G M, Geetha M, Asthana S N and Singh H 2005 Hexanitrohexaazaisowurtzitane (CL-20) and CL-20-based formulations (review) *Combust Explos Shock Waves* **41** 121–32
[66] Nair U R, Asthana S N, Subhananda Rao A and Gandhe B R 2010 Advances in high energy materials *Def. Sci. J.* **60** 137–51
[67] MSDS CL-20 High explosive interim 1.1, dot class A—proprietary—CAS (chemcas.com).
[68] Safety-Data-sheet-NTO-eng.pdf (https://multihost.no/).
[69] Latypov N V, Johansson M, Holmgren E, Sizova E V, Sizov V V and Bellamy A J 2007 On the synthesis of 1,1-diamino-2,2-dinitroethene (FOX-7) by nitration of 4,6-dihydroxy-2-methylpyrimidine *Org. Process Res. Dev.* **11** 56–9
[70] FOX-7 [Eurenco MSDS].pdf.
[71] Badgujar D M, Wagh R M, Pawar S J and Sikder A K 2014 Process optimization for synthesis of guanylurea dinitramide (GUDN) *Propellants Explos. Pyrotech.* **39** 658–61
[72] Östmark H, Helte A and Carlsson T *Guanylurea-dinitramide (FOX-12): a new extremely insensitive energetic material for explosives applications* (https://researchgate.net/publication/268046924_Guanylurea-dinitramide_FOX-12_A_new_extremely_insensitive_energetic_material_for_explosives_applications)
[73] AJB 38 CEJEM July 08 Bellamy (cranfield.ac.uk)
[74] Zhao X and Liu Z 2013 An improved synthesis of 2,6-diamino-3,5-dinitropyrazine-1-oxide *J. Chem. Res.* **37** 425–26
[75] Jing S, Liu Y, Liu D and Guo J 2016 Research on a new synthesis of LLM-105 using N-nitroso-bis(cyanomethyl)amine *Cent. Eur. J. Energ. Mater.* **13** 21–32
[76] Highsmith T K, Wayne Edwards W and Wardle R B 1996 Synthesis of 4,10-dinitro-2,6,8,12-tetraoxa-4,10-diazatetracyclo[5.5.0.05,903,11]dodecane *Patent Number: 5,498,711*, Thiokol Corporation. 1498386766339565799-5498711 (storage.googleapis.com)
[77] Karaghiosoff K, Klapoetke T M, Michailovskia A and Hollb G 2002 4,10-Dinitro-2,6,8,12-tetraoxa-4,10-diazaisowurtzitane (TEX): a nitramine with an exceptionally high density *Acta Cryst.* **58** 580–81
[78] Xiang D and Zhu W 2017 Thermal decomposition of isolated and crystal 4,10-dinitro-2,6,8,12-tetraoxa-4,10-diazaisowurtzitane according to ab initio molecular dynamics simulations *RSC Adv.* **7** 8347–56

[79] Francois E G and Hanson K 2010 *Task 09-2-29 insensitive explosive formulations containing diaminoazoxyfurazan (DAAF) as replacements for PBXN-7* LA-UR-10–07319, Los Alamos National Laboratory
[80] Koch E-C 2016 Insensitive high explosives II: 3,3′-diamino-4,4′-azoxyfurazan (DAAF) *Propellants Explos. Pyrotech.* **41** 526–38
[81] TNAZ—energeticchemical (google.com)
[82] Sikder N, Sikder A K, Bulakh N R and Gandhe B R 2004 1,3,3-Trinitroazetidine (TNAZ), a melt-cast explosive: synthesis, characterization and thermal behaviour *J. Hazard. Mater.* **113** 35–43
[83] Zhang M-X, Eaton P E and Gilardi R 2000 Hepta- and octanitrocubanes *Angew. Chem. Int. Ed.* **39** 401–04
[84] Sikder A and Sikder N 2004 A review of advanced high performance, insensitive and thermally stable energetic materials emerging for military and space applications *J. Hazard. Mater.* **112** 1–15
[85] Klapötke T M, Fischer N, Piercey D G, Stierstorfe J and Reymann M 2003 *Patent* DE 102011081254 A1
[86] Fischer N, Fischer D, Klapötke T M, Piercey D G and Stierstorfer J 2012 Pushing the limits of energetic materials – the synthesis and characterization of dihydroxylammonium 5,5′-bistetrazole-1,1′-diolate *J. Mater. Chem.* **22** 20418
[87] Klapötke T M Original Pres (1).pdf
[88] Li N, Zhao F, Luo Y, Gao H X, Yao E, Zhou Z, Xiao L, Hu R and Jiang H 2015 Dissolution thermokinetics of dihydroxylammonium 5,5′-bistetrazole-1,1′-diolate in dimethyl sulfoxide**122** 1023
[89] Yu Y, Chen S, Li T, Jin S, Zhang G, Chenb M and Li L 2017 Study on a novel high energetic and insensitive munitions formulation: TKX-50 based melt cast high explosive *RSC Adv.* **7** 31485–1492
[90] Ang G G and Sreekumar P 2012 Polymers as binders and plasticizers—historical perspective *Energetic Polymers*—Binders and Plasticizers for Enhancing Performance 510
[91] Turner N 2012 *Qualification and Characterisation of Nitrocelluloce* (London: Defence Ordnance Safety Group (DOSG))
[92] Desai H J, Cunliffe A V, Lewis T, Millar R W, Paul N C and Stewart M J 1996 Synthesis of narrow molecular weight, ω-hydroxy telechelic poly(glycidyl nitrate) and estimation of theoretical heat of explosion *Polymer* **37** 3471–76
[93] Chen J, Jin B, Luo G, Liu H, Zhang Q, Huang Q and Peng R 2019 Thermodynamics and kinetics of polyglycidyl nitrate-based urethane network formation by microcalorimetry *J. Chem. Thermodyn* **132** 397–404
[94] Paraskos A J, Dewey M A and Edwards W 2006 One pot procedure for poly (glycidyl nitrate) end modification, Northrop Grumman Systems Corp. *US Patent* 7,714,078 B2
[95] Wang X, Shu Y, Lu X, Mo H and Xu M 2018 Synthesis and characterization of PolyNIMMOHTPB-polyNIMMO triblock copolymer as a potential energetic binder *Cent. Eur. J. Energ. Mater.* **15** 456–67
[96] Kawamoto A M, Oliveira J I S, de Cássia R and Dutra L *et al* 2009 Synthesis and characterization of energetic thermoplastic elastomers for propellant formulations *J. Aerosp. Technol. Manag.* **1** 1
[97] Talawar M B, Sivabalan R, Anniyappan M, Gore G M, Asthana S N and Gandhe B R 2007 *Combust. Explos. Shock, Waves* **43** 62–72

[98] EURENCO 2022 New energetics http://eurenco.com/en/high_explosives/new_energetic.html
[99] Kubota N 2007 *Propellants and Explosives: Thermochemical Aspects of Combustion* 2nd edn (New York: Wiley)
[100] Florczak B, Bogusz R, Skupinski W, Chmielarek M and Dzik A 2015 Study of the effect of nitrated hydroxyl-terminated polybutadiene (NHTPB) on the properties of heterogeneous rocket propellants *Cent. Eur. J. Energetic Mater.* **12** 841–54
[101] Ruebner A, Statton G L and Consaga J P 2003 Polymeric cyclodextrin nitrate esters.Mach I, Inc. *US Patent* 6527887
[102] Consaga J P and Collignon S L 1992 Energetic composites of cyclodextrin nitrate esters and nitrate *US Patent* 5114506
[103] Oyler K D 2014 Green primary explosives *Green Energetic Materials* ed T Brinck (Chichester: Wiley) ch 5
[104] Benge R J 1979 *Potassium 4,6-dinitobenzofuroxan, development, manufacture and properties of RD1372* 41 PERME Technical Report
[105] Sabatini J J and Oyler K D 2016 Recent advances in the synthesis of high explosive materials *Crystals* **2016** 5
[106] Fronabarger J W, Williams M D, Sanborn W B, Parrish D A and Bichay M 2011 Environmentally acceptable alternatives to existing primary explosives *Propellants Explos. Pyrotech.* **36** 459–70
[107] Dušek K 1987 Theory of network formation by additional crosslinking of polyurethanes due to biuret and allophanate formation *Polym. Bull.* **17** 481–88
[108] Chen K-S, Chen Y-S, Yu T L and Tsai C-L 2002 Physical properties of tri-isocyanate crosslinked polyurethane *J. Polym. Res.* **9** 119–28
[109] Sekkar V, Gopalakrishnan S and Ambika D K 2003 Studies on allophanate-urethane networks based on hydroxyl terminated polybutadiene: effect of isocyanate type on the network characteristics *Eur. Polym. J.* **39** 1281–90
[110] Daniel M A 2006 Polyutherane binder systems for polymer bonded explosives *DSTO-GD-0492* (Australia: Defence Science and Technology Organisation)
[111] Williams R 2012 *MSc Thesis* Cranfield University
[112] Dossi E, Akhavan J, Williams R G, Doe W J and Gaulter S 2018 Cross-linking of hydroxyl terminated polyols with triethyleneglycol diglycidyl ether: an alternative to toxic isocyanates *Propellants Explos. Pyrotech.* **43** 241–50
[113] Yang B, Huang W M, Li C and Li L 2006 Effects of moisture on the thermomechanical properties of a polyurethane shape memory polymer *Polymer* **47** 1348–356
[114] Broughton E 2005 The Bhopal disaster and its aftermath: a review *Environ. Health* **4** 6
[115] U.S. Environmental Protection Agency *Toluene Diisocyanate (TDI) and Related Compounds Action Plan* (https://epa.-gov/assessing-and-managing-chemicals-under-tsca/toluenediisocyanates-tdi-action-plan)
[116] Bello D, Herrick C A, Smith T J, Woskie S R, Streicher R P, Cullen M R, Liu Y and Redlich C A 2007 Skin exposure to isocyanates: reasons for concern *Environ. Health Perspect.* **115** 328–35
[117] Doe J E and Hoffmann H D 1995 Toluene diisocyanate: an assessment of carcinogenic risk following oral and inhalation exposure *Toxicol. Ind. Health* **11** 13–32
[118] Pradyot P 2007 *A Comprehensive Guide to the Hazardous Properties of Chemical Substances* 3rd edn ed P Pradyot (Hoboken, NJ: Wiley)
[119] Substances restricted under REACH—ECHA (https://european-union.europa.eu/index_en)

[120] Zabalova M V, Levinaa M A and Tiger R P 2019 Polyurethanes without isocyanates and isocyanates without phosgene as a new field of green chemistry: mechanism, catalysis, and control of reactivity *Russ. J. Phys. Chem.* B **13** 778–88
[121] msds (https://www.fishersci.com/us/en/home.html)
[122] Swensson A and Andersson K 1985 Nordiska expertgruppen for gransvarde dokumentation 58 Diiscoyanater *Nordic Expert Group* (http://hdl.handle.net/ 2077/4059)
[123] Guam J *et al* 2011 Progress in study of non-isocyanate polyurethane *Ind. Eng. Chem. Res.* **50** 6517–27
[124] Figovsky O L and Shapovalov L D 2002 Features of reaction amino-cyclocarbonate for production of new type non isocyanate polyurethane coatings *Macromol. Symp.* **187** 325–32
[125] Nanotech enters into agreement for distribution/manufacturing rights of green polyurethane (TM) in the Russian Federation (https://www.nbcnews.com/)
[126] Kar S and Banthia A K 2005 Synthesis and evaluation of liquid amine-terminated polybutadiene rubber and its role in epoxy toughening *J. Appl. Polym. Sci.* **96** 2446–53
[127] Cunuffe V C, Tod D A and Torry S A 2011 Curing method for polyether *Patent* EP2283061 A1 UK
[128] Luppi F, Kister G, Carpenter M and Dossi E 2019 Thermomechanical characterisation of cross-linked β-cyclodextrin binders *Polym. Test.* **73** 338–45
[129] Szejtli J 1998 Introduction and general overview of cyclodextrin chemistry *Chem. Rev.* **98** 1743–754
[130] Zheng M, Endo T and Zimmermann W 2002 Enzymatic synthesis and analysis of large-ring cyclodextrins *Aust. J. Chem.* **55** 39–48
[131] Del Valle E M M 2004 Cyclodextrins and their uses: a review *Process Biochem.* **39** 1033–046
[132] Das S K, Rajabalaya R, David S, Gani N, Khanam J and Nanda A 2013 Cyclodextrins-the molecular container *Res. J. Pharm., Biol. Chem. Sci.* **4** 1694–720
[133] Redenti E, Szente L and Szejtli J 2000 Drug/cyclodextrin/hydroxy acid multicomponent systems. Properties and pharmaceutical applications *J. Pharm. Sci.* **89** 1–8
[134] Kono H, Onishi K and Nakamura T 2013 *Carbohydr. Polym.* **98** 784–92
[135] Szejtli J 2010 Introduction and general overview of cyclodextrin chemistry *ChemInform* **29**

IOP Publishing

Energetics Science and Technology
An integrated approach
Adam S Cumming

Chapter 3

Crystal form and morphology

Ruth M Doherty

3.1 Introduction

Most of the substances that are used in energetic materials (EMs), i.e. propellants, explosives, and pyrotechnics, explicitly for their energy contribution are used as crystalline solids. This includes small organic compounds, such as cyclo-1,3,5-trinitro-1,3,5-triazinane (RDX), 1,3,5,7-tetranitro-1,3,5,7-tetrazocane (HMX), 1,3,5-triamino-2,4,6-trinitrobenzene (TATB), and pentaerythritol tetranitrate (PETN), the structures of which are shown in figure 3.1. Other commonly used crystalline solids are oxidizers, e.g. ammonium nitrate (AN), ammonium dinitramide (ADN), and ammonium perchlorate (AP), and fuels, e.g. aluminum. The characteristics of the solid particles used in a formulation have a profound impact on the processability, safety, performance, and aging properties of the materials in which they are incorporated.

Crystalline solids are those solids in which the molecules or ions in the particles of the solid are arranged in a small repeating volumetric unit, referred to as the unit cell, throughout the whole of the particle. The possible arrangements of the structural elements (molecules or ions) within the unit cell are characterized by the shape of the unit cell and the symmetrical relationships between the molecules or ions within the unit cell. These arrangements are the allowed crystallographic space groups. The space-filling stacking of unit cells in three dimensions gives a crystal lattice that exhibits a unique pattern of diffraction of x-rays that can be used to identify the arrangement of the atoms within the crystal [1].

The properties of interest in consideration of the physical form of the solids used can be divided into two categories: (1) the intrinsic properties of the particles being used, which depend on the crystallographic form of the particle, and (2) the controllable properties that can be varied to produce desired properties in the energetic composition. The first category includes such properties as theoretical maximum density, solubility, and, to some extent, reactivity or sensitivity to initiation by mechanical stimuli. The second category, i.e. those properties that

doi:10.1088/978-0-7503-3943-8ch3

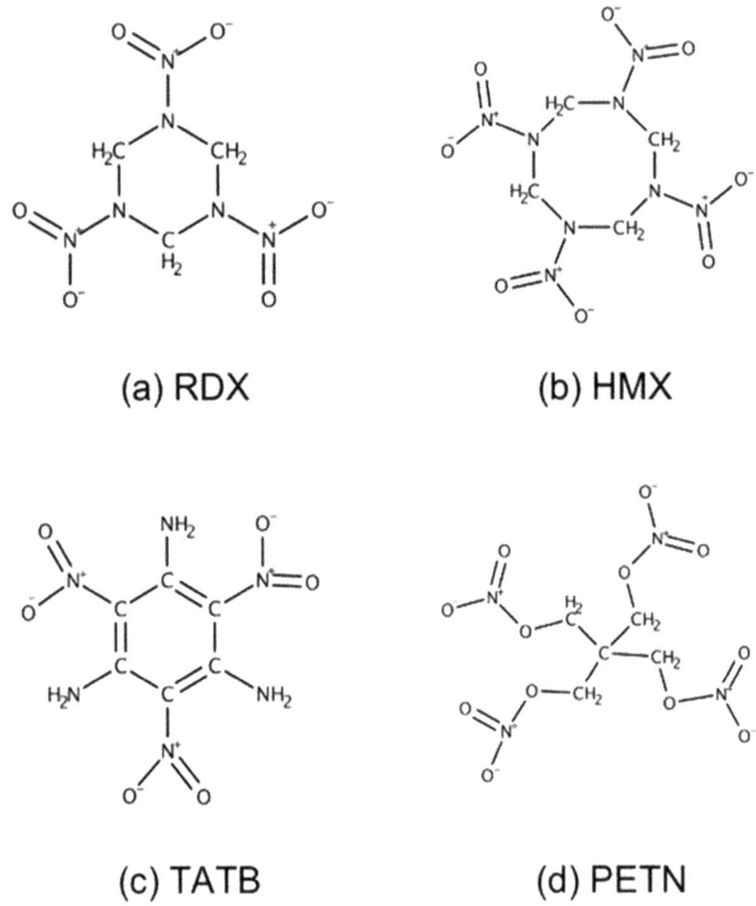

Figure 3.1. Structures of common energetic compounds. (a) Cyclo-1,3,5-trinitro-1,3,5-triazinane (RDX), (b) 1,3,5,7-tetranitro-1,3,5,7-tetrazocane (HMX), (c) 1,3,5-triamino-2,4,6-trinitrobenzene (TATB), (d) pentaerythritol tetranitrate (PETN).

can be controlled or varied, includes particle size and particle size distribution, morphology, and degree of perfection, all of which influence the processability of mixtures containing the ingredient. These properties may also affect the reactivity or sensitivity of an EM.

Several energetic compounds that are widely used in EMs will serve as examples to illustrate the importance of understanding the crystalline nature and variability of properties of solid materials as we consider how to take advantage of their contributions to explosives, propellants, and pyrotechnics.

3.2 Polymorphism and phase diagrams

Many of the commonly used ingredients in energetic compositions have phase diagrams that are characterized by the existence of multiple crystalline forms (polymorphs) within the range of temperatures and pressures normally encountered in EM applications. An example of this behavior is illustrated in the phase diagram of RDX, as shown in figure 3.2 [2]. The phase diagram displays the regions of thermodynamic stability of different solid forms of RDX, though it has been noted that 'ε-RDX has limited chemical stability under the pressure and temperature conditions where it is produced.' [3]

As defined by Haleblian and McCrone [4], a polymorph is 'a solid crystalline phase of a given compound resulting from the possibility of at least two different arrangements of the molecules of that compound in the solid state.' The specification of the phase being *crystalline* conveys the idea that a space-filling arrangement of some small number of the molecules (the unit cell) is repeated throughout the whole of the material.

In the case of small organic compounds, polymorphism (the existence of multiple distinct solid phases) can result from differences in the way in which molecules with the same conformation (spatial arrangement of the atoms in the molecule relative to one another) are packed into a crystal lattice or from differences in the conformation assumed by the molecules in the lattice [5]. This distinction is illustrated by one of the most widely studied EMs of the past 30 years, 2,4,6,8,10,12-hexanitro-2,4,6,8,10,12-hexazazisowurtzitane (HNIW),

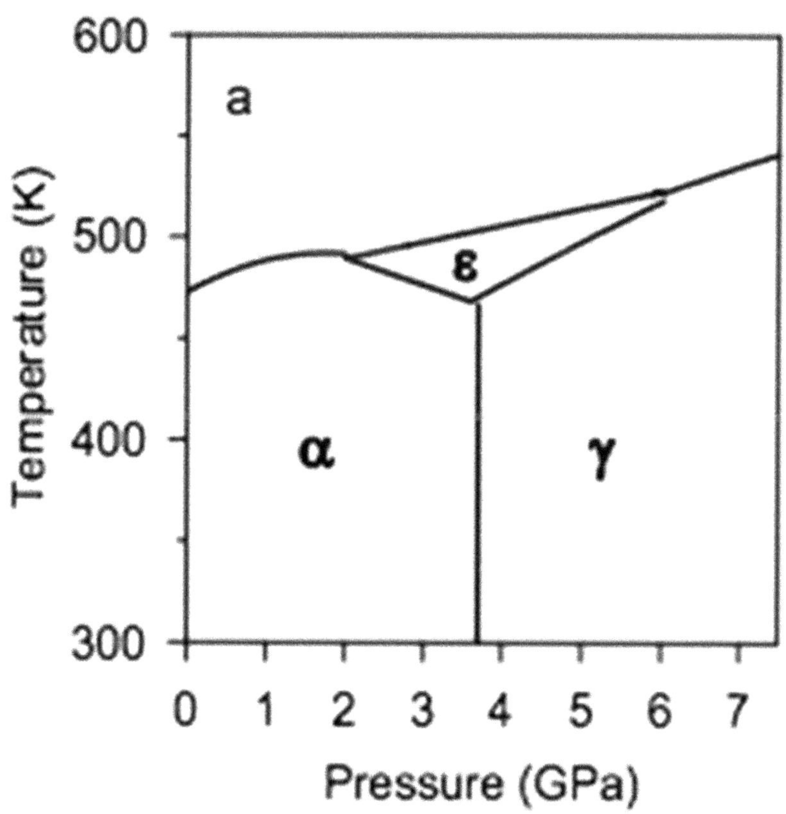

Figure 3.2. Phase diagram of RDX [2]. IOP Publishing. Reproduced with permission.

also known as CL-20. The conformational polymorphism of HNIW is discussed below.

The different arrangements of molecules within the basic unit cell are associated with differences in the strength of the interactions between adjacent molecules in the solid phase. Consequently, different polymorphic forms of a given compound exhibit differences in melting point, solubility, enthalpy of formation (ΔH_f), and other physical properties. It should be noted that multiple polymorphic forms may be obtained at a given temperature and pressure, through the use of different solvents or conditions under which nucleation and growth of the crystals takes place (e.g. degree of supersaturation, agitation, presence of seeds or other materials to promote crystallization) [6]. Polymorphic forms that are not thermodynamically stable (i.e. the lowest-energy form at a given temperature and pressure) may still persist if there is no energetically available pathway for a transition to a lower-energy form. Such persistent forms are categorized as metastable.

Distinguishing the regions of thermodynamic stability can be difficult due to the formation and persistence of polymorphic forms in regions of temperature and pressure where a lower-energy form can also be observed. This persistence over an extended period of time is sometimes referred to as kinetic (as opposed to thermodynamic), and is dependent on the existence of an energy barrier between the metastable (kinetically stable) form and the thermodynamically stable form. As shown in figure 3.3, at a given temperature and pressure, a higher-energy form must overcome the energy barrier to transform to the lower-energy, thermodynamically stable form. The availability of an energetically accessible pathway for the transformation determines the degree of persistence of the higher-energy form.

There has been some discussion of whether solvates or hydrates should be considered as polymorphs or 'pseudopolymorphs'. [7, 8] An example of this can be found in the treatment of HNIW, one crystalline form of which is a hydrate, with the water content varying from 0.17 to 0.50 mol H_2O per mol HNIW [9, 10]. Those forms are different from a compositional as well as a structural point of view, and will not be considered as polymorphs in the following discussion.

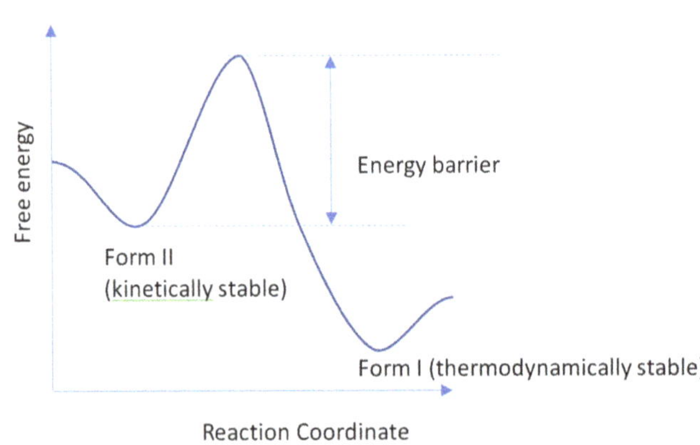

Figure 3.3. Transformation of one polymorphic form to another.

In compounds that exhibit polymorphism, it is very common for the first form isolated to be one of the metastable forms. This phenomenon was articulated in the well-known Ostwald Rule of Stages [11]: '… in the course of transformation of an unstable (or metastable) state into a stable one, the system does not go directly to the most stable conformation (corresponding to the modification with the lowest free energy) but prefers to reach intermediate stages (corresponding to other possible metastable modifications) having the closest free energy difference to the initial state.' [12] This can be understood as underscoring the importance of kinetics over thermodynamics in the early stages of crystal formation and growth, as discussed below.

3.2.1 Differences in physical properties of polymorphs

An example of the differences in physical properties of the different polymorphic forms of a compound is found with the pharmaceutical carbamazepine (CBZ, 5H-dibenzo[b,f]azepine), a tricyclic compound (see figure 3.4) that is used as an anticonvulsant in the treatment of epilepsy [13]. There are many analogies that may be drawn between compounds that are used as active pharmaceutical ingredients (APIs) and those that are used in EMs, so CBZ can serve as a useful material to illustrate the influence of polymorphic form on the utility of a material. Many of the compounds of importance as APIs or EMs are small organic molecules that can adopt multiple conformational arrangements. In the case of APIs, bioavailability is a metric that can be used to evaluate a new material, and in the case of EMs, energy content and sensitivity are used. Both of these types of metrics are related to the characteristics of the solid forms of the materials.

CBZ is known to exist in five polymorphic forms [14], as well as some hydrates. Form I (triclinic form) and Form III (primitive monoclinic form, $P2_1/n$) constitute an enantiotropic pair, i.e. they can undergo a reversible polymorphic phase change [15].

- Form III, which is the commercially available form of the drug, is the thermodynamically stable form below 71 °C. Above 71 °C Form I is the thermodynamically stable form [16].
- The melting points of the two polymorphs are reported as ~171 °C for Form III and 191 °C for Form I [17].
- The solubility in methanol at 22 °C is 0.1221 g CBZ/g CH_3OH for Form I, but only 0.0896 for Form III [14].

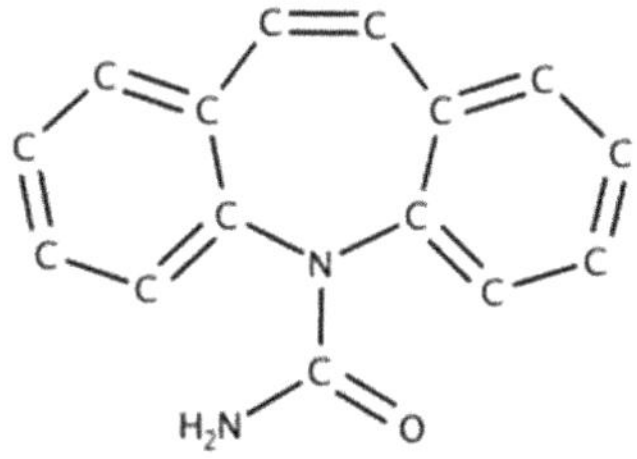

Figure 3.4. Carbamazepine (CBZ).

- The crystal densities of forms I and III are very similar, but the density of Form III is slightly higher (1.35 vs. 1.31 g cm^{-3}) [18, 19]. The densities of two other polymorphs are lower (1.23–1.27 g cm^{-3}) [20].

O'Mahony *et al* noted that it was difficult to measure the solubility of Form I accurately due to the fact that the thermodynamically stable form (Form III) has a lower solubility. Consequently, and when the concentration of CBZ in solution reached a high enough level, Form III began to precipitate: 'The maximum achievable solution concentration during such a process is governed by both dissolution of the metastable phase and growth of the stable phase and this can establish a kinetic steady-state for the solution concentration.' [14] Measuring the solubility of the stable form is more straightforward since this solvent-mediated phase transformation does not occur.

As a rule, the polymorph with the lowest solubility is the one that is thermodynamically stable at the temperature conditions of the measurement [21]. For some pharmaceuticals, this may mean that the preferred form of a drug is not the stable polymorph but rather a metastable form. Since the effectiveness of a drug is related to its bioavailability, compounds that have low solubility in biological systems may be more effective if used in their metastable form, which has a higher solubility than the stable form [22].

In the case of EMs, which are often formulated with plasticized binders and stored for many years with the requirement that performance and sensitivity will not be adversely affected by aging, the caveat regarding a polymorphic change associated with the use of a metastable polymorph must be kept in mind. The plasticizer in a polymeric binder may provide a pathway for solvent-mediated phase transformation, possibly exacerbated by temperature cycling in storage. Such a transformation might result in development of cracks or voids and consequent sensitization of the EM to accidental initiation.

3.2.2 Prediction of crystal structures

The existence of multiple polymorphic forms of a material has been known since the observations of Liebig and Wohler were reported in the first half of the 19th century [23]. The question of the ubiquity of the phenomenon has been debated, but increasingly it has been accepted as very common. A 2015 report identified that 37% of the single-component structures in the Cambridge Structural Database exhibited polymorphism, while well over 50% of the compounds studied in two pharmaceutical databases did so [24].

The ability to predict whether a given compound might exist in multiple polymorphic forms is particularly important in the field of pharmaceuticals, where bioavailability and stability are of paramount importance, as illustrated by the history of ritonavir, an API that was introduced in 1996. The polymorphic form that was used turned out not to be thermodynamically stable. When the stable form made an appearance in 1998, its lower solubility and consequent lower bioavailability caused the oral capsule to be removed from the market [25].

Polymorphic stability is also important in the field of EMs, as the required service lifetimes of some items containing EMs may be decades long. The environment surrounding the particulates used in energetics often involves the presence of plasticizers, in which the energetic compounds may have non-negligible solubility. Oscillations in temperature and humidity can also affect the condition of the EMs. One of the most widely publicized failures of an EM in recent years involved the use of phase-stabilized ammonium nitrate (PSAN) formulations in automobile airbags (see section 3.3.3).

Progress in the ability to predict crystal structures has been assessed periodically since 1999 under the auspices of the Cambridge Crystallographic Data Centre. The results of the most recent assessment in a blind crystal structure prediction (CSP) assessment, the sixth in the series, were reported in 2016 [26]. The participants in the comparisons were given '… two-dimensional chemical diagrams and crystallization conditions of a set of target systems where the experimental structure had been determined but not yet reported.' [26] In the 2016 study five molecules representing different sets of structural features were examined. Compound XXIII, the compound that most closely resembles the kinds of molecules used in EMs, was known to exist in five polymorphic forms.

As shown in figure 3.5, compound XXIII comprises 26 non-hydrogen atoms, several bonds around which rotation is free to occur, some hetero-atoms (N, O, Cl), and the possibility of hydrogen bonding. The details of the approaches taken by the participants may be found in [26].

The aim of the comparison of predictive methods was not only to see whether the participants could predict the known polymorphs of the target materials, but also to examine the structure of the potential energy surfaces to assess where the local minima were located. For the predictions of the structure of compound XXIII that were submitted, 14 crystal structures were submitted for polymorphs A, B, and D, and three structures each for polymorphs C and E. After candidate conformations were identified, the energies of the proposed crystal structures were generated and the candidate crystal structures were then ranked in order of stability or likelihood of occurrence. Of the structures submitted, the actual structure was predicted in the list of the top candidates for four of the five polymorphs. In only one case, for Form B, was the thermodynamically stable structure ranked number 1. The authors of the study ultimately concluded 'state-of-the-art CSP calculations are now applicable to a

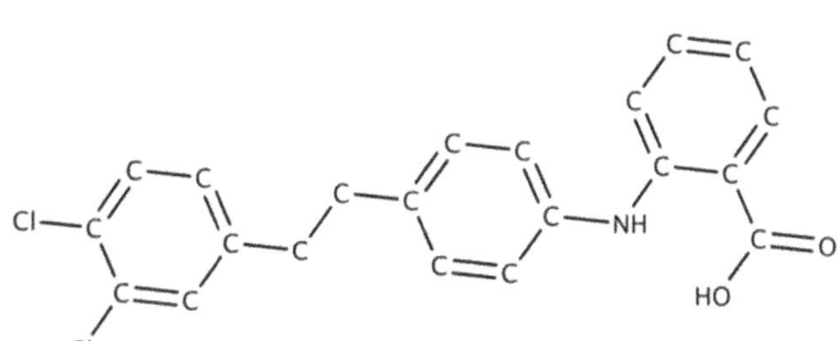

Figure 3.5. Compound XXIII in blind crystal structure prediction [26].

wider range of solid forms, such as salts and hydrates, as well as larger more flexible molecules' [26].

The complexity of the potential energy surfaces, with many local minima found in the calculations, suggests that there may be other polymorphic forms that might be isolable under some set of conditions. The fact that many compounds are known in only a single crystalline form was attributed by McCrone to a low level of investment in searching for polymorphs [27]. Price has identified several possible explanations for the failure to find more polymorphs [28]. Among the explanations are the use of approximations, such as neglecting thermal effects; failure to do experiments under conditions that would produce a predicted polymorph; and lack of analytical methods that would allow the detection and characterization of a new polymorph.

3.3 Polymorphism in energetic materials

The most commonly used ingredients in EMs are all known to exhibit polymorphism, as shown in table 3.1. In addition to the compounds shown in table 3.1, TATB is reported to undergo a high-pressure phase transition above 4 GPa from the known structure (space group P-1) to a monoclinic phase, but the crystal structure determination of that phase has not been reported [29]. These compounds are typically used in a polymorphic form that is stable at ambient temperature and pressure, and often the particular form is not identified in a material specification, such as the Detail Specification for RDX [30], though properties characteristic of the desired polymorph (e.g. melting point) may be included. One exception to this general rule is found in the specification for HMX, in which the β polymorph is identified as the form to be used, and there is a limit on the allowable quantity of the α polymorph (maximum 0.10 wt%) [31]. The α form is stable at high temperature but metastable at ambient temperature. It is also more sensitive than the β form to mechanical stimuli such as impact and friction [32].

As illustrated by the example of HMX above, the relationship between polymorphic form and susceptibility to accidental initiation (sensitivity) is an important concern in EM technology. The benchmark for insensitivity in commonly used energetic ingredients is TATB. Its profound insensitivity to initiation by impact is related not only to its relatively low energy content ($\Delta H_f = -541.3$ kJ kg^{-1}) and low oxygen balance (−55.8%) [33], but also to the arrangement of the molecules in the crystal lattice. The presence of layering of planar or nearly planar molecules and the possibility of intermolecular hydrogen bonding have been identified as factors contributing to insensitivity [34, 35].

3.3.1 Hexanitrohexaazaisowurtzitane (HNIW, CL-20)

HNIW is an energetic compound with a polycyclic structure, shown in figure 3.6, viewed from the side and from the top. The arrangement of the nitro groups relative to the ring system is not fixed in the liquid phase or in solution, due to facile interconversion of the possible geometries about the nitrogen atoms in the ring. However, once HNIW crystallizes, the nitro group orientations about those nitrogen

Table 3.1. Polymorphism of common energetic material ingredients.

Ingredient	Number of polymorphs‡	Stable form† (1 atm, 298 K)	Comments
TNT	3	Monoclinic ($P2_1/a$) [36]	Orthorhombic ($Pca2_1$) has no identified region of thermodynamic stability [36]. No x-ray structure for third reported polymorph [37].
RDX	5	α (Pbca) [38]	High P form: γ ($Pca2_1$) [39]. High P, high T form: ε [3]. β ($Pca2_1$) [40] has no region of thermodynamic stability [41].
HMX	4	β ($P2_1/c$) [42]	High T forms: α(Fdd2) [43], δ($P6_122$); γ metastable at high T [42].
PETN	4	PETN-I [44]	High P form: PETN-III [45]. High P, T form: PETN-IV [46]. Form II has no identified region of thermodynamic stability.
HNIW (CL-20)	4	ε ($P2_1/n$)	High P form: ζ [47]. High T form: γ ($P2_1/c$). β has no identified region of thermodynamic stability.
DNAN	6	Form I ($P2_1/n$) [48]	Low T form: III ($P2_1/n$) [49].
NTO	2	α (P1)	β ($P2_1/c$) [50].
AN	5	AN-IV (Pmmn) [51]	High T forms: I, II, III, VI. Low T forms: V and VII.
AP	6	Form I (Pnma) [52]	High P forms: II (Pnma), III, IV, VI. High T form: V (F4bar3m) [52].

‡ Excludes solvates/hydrates.
† Crystallographic space group in parentheses.

atoms are fixed. Consequently, multiple polymorphic forms of HNIW are observed, corresponding to different arrangements of the nitro groups about the ring nitrogen atoms. These fit into the subset of polymorphs referred to as conformational polymorphs. The nitro groups on each of the ring nitrogen atoms (atoms 2, 6, 8, and 12 in figure 3.6(a)) can be oriented away from the center of the molecule (*exo* position, as shown in figure 3.6(b)) or toward the center of the molecule (*endo* position, as shown in figure 3.6(c)). Furthermore, the nitro groups on atoms 4 and 10

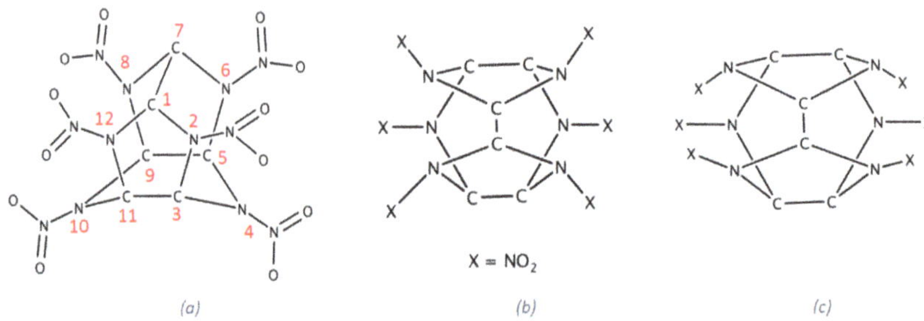

Figure 3.6. Structure of HNIW, side (a) and top (b and c) views. Note the variation of possible arrangement of the nitro groups as illustrated in the top views. Numbering of atoms in the rings is given in red.

Table 3.2. Possible orientations of nitro groups in HNIW [53].

	Orientation (*endo*/*exo*) about ring nitrogen atoms					
	5-Membered rings				6-Membered ring	
Polymorph	N2	N6	N8	N12	N4	N10
α	Exo	Exo	Endo	Exo	Exo	Exo
β	Exo	Endo	Exo	Endo	Exo	Exo
γ	Exo	Exo	Endo	Exo	Exo	Exo
ε	Endo	Exo	Exo	Endo	Endo	Exo

can be oriented toward (endo) or away from (exo) the rest of the molecule. The possible orientations of the nitro groups in the HNIW molecule have been summarized by Foltz [53], as shown in table 3.2.

There are four well-known solid forms of HNIW at ambient pressure (designated α, β, γ, and ε) and one form (ζ) that exists at high pressure [47]. Of these, the α form exists as a hydrate with variable water content. A fifth form, δ, was reported but no crystallographic data exists for such a form [54]. Transitions from either β-, γ-, or ε-HNIW to ζ-HNIW all take place at high pressure, but not all are reversible. No crystal structure is available for the ζ form of HNIW.

As can be seen from the progression of the names of the polymorphs of HNIW, the thermodynamically stable ε form was not one of the first forms to be isolated and characterized. The first named form of HNIW, the α form, is a hydrate. The second form, the β polymorph, is not known to have any region of temperature and pressure in which it is the thermodynamically stable form. The recovery of this form in recrystallizations is an example of Ostwald's Rule of Stages (*vide supra*). As-received samples of the stable ε polymorph have been found to be contaminated with small amounts of the β polymorph. Exposure of ε samples to elevated temperature in the presence of a solvent can lead to disappearance of the β crystals [55].

The high solubility of HNIW in common solvents, including some materials used as plasticizers in energetic compositions, differs from that of other common nitramines such as RDX and HMX. For example, whereas HMX is only slightly soluble in acetone (0.1–5 g/100 ml at room temperature [56]), the solubility of HNIW in acetone at 30°C is 75 g/100 ml [55]. This relatively high solubility can facilitate the process known as Ostwald ripening, i.e. a change in particle size distribution as a result of dissolution of smaller particles and growth of larger particles. Liu *et al* have shown that there is 'a strong effect of solute solubility on Ostwald ripening' [57], with higher solubility accelerating the process. The implications of this for the use of nanoscale particles of HNIW to mitigate the sensitivity problems associated with that material have not yet been assessed.

3.3.2 2,4-Dinitroanisole

Interest in the use of EMs that are less sensitive to accidental initiation in the latter part of the 20th century led to the evaluation of 2,4-dinitroanisole (DNAN) as a potential replacement for 2,4,6-trinitrotoluene (TNT) in many explosive formulations [58]. DNAN was used in some TNT compositions, such as Amatol (which is a mixture of AN and TNT), during World War II when supplies of TNT were low [59], and it has also been used as an insecticide [60] and in the production of azodyes [61]. The main reason for the renewed interest in DNAN for EMs is its insensitivity. The performance of DNAN as an explosive is inferior to that of TNT, and it has a density less than that of TNT, but its melting point is sufficiently low (<100°C) for it to be considered as a matrix material for melt-cast explosives analogous to Composition B (mixture of RDX and TNT) or Octol (mixture of HMX and TNT). Evidence of irreversible volumetric expansion (irreversible growth) was observed in items loaded with explosive compositions based on DNAN that were subjected to thermal cycling. Investigations into this phenomenon revealed complex polymorphic behavior [62].

The first crystal structure reported for DNAN showed that it crystallizes in a monoclinic space group, $P2_1/n$, with a measured density of 1.56 g cm^{-3} at 298 K [48]. The melting point of this form of DNAN (referred to as Form I or α) has been reported to be in the range of 94.6 to 96.99°C in various sources [48, 63, 64]. In 2007 Xue *et al* reported a second monoclinic crystal structure for DNAN (Form II or β), also in space group $P2_1/n$, but with slightly different cell parameters [65]. The wide range of values reported for the melting point of DNAN may be due to the cocrystallization of forms I and II under some conditions of crystallization [64].

DNAN is known to exist in six polymorphic forms [49]. Forms I and II are known to exist at ambient temperature (around 298 K) and pressure. A third polymorph (DNAN-III) is observed at low temperatures, below 266 K. It is observed in the transformation from DNAN-II, in which the nitro groups are disordered, to Form III, in which the nitro groups are ordered. Forms II and III both exist in the crystallographic space group $P2_1/n$, with the difference between the forms being one of the degrees of order of the nitro groups in the crystal lattice. The II → III transition is also observed when the pressure is increased to ~0.1 GPa. Both a decrease in temperature and an increase in pressure result in less space for the disordered arrangement and provide the driving force for the transition to an ordered state. This change at ~266 K has also been interpreted as a 'low-temperature phase transition' of Form II characterized by negative thermal expansion [66] rather than the emergence of a distinct polymorphic form. The differently ordered arrangements are sometimes referred to as 'two β forms'. [64]

The strong similarities in polymorphic forms I and II of DNAN result in interesting crystallization behavior. When DNAN-I is heated above its melting point and then cooled, the melt undergoes supercooling, and crystallization does not occur until the temperature has dropped to ~331 K. Subsequent heating shows an endotherm at 356 K, indicating that the DNAN has crystallized as DNAN-II [64].

When DNAN crystallizes from solution, both the polymorphic form and the morphology of the crystals formed are strongly dependent on the conditions of the crystallization: the choice of solvent, the time over which the crystallization occurs, and the concentration of DNAN in the solvent. DNAN dissolved in ethanol at 313 K, allowed to cool to room temperature, and crystallized slowly over three days by evaporation of the solvent produced block-shaped crystals of Form I. Heating the solution to a higher temperature (343 K), followed by one day of crystallization, produced Form II [66]. Denekamp *et al* found that a solution of DNAN heated to 343 K, filtered, and cooled to room temperature produced a mixture of forms I and II. If left in the supernatant liquid, the mixture converted entirely to Form I (the thermodynamically stable form at room temperature) [64].

In order to avoid the problems associated with irreversible growth of DNAN-based formulations during low-temperature thermal cycling, Ward *et al* sought to suppress the phase transition at low temperature between DNAN-II and DNAN-III. They examined the influence of several structurally similar compounds on the crystallization of DNAN [49]. As shown in figure 3.7, 2, 4-dinitrotoluene and 2,4-dinitrobenzene have the same structure as DNAN, with the exception of the group at the top of the ring in the structures below. The deliberate addition of a small amount of an impurity, referred to as a dopant, in the mixture to be crystallized can result in the incorporation of the dopant in the crystal lattice. The presence of the dopant, with smaller substituents (methyl or hydrogen in place of methoxy) on the carbon adjacent to one of the nitro groups can change the system enough to suppress or eliminate the polymorphic transformation at low temperature. In this case, with 5% 2,4-dinitrotoluene (DNT), the polymorphic transition is moved from 266 K to below 240 K. Increasing the amount of DNT dopant to 10% further suppresses the temperature of the phase transformation to 210–190 K. Similar phase transition suppression results were obtained using 1,3-dinitrobenzene (DNB) dopant at levels of 5% and 10%.

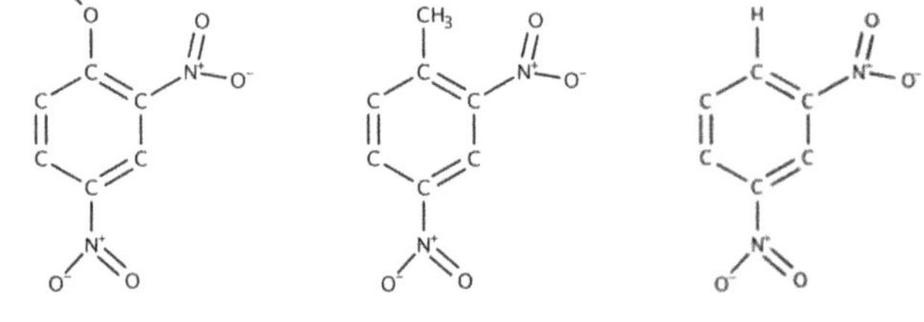

Figure 3.7. Nitroaromatic compounds: (a) DNAN, (b) DNT, (c) DNB.

3.3.3 Ammonium nitrate

Ammonium nitrate (AN) is used in explosive mixtures mainly as an ingredient of industrial explosives such as ammonium nitrate-fuel oil (ANFO), though it has been used in military explosives such as Amatol. It is also widely used as a fertilizer. With an oxygen balance of +20%, it is included as an oxidizer for some other component, such as fuel oil in ANFO, and is not used as the main energetic compound. Its contribution to the energy of such mixtures derives from its oxygen content.

AN itself is assigned to Hazard Class and division 5.1 (Oxidizing Substances) [67] if it contains less than a minimal amount (<0.1%) of an additive or impurity that may serve as fuel for the AN oxidizer. AN-containing blasting agents, such as ANFO, are assigned to Hazard Class 1 (Explosives). Nonetheless, AN has been involved in several explosive incidents of note. Among the most spectacular in terms of damage are the 1921 explosion in Oppau, Germany [68], the 1947 explosion of AN aboard a ship in Texas City, Texas [67], the 2001 explosion of fertilizer-grade AN (FGAN) in a warehouse in Toulouse, France [67], and most recently the 2020 explosion in Beirut, Lebanon [69].

Two grades of AN are defined based on the porosity of the AN prills: (1) technical grade (TGAN), with relatively high porosity in the prills, and (2) fertilizer grade (FGAN), with relatively low porosity in the prills. Zygmunt and Buczkowski [70] discussed a method by which FGAN can be converted to a more porous form (TGAN) through heating. In most of the cases cited above, there was a fire that preceded the explosion, leading not only to polymorphic changes but also to decomposition of some of the AN. The presence of contaminants and the very large masses of AN involved may also have contributed to the catastrophic nature of these incidents.

The phase diagram of AN is very complex, involving up to seven polymorphic forms (AN-I through AN-VII [71]), and including phase boundaries at ambient pressure that fall within the normal temperature range for operations (−40°C to +60°C). The phase diagram for the commonly used solid solution of AN and potassium nitrate (KN) is shown in figure 3.8. The AN phase that is stable at room temperature (around 20°C–25 °C) is AN-IV. It transforms to a higher-temperature polymorph, AN-III, around 32°C, and to a low-temperature form, AN-V, below −16°C [72]. The addition of KN causes the temperatures at which the transformations occur to shift, providing some phase stability [73]. The exact boundary location depends on humidity as well.

Two problems that have been encountered in the use of AN in energetic compositions are related to the polymorphic phase instability of AN. These problems, the introduction of porosity and disaggregation of the materials, are due to polymorphic changes that may occur in the course of diurnal temperature cycling [74]. Thermal cycling leads to increased porosity and in some cases decrepitation of the material, which can result in unexpectedly rapid and uncontrolled release of energy in a propellant and increased sensitivity to accidental initiation in explosives. Degradation over time is especially problematic in climates with high temperature and high absolute humidity. Diverse efforts to suppress the

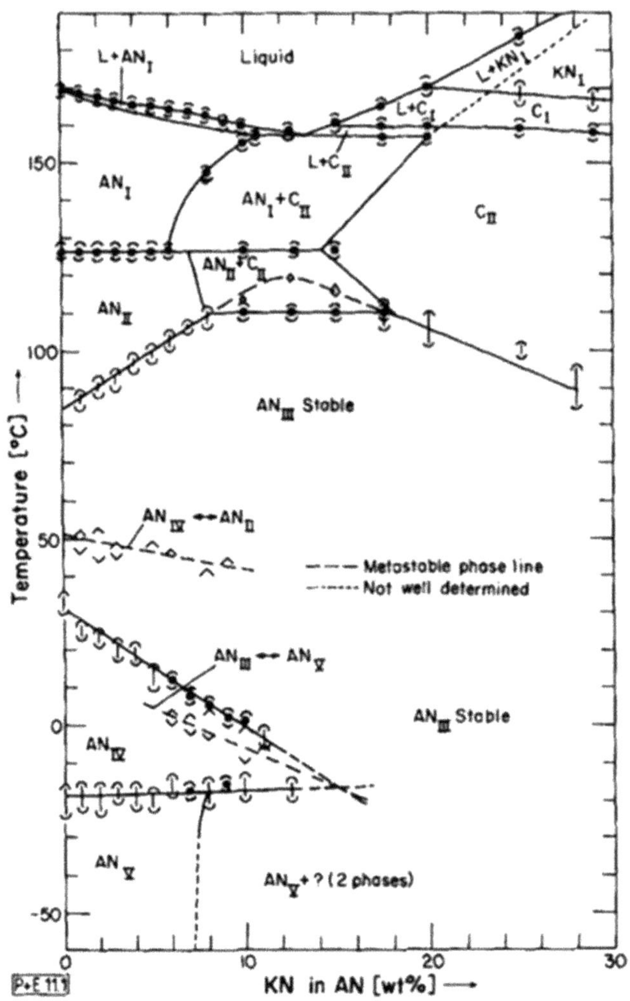

Figure 3.8. Phase diagram of AN-KN system [73]. Reproduced with permission from Wiley.

phase transitions have been pursued, many of them focused on the use of potassium nitrate to produce a solid solution with AN that is referred to as phase-stabilized ammonium nitrate (PSAN).

In the past 15 years, PSAN used in automotive airbag inflators was implicated in failures of the devices that led to the confirmed deaths of 13 people in the US as a result of improper functioning of vehicle airbags. The National Highway Traffic Safety Administration report on the state of the 'Takata airbag recalls' summarizes the cause of the failures as follows: 'The precise cause of the chemical propellant's accelerated burn rate in defective Takata inflators is the tendency of the chemical—phase-stabilized ammonium nitrate ("PSAN")—to degrade over time when exposed to high absolute humidity and high temperature thermal cycling.' [75] Degradation of the propellant in the airbags resulted in uncontrolled burning that caused fragmentation of the metal case, with fragments of metal shooting into the vehicle. Takata recalled the PSAN-based airbag inflators that did not contain a desiccant to absorb ambient moisture and mitigate the degradation of the PSAN.

3.4 Physical particle characteristics

The utility of a compound as an ingredient in an energetic formulation is determined by many factors that can be controlled to some extent. While the intrinsic properties of a particular material, such as the theoretical maximum density (based on the crystal structure) of a given polymorph at a temperature and pressure of interest cannot be modified, there are several attributes that are controllable. Processing techniques that can be used to produce the particles with the desired properties include the following:

- Crystallization from solution under controlled conditions (choice of solvent, temperature, and concentration) [76].
- Mechanical means of particle size reduction, e.g. milling or grinding.
- Spray crystallization [77].
- Crystallization through rapid expansion of supercritical solutions [78].
- Spheroidization to produce smoothed particles, and remove surface defects [79].

In addition to the mean particle size, other attributes that are important to the processability of a material include the particle size distribution, the crystal habit and morphology, and the quality of the crystals.

3.4.1 Crystal size and morphology

The processability of a formulation, i.e. the degree to which it can be mixed and loaded into suitable hardware, depends on particle size, particle size distribution, and shape (morphology or crystal habit) of the particles. In order to be of practical use in an energetic composition, a material must not only be available in sufficient quantities at an acceptable price, but also in a range of particle sizes in a shape that is amenable to mixing with other ingredients. To some extent the innate solubility of a material may limit the size of the crystals that are likely to be formed, as in the case of TATB, which is insoluble in most common solvents and hence less amenable to conventional methods of recrystallization [80]. The quality or degree of perfection of the crystals is also an important factor in determining the sensitivity to initiation by various mechanical stimuli [81, 82].

These important characteristics of particles can be controlled by the way in which the particles are formed in the initial crystallization process, and can be modified to improve the processability of a material [83]. The morphology of the particles in particular can be controlled by the choice of solvent and crystallization conditions. As can be seen in figure 3.9, the morphology of ADN crystals obtained from methanol solution is acicular (needle-like), whereas the crystals obtained from trifluoroethanol have a more plate-like morphology.

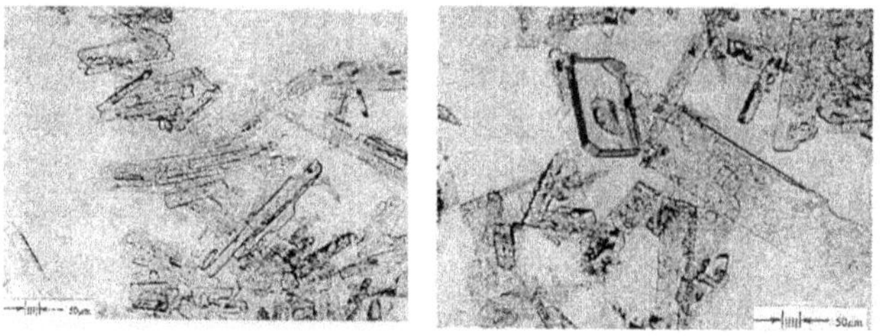

Figure 3.9. Influence of solvent on morphology of ADN.

It should be pointed out that the existence of a material in several morphologies—e.g. plates, needles, and prismatic particles—is not by itself evidence of distinct polymorphs. The angles formed by the intersection of the faces in a faceted crystal do reflect the underlying arrangement of the molecules or ions in the crystal lattice, but the rates at which additional material is added to the growing faces of a particle are dependent on the crystallization conditions as well as the energies associated with transferring a molecule from solution to the growing particle. Both the polymorphic form and the morphology are influenced by the choice of conditions for crystallization from solution [7, 84]. The polymorphic form that results from crystallization from solution can be guided by the presence of a small amount of seed crystals of the preferred polymorph, or through the use of other impurities that may provide a surface upon which the growth of crystals may be initiated [85].

The formation of crystals from solution is one of the most common methods of producing particles of organic compounds. The process of crystallization can be considered in successive steps, commonly referred to as nucleation and growth. As noted by Bernstein, 'once the growth process has begun the structure of the crystal form has essentially been determined.' [7] It is the nucleation step—the formation of the first aggregates of molecules to which additional molecules will be added in the growth stage—that lays the foundation for the polymorph that is to be formed under the prevailing conditions. Modeling has shown that kinetics plays a major role in the nucleation step, as well as in the subsequent growth of the particles [86].

The ideal particle shape for achieving high solids loading in an EM is roughly spherical, with a smooth surface [87, 88]. Unfortunately, it is often the case that both old and new materials commonly crystallize in a morphology that is unfavorable for processing [89]. Acicular (needle-like) crystals are particularly difficult to process, but are very common in crystallization from solution. For some materials with a propensity for crystallizing in an acicular form, such as ammonium nitrate or ammonium dinitramide [90], it is possible to obtain a more useful morphology by prilling [91]. The prilling process involves the solidification of a molten material or heated, highly concentrated solution as droplets of the melt or solution fall from a reservoir of material through a cooling atmosphere. As the droplets cool, they form nearly spherical solid particles, referred to as prills [92]. The nearly spherical shape makes the material easier to process than would be the case for a material with an acicular habit.

Loading EMs into test items or other hardware is commonly done by pouring the material after the solid ingredients have been added to a fluid binder phase and mixed. This process is also referred to as casting. The fluid phase may be a compound that is processed above its melting point, then cooled to allow it to solidify after the other ingredients have been incorporated. This is referred to as melt casting. Examples of ingredients used in this way are TNT and DNAN. Alternatively, a liquid polymeric binder may be used. After the incorporation of the particulate ingredients, the binder is cured through a chemical reaction to give a solid material.

The use of conventional vertical mixers to perform the mixing step imposes a limit on the volumetric solids loading for compositions processed in this way. The mixture of ingredients must have a viscosity low enough to allow a uniform distribution of ingredients in the fluid phase and to allow loading into hardware but high enough to prevent settling of the dispersed solid fill while the mixture is still fluid (i.e. before cooling and solidification of melt-cast mixtures or curing of polymeric binders)[1]. Other processing technologies, such as resonant acoustic mixing [93, 94] or extrusion, can be used when the performance requirements require higher solids loading than can be accommodated by conventional mixers and loading a viscous material into the final hardware is feasible.

3.4.2 Crystal quality

The susceptibility of EMs to unplanned initiation is characterized by various types of tests intended to quantify the sensitivity of those materials to various stimuli (impact, friction, electrostatic discharge, shock) [95]. The emphasis in the last quarter of the 20th century on producing forms of EMs with low levels of sensitivity resulted in new versions of known materials such as RDX and HMX, in which the response to low-level stimuli was less violent than in previously available forms.

In 2000, reports were beginning to be published on the subject of a less-sensitive form of RDX, dubbed 'insensitive RDX' or I-RDX® by the manufacturer (SNPE, now EURENCO) [96]. Later the designation 'reduced sensitivity' (RS) rather than 'insensitive' was adopted to describe such forms [97], with the 'RS' designation also being applied to other EMs such as HMX [98] and HNIW [99]. The direct nitration (Woolwich) process for synthesizing Type I RDX [30] is used by Eurenco and results in a material with very little HMX as a byproduct. In contrast to this, the acetic anhydride (Bachmann) process is used by other manufacturers to produce Type II RDX [30] with HMX present as an impurity. The amount of HMX allowed in Type II RDX according to the US specification for RDX is 4–17 wt% [100]. Some of the forms of RS-RDX produced within the past 20 years are based on Type II RDX purified to remove most of the HMX impurity [101].

[1] One notable exception to this is the so-called 'sedimentation-cast' octol [132]. In this case, the settling of HMX to give a density and composition gradient is intentional, and is facilitated by maintaining the temperature of the loaded items above the melting point of TNT (the fluid phase), while the loaded cases are subjected to vibration.

The differences in sensitivity of various types of RDX were attributed to differences in crystal quality, as reflected in the presence of internal defects (intragranular voids and solvent inclusions) as well as impurities such as HMX, which is commonly found in RDX produced by the Bachmann process [102, 103]. I-RDX and some other RS variants of RDX are produced by proprietary recrystallization processes that result in fewer defects in the crystals, as shown in figure 3.10. In an interlaboratory comparison of seven lots of RDX produced by five manufacturers, it was found that material lots with higher crystal quality were less sensitive to shock initiation than were lower-quality lots. However, there was no discernable difference in sensitivity as evaluated in a small-scale impact test [104].

In the past decade work on the production of less-sensitive forms of known EMs has included studies on the choice of solvent and crystallization conditions [105–107], use of additives [108, 109], and microencapsulation [110]. The use of particles with sizes in the nanometric range (10s to 100s of nanometers in diameter) has been touted as a way of reducing the sensitivity of the particles to mechanical stimuli [111]. However, the opposite behavior has been found in some cases, with the nanometric particulates being more sensitive than the corresponding micron-sized particles [112].

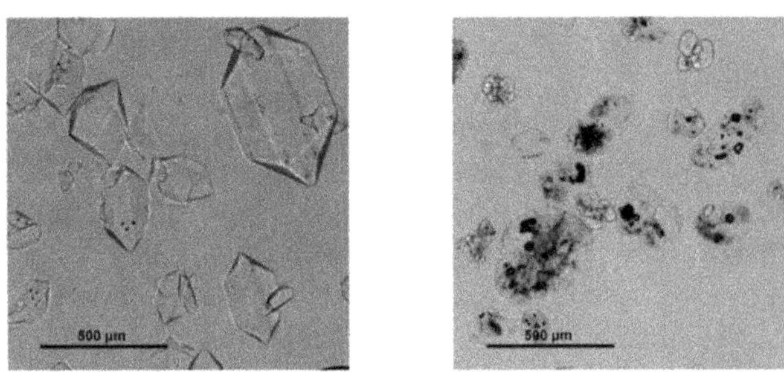

Figure 3.10. RDX crystals optical microscopy with matching refractive index. Recrystallized (left) and as received (right) [103]. Reproduced with permission from Wiley.

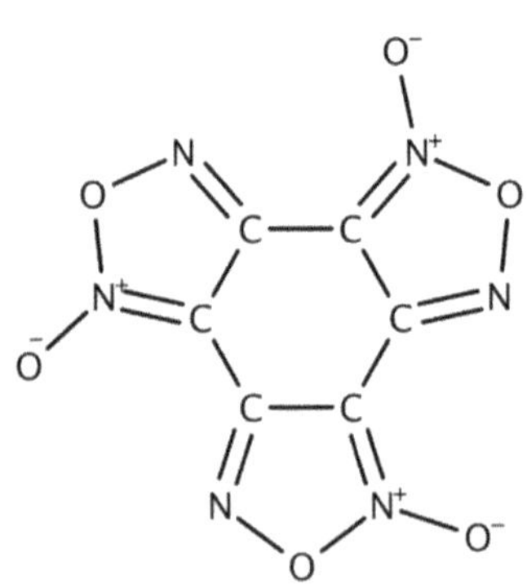

Figure 3.11. Structure of benzotrifuroxan (BTF).

3.4.3 Cocrystallization

The demands placed on energetic compounds intended for use in EMs present challenges for those who would synthesize new materials to be used in EM applications [113]. In addition to the obvious need for a pathway by which to make a new compound, there are other constraints relating to performance (relative to current energetic compounds such as HNIW), safety, long-term stability, availability in significant quantities from a reliable source, and cost.

One approach that has received considerable attention since 2010 is the formation of energetic cocrystals based on compounds that are currently available. Cocrystals, in the most general definition, are crystalline materials comprising more than one molecular species [114]. By this definition, hydrates and other solvates may be considered to be cocrystals, though other definitions restrict the use of the term to multicomponent systems in which each component exists as a crystalline solid by itself [115]. The two components may interact with one another in a variety of ways to form a new crystalline material with a specific stoichiometric ratio of the components. The interactions may include hydrogen bonding, π–π stacking, and amine-nitro group attractions [116].

The desired advantages resulting from the use of cocrystals, as opposed to use of the components as discrete ingredients in EMs, are usually presented as the following:

- A reduction in sensitivity of a powerful but highly sensitive compound, such as HNIW [117].
- An improvement in performance over commonly used ingredients by changing the oxygen balance of the energetic fill in a formulation [118].
- Improved thermal stability [119].

Reduction in sensitivity is a common theme in energetic research, and the production of cocrystals to accomplish this end must be viewed in the context of the other approaches available to achieve the same degree of desensitization. The production of high-quality crystals, as in the case of RS-RDX, and the use of submicron particles have been studied extensively as methods of reducing the sensitivity of the ingredients themselves. The use of polymeric binders offers a means of producing complete formulations with acceptable sensitivity. Each of these approaches has limitations. High-quality crystals are preferable to highly defected crystals in terms of sensitivity, but the intrinsic stability of the compound is not changed. Nanocrystalline materials contain fewer large defects than larger particles, but present challenges in many common processing technologies due to the high viscosity of mixtures containing nanoparticulates. The use of non-energetic binders in a polymer-bonded explosive (PBX) dilutes the energy of the energetic fill. Similar considerations apply to cocrystals.

Two properties that are often associated with the performance of explosives are density and oxygen balance. One often encounters statements to the effect that both higher density and higher oxygen balance are associated with higher detonation pressure and detonation velocity. All other things being equal, this is generally correct, but it is rare that all other things are equal. Given that the detonation pressure is related to the square of the density and the detonation velocity is linearly related to the density [120], one can expect that similar compositions with higher density are to be preferred. For compositions containing the same set of elements (typically C, H, N, and O) higher oxygen balance generally translates to high performance. However, both density and oxygen balance can be increased by the incorporation of heavy elements such as Cl, Br, and I, without increasing performance.

The density of cocrystals most often lies between the densities of the two coformers. In an early study of energetic cocrystals, Landenberger and Matzger reported the densities of 17 cocrystals in which TNT was one of the components [121]. The densities of the cocrystals were determined from x-ray diffraction data collected at 95 K. The densities of the all but one of the coformers were lower than that of TNT at 123 K, which has ρ = 1.704 g cm^{-3} at that temperature. The highest density of the coformers was that of tetrathiofulvalene, with ρ = 1.75 g cm^{-3} at 98 K. In 16 of the 17 cases, the density of the cocrystal was between that of the highest-density polymorphs of TNT and the coformer. The cocrystal with the highest density (ρ = 1.737 g cm^{-3} at 95 K) involved 1-bromonaphthalene, which is a liquid at room temperature. As noted by the authors, the high density of this cocrystal does not translate to expected improvement in performance.

To date, the most powerful energetic ingredient is ε-HNIW, with a density of 2.03 g cm^{-3} at 298 K. In addition to being very powerful, HNIW is quite sensitive, as assessed in the common small-scale safety tests for impact and friction. Cocrystallization with another energetic compound has been explored as a means of mitigating the problems associated with the sensitivity of HNIW while maintaining the performance benefits associated with its heat of formation, oxygen balance, and density [122]. Properties of the cocrystals that have been reported to date are given in table 3.3.

As can be seen in table 3.3, the cocrystals containing HNIW all have densities that lie between those of the two components of the cocrystals, if the thermodynamically stable ε polymorph of HNIW is used for the comparison. More than half have densities that are below the density of the β polymorph of HMX. The calculated detonation velocities of all but the HNIW-HMX cocrystal are similarly unexceptional.

The reduction in sensitivity and increase in thermal stability that often accompanies cocrystallization may be exploited to achieve more powerful but less sensitive explosive compositions. The typical metrics used in the early stages of development of a new material are small-scale sensitivity tests that are, at best, blunt tools to evaluate overall sensitivity [131]. Further investigation of the response of cocrystals to various mechanical and thermal stimuli is needed to provide a more complete picture of the advantages available through formation of cocrystals.

Table 3.3. Cocrystals of HNIW with energetic coformers.

Coformer	Coformer density (g cm^{-3})	Stoichiometry (HNIW:coformer)	Cocrystal density (g cm^{-3})	Calc. det. velocity (m s^{-1})	References
TNT	1.654	1:1	1.84	8402	[121]
HMX	1.906	2:1	1.945	9484	[123]
BTF	1.901	1:1	1.918	8969	[124]
DNB	1.575	1:1	1.880	8434	[125]
MDNT	1.676	1:1	1.883 (95 K)	7760*	[126, 127]
DNDAP	1.55	2:1	1.871	8997	[128]
DNG	1.39	1:1	1.750	8327	[129]
TFAZ	1.87	1:1	1.932	9103	[130]

*Measured at density of 1.40 g cm^{-3}.

3.5 Summary

The particle characteristics of many materials utilized in energetic compositions are critically important to the performance and safety of those compositions. The energy available from a given compound will largely be determined by its molecular bonding, but the specific details of the solid phase in which it exists will strongly influence properties such as density that are also important in the performance of an EM. Both the phase diagram and the particle properties that can be tailored to a desired application are important in determining the utility of a compound as an ingredient in EMs.

Abbreviations

ADN	Ammonium dinitramide
AN	Ammonium nitrate
ANFO	Ammonium nitrate/fuel oil
API	Active pharmaceutical ingredient
CL-20	2,4,6,8,10,12-Hexanitro-2,4,6,8,10,12-hexaazaisowurtzitane
CSP	Crystal structure prediction
DNAN	2,4-Dinitroanisole
DNB	1,3-Dinitrobenzene
DNDAP	2,4-Dinitro-2,4-diazapentane
DNG	3,5-Dinitro-3,5-diazaheptane
DNT	2,4-Dinitrotoluene
EM	Energetic material
FGAN	Fertilizer-grade ammonium nitrate
HMX	Cyclo-1,3,5,7-tetramethylene-2,4,6,8-tetranitramine
HNIW	Hexanitrohexaazaisowurtzitane
MDNT	1-Methyl-3,5-dinitro-1,2,4-triazole
PETN	Pentaerythritol tetranitrate
PSAN	Phase-stabilized ammonium nitrate
RDX	Cyclo-1,3,5-trimethylene-2,4,6-trinitramine
SNPE	Société Nationale des Poudres et des Explosifs
TATB	1,3,5-Triamino-2,4,6-trinitrobenzene
TFAZ	7H-Trifurazano[3,4-b:3′,4′-f:3″,4″-d]azepine
TGAN	Technical grade ammonium nitrate
TNT	2,4,6-Trinitrotoluene

References

[1] Massa W 2004 *Crystal Structure Determination* (Berlin: Springer)

[2] Dreger Z A 2012 Energetic materials under high pressures and temperatures: stability, polymorphism and decomposition of RDX *J. Phys.: Conf.* **377** 102047

[3] Dreger Z A and Gupta Y M 2010 Phase diagram of hexahydro-1,3,5-trinitro-1,3,5-triazine crystals at high pressures and temperatures *J. Phys. Chem.* A **114** 8099–105

[4] Haleblian J and McCrone W 1969 Pharmaceutical applications of polymorphism *J. Pharm. Sci.* **58** 911–29

[5] Cruz-Cabeza A J and Bernstein J 2014 Conformational polymorphism *Chem. Rev.* **114** 2170–91
[6] Desiraju G R 2007 Crystal engineering: a holistic view *Angew. Chem. Int. Ed.* **46** 8342–56
[7] Bernstein J 2011 Polymorphism—a perspective *Cryst. Growth Des* **1** 632–50
[8] Desiraju G R 2008 Polymorphism: the same and not quite the same *Cryst. Growth Des.* **8** 3–5
[9] Nielsen A T, Chafin A P, Christian S L, Moore D W, Nadler M P, Nissan R A and Vanderah D J 1998 Synthesis of polyazapolycyclic caged polynitramines *Tetrahedron* **54** 11793–812
[10] Sysolyatin S V, Lobanova A A, Chernikova Y T and Sakovich G V 2005 Methods of synthesis and properties of hexanitrohexaazaisowurtzitane *Russ. Chem. Rev.* **74** 757–64
[11] Ostwald W Z 1897 Studies on formation and transformation of solid materials *Z. Phys. Chem.* **22** 289–330
[12] Tahri Y, Gagniere E, Chabanon E, Bounahmidi T, Kozisek Z, Candoni N, Veesler S, Boukerche M and Mangin D 2019 Multiscale experimental study and modeling of L-glutamic acid crystallization: emphasis on a kinetic explanation of the Ostwald Rule of Stages *Cryst. Growth Des* **19** 3329–37
[13] National Library of Medicine *Pubchem* (https://pubchem.ncbi.nlm.nih.gov/compound/2554)
[14] O'Mahony M A, Maher A, Croker D M, Rasmuson A C and Hodnett B K 2012 Examining solution and solid state composition for the solution-mediated polymorphic transformation of carbamazepine and piracetam *Cryst. Growth Des* **12** 1925–32
[15] Bernstein J 2020 *Polymorphism in Molecular Crystals* 2nd edn (Oxford: Oxford University Press)
[16] Behme R J and Brooke D 1991 Heat of fusion measurement of a low melting polymorph of carbamazepine that undergoes multiple-phase changes during differential scanning calorimetry analysis *J. Pharm. Sci.* **80** 986–90
[17] Sethia S and Squillante E 2004 Solid dispersion of carbamazepine in PVP K30 by conventional solvent evaporation and supercritical methods *Int. J. Pharm.* **272** 1–10
[18] Reboul J P, Cristau B and Soyfer J C 1981 5H-Dibenz[b,f]azepinecorboxamide-5 (Carbamazepine) *Acta Crystallogr.* **B37** 1844–8
[19] Ceolin R, Toscani S, Gardette M-F, Agafonov V N, Dzyabchenko A V and Bachet B 1997 X-ray characterization of the triclinic polymorph of carbamazepine *J. Pharm. Sci.* **86** 1062–5
[20] Lang M, Kampf J W and Matzger A J 2002 Form IV of carbamazepine *J. Pharm. Sci.* **91** 1186–90
[21] Threlfall T L 1995 Analysis of organic polymorphs: a review *Analyst* **120** 2435–60
[22] Nicoud L, Licordari F and Myerson A S 2018 Estimation of the solubility of metastable polymorphs: a critical review *Cryst. Growth Des* **18** 7228–37
[23] Liebig J and Wohler F 1832 *Ann. Pharm. III* **249** 514
[24] Cruz-Cabeza A J, Reutzel-Edens S M and Bernstein J 2015 Facts and fictions about polymorphism *Chem. Soc. Rev.* **44** 8619–35
[25] Morissette S L, Soukasene S, Levinson D, Cima M J and Almarsson O 2003 Elucidation of crystal form diversity of the HIV protease inhibitor ritonavir by high-throughput crystallization *Proc. Natl Acad. Sci.* **100** 2180–4
[26] Reilly A M *et al* 2016 Report on the sixth blind test of organic crystal structure prediction methods *Acta Cryst.* B **B72** 439–59

[27] McCrone W 1965 *Physics and Chemistry of the Organic Solid State* vol 2 (New York: Wiley Interscience)

[28] Price S L 2013 Why don't we find more polymorphs? *Acta Crystallogr.* **B69** 313–28

[29] Steele B A *et al* 2019 Pressure-induced phase transition in 1,3,5-triamino-2,4,6-trinitrobenzene (TATB) *Appl. Phys. Lett.* **114** 191901

[30] U. S. Army ARDEC 1996 Detail specification RDX (Cyclotrimethylenetrinitramine) *U. S. Army ARDEC, Picatinny Arsenal*

[31] U.S. Army ARDEC 1999 Detail specification HMX (Cyclotetramethylenetetranitramine) *U.S. Army ARDEC, Picatinny Arsenal*

[32] Peterson P D, Lee K-Y, Moore D S, Scharff R J and Avilucea G R 2007 The evolution of sensitivty in HMX-based explosives during the reversion from delta to beta phase *Proc. 15th APS: Shock-Compression of Condensed Matter (Waikoloa, Hawaii)*

[33] Meyer R, Kohler J and Homburg A 2007 *Explosives* 6th edn (Weinheim: Wiley-VCH)

[34] Ma Y, Zhang A, Zhang C, Jiang D, Zhu Y and Zhang C 2014 Crystal packing of low-sensitivity and high-energy explosives *Cryst. Growth Des* **14** 4703–13

[35] Michalchuk A A L, Rudic S, Pulham C R and Morrison C A 2021 Predicting the impact sensitivity of a pollymorphic high exploaive: the curious case of FOX-7 *Chem. Commun.* **57** 11213–6

[36] Vrcelj R M, Sherwood J N, Kennedy A R, Gallagher H G and Gelbrich T 2003 Polymorphism in 2,4,6-trinitrotoluene *Cryst. Growth Des* **3** 2017–1032

[37] Golovina N I, Titkov A N, Raevskii A V and Atovmyan L O 1994 Kinetics and mechanism of phase transitions in the crystals of 2,4,6-trinitrotoluene and benzotrifuroxane *J. Solid State Chem.* **113** 229–38

[38] Hakey P, Ouellette W, Zubieta J and Korter T 2008 Redetermination of cyclo-trimethylenetrinitramine *Acta Crystallogr. Sec.* E **64** 1428

[39] Davidson A J, Oswald I D H, Francis D J, Lennie A R, Marshall W G, Millar D I A, Pulham C R, Warren J E and Cumming A S 2008 Explosives under pressure—the crystal structure of γ-RDX as determined by high-pressure x-ray and neutron diffraction *Cryst. Eng. Commun.* **10** 162–5

[40] Millar D I, Oswald I D, Francis D J, Marshall W G, Pulham C R and Cumming A S 2009 The crystal structure of β-RDX—an elusive form of an explosive revealed *Chem. Commun.* **2009** 562–4

[41] Goldberg I G and Swift J A 2012 New insights into the metastable B form of RDX *Cryst. Growth Des* **12** 1040–5

[42] Cady H H and Smith L C 1962 *Studies on the Polymorphs of HMX* (Los Alamos, NM: Los Alamos Scientific Laboratory)

[43] Cady H H, Larson A C and Cromer D T 1963 The crystal structure of alpha-HMX and a refinement of the structure of beta-HMX *Acta Crystallogr.* **16** 617

[44] Booth A D and Llewellyn F J 1947 The crystal structure of pentaerythritol tetranitrate *J. Chem. Soc.* **1947** 837–46

[45] Tschauner O, Kiefer B, Lee Y, Pravica M, Nicol M and Kim E 2007 Structural transition of PETN-I to ferroelastic orthorhombic phase PETN-III at elevated pressures *J. Chem. Phys.* **127** 094502

[46] Dreger Z A and Gupta Y M 2013 High pressure-high temperature polymorphism and decomposition of pentaerythritol tetranitrate (PETN) *J. Phys. Chem.* A **117** 5306–13

[47] Russell T P, Miller P J, Piermarini G J and Block S 1993 Pressure/temperature phase diagram of hexanitrohexaazaisowurtzitane *J. Phys. Chem.* **97** 1993–7

[48] Nyburg S C, Faerman C H, Prasad L, Palleros D and Nudelman N 1987 Structures of 2,4-dinitroanisole and 2,6-dinitroanisole *Acta Crystallogr.* **C43** 686–9

[49] Ward D W, Coster P L, Hope K S and Pulham C R 2015 Controlling a polymorphic transition in 2,4-dinitroanisole using crystal doping *New Trends in Research of Energetic Materials (NTREM) (Pardubice, Czech Republic)*

[50] Lee K Y and Gilardi R 1992 NTO polymorphs *Proc. MRS Online* **296** 237–42

[51] Chellappa R S, Dattelbaum D M, Velisavlyevic N and Sheffield S 2012 The phase diagram of ammonium nitrate *J. Chem. Phys.* **137** 064504-1–15

[52] Dunuwille M and Yoo C-S 2016 Phase diagram of ammnoium perchlorate: Raman spectroscopic constrains at high pressures and temperatures *J. Chem. Phys.* **144** 244701

[53] Foltz M F, Coon C L, Garcia F and Nichols A L 1994 The thermal stability of the polymorphs of hexanitrohexaazaisowurtzitane, part I *Propellants Explos. Pyrotech.* **19** 19–25

[54] Nielsen A T, Chan M L, Kraeutle K J, Lowe-Ma C K, Hollins R A, Nadler M P, Nissan R A, Norris W P, Vanderah D J and Yee R Y 1989 *Polynitropolyaza caged explosives—part 7* (Naval Weapons Center, China Lake, CA)

[55] von Holtz E, Ornellas D, Foltz M F and Clarkson J E 1994 The solubility of ϵ-CL-20 in selected materials *Propellants Explos. Pyrotech.* **19** 206–12

[56] Dobratz B M 1981 *LLNL Explosives Handbook: Properties of Chemical Explosives and Explosive Simulants* (Livermore, CA: Lawrence Livermore National Laboratory)

[57] Liu Y, Kathan K, Saad W and Prud'homme R K 2007 Ostwald ripening of beta-carotene nanoparticles *Phys. Rev. Lett.* **98** 036102

[58] Nicolich S, Niles J, Ferlazzo P, Doll D, Braithwaite P, Rausmussen N and Ray M 2003 Recent developments in reduced sensitivty melt pour explosives *Energetic Materials: Reactions of Propellants, Explosives and Pyrotechnics (Karlsruhe, Germany)*

[59] Fedoroff B 1958 *Dictionary of Explosives, Ammunion and Weapons (German Section)* (Dover, NJ: Picatinny Arsenal)

[60] Nelson A A 1948 Toxicological investigations of compounds proposed for use as insect repellents. C. Pathological examination *J. Pharm. Expl. Therap.* **93** 26–39

[61] Chudgar R J 2000 Azo dyes *Kirk-Othmer Encyclopedia of Chemical Technology* (New York: Wiley)

[62] Samuels P 2012 Irreversible growth of DNAN based formulations *NDIA IM/EM Symp. (Las Vegas, NV)*

[63] Davies P J and Provatas A 2006 Characterisation of 2,4-dinitroanisole: an ingredient for use in low sensitivity melt cast formulations *Weapons Systems Division* (Edinburgh, South Australia: Defence Science and Technology Organisation)

[64] Denekamp C, Meikler O, Zelner M, Suwinska K and Eichen Y 2018 Controlling the crystal morphology and polymorphism of 2,4-dinitroanisole *Cryst. Growth Des* **18** 1350–7

[65] Xue G, Gong C-R and Chen H-Y 2007 Crystal structure of 2,4-dinitroanisole, C7H6N2O5 *Z. Kristallogr.* **222** 321–2

[66] Takahashi H and Tamura R 2015 Low temperature phase transition induced biaxial negative thermal expansion of 2,4-dinitroanisole *CrystEngComm* **17** 8888–96

[67] US EPA 2013 *Chemical Advisory: Safe Storage, Handling and Management of Solid Ammonium Nitrate Prills* EPA 55-S-13-001 United States Environmental Protection Agency

[68] Robertson R, Thomas H H, Hallimond A F, Bragg W and Threflall R 1924 Investigation on the chemical and physical properties of Oppau ammonium sulphate-nitrate at the Government Laboratory *Trans. Faraday Soc.* **20** 46

[69] Guglielmi G 2020 Nature: Why Beirut's ammonium nitrate blast was so devastating (10 August 2020) https://doi.org/10.1038/d41586-020-02361-x

[70] Zygmunt B and Buczkowski D 2007 Influence of ammonium nitrate prills' properties on detonation velocity of ANFO *Propellants Explos. Pyrotech.* **32** 411–4

[71] Theoret A and Sandorfy C 1964 Infrared spectra and crystalline phase transitions of ammonium nitrate *Can. J. Chem.* **42** 57–62

[72] Sowell R R, Karnowsky M M and Walters L C 1971 The transitions in phases II-III-IV in high purity ammonium nitrate *J. Therm. Anal.* **3** 119–29

[73] Cady H H 1981 The ammonium nitrate-potassium nitrate system *Propellants Explos.* **6** 49–54

[74] Boyars C, Holden J R and Bertram A L 1973 *Minol IV, a New Explosive Composition Containing Ammonium Nitrate-Potassium Nitrate Solid Solution. Part I* (White Oak, Silver Spring, MD: Naval Ordnance Laboratory)

[75] Buretta J D 2017 *The State of the Takata Airbag Recalls* (Washington, DC: National Highway Traffic Safety Administration)

[76] Parmar M M, Khan O, Seton L and Ford J L 2007 Polymorph selection with morphology control using solvents *Cryst. Growth Des* **7** 1635–42

[77] Teipel U 1999 Production of particles of explosives *Propellants Explos. Pyrotech.* **24** 134–9

[78] Stepanov V, Krasnoperov L N, Elkina I B and Zhang X 2005 Production of nano-crystalline RDX by rapid expansion of supercritical solutions *Propellants Explos. Pyrotech.* **30** 178–83

[79] van der Heijden A E D M and Bouma R H B 2004 Crystallization and characterization of RDX, HMX and CL-20 *Cryst. Growth Des.* **4** 999–1007

[80] Hoffman D M and Fontes A T 2008 *Density Distributions in TATB Prepared by Various Methods* (Livermore, CA: Lawrence Livermore National Laboratory)

[81] van der Heijden A E D M and Bouma R H B 2004 Crystallization and characterization of RDX, HMX and CL-20 *Cryst. Growth Des.* **4** 999–1007

[82] Johansen O H, Kristiansen J D, Gjersoe R, Berg A, Halvorsen T, Smith K-T and Nevstad G O 2008 RDX and HMX with reduced sensitivity towards shock initiation—RS-RDX and RS-HMX *Propellants Explos. Pyrotech.* **33** 20–4

[83] Mullin J W 2001 *Crystallization* (Woburn, MA: Butterworth-Heinemann)

[84] LLinas A and Goodman J M 2008 Polymorph control: past, present and future *Drug Discov. Today* **13** 198–210

[85] Beckmann W 2000 Seeding the desired polymorph: background, possibilities, limitations and case studies *Org. Process Res. Dev.* **4** 372–83

[86] Desgranges C and Delhommelle J 2006 Insights into the molecular mechanism underlying polymorph selection *J. Am. Chem. Soc.* **128** 15104–5

[87] Anderson E W 1993 Tactical missile warheads *Progress in Astronautics and Aeronautics* vol 155 (Washington, DC: American Institute of Aeronautics and Astronautics)

[88] Sarangapani R, Ramavat J, Reddy T S, Patil R S, Gore G M and Sikder A K 2014 Effect of particle size and shape of NTO on micromeritic characteristics and its explosive formula-tioins *Powder Technol.* **253** 276–83

[89] Koch E-C 2019 Insensitive high explosives: III. Nitroguanidine—synthesis—structure—spectroscopy—sensitiveness *Propellants Explos. Pyrotech.* **44** 267–92
[90] Ostmark H, Bemm U, Langlet A, Sanden T and Wingborg N 2000 The properties of ammonium dinitramide (ADN): part 1, basic properties and spectroscopic data *J. Energetic Mater.* **18** 123–8
[91] Heintz T, Pontius H, Aniol J, Birke C, Leisinger K and Reinhard W 2009 Ammonium dinitramide—prilling, coating and characterization *Propellants Explos. Pyrotech.* **34** 231–8
[92] Saleh S N, Ahmed S M, Al-mosuli D and Barghi S 2015 Basic design methodology for a prilling tower *Can. J. Chem. Eng.* **93** 1403–9
[93] Hope K S, Lloyd H J, Ward D, Michalchuk A A L and Pulham C R 2015 Resonant acoustic mixing: its applications to energetic materials *New Trends in Research of Energetic Materials (Pardubice, Czech Republic)*
[94] ResoDyn Corporation Mixing energetic materials 15 January 2019 (www.resodynmixers.com/download/mixing-energetic-materials)
[95] United Nations 2019 *Manual of Tests and Criteria, Seventh Revised Edition* (New York: United Nations)
[96] Freche A, Aviles J, Donnio L and Spyckerelle C 2000 Insensitive RDX (I-RDX) *Insensitive Munitions and Energetic Materials Symp. (San Antonio, TX)*
[97] Lochert I J, Franson M D and Hamshere B L 2003 *Reduced Sensitivity RDX Part I: Literature Review and DSTO Evaluation* (Edinburgh, South Australia: DSTO Systems Science Laboratory)
[98] Halvorsen T, Kristiansen J D, Johansen O H, Smith K-T, Berg A, Gjersoe R, Christensen M and Nevstad G 2005 RS-RDX and RS-HMX: ageing and shock sensitivity *36th Int. Annual Conf. of ICT (Karlsruhe, Germany)*
[99] Elbeih A, Husarova A and Zeman S 2011 Path to ε-HNIW with reduced impact sensitivity *Cent. Eur. J. Energetic Mater.* **8** 173–82
[100] U. S. Department of Defense 1996 *Detail Specification—RDX (Cyclotrimethylenetrinitramine)* (Picatinny Arsenal, NJ: US Army ARDEC)
[101] Halvorsen T 2002 Improved RDX, properties and processability *33rd Int. Annual Conf. of ICT (Karlsruhe, Germany)*
[102] Spyckerelle C, Eck G, Sjoberg P and Amneus A-M 2008 Reduced sensitivity RDX obtained from Bachmann RDX *Propellants Explos. Pyrotech.* **33** 14–9
[103] Borne L, Mory J and Schlesser F 2008 Reduced sensitivity RDX (RS-RDX) in pressed formulations: respective effects of intra-granular pores, extra-granular pores and pore sizes *Propellants Explos. Pyrotech.* **33** 37–43
[104] Doherty R M and Watt D S 2008 Relationship between RDX properties and sensitivity *Propellants Explos. Pyrotech.* **33** 13
[105] Ramavat V, Sarangapani R, Reddy S T, Patil R S, Gore G M and Sikder A K 2013 Studies on the tailoring of particle size and micromeritic properties of reduced shock sensitivity RDX (RSS-RDX) *Cent. Eur. J. Energetic Mater.* **10** 581–592
[106] Cooper J, Borne L and Coquerel G 2020 Antisolvent additino: an effective method of controlled fluid inclusion formation in RDX crystals *Cryst. Growth Des.* **20** 7120–8
[107] Wang Y, Li X, Chen S, Ma X, Yu Z, Jin S, Li L and Chen Y 2017 Preparation and characterization of cyclotrimethylenetrinitramine (RDX) with reduced sensitivity *Materials* **10** 974

[108] Zhang X-x, Xue Z-H, Wang Z, Chen S, Yang Z-J and Yan Q-L 2020 Hybrid RDX crystals assembled under constraint of 2D materials with largely reduced sensitivity and improved energy density *J. Hazard. Mater.* **398** 122842

[109] Qiu H, Stepanov V, Di Stasio A R, Chou T and Lee W Y 2011 RDX-based nanocomposite microparticles for significantly reduced shock sensitivity *J. Hazard. Mater.* **185** 489–93

[110] Yang Z, Ding L, Wu P, Liu Y, Nie F and Huang F 2015 Fabrication of RDX, HMX and CL-20 nbased microcapsules via *in situ* polymerization of melamine-formaldehyde resins with reduced sensitivity *Chem. Eng. J.* **268** 60–6

[111] Blas L, Klaumunzer M, Pessina F, Braun S and Spitzer D 2015 Nanostructuring of pure and composite-based K6 formulations with low sensitivies *Propellants Explos. Pyrotech.* **40** 938–44

[112] Risse B, Schnell F and Spitzer D 2014 Synthesis and desensitization of nano-beta-HMX *Propellants Explos. Pyrotech.* **39** 397–401

[113] Sabatini J J and Oyler K D 2016 Recent advances in the synthesis of high explosive materials *Crystals* **6** 5

[114] Friscic T and Jones W 2009 Recent advances in understanding the mechanism of cocrystal formation via grinding *Cryst. Growth Des.* **9** 1621–37

[115] Almarsson O and Zaworotko M J 2004 Crystal engineering of the composition of pharmaceutical phases. Do pharmaceutical co-crystals represent a new path to improved medicines? *Chem. Commun.* **2004** 1889–96

[116] Kennedy S R and Pulham C R 2018 Co-crystallization of energetic materials *Co-crystals: Preparation, Characterization and Applications* (London: Royal Society of Chemistry)

[117] Zhang J and Shreeve J M 2016 Time for pairing: cocrystals as advanced energetic materials *Cryst. Eng. Comm.* **18** 6124–33

[118] Bellas M K and Matzger A J 2019 Achieving balanced energetics through cocrystallization *Angew. Chem. Int. Ed.* **58** 17185–8

[119] Bennion J C, SIddiqi Z R and Matzger A J 2017 A melt castable energetic cocrystal *Chem. Commun.* **53** 6065–8

[120] Kamlet M J and Jacobs S J 1968 Chemistry of detonations. I. A simple method for calculating detonation properties of C-H-N-O explosives *J. Chem. Phys.* **48** 23–35

[121] Landenberger K B and Matzger A J 2010 Cocrystal engineering of a prototype energetic material: supramolecular chemistry of 2,4,6-trinitrotoluene *Cryst. Growth Des.* **10** 5341–7

[122] Bennion J C and Matzger A J 2021 Development ABD evolution of energetic cocrystals *Acc. Chem. Res.* **54** 1699–710

[123] Bolton O, Simke L R, Pagoria P F and Matzger A J 2012 High power explosive with good sensitivity: A 2:1 cocrystal of CL-20:HMX *Cryst. Growth Des.* **12** 4311–4

[124] Yang Z, Li H, Zhou X, Zhang C, Huang H, Li J and Nie F 2012 Characterization and properties of a novel energetic-energetic cocrystal explosive composed of HNIW and BTF *Cryst. Growth Des.* **12** 5155–8

[125] Wang Y, Yang Z, Li H, Zhou X, Qi Z, Wang J and Liu Y 2014 A Novel cocrystal explosive of HNIW with good comprehensive properties *Propellants Explos. Pyrotech.* **39** 590–6

[126] Ravi P, Badgujar D M, Gore G M, Tewari S P and Sikder A K 2011 Review on melt cast explosives *Propellants Explos. Pyrotech.* **36** 3930403

[127] Vuppuluri V S, Samuels P J, Caflin K C, Gunduz I E and Son S F 2018 Detonation performance characterization of a novel CL-20 cocrystal using microwave interferometry *Propellants Explos. Pyrotech.* **43** 38–47
[128] Liu N, Duan B, Lu X, Mo H, Xu M, Zhang Q and Wang B 2018 Preparation of CL-20/DNDAP cocrystals by a rapid and continuous spray drying method: an alternative to cocrystal formations *Cryst. Eng. Comm.* **20** 2060–7
[129] Goncharov T K, Aliev Z G, Aldoshin S M, Dashko D V, Vasil'eva A A, Shishov N I and Milekhin Y M 2015 Preparation, structure and main properties of bimolecular crystals CL-20-DNP and CL-20-DNG *Russ. Chem. Bull., Int. Ed.* **64** 366–74
[130] Liu N, Duan B H, Lu X M and Want B Z 2020 Investigation of CL-20/TFAZ cocrystal: preparation, structure and performance *J. Phys. Conf. Ser.* **1721** 012005
[131] Doherty R M 2018 Making sense of sensitivity data *Proc. 21st Seminar on New Trends in Research of Energetic Materials (Pardubice, Czech Republic)*
[132] Hallstrom S and Horman L 1985 *Method and means for sedimentation casting of explosive charges* Canada Patent CA 1185094A

Chapter 4

Application of machine learning and artificial intelligence methods to energetics science and technology

William H Wilson, Peter W Chung, Zois Boukouvalas and Elan Moritz

4.1 Introduction

Today, one sees many headlines involving big data, machine learning, artificial intelligence, robotics, analytics, self-driving cars, the Internet of Things, and so on. At their core, all these technologies are based on the rapid advances being made in powerful computing hardware, accompanied by intense, well-funded, and widespread activity to develop a rapidly growing reservoir of sharable data-processing algorithms. Machine learning (ML) and artificial intelligence (AI) methodologies have advanced significantly in the last two decades. ML fundamentally enables computers to learn from data and to draw out relationships for which a firm scientific basis may not yet be known (LeCun *et al* 2015) [1]. The recent pairing of AI algorithms with the ever-increasing power of computer technologies has allowed greater consideration of massive data, and has resulted in many sensational demonstrations [2]. A partial listing includes the Netflix Prize [3], IBM Watson [4], deep learning and convolutional neural networks [5], and AlphaGo [6], among many others. In recent years, this convergence of fast and ever more powerful high-capacity computers with the compounding evidence that AI/ML approaches actually perform well (in a surprising variety of disparate problem types encountered in seemingly unrelated fields), has led to a sea change in the way researchers and scientists view previously intractable problems. Increasingly, researchers from a growing number of fields are considering, and pursuing, the application of AI/ML and data science methods to rapidly address and solve problems previously seen as either too hard, too expensive, or too time consuming to practically address using more traditional approaches. In fact, increasingly, thought leaders from a growing number of fields see inclusion of AI/ML considerations in their research and analysis 'tool box' as *de rigueur*. As the acceptance of AI/ML approaches has grown, the

doi:10.1088/978-0-7503-3943-8ch4

range of AI/ML applications has diffused from the early concentration in what could perhaps be thought of as issues related mostly to social science and human behavior fields (e.g. marketing research, trend analysis, finance, advertising, etc) and moved into fields more in the 'hard' sciences and technology. An early instance of that trend has been the successful use of AI/ML approaches in medical diagnostics and then in the pharmaceutical field. Pharmaceutical companies in the early 2010s time frame partnered with AI startups and academics for internal research and development programs to predict the pharmacological properties of small molecules, and to generate novel chemistry through deep learning techniques. The range of applications of ML and deep learning techniques has now permeated into many areas of drug discovery. Similarly, research leaders in the field are now developing significant evidence that the energetics field may also be a beneficiary of AI/ML, in which molecule design is but just one area. But even for the design of molecules, contrary to popular belief, the advances in pharmacology leave many important research questions unique to energetics unanswered. Meanwhile, the use of AI/ML approaches has been growing in other fields, in search of solutions to previously intractable problems in theoretical mathematics, physics, materials science, and chemistry. Thus, it is perhaps inevitable that a growing number of researchers from the broad energetic materials field have also begun to explore creative ways to adopt, adapt, and apply AI/ML methods to various energetic materials science and technology areas.

As with any popular 'next great thing', however, grandiose expectations should be tempered with caution. Skeptics of the potential value of the use of AI/ML approaches to energetics point to the Gartner hype cycle [7] for a dose of sobriety. Though evidence of practical technical problem solving using AI/ML methods has been compelling for finding patterns and correlations, the same cannot be said for the *understanding* of underlying causalities that emerges from the work. Yet, fundamental understanding is central in many problems of scientific interest. The relationship between cause and effect is a necessary key to understanding the robustness—or the sensitivity to perturbations—of a system. Many in the energetics research community have therefore been understandably measured in their support of any adoption of AI/ML. Perhaps due to the inherent challenges of working with explosive materials, the energetics community is often slow to jump in with both feet when a new technique or methodology, ostensibly of value to them, first comes along. Variations introduced at the molecular or system-design levels are part and parcel to all downstream-engineered products. Confidence that small perturbations in the design, processing, and/or integration steps will not significantly hinder the functions of the overall system is a foundational requirement in design, fabrication, and systems-integration practices of present-day energetics and weapons systems. So, it is not surprising that great enthusiasm is not universally elicited within the energetics field for new modeling methods with unknown sensitivities, particularly ones that do not always provide an easily followed route between the data they ingest and the solutions they generate.

Another obstacle to applying AI/ML methods to problems of energetic material science and technology lies in the heavy reliance of most current AI/ML methodology on vast amounts of data. As will be discussed in more detail below, data availability in large quantities is one of three critical factors generally required in AI/ML methodologies. While large amounts of data can either be found or fairly easily generated in many other fields, energetics experimental data by contrast is typically sparse, in part due to the safety considerations inherent in its generation, as well as due to security considerations that often limit its availability. Furthermore, the energetics science and technology fields are, by comparison to much larger applications such as structural materials or chemical discovery and development for pharmacological applications, rather niche areas, with orders of magnitude smaller pools of resources and interest. Thus, the rate of experimental data generation in energetics is quite small by comparison to these other domains. We will also describe below efforts, currently arising from research in using AI/ML methods for energetics applications, to develop new algorithms and methods for adapting AI/ML to more robustly handle scarce data problems while still providing beneficial AI/ML solutions. These research efforts have been motivated by the scarce data 'fact of life' problem in energetics, and are allowing the niche field of energetics research to not only borrow methods from the large body of previous AI/ML work in other fields, but also to identify the unique scientific challenges associated with scarce data.

Data for any field can exist in many forms, including tabulated measurements, results from experiments, graphical representations, computational results, and so on. Data can reside in many places, including textbooks, lab notebooks, film records, journal articles, and so on. Subtly different than the problems of data scarcity that are particularly significant for the field of energetics, the many forms and locations of data pose different sorts of difficulties. Data may exist but still be considered scarce if the costly tasks of curation— including, but not limited to, creating, collecting, digitizing, structuring, indexing, organizing, and maintaining—have not been undertaken at a community-level. This is more of a problem of data availability in a format suitable for use in AI/ML algorithms. It is also important to realize that, to most aggressively pursue and benefit from the emerging power of AI and ML methods, potential sources of data must be inclusively defined. Much of the promise being shown for modern AI methods lies not just in its use to 'learn' from data so as to predict future outcomes, but to also learn, manipulate, and apply facts from broad and diverse sources of data about a subject. That is, data may be tabulated results from numerical measurements, but useful and actionable information is also more broadly contained in factual knowledge or behavior about a field of interest. By illustration, the statement 'if one mixes sufficient quantities of hydrogen gas with oxygen gas and introduces an electrical discharge to the mixture, then a rapid combustion of the two gases will occur, releasing a large quantity of heat energy' can still be, for certain purposes of ML, useful input information, despite the lack of any numerical quantities being provided in the statement. In this interpretation, concepts and facts from what may be traditionally and broadly called 'knowledge' can also be useful inputs for AI/ML use. Textbooks and journal articles are just two examples of repositories of both data and knowledge: they typically contain numerical or experimental or measurements data closely intertwined with fact or behavior or knowledge.

One might wonder: How does this broader pool of information useful for AI/ML applications affect the issue of data scarcity for energetics, as discussed above? Paradoxically, while the field of energetics suffers from a scarcity of traditional numerical, experimental, and measurement data types compared to most other application fields for AI/ML, the amount of factual and behavioral and knowledge information about energetics, like essentially all science and technology fields, is daunting, even if one limits consideration just to the information in textbooks and journal articles. Energetics knowledge is available broadly across many openly available documents. The size of this set is likely immense and grows unabated each year. During the period between 1990 and 2011, it is estimated that roughly 17,000 scholarly papers, identifiable as energetics relevant, were published worldwide [8]. Based on the estimate that global scientific output doubles every 9 years, or the equivalent of around 9% increase each year [9], over a thousand new publications with a self-identifying focus on energetics are likely now appearing annually. Accounting also for tangential and adjacent fields in, for instance, pharmaceutical research or chemical engineering areas, the number of documents that are potentially relevant for energetics innovation constitute a set that is too large to count, let alone for one human, or even a large team of humans, to read, comprehend, and evaluate. The question and challenge of capturing and 'ingesting' all, or at least most, science and technology-relevant information will not just persist but will grow. The plain fact is that there are many more sources of useful, if not vitally critical, data, information, and knowledge, than individuals or collections of individuals can reasonably process at a sustained rate. Detailed awareness of important emerging energetics developments, and, more importantly, knowledge that could inspire insights that would lead to new energetics capabilities, in all probability lie buried in this glut of documents. But given the amount of information this is expected to entail, manually achieving this is simply intractable. A manual effort to do this by hand with the intent of enabling inference and deduction with subject matter expertise is neither technically nor economically feasible. Yet the inferable knowledge alluded to in such analyses is strongly believed to exist. The question is not about its existence but rather about how to gain access. If the processes required to reach this information can be automated and accelerated, then the desired knowledge can be obtained. Approaches addressing the problems of automating the access to knowledge-producing information, and of automating the extraction of connections and inferences from the collection of reference sources, lie in the AI/ML realms of natural language processing (NLP) and literature-based discovery (LBD). In recent years, researchers have begun adapting many of the NLP and LBD methods and customizing them specifically to address energetics science and technology knowledge source materials. We will describe some of those efforts in more detail below.

In summary, despite the caution needed to temper expectations, the skepticism about just how beneficial the use of AI/ML will be for energetics technology, the issues of scarce energetics experimental data, the daunting problems of ingesting literature-based energetics knowledge, and the recognition that the current state of art in AI/ML as yet lacks a robust capability to produce scientific understanding of underlying fundamental cause-and-effect mechanisms, interest and work in applying AI/ML and NLP methods and approaches to energetic material research is growing. Early work, started only around 2015, is already providing encouraging results. Some of these results will be described in more detail in this chapter, while others will be summarized and referred to mainly in passing.

4.2 Background

In general (not just in the field of energetics) AI/ML requires three factors in a 'package': (a) the *data*, (b) *models*, and (c) a *problem for analysis*[1]. The mutual relationship of these three factors ussd in AI/ML is illustrated in figure 4.1. From the point of view of proposing or assessing an AI/ML capability, one must address and define what will be used for all three factors. That is, an AI/ML research problem may entail the development of data, creation of new models, or performance of an analysis to seek answers to a specific question, but it cannot do one without knowing how it will use the other two. Thus, the three factors are helpful in contextualizing the significance of an AI/ML innovation.

It is important to recognize that, in general, AI/ML is not only about the modeling methods: the growth in interest has as much to do with recent advances to make possible the greater availability of data, the creative articulation of analysis questions, and the technological advances that have happened elsewhere. Put another way, many of the modeling methods have existed for decades; it would not be unreasonable to

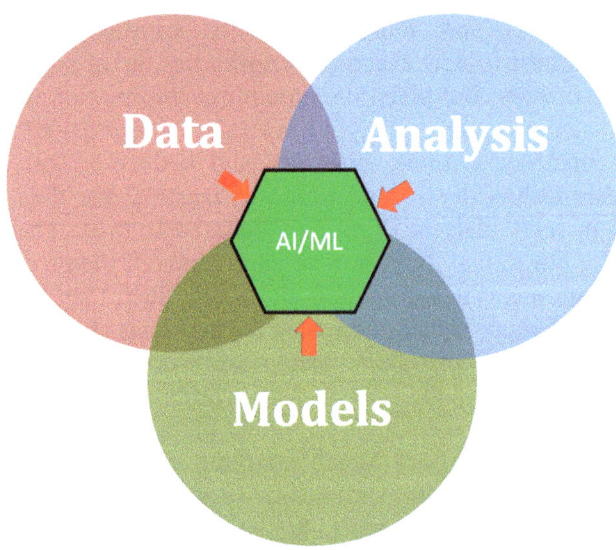

Figure 4.1. Three equally important factors needed in ML methods.

[1] Usually, an implied fourth factor is also included but not as often discussed regarding the availability of computational resources; the furtherance of AI/ML would not have been possible without modern graphics processors.

argue that algorithms written in the 1970s are the initial bases for the most fantastical demonstrations seen of late. What is different is that concerted efforts have developed sharable data and, with time, the workforce was educated on methods to enable the articulation of interesting problems that have been refined with each new generation of students. This was helped along by innovations in computer hardware capabilities, curiously thanks in part to the interest in computer video games.

As AI/ML approaches increasingly are applied to energetics problems, many of the future directions will no doubt revolve around questions of how to make the three factors of data, models, and appropriate problem selection relevant to energetic material systems. For instance, in order to enable analyses based on AI/ML for energetics, one must ask, 'Are the models well suited to these new problems?' and 'Are the data sufficient to use the models?', and 'What problems can be analyzed, and how should they be posed?'

In the energetics science and technology community especially, significant interesting problem areas are ripe for early fruits from application of AI/ML. However, a famous quote attributed to the great statistical mathematician John Tukey goes like this: 'An approximate answer to the right problem is worth a good deal more than an exact answer to an approximate problem.' This is to say that the Universe is comprised of problems that are subject to unknowable vagaries and therefore are impossible to model accurately with single-point solutions. Instead, the right approach is to understand that solutions are in fact sampled from an underlying statistical distribution possibly containing an infinite number of possible single-point answers. Among the infinite possible solutions, some may be more probable than others. This is what makes AI/ML models so different from most physics-based models. For one, AI/ML models are not necessarily physics-based. One might surmise that this is a significant flaw in AI/ML. However, by the same token, AI/ML is also not limited by *unknown* physics. Many mechanisms and scientific foundations of the synthesis, processing, formulation, initiation, detonation, reaction, and ignition in energetics are still beyond the frontiers of current scientific understanding. The lack of a theoretically founded mathematical model of any physical or natural system *necessarily precludes* modeling if one's methodology is confined to using only physics-based modeling approaches.

The distinction between statistical models based on AI/ML methods and deterministic physics-based models could not be more important. The models developed through AI/ML methods are often referred to as being *predictive*, which contrasts with the instructive nature of most physics-based models. This nuance is profound since the former holds a 'mathematical guarantee' that the results of the model will agree with reality, but only to within a definable probability, while the other is associated with no such probability. In fact, strictly speaking, the probability of a physics-based model disagreeing with actual experiment is 100%! While physics-based models may have their foundations in simple-to-understand first principles, there is no clear assurance that the solution to those models will accurately represent, say, detonation, when it occurs in a natural environment. In contrast, AI/ML models can be trained using data from actual detonation events, provided the measurements are detailed enough. As such, there is an assurance that if the data are obtained in such a way that the vagaries of interest are captured, then the model can be developed to represent the underlying uncertainties in the solution. But though the AI/ML model's value lies in the ability to capture and represent real effects, it lacks the ability to survive the reductionist's scrutiny into first principles.

Despite these distinctions that separate the capabilities of AI/ML from other classical modeling approaches, it should be noted that nothing within AI/ML makes its methods fundamentally incompatible with working with other models, even those that are physics based. Creativity and imagination are enabling researchers to deploy AI/ML within, for instance, atomistic modeling to overcome calculation bottlenecks and allow researchers to reach beyond current scientific limitations. Others are combining multiple types of AI/ML models to recognize deficiencies in available data to generate remedial data, developing models that are predisposed to preserving some basic physical or mathematical properties, or developing models that can guide how new experiments should be performed.

It is often said, somewhat dismissively, that AI/ML is but a fancy form of regression. There is a good amount of truth to this characterization. But what is often difficult to convey in addition is the fact that the regression approaches used in AI/ML are typically not over a single variable and have no restriction to being a linear one. A highly nonlinear, high-dimensional multivariate regression, packaged with underlying uncertainties, should in principle be able to estimate the behavior of any system with arbitrary complexity. The evidence of this grows daily as the methods are finding success in applications far beyond just energetics, including control over autonomous swarms, high-throughput material discovery and screening, visual pattern recognition, and encryption/decryption technologies, among others.

Although application of AI/ML analysis methods is characterized by some as a relatively recent and new development within the energetics community, in fact the introduction of these approaches should more accurately be seen as a logical progression from previous traditional analysis methods that have long been used with good success for practical solution to energetics problems. The recent interest in AI/ML for energetics is indebted to historical developments in energetics analysis separate from contemporary AI/ML advances. Much early work in energetics was based on statistical techniques that sought correlated relationships using regression-based analyses and a combination of experimental and synthetic data, sometimes dismissively described as engineering curve-fitting. The introduction in the paper by Rice and Hare [10] provides a brief but exhaustive summary of such earlier correlation works in the energetics domain starting with the efforts of Mortimer Kamlet in the 1970s. Remarkably, the work of Rice and Hare marks a turning point where methods developed in the pharmaceutical industry began taking on larger roles in the research of energetics. Much work followed. For instance, see references [11–15].

These early developments borrowed heavily from the field of quantitative structure property relationship (QSPR) and quantitative structure activity relationship (QSAR) modeling [10]. These are systematic methods for developing empirical, statistics-based quantitative relationships between chemical structures and the properties of molecules. Their starts are believed to have been in classical chemistry and drug research in 1868 [16]. Due to their reliance on statistics, these models are often termed 'predictive' due to the quantifiable likelihood that the naturally occurring answer will be found among all the possible answers held within the statistical distributions. Contrast this to deterministic models, often through physics-based simulations, that provide a single answer or a small set of possible answers that may or may not represent the naturally occurring behavior, since the focus may not have been on the categorical accounting of all sources contributing to variability in the answers. The experiences gained from both schools of thought run deep in the current energetics workforce, but the increasing use of statistical methods in QSAR/QSPR are arguably the first instances of AI/ML-like techniques showing value for energetics research.

With the then growing reliance on deterministic modeling approaches and the increasing complexity of those models, particularly those solved using supercomputers, interest grew in the late 1990s to recognize the importance of quantifying uncertainty in the models, ranging from those used in basic research simulations to the models used in the certification and testing of weapons platforms [17]. Verification, validation, and uncertainty quantification (VVUQ) seeks to reduce the reliance on qualitative measures of evaluation and place significant attention on mathematical foundations to enable reproducibility and assess the potential for validation of large complex models [18]. The central idea of VVUQ is to identify significant sources of uncertainty that would cause a model to deviate from the physical system and aggregate the uncertainties that arise from the different sources. An analyst can then develop emulation or reduced-order models, perform sensitivity analyses, or process information obtained from multiple models to learn how to make inferences about the real physical system from an ensemble of information sources.

Unlike QSAR/QSPR, VVUQ is not necessarily a single calculation procedure, model, or methodology. It is a way of thinking primarily as it relates to solutions of partial differential equations—the basis for most deterministic physics-based models. To use VVUQ effectively requires a deep understanding of these models, the physical (or experimental) problem of interest, and statistical methods. In many cases, it leads to new models that work in tandem with the original mathematical representation to provide additional information not otherwise available. Among energetics areas, VVUQ is an established and prominent field of applied mathematics, such as polynomial chaos expansion [19] and stochastic partial differential equations, and in ensemble modeling for complex multiphysics simulation codes [20].

But then in the last 15 years or so, with the development of progressively faster computing technologies and high-density data storage, long-known statistical methods algorithms could finally be implemented to show breathtaking automated capabilities, where computers can be taught how to learn and then answer specific types of questions. A listing of key developments will be necessarily remiss and quickly outdated, but as discussed in the introduction, it includes the Netflix Prize [3], IBM Watson [4], deep learning and convolutional neural networks [21], AlphaGo [22], among many others. While still reliant on regression of sorts, these breakthroughs illustrate capabilities far beyond the single and multiple linear/nonlinear regression baselines of QSAR and the re-derivation of stochasticity in the applied mathematics of VVUQ. They point to translations of capabilities that have long been used in computer science and electrical engineering fields but have rarely found acceptance in other applied science fields that could benefit from them.

Nondeterministic correlation model efforts to predict difficult-to-simulate behaviors of energetics began appearing nearly three decades ago, motivated by the difficulty in finding mathematically perfect relationships between chemical constituents and the final performance properties of energetic weapon systems [23, 24]. By the 1990s, the emergent global demand for insensitive munitions became manifest in the open literature, with much effort focused on overcoming the great number of chemical and physical uncertainties associated with characterizing impact sensitivity [25–30]. The modeling techniques that were used began to experience wider acceptance later in the same decade and thereafter when other works began to appear where such statistics-based mathematical techniques were showing successful applications in problems accompanying a greater level of physics, chemistry, and materials science sophistication [31–37]. Once the ML revolution had firmly taken root by the 2010s, other types of properties such as detonation velocity and detonation pressure were used as target problems of ML studies of energetics [38–46]. Relative to the number of studies focused on molecule and material property prediction, the use of AI/ML for energetics detection has been fewer in number [47–50], but their appearance shows convincingly the potential diversity of energetics AI/ML applications based on the same set of underlying mathematical and computational principles. Any process that involves 'humans-in-the-loop', and any science and technology development field supporting energetics can potentially exploit AI/ML to vastly accelerate workflows or improve the accuracy or efficiency of repetitive tasks.

4.3 Machine learning for new energetics

During the past few decades, interest in discovery of new energetic materials has remained strong, motivated by a desire to find materials that would have improved performance, thermodynamic stability, or safety characteristics compared to the handful of energetics that had been used almost exclusively since the mid-20th century: HMX, RDX, TNT, PETN, and TATB [51]. Generally, the traditional approaches to the discovery and development of new energetic materials has required expensive and time-consuming synthesis and characterization trial-and-error discovery loops, with many synthesis experiments leading to dead ends and/or yielding little useful or fruitful results. As successes from other fields emerged, interest grew in exploring if computational screening and lead generation methods using AI/ML approaches could speed up the pace of new energetic material discovery and development. Traditionally, screening has been done using either *ad hoc* rules of thumb, which are usually limited in their domain of applicability, or by running many expensive quantum chemistry calculations, which require significant supercomputing time. ML from data, on the other hand, holds the promise of allowing for rapid screening of materials at a much lower computational cost. A properly trained ML model can make useful predictions about the properties of a candidate material in milliseconds rather than hours or days [52].

In energetics research areas, the use of emergent AI/ML approaches at first developed gradually, starting about 15 years ago. Early literature can be found before the AI/ML successes in other fields renewed the interest for energetics. Not surprisingly, we find a small number of energetics themes consistently appearing among these early efforts. These can be categorized as being either 'Prediction' or 'Detection'. Although the two are not unrelated, the former can be thought of as regression based, where a computable relationship is sought to enable predictions. The goal of detection, on the other hand, is to determine through ML approaches whether energetic materials are in a location or vicinity through indirect measurement techniques. Its appearance in the literature follows significant public interest in detection approaches for explosives in the name of public safety. Contemporary ML methods often used in prediction include regression, neural networks, and deep learning. Detection focuses on identifying the appearance or existence of a recognizable structure either in an image or in raw data. ML methods suited for this type of objective include classification. Examples of these efforts that can be thought of as early applications of ML to energetics include the following:

- Prediction:
  - A mixed regression model for the analysis of explosives performance data [23].
  - Application of rigorous nonlinear regression to the decomposition kinetics of explosives [53].
  - Impact sensitivity prediction by the means of feed-forward multilayer neural networks [54].
  - Prediction of the impact sensitivity by neural networks [20, 55].
  - Application of nonparametric binary regression to evaluate the sensitivity of explosives [56].
  - Neural network correlations of detonation properties of high energy explosives [57].
  - A new efficient approach for variable selection Based on multiregression: prediction of gas chromatographic retention times and response factors [58].
  - GMDH-type neural network modeling of explosive cutting process of plates using singular value decomposition [59].
  - Optimization of neural networks architecture for impact sensitivity of energetic molecules [60].
  - The application of neural networks to the assessment of impact testing on explosives [61].
  - Investigation of the various structure parameters for predicting impact sensitivity of energetic molecules via artificial neural network [62].
  - Prediction of impact sensitivity of nonheterocyclic nitroenergetic compounds using genetic algorithm and artificial neural network [63].
  - Prediction of detonation pressure of aluminized explosive by artificial neural network [64].
  - Development of simple QSPR models for the impact sensitivity of nitramines [65].
  - Impact sensitivity prediction by neural network [66].

  - Artificial neural network approach to predict blast-induced ground vibration, airblast, and rock fragmentation [41].
  - Applying machine learning techniques to predict the properties of energetic materials [67].
  - Determination of the velocity of detonation of primary explosives using genetically optimized support vector regression [68].
  - Prediction of detonation velocity and N–O composition of high energy C–H–N–O explosives by means of artificial neural networks [69].
  - Machine-learning assisted screening of energetic materials [70].
  - Deep learning for synthetic microstructure generation in a materials-by-design framework for heterogeneous energetic materials [71].
- Detection
  - Use of artificial neural networks for explosives detection through tomographic images with thermal neutrons [72].
  - Optimization of the separation of organic explosives by capillary electrophoresis with artificial neural networks [73].
  - Study of artificial neural network on explosive detection with PFTNA method [74].
  - Detection of nitro-aromatic explosives with parametric regression modeling approach [75].

More recently, as the power and sophistication of AI/ML methods have been rapidly advanced in other application areas, interest in adapting those emerging capabilities to the energetics field has only increased. One of the earliest objectives from the energetics field, which has received growing attention recently for application of AI/ML-specific methods, has been development of AI/ML approaches to generate (or 'discover') new target energetic molecules with desirable or improved energetics properties, compared to the best energetic materials currently available. In terms of what we have called the three necessary factors for any application of AI/ML methods, namely data, models, and a problem for analysis, as introduced and briefly described above, the energetics 'problem for analysis' chosen in this case could be summarized as *to generate new energetic molecule candidates having desired properties.* Actually, this is a rather simplified or reduced statement of the problem, which is in fact a many-faceted, complex problem, if it is to be productive and workable. Upon reflection, it becomes desirable to include within that problem another more preliminary ML problem, which is *to predict the energetic properties of energetic molecules.* For such problem statements, to use ML methods, data that can be used to 'train' the ML models are needed. Determination of the details of what sort of energetic molecule data would be needed to tackle such energetics problems ideally should depend on which ML algorithms and models are going to be used. But, conversely, in practical terms, which ML models will be the most effective for the problem can be highly dependent upon what kinds of data are actually available, and how much such data can be found. As has been mentioned, molecular property data for energetic molecules is quite sparse compared to most standard AI/ML applications in other fields. Thus, for much of the early work for energetics problems such as these, much of the effort has been to explore various ML approaches and models to determine which ones seem to show the most promise in the context of what data are actually available. Some of the earliest such efforts were reported in papers by Elton *et al* ('Applying machine learning techniques to predict the properties of energetic materials') [67], and by Barnes *et al* ('Machine learning of energetic material properties') [76]. These papers describe proof-of-concept efforts to show that ML techniques could be used to predict the properties of CNOHF energetic molecules from their molecular structures. The data used in these efforts were taken from a small but diverse data set consisting of 109 molecular structures spread across 10 compound classes, found in a paper by Huang and Massa [77]. The papers by Elton *et al* and Barnes *et al* describe a comprehensive comparison of approximately 10 different ML models and several different molecular featurization methods, including *sum over bonds*, *custom descriptors*, *Coulomb matrices*, *bag of bonds*, and *fingerprints*. For instance, the Coulomb matrix featurization process, in which the spatial coordinates and nuclear charges of a molecule are transformed into a eigenvalue vector, is illustrated in figure 4.2.

The ML models used included *kernel ridge regression*, *ridge regression*, *support vector regression*, *random forest*, *k-nearest neighbors*, and a *take-the-mean* dummy predictor. Details of each of these featurization methods and ML models can be found in the paper by Elton *et al*. For the particular molecular data types that they had available, Elton *et al* concluded that the best featurization found was *sum over bonds* (bond counting), and the best model was *kernel ridge regression*. Despite having a small data set, Elton *et al* showed that they obtained acceptable errors and Pearson correlations for the prediction of detonation pressure, detonation velocity, explosive energy, heat of formation, density, and several other properties. The speed of these predictions, compared to typical physics-based code simulations, is more than an order of magnitude faster, and these predictions typically could be run on a laptop computer. By then including an additional data set with ≈300 additional molecules in their training set, they showed that the error could be pushed still lower, although convergence with number of molecules was slow.

In addition to being a proof of concept, the results of the ML work of Elton *et al* and of Barnes *et al* also challenged the typical ML 'rule' that large data sets are necessary for ML to be useful. These efforts were among the very first comprehensive explorations of ML methods to predict energetics performance data. An important aspect contributing to the success of that work was that although the data set was small, it was diverse. That is, the data used for the model training contained molecules from 10 distinct compound classes. This helped to increase the ability of

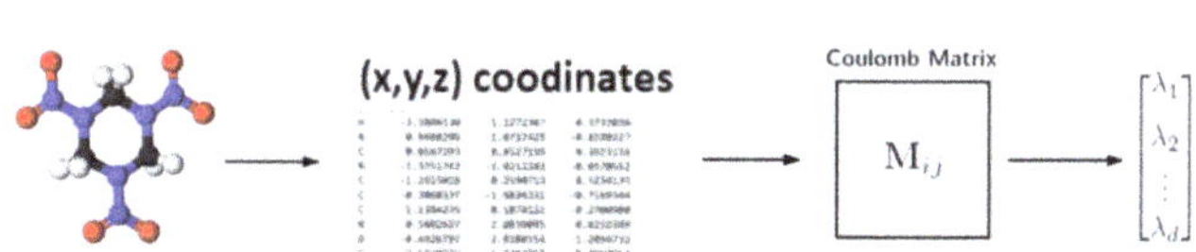

Figure 4.2. Transformation of (x, y, z) coordinates and nuclear charges to the Coulomb matrix eigenvalue spectra representation of a molecule.

the models trained on that data set to be relatively general in their applicability. The work of Elton *et al* and Barnes *et al* helped pave the way for later applications of ML in the energetics domain, including automated lead generation and interpreting ML models to obtain novel chemical insights.

With good progress made for the energetics ML problem *to predict the energetic properties of energetic molecules*, cited above, these results could then be extended to the more general energetics ML problem of *to generate new energetic molecule candidates having desired properties.* An important paper describing the foundations and background needed for this next energetics ML application is another paper by Elton *et al*, 'Deep learning for molecular design—a review of the state of the art' [78]. This paper is an important reference not only because it reports initial efforts to develop methods to use ML for energetic molecular discovery by generating a pool of candidates for computational study and possible selection for synthesis efforts, but also because it provides a fairly comprehensive overview of several AI/ML methods and practices for molecular generation generally, and how they might be adapted and applied for use in energetics applications. Thus, that paper serves somewhat as a primer for any that may want to apply AI/ML methods specifically to energetic molecule generation. The focus of their review is on the use of deep learning techniques for the targeted generation of molecules and guided exploration of a virtual chemical space in which the molecules can be thought to reside. The paper gives a broad introduction to the various methods that are being used to represent molecules within computer models, including molecular graphs and string-based representations (e.g. simplified molecular-input line-entry system, or SMILES), image-based representations, tensor representations, and 3D geometry representations, and discusses some advantages and difficulties in the use of each. Next, various deep learning architectures are discussed, including recurrent neural networks, autoencoders, and generative adversarial networks (GANs). (In review of the approaches using GANs, it becomes apparent why the solution of the ML problem of *to predict the energetic properties of energetic molecules* was first needed for the new molecule generation problem, since the property prediction capability is needed to select and optimize the properties of the candidate molecules as they are generated, in various feedback loops.) The paper also describes how metrics and so-called 'reward functions' can be used in the molecular generation process, to steer the generation toward particular desired features or properties. Figure 4.3 illustrates several functional variations on how these GANs use SMILES strings in generating molecular candidates.

Building upon and extending the ML methods and approaches outlined by Elton *et al*, a paper by Balakrishnan *et al* ('Locally optimizable joint embedding framework to design nitrogen-rich molecules that are similar but improved') [79] describes use of a variational autoencoder in combination with the property prediction model developed earlier to generate new energetic molecule candidates, followed by application of a joint embedding optimization framework to maximize the desired performance property of the molecular candidates. The optimization used is a local method based on a gradient ascent approach, so it is not global in this case; that is, the optimization takes place only in the near neighborhood of the starting candidate molecule, in the chemical 'space' of all possible molecules.

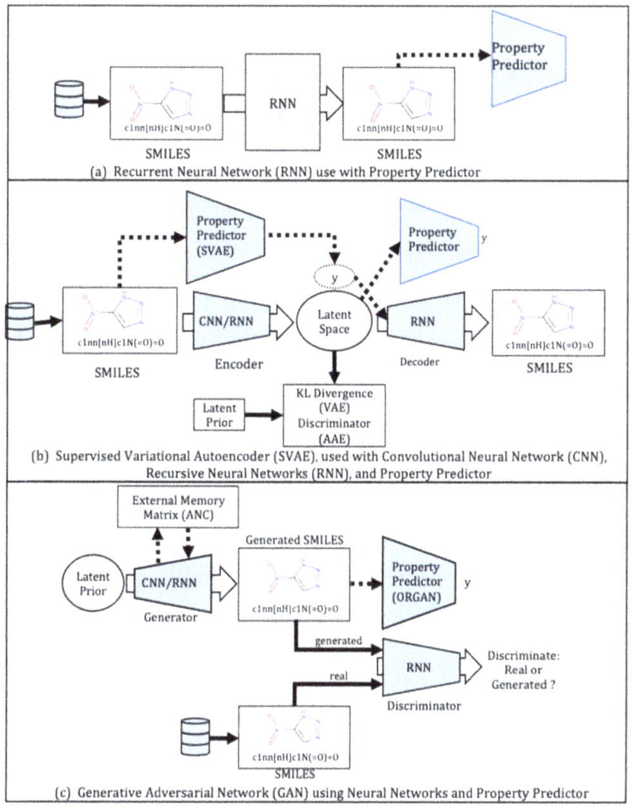

Figure 4.3. Diagrams of three example frameworks used to generate and optimize candidate molecules.

4.4 Energetics applications of natural language processing

As mentioned in section 4.2, the task of keeping up with sources of knowledge of importance or relevance to the field of energetics, like all technical fields, is a daunting one that could greatly benefit from automated approaches that have been developed in the AI/ML area known as NLP. In the last few years, good progress has also been made in applying NLP capabilities to the energetics field. A paper presented by Bouklouvavlas at the 22nd International Seminar in New Trends in Research of Energetic Materials describes some of this early progress ('Using natural language processing techniques to extract information on the properties and functionalities of energetic materials from large text corpora') [80]. This paper describes techniques for using word-embedding methods to create vectors for each word within a text, with emphasis on the mathematical details of two popular methods known as *word2vec* [81, 82] and *GloVe* [83]. Both *word2vec* and *GloVe* learn geometrical encodings (vectors) of words from their co-occurrence information (i.e. how frequently they appear together in large text corpora). They differ in the method they learn: *Word2Vec* uses a neural network (liner model) approach to train, using co-occurrence within a local context (i.e. neighboring words), whereas *GloVe* ('global vector') uses matrix factorization techniques (nonlinear model) to train, using global word-to-word co-occurrence counts. The *word2vec* method is actually an umbrella term for a closely related family of word-embedding algorithms. After describing their use of these methods in creating the word embeddings, the authors then describe the specifics of 'training' these embedding approaches specifically for energetics applications, and approaches for determining word-embedding similarities. The end goal of these NLP efforts is to autonomously extract application and property information from documents. The process can be somewhat summarized in a sequence of steps that are followed in a so-called 'pipeline'. From their paper, these pipeline steps are listed as follows:

1. Convert the text from the source materials into plain text.
2. Process the text to remove information that causes unnatural breaks in the narrative.
3. Process the text to remove unimportant words—so-called stop words—from sentences.
4. Remove text that can produce artifactual tokens such as abbreviations or numerical values.
5. Identify the chemical entities using chemical entity recognition tools such as *ChemDataExtractor*.
6. Normalize the chemical entities by performing synonym matching and replacement.
7. Use a ML-based classifier to classify energetic chemical names.
8. Remove spaces and dashes from chemical names in text.
9. Segment sentences using Natural Language Toolkit *NLTK sent_tokenize.*
10. Convert all text to lower case and tokenize sentences using Natural Language Toolkit *NLTK simple_preprocess.*

These steps for an NLP pipeline as used in this work are illustrated in figure 4.4.
While this effort was an early foray into use of NLP to automate the extraction of knowledge from text sources, it showed interesting and promising results. One of the

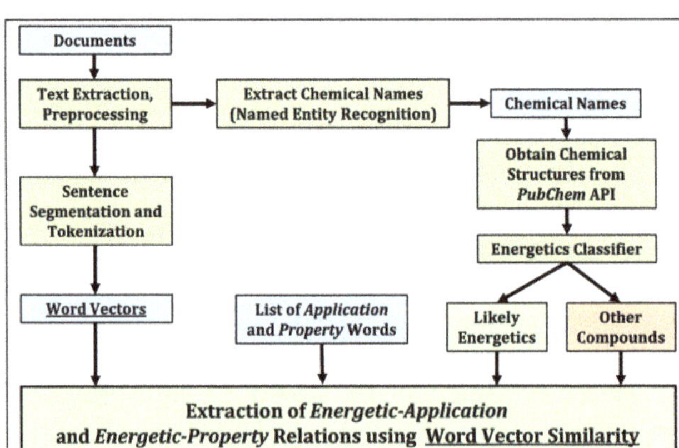

Figure 4.4. NLP pipeline for energetics text processing

first things the authors observed was that words ‘clustered’ within a notional word-space according to the word’s broad characteristics, for instance whether the word comes from, say, an ‘energetic’, ‘binder’, or ‘element’ category. Next, they found that they could use measures of word positions and abstract distances within the word-space to infer relationships between such concepts as type of material and type of application. Another thing the authors looked for was to see how well the NLP pipeline could detect common energetics synonyms. For instance, they found that synonyms often appeared in what they defined as top 5 similarity rankings for the most similar words. For instance, when using the *word2vec* tool for ‘RDX’, the third most similar word (after ‘HMX’ and ‘TNT’) was an older name for RDX, ‘hexogen’. Similarly, for PETN, the third most similar word was its full name, ‘pentaerythritol tetranitrate’.

In summary, this early energetics NLP work explored a proof of concept for NLP methods to extract chemical–chemical and application–chemical relationships from large text corpora, without the need for hand labeling (i.e. without the need for human subject matter experts to provide examples of such relationships). This early demonstration work used only a somewhat limited in size corpus of energetics-relevant text comprised of open source energetics-relevant documents obtained from US patents, and proceedings from papers presented at several technical conferences. Nevertheless, the work showed good promise for NLP methods to extract chemically meaningful associations, and for NLP to provide novel insights into how chemicals are used in energetic materials research and energetic systems. As appreciation for the potential benefits of such NLP approaches grows, more research will provide construction of much larger and more inclusive text corpora, and the power of the inferencing capability will no doubt grow.

As use of NLP methods for extraction of knowledge from text sources has grown, interest has naturally arisen regarding how specific topics of interest are identified from within the text, and the so-called ’explainability’ of results arrived at by these automated NLP processes. A recent paper by Puerto *et al* [84] addressed some of these topic identification and explainability concepts for the energetics field. In [84], the authors proposed an automatic explainable energetics topic identification process consisting of creating document representations and performing energetics document classifications. In this work the authors described use of three complex document representation methods based on Latent Dirichlet Allocation (LDA) [85], *word2vec*, and Bidirectional Encoder Representations (BERT) [86]. Based on those three methods they presented a classification algorithm to predict the topic of an energetics document. Moreover, they introduced local interpretability model-agnostic explanations (LIME) [87] to provide a localized understanding of each document classification. The objective is to provide interpretations that indicate to subject matter users that the NLP-produced topic predictions are understandable and can be trusted. This study was carried out using a novel annotated energetics dataset created and validated by a team of energetics experts.

The main components of the NLP explainable classification pipeline used in this recent work consisted of abstract collection performed by chemistry experts, followed by data labeling (in this case, topic identification) performed by the same chemistry experts, then creation of three distinct document embedding datasets. Next, the NLP classifier was fitted with these embeddings individually, and finally the automated local interpretability model-agnostic explanations (LIME) outputs were obtained and compared with the chemistry experts' topic identifications to see how closely they matched. In particular, the energetics abstracts were labeled by the experts into topics such as energetics property characterization, modeling, processing, and synthesis. Using the labeled data, document embeddings were created. Then, based on these embeddings, document representations were constructed and used to train a random forest classification model. After the classification model was trained, the LIME observations were extrapolated using the LIME package with the fitted model for each testing observation. Finally, prediction performance of the model was assessed, as well as the of agreement between the LIME output and the chemistry expert that aided in the manual labeling.

4.5 Approaches for 'small data': data fusion based on independent vector analysis

Recent ML advances as presented in earlier sections have shown significant promise for the prediction of energetic molecules; however, the problem remains a big challenge due to a number of key limitations. These limitations are related to the use of multimodal data, i.e. information collected about the same sets of molecules but using different ways of characterizing this set of molecules, as well as the explainability of a system geared at the prediction of molecular properties. With explainability we refer to the ability of a model to summarize the causes of its decisions in an efficient manner, illuminate various connections between high-level molecular features and low-dimensional joint representations, and hence gain the trust of its users. Data-fusion methods based on the independent vector analysis model, as demonstrated in two papers by Boukouvalas *et al* [88, 89], have responded to these complex sets of requirements by exploiting alignments of data fragments where there is a common underlying molecular space.

In both of these papers, the data-fusion approach takes an arbitrary number of individual feature vectors and generates joint low-dimensional features, i.e. molecular signatures, which are used for prediction of molecular properties and for knowledge discovery. It is worth mentioning that the data-fusion method is completely unsupervised, parameter free, and computationally attractive.

4.6 Conclusions

Recently, AI/ML approaches, capabilities, and techniques suitable for the energetics field have emerged and are gaining traction in application to many facets of energetics science and technologies. Early successes have shown that AI/ML methodologies can be used to predict the properties of CNOHF energetic molecules from their molecular structures with surprising speed and accuracy. In short order, it seems probable that these capabilities will grow, following a similar path taken by AI/ML applications for general chemical molecules and pharmaceuticals. Current evidence also shows that, with time, the modeling capabilities will grow into significant, if not revolutionary, new tools for use by the energetics communities to discover, innovate, and develop energetic materials and systems. Emergent capabilities for energetic molecules/materials and for successful use, even given the limits of scarce data, have already begun to appear. Early studies are demonstrating the promise of deeper and more significant developments to come. The emergence of many new types of AI/ML methods will likely have an impact not limited to just the energetic material: the complexity found in the science and technology of energetic molecular and formulation materials is echoed in the systems that produce them and the systems into which energetics are integrated. The design and optimization of these system applications will also benefit from the use of AI/ML approaches. The data needs are indeed acute for researching the suitability of novel AI/ML mathematical and computational approaches for energetics and defence sciences, but they are nevertheless now being addressed. To meet the challenges of AI/ML integration into the field of energetics technology, it will be vital to train an AI/ML conversant energetics workforce of the future: a workforce capable of world-leading innovations must be equipped with AI/ML skills, data, and methods.

References

[1] LeCun Y, Bengio Y and Hinton G 2015 Deep learning *Nature* **521** 436–44 http://refhub.elsevier.com/S2211-3835(18)30282-X/sbref19

[2] Marx V 2019 Machine learning, practically speaking *Nat. Methods* **16** 463–67

[3] Netflix Prize 2009 Netflix Prize. Available at: https://netflixprize.com/

[4] IBM 2022 IBM Watson. Available at: https://ibm.com/watson

[5] LeCun Y, Boser B and Denker J S *et al* 1989 Backpropagation applied to handwritten zip code recognition *Neural Comput* **1** 541–51

[6] Silver D, Schrittwieser J and Simonyan K *et al* 2017 Mastering the game of go without human knowledge *Nature* **550** 354–59

[7] https://en.wikipedia.org/wiki/Hype_cycle

[8] Kaczmarck R. E. (ed) 2013 *Topics in Energetics Research and Development* vol 1 (College Park, MD: CALCE ESPC)

[9] Nature 2014 Blog: Global scientific output doubles every nine years http://blogs.nature.com/news/2014/05/global-scientific-output-doubles-every-nine-years.html

[10] Rice B and Hare J 2002 A quantum mechanical investigation of the relation between impact sensitivity and the charge distribution in energetic molecules *J. Phys. Chem.* A **106** 1770–83

[11] Gao C and Gao S 2007 Two dominant factors influencing the impact sensitivities of nitrobenzenes and saturated nitro compounds *J. Phys. Chem.* B **111** 12399–402

[12] Keshavarz M and Pouretedal 2005 Simple empirical method for prediction of impact sensitivity of selected class of explosives *J. Hazard. Mater.* **124** 27–33

[13] Badders N, Wei C, Aldeeb A, Rogers W and Mannan M 2006 Predicting the impact sensitivities of polynitro compounds using quantum chemical descriptors *J. Energetic Mater.* **24** 17–33

[14] Morrill J and Byrd E 2008 Development of quantitative structure–property relationships for predictive modeling and design of energetic materials *J. Mol. Graph. Model.* **27** 349–55

[15] Li G and Zhang C 2020 Review of the molecular and crystal correlations on sensitivites of energetic materials *J. Hazard. Mater.* **395** 122910

[16] Crum-Brown A and Fraser T R 1868 On the connection between chemical sonstitution and physiological action. Part 1. On the physiological action of the ammonium bases, derived from Strychia, Brucia, Thebaia, Codeia, Morphia and Nicotia *Trans. R. Soc.* **25** 151–203
[17] American Institute for Aeronautics and Astronautics (AIAA) 1998 *Guide for the Verification and Validation of Computational Fluid Dynamics Simulations* (Reston, VA: AIAA)
[18] National Research Council 2012 Assessing the reliability of complex models: mathematical and statistical foundations of verification *Validation, and Uncertainty Quantification* (Washington, DC: The National Academies Press)
[19] Xiu D and Karniadakis G 2002 The Wiener–Askey polynomial chaos for stochastic differential equations *SIAM J. Sci. Comput.* **24** 619–44
[20] LLNL 2022 Weapons and Complex Integration Website, Uncertainty Quantification, Lawrence Livermore National Laboratory. https://wci.llnl.gov/simulation/computer-codes/uncertainty-quantification
[21] LeCun Y, Boser B, Denker J, Henderson D, Howard R, Hubbard W and Jackel L 1989 Backpropagation applied to handwritten zip code recognition *Neural Comput.* **1** 541–51
[22] Silver D *et al* 2017 Mastering the game of Go without human knowledge *Nature* **550** 354–59
[23] McDonald W A 1982 *Mixed Regression Model for the Analysis of Explosives Performance Data*. NSWC Technical Report 82-74, Naval Surface Weapons Center, Silver Spring, MD
[24] Pickard J M 1989 Application of rigorous nonlinear regression to the decomposition kinetics of explosives *Thermochim. Acta* **149** 301–15
[25] Nefati H, Legendre J J and Michot C 1995 Impact sensitivity prediction by the means of feed-forward multilayer neural networks *J. Hazard. Mater.* **330** 562–67
[26] Nefati H, Cense J M and Legendre J J 1996 Prediction of the impact sensitivity by neural networks *J. Chem. Inf. Comput. Sci.* **36** 804–10
[27] Efromovich S and Thomas E V 1996 Application of nonparametric binary regression to evaluate the sensitivity of explosives *Technometrics* **38** 50–8
[28] Prana V, Fayet G, Rotureau P and Adamo C 2012 Development of validated QSPR models for impact sensitivity of nitroaliphatic compounds *J. Hazard. Mater.* 169–77
[29] Devinyak O, Havrylyuk D and Lesyk R 2014 3D-Morse descriptors explained *J. Mol. Graph. Model.* 194–203
[30] Xu J, Zhu L and Fang D *et al* 2012 QSPR studies of impact sensitivity of nitro energetic compounds using three-dimensional descriptors *J. Mol. Graph. Model.* **36** 10–19
[31] Chen D S and Wong D S H 1998 Neural network correlations of detonation properties of highenergy explosives *Propellants Explos. Pyrotech.* **23** 296–301
[32] Lučić B, Trinajstić N, Sild S, Karelson M and Katritzky A R 1999 A new efficient approach for variable selection based on multiregression: prediction of gas chromatographic retention times and response factors *J. Chem. Inf. Comput. Sci.* **39** 610–21
[33] Darvizeh A, Nariman-Zadeh N and Gharababaei H 2003 GMDH-type neural network modelling of explosive cutting process of plates using singular value decomposition *Systems Analysis Modelling Simulation.* **43** 1383–97
[34] Cho S G, No K T and Goh E M *et al* 2005 Optimization of neural networks architecture for impact sensitivity of energetic molecules *Bull. Korean Chem. Soc.* **26** 399–408
[35] Smith P P, Nolan P F and Wharton R K 2006 The application of neural networks to the assessment of impact testing on explosives *J. Energetic Mater.* **16** 129–46

[36] Keshavarz M H and Jaafari M 2006 Investigation of the various structure parameters for predicting impact sensitivity of energetic molecules via artificial neural network *Propellants, Explosives and Pyrotechnics* **31** 216–25
[37] Rui W, Juncheng J and Pan Y 2012 Prediction of impact sensitivity of nonheterocyclic nitroenergetic compounds using genetic algorithm and artificial neural network *J. Energetic Mater.* **30** 135–55
[38] Liu Y G, Tian X, Jiang Y Q, Li G B and Li Y Z 2013 Prediction of detonation pressure of aluminized explosive by artificial neural network *Advanced Materials Research.* **641–642** 460–3
[39] Fayet G and Rotureau P 2014 Development of simple QSPR models for the impact sensitivity of nitramines *J. Loss Prev. Process Ind.* **30** 1–8
[40] Rim O K and Wk L 2015 Impact sensitivity prediction by neural network *Science and Technology Analytical.* **28** 430–5
[41] Tiile R 2016 Artificial neural network approach to predict BLASt-induced ground vibration, airblast and rock fragmentation *MS Thesis* Missouri University of Science and Technology
[42] Elton D C, Boukouvalas Z, Butrico M S, Fuge M D and Chung P W 2018 Applying machine learning techniques to predict the properties of energetic materials *Sci. Rep.* **8** 9059
[43] Owolabi T O 2019 Determination of the velocity of detonation of primary explosives using genetically optimized support vector regression *Propellants Explos. Pyrotech.* **44** 1282–92
[44] Chandrasekaran N, Oommen C, Sanal Kumar V R, Lukin A N, Abrukov V S and Anufrieva D A 2019 Prediction of detonation velocity and N–O composition of high energy C–H–N–O explosives by means of artificial neural networks *Propellants Explos. Pyrotech.* **44** 579–87
[45] Kang P, Liu Z, Abou-Rachid H and Guo H 2020 Machine-learning assisted screening of energetic materials *J. Phys. Chem.* A **124** 5341–51
[46] Chun S, Roy S, Nguyen Y T, Choi J B, Udaykumar H S and Baek S S 2020 Deep learning for synthetic microstructure generation in a materials-by-design framework for heterogeneous energetic materials *Sci. Rep.* **10** 13307
[47] Ferreira F J O, Crispim V R and Silva A X 2009 Use of artificial neural networks in drug and explosive detection through tomorgraphic images with thermal neutrons *2009 Int. Nuclear Atlantic Conf. (INAC) (Rio de Janeiro, Brazil)*
[48] Casamento S, Kwok B, Roux C and Dawson M 2003 Optimization of the separation of organic explosives by capillary electrophoresis with artificial neural networks *J. Forensic Sci.* **48** 1075–83
[49] Wang H, Li Y, Yang Y, Hu S, Chen B and Gao W 2005 Study of artificial neural network on explosive *2005 IEEE Nuclear Science Symposium Conf. Record* 471–3
[50] Ramdasi D and Mudhalwadkar R 2018 Detection of nitro-aromatic explosives with parametric regression modeling approach *Int. J. Control Autom.* **11** 13–24
[51] National Research Council 2004 *Advanced Energetic Materials* (Washington, DC: National Academies Press)
[52] Gilmer J, Schoenholz S S, Riley P F, Vinyals O and Dahl G E 2017 Neural message passing for quantum chemistry arXiv e-prints 1704.01212
[53] Pickard J 1989 Application of rigorous nonlinear regression to the decomposition kinetics of explosives *Thermochim. Acta* **149** 301–15
[54] Nefati H, Legendre J and Michot C 1995 Impact sensitivity prediction by the means of feed-forward multilayer neural networks *J. Hazard. Mater.* **330** 562–7
[55] Nefati H, Cense J-M and Legendre J-J 1996 Prediction of the impact sensitivity by neural networks *J. Chem. Inf. Comput. Sci.* **36** 804–10

[56] Efromovich S and Thomas E 1996 Application of nonparametric binary regression to evaluate the sensitivity of explosives *Technometrics* **38** 50–8
[57] Chen D S and Wong D S H 1998 Neural network correlations of detonation properties of high energy explosives *Propellants Explos. Pyrotech.* **23** 296–301
[58] Lučić B, Trinajstić N, Sild S, Karelson M and Katritzky A 1999 A new efficient approach for variable selection based on multiregression: prediction of gas chromatographic retention times and response factors *J. Chem. Inf. Comput. Sci.* **39** 610–21
[59] Darvizeh A, Nariman-Zadeh N and Gharababaei H 2003 GMDH-type neural network modelling of explosive cutting process of plates using singular value decomposition *Syst. Anal. Modell. Sim.* **43** 1383–97
[60] Cho S, No K, Goh E, Kim J, Shin J, Joo Y and Seong S 2005 Optimization of neural networks architecture for impact sensitivity of energetic molecules *Bull. Korean Chem. Soc.* **26** 399–408
[61] Smith P, Nolan P and Wharton R 2006 The application of neural networks to the assessment of impact testing on explosives *J. Energetic Mater.* **16** 129–46
[62] Keshavarz M and Jaafari M 2006 Investigation of the various structure parameters for predicting impact sensitivity of energetic molecules via artificial neural network *Propellants Explos. Pyrotech.* **31** 216–25
[63] Wang R, Jiang J and Pan Y 2012 Prediction of impact sensitivity of nonheterocyclic nitroenergetic compounds using genetic algorithm and artificial neural network *J. Energetic Mater.* **30** 135–55
[64] Liu Y G, Tian X, Jiang Y Q, Li G B and Li Y Z 2013 Prediction of detonation pressure of aluminized explosive by artificial neural network *Advanced Materials Research* **641–642** 460–3
[65] Fayet G and Rotureau P 2014 Development of simple QSPR models for the impact sensitivity of nitramines *J. Loss Prev. Process Ind.* **30** 1–8
[66] Rim O K and Lim W-k 2015 Impact sensitivity prediction by neural network *Sci. Tech. Anal.* **28** 430–5
[67] Elton D, Boukouvalas Z, Butrico M, Fuge M and Chung P 2018 Applying machine learning techniques to predict the properties of energetic materials *Sci. Rep.* **8** 9059
[68] Owolabi T 2019 Determination of the velocity of detonation of primary explosives using genetically optimized support vector regression *Propellants Explos. Pyrotech.* **44** 1282–92
[69] Chandrasekaran N, Oommen C, Sanal Kumar V, Lukin A, Abrukov V and Anufrieva D 2019 Prediction of detonation velocity and N–O composition of high energy C–H–N–O explosives by means of artificial neural networks *Propellants Explos. Pyrotech.* **44** 579–87
[70] Kang P, Liu Z, Abou-Rachid H and Guo H 2020 Machine-learning assisted screening of energetic materials *J. Phys. Chem.* A **124** 5341–51
[71] Chun S, Roy S, Nguyen Y, Choi J, Udaykumar H and Baek S 2020 Deep learning for synthetic microstructure generation in a materials-by-design framework for heterogeneous energetic materials *Sci. Rep.* **10** 13307
[72] Ferreira F, Crispim V and Silva A 2009 Use of artificial neural networks in drug and explosive detection through tomorgraphic images with thermal neutrons *2009 Int. Nuclear Atlantic Conf. (INAC) (Rio de Janeiro, Brazil)*
[73] Casamento S, Kwok B, Roux C and Dawson M 2003 Optimization of the separation of organic explosives by capillary electrophoresis with artificial neural networks *J. Forensic Sci.* **48** 1075–83

[74] Wang H, Li Y, Yang Y, Hu S, Chen B and Gao W 2005 Study of artificial neural network on explosive *2005 IEEE Nuclear Science Sympos. Conf. Record (Fajardo, Puerto Rico)*

[75] Ramdasi D and Mudhalwadkar R 2018 Detection of nitro-aromatic explosives with parametric regression modeling approach *Int. J. Control Autom.* **11** 13–24

[76] Barnes B C, Elton D C and Boukouvalas Z *et al* 2018 Machine learning of energetic material properties arXiv:1807.06156

[77] Huang L and Massa L 2013 Applications of energetic materials by a theoretical method (discover energetic materials by a theoretical method) *Int. J. Energ. Mater. Chem. Propul.* **12** 197–262

[78] Elton D C, Boukouvalas Z, Fuge M D and Chung P W 2019 Deep learning for molecular design—a review of the state of the art *Mol. Syst. Des. Eng.* **4** 828–49

[79] Balakrishnan S, VanGessel F, Boukouvalas Z, Barnes B, Fuge M and Chung P W 2021 Locally optimizable joint embedding framework to design nitrogen-rich molecules that are similar but improved *Mol. Informat.* **40** 2100011

[80] Elton D C, Turakhia D, Boukouvalas Z, Doherty R M, Reddy N, Fuge M D and Chung P W 2019 Using natural language processing techniques to extract information on the properties and functionalities of energetic materials from large text corpora *22nd Int. Seminar in New Trends in Research of Energetic Materials*

[81] Mikolov T, Chen K, Corrado G and Dean J 2013 Efficient estimation of word representations in vector space arXiv:1301.3781

[82] Mikolov T, Sutskever I, Chen K, Corrado G S and Dean J 2013 Distributed representations of words and phrases and their compositionality *Proc. 26th Int. Conf. Neural Information Processing Systems* **2** 3111–9 Available at https://proceedings.neurips.cc/paper/2013/file/9aa42b31882ec039965f3c4923ce901b-Paper.pdf

[83] Pennington J, Socher R and Manning C 2014 GloVe: global vectors for word representation *Proc. 2014 Conf. on Empirical Methods in Natural Language Processing (EMNLP)* 2014 *(Doha, Qatar)* 1532–43

[84] Puerto M *et al* 2022 Assessing the trade-off between prediction accuracy and interpretability for topic modeling on energetic materials corpora arXiv: 2206.00773

[85] Blei D M, Ng A Y and Jordan M I 2003 Latent dirichlet allocation *J. Mach. Learn Res.* **3** 993–1022

[86] Devlin J *et al* 2022 2018 Bert: Pre-training of deep bidirectional transformers for language understanding arXiv: 1810.04805

[87] Ribeiro M T, Singh S and Guestrin C 2016 Why should I trust you?" Explaining the predictions of any classifier Proc. of the 22nd ACM SIGKDD Int/ Conf. on Knowledge Discovery and Data Mining

[88] Boukouvalas Z, Puerto M, Elton D C, Chung P W and Fuge M D 2021 Independent vector analysis for molecular data fusion: application to property prediction and knowledge discovery of energetic materials *2020 28th European Signal Processing Conf. (EUSIPCO)* 2021 (Amsterdam, the Netherlands) 1030–4

[89] Boukouvalas Z, Elton D C, Chung P W and Fuge M D 2018 Independent vector analysis for data fusion prior to molecular property prediction with machine learning arXiv: 1811.00628

Chapter 5

The impact of resonance acoustic mixing on the production of solid propellants and explosives

Ning Ma, Song Chen, Zhe Zhang, Xiaopeng Sun, Zhongyuan Xie, Weiqiang Pang[1] and Guangbin Zhang

Resonance acoustic mixing (RAM) technology is a system that performs high-speed mixing by vibration through the control of acceleration and frequency. RAM is considered to have potential advantages over traditional, high-shear processing methods for explosives and propellants. Because propellants and explosives are hazardous materials, as a new processing technology used for energetic materials, the task of the safety evaluation of RAM is essential to guide its operation. This chapter focuses on testing and evaluation of impact, including impact forces and impact force rates, by a 'bridge' equivalent testing device which relates impact force to impact sensitivity. The work is divided into three parts. In the first part, the impact forces applied by the top, bottom, and sidewall of a container are tested. The second part provides a method for the equivalence of impact force in RAM and impact sensitivity data, which can convert the height data (in centimetres) of the impact sensitivity into the corresponding impact stimulus force (megapascals or MPa μs^{-1}). In the third part, the impact forces obtained in RAM and impact forces obtained from an impact sensitivity testing course are compared to evaluate the safety of the process. The purpose of the test combining with lower limit sensitivity H_0 of typical explosives is to evaluate the process safety. The results show that the impact forces in RAM are only about 1/100th of those corresponding to sensitivity testing, and the impact force rate in RAM is only about 1/1500th that corresponding to sensitivity testing, which indicates that the impact stimulus in RAM is nearly negligible compared to that in the sensitivity testing and thus that RAM is a safe processing method for most explosives.

5.1 Background

Many commercial processes involve the mixing of two or more materials. For instance, one may consider a convolution of chemical resins or the combination of a

[1] Author to whom any correspondence should be addressed.

doi:10.1088/978-0-7503-3943-8ch5

viscous liquid and a finely divided powder. The constituents of the mixture are regarded as nonreactive for now. High-shear mixing is a very familiar technique. Generally, it involves the axial rotation of an array of metal or plastic blades through a volume of material in question [1]. High-shear mixing is effective and simple, but it does engender some deleterious side effects. The action of viscosity causes friction, which dissipates mechanical energy into heat. In certain cases, the material being mixed may reach high temperatures. For energetic materials, this heating is not welcome. It is thought that the magnitude of the temperature rise during mixing may be mitigated if one avoids the generation of high shear. Resonant acoustic mixing (RAM) may provide a solution to this problem [2, 3].

RAM, an innovative type of mixing technology, has been expanded in its use and application over recent years. It is reliant on both bulk mixing and micro-mixing techniques [4]. Even though there is only one company involved in the manufacture of the apparatus, the number of applications and publications associated with the technology is constantly increasing [5]. Moreover, the number of energetics-related publications has also slowly increased as the technology has become accepted by the community over the last 10 years.

RAM was first addressed in a paper published in 2007, unveiling the new mixer [6]. Since then, significant research has been undertaken, initially focused on understanding RAM's mechanisms and the effect of equipment parameters on the mixing performance of materials. More recently, this has progressed into studies using RAM for mixing but not as the sole focus of the research. Although RAM is still a relatively new technology, this shows how it is already being used for applications in industry, primarily the pharmaceutical and energetics industries. Moreover, RAM has already gained considerable industrial interest and is expected to grow with an expanding product range to meet specific industry needs and as more research is published.

Instead of mixing blades, a Resonant Acoustic® Mixer (RAM) consists of a vertically vibrating spring-mounted platform to which a mixing vessel is attached. Oscillations occur at high acceleration (up to 100 G) and low amplitude (up to 14 mm) at the mechanical resonance of the system (approximately 60 Hz) [7]. The principle of RAM is illustrated in figure 5.1. The RAM mixing system consists of a

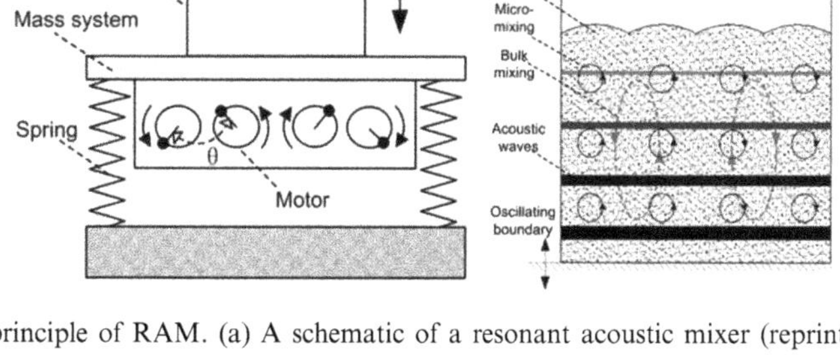

Figure 5.1. The principle of RAM. (a) A schematic of a resonant acoustic mixer (reprinted from [43] with permission from the *Chinese Journal of Explosives and Propellants*). (b) The resonant acoustic mixer uses acoustic energy to provide both bulk and micro-mixing.

mass system, a spring assembly, and a loaded mixing vessel. A motor that subjects the mixing media to a reciprocating, agitating movement controls the spring assembly, and the RAM mixing system attains resonance when the stored forces in the spring and the inertia forces form a mass equal each other.

Since its development, a large proportion of the published research has tried to understand the mixing process, but further development is needed before it can be validated and utilised in industry. Resodyn's product range has advanced since the first laboratory RAMs (LabRAM) were demonstrated in 2006. The products range in both payload capacity and in application. The 'H' ranges are for energetic materials and meet the standards required for safe processing of sensitive and hazardous materials, and the mixing capacity ranges from 0.5 kg to 420 kg, as described by the following model nomenclature: LabRAM I (0.5 kg), LabRAM II (1.0 kg), OmniRAM (5.0 kg), RAM5 (36 kg), RAM 55 (420 kg), respectively [8].

5.2 Reliability and challenges of RAM mixtures

Solid propellants and explosives combine an energetic binder or non-energetic binder with an oxidiser such as ammonium perchlorate (AP). Other ingredients include, but are not limited to, metal fuels, plasticisers, curing agents, bonding agents, and burning rate modifiers. The manufacturing is usually done by a batch process, with multiple stages that are potentially hazardous for the operator and thus are controlled and monitored remotely. The mixing and casting processes are considered the most complex of these stages as they have the greatest impact on the quality and performance of the propellants and explosives. The most common method is to mix the material in a batch mixer, such as a sigma-blade or planetary mixer, then to cast (pour) it into moulds, before then curing it [9]. The maximum oxidiser content is desirable to increase the performance of the propellants or explosives during their design and manufacturing process. Increasing the oxidiser solid loading, reducing particle size, and combining multiple particle sizes for better packing to increase density are all potential methods to increase the oxidiser content and improve performance. Bi-modal and tri-modal packing use two or three different particle sizes, respectively, to increase the theoretical maximum packing fraction. The optimum theoretical concentration of AP oxidiser would be about 90%–93% w/w, but this cannot be achieved with conventional mixers. Solid loading (oxidiser, metal, and other ingredients) above 90% w/w increases the viscosity to an extent where the slurry does not flow sufficiently to mix the ingredients to the required standard [10].

It has been established through research carried out to date that RAM technology offers several advantages over traditional mixing techniques, such as reduced mixing time, reduced energy cost, reduced waste, and being able to be mixed in-case, especially where energetics are concerned [11]. The main advantage lies in the relatively gentle mechanism of the RAM technique, where only minor damage occurs to sample particulates, with no blades or impellers being used in the process and thus the lives of the operators being protected. In the research field of explosives and solid propellants, RAM has been widely studied [12–14]. Regarding energetics-related publications, the topics within the field that have been tackled are diverse, ranging from aluminium/bismuth (III) oxide (Al/Bi_2O_3) nanothermites [15] to co-crystallisation [16, 17], castable propellants [18] to pyrotechnics [19], hydroxyl terminated polybutadiene (HTPB)/ammonium nitrate (AN) propellants [20], and Al-ice suspensions [21]. For instance, a comparative study of the Al/Bi_2O_3 nanothermite mixed by ultrasonics and LabRAM has been reported [15]. Ultrasonic mixing was limited to a very low solids mixture of 0.6% v/v and RAM up to 50 vol%. Comparison and consistency of electrostatic discharge (ESD) times were used to determine the quality of the products. The ignition delay, for 10 experiments, for thermite ultrasonicated in hexanes was 106 ± 7 μs, compared to 95 ± 3 μs for thermite with N,N-dimethylformamide (DMF) at 40% v/v using RAM. The 30%–50% v/v solid-loaded RAM mixtures had the most consistent ignition delays, and smallest aggregates (2.3 μm nom. aggregate size), overall producing an improved mixing quality. In both cases, RAM was able to produce mixtures beyond the capability of a conventional mixer. At the same time, a resonant acoustic mixer was used to intimately mix active pharmaceutical ingredient compounds and co-former at high frequency in the presence of a small amount of solvent, to induce conversion to co-crystals with no grinding media required. Carbamazepine (CBZ) and nicotinamide (NCT) were used as a model system for successfully producing CBZ:NCT co-crystals [16]. It was shown that RAM provides the mixing intensity required of lab-scale mechanochemical methods, and it in general has been demonstrated to be scalable to volumes greater than 200 l and thus affords a potential new platform for co-crystallisation processes.

Another example, the density, ballistic properties, and homogeneity of RAM and cast-cured rocket propellant of HTPB, AN, and an isocyanate-based curing agent, with an overall solid loading of 81% w/w, were compared [20]. The densities of samples were found to be within 1% using helium gas pycnometry. A chimney burner determined that the RAM-mixed propellant was within 5% of the burning rate (2–10 MPa pressure) for the Baker Perkins–mixed propellant. Overall, there were no discernible differences between the two products.

The ultimate tensile strength of the product could be reduced by using RAM compared to when using a conventional mechanical mixer. For example, an improved mixing quality of epoxy resin and titanium oxide powder with LabRAM compared to an unspecified mechanical mixer were carried out [22]. The average relative standard deviation (RSD) of the ultimate tensile strength was found to be reduced from 18.6% (conventional mixer) to 5.1% (RAM). The reduction in RSD shows good reliability and reproducibility of RAM as a mixing technique, and further shows that it produces a higher quality product. However, it is difficult to validate this experiment due to the lack of detail on the exact mixture and mechanical mixer used.

NiO-yttria-stabilised zirconia anode supports for solid oxide fuel cells were successfully fabricated using LabRAM [23]. Conventional ball milling (BM), addition of plasticiser and binder, and de-airing the slurry would comprise a >72 h process, whereas it took about 30 min with RAM (×144 faster). Quantification of the product was conducted using a 3D reconstruction technique and the electrochemical performance was tested, both finding that the mix was highly homogeneous and statistically identical to BM. The product also showed good long-term stability of over 300 h, making RAM a desirable alternative to BM in this field.

RAM was used in the preparation of Composition B, containing TNT and RDX mixture, at a mass ratio of x% and (100–x)%, and polymer-bonded explosives (PBXs) in the laboratory scale. The efficiencies were increased by 36% and 114%, respectively, compared to traditional mixing methods such as a vertical kneading machine [24]. It was found that the most consistent and highest quality samples were obtained at 55 g within 20 min by the preparation of PBX explosives, at BAE Systems Land UK (BEAS) [25]. Furthermore, the co-crystalline product from the RAM preparation was consistent with the product from solution crystallisation [26]. For example, an energetic–energetic co-crystal comprised of CL-20 and 1,3,5,7-tetranitro-1,3,5,7-tetracyclooctane (HMX) in a 2:1 mol ratio were prepared in a resource-efficient manner by using RAM. It was found that the novel RAM method of producing energetic materials greatly increases safety and is scalable for manufacturing of large quantities of energetic co-crystals and their formulations. However, future work should include evaluations of various new co-crystals as potential routes to meeting insensitive munitions (IMs) challenges.

Using the same material type and similar container proportion, the RAM mixing time is almost unchanged for different quantities of mixed materials. For example, the mixing of PBXs was investigated, with detailed compositions of mixed PBXs, as shown in table 5.1 [27], with 0.2 kg, 20 kg, and 200 kg samples for 12, 12, and 14 min, respectively; it was found that the tensile deformation rates were 2%, 3.6%, and 3.5%, respectively, indicating that the quality of the mixed products decreased with an increase in the order of magnitude under the same mixing time (figure 5.2).

In order to improve the solids loading to obtain high operational performance and improve the shock sensitivity for better IM performance, a comparative study on a 125 g hydroxyl-terminated caprolactone ether binder, AP, and Al mixture, made by LabRAM and Baker Perkins (BP) mixers, were conducted. Two explosive compositions with 0.165 $m^2\ g^{-1}$ and 1.65 $m^2\ g^{-1}$ explosive particle surface area were used, and compared to the baseline mixture with 0.021 $m^2\ g^{-1}$ explosive particle surface area; the viscosity, burning rate, and tensile properties of the final products

Table 5.1. Compositions of mixed PBXs.

Ingredients	Characteristics	Mass fraction/%
HTPB	Binder	7.91
Lecithin	Thickening agent	0.30
DOA	Plasticiser	7.91
Plaster	Energetic materials simulant	41.49
Granulated sugar	Energetic materials simulant	41.49
Blue pigment	Inert materials	0.05
DBTDL	Catalyst	0.03
IPDI	Curing agent	0.71

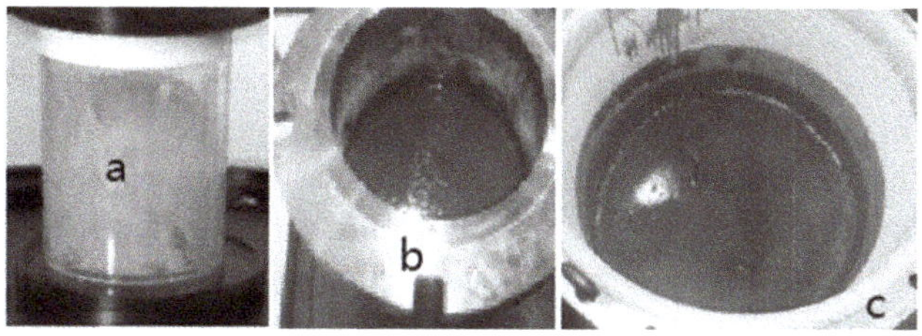

Figure 5.2. Photographs of mixed samples with different inventories: (a) 0.2 kg, (b), 20 kg, (c) 200 kg [27].

Figure 5.3. Photographs of mixed samples with different particle surface areas. Top: 0.165 $m^2\ g^{-1}$, bottom: 1.65 $m^2\ g^{-1}$. Reprinted from [28] with permission.

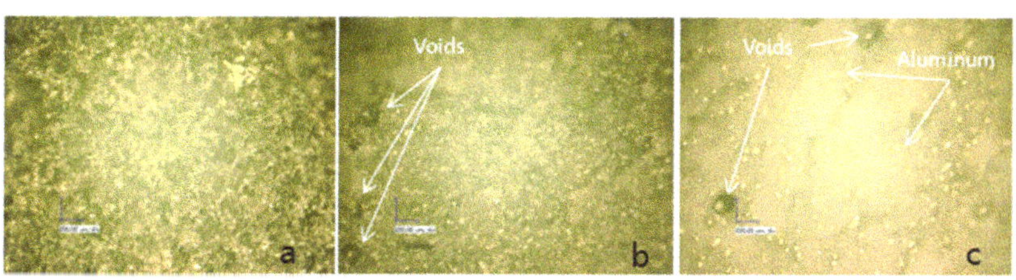

Figure 5.4. Optical microscopies of two compositions, compared to the baseline. (a) Baseline, 0.021 $m^2\ g^{-1}$, (b) 0.165 $m^2\ g^{-1}$, (c) 1.65 $m^2\ g^{-1}$. Reprinted from [28] with permission.

were then compared [28]. Photographs of example mixed samples with different particle surface areas are shown in figures 5.3 and 5.4. It was found that RAM can be used to effectively mix extremely viscous explosive formulations, and there is no evident particles agglomeration in Resodyn-mixed explosives. The end viscosity was significantly greater for RAM (21.5 kPs at 35.5 °C), than BP (7.5 kPs at 49.3 °C). The reason for the lower viscosity for the BP-produced material was not stated but may be due to increased processing temperature. It was noted that the mixing time was shorter for RAM – 30 min compared to 100 min for BP – and that there were no agglomerations seen in the RAM mixture. It is possible/likely that the mix process and time for the RAM was not optimised and could have been considerably faster. The mix time for the BP mixer seems quite short, suggesting that the composition should be easy to formulate.

In addition, we examined the uniformity of RAM-mixed materials in stimulant explosives and propellants (HTPB:Al:$CaCO_3$:ethyl acetate = 1:1.5:3:1, the average diameter of Al and $CaCO_3$ being 29 μm and 1–5 μm, respectively); the scanning electron microscopy results are shown in figure 5.5. It was demonstrated that the mixing efficiency increases with time and amplitude within a certain amplitude range [29].

However, as an innovative technology, the scale-up applications of RAM in solid propellants and explosives needs to meet and complete process safety testing and evaluation. Because of the large amplitude and acceleration characteristics, the main hazardous stimuli in RAM are impact collision, friction, ESD, mixture temperature, and so on [28]. One example, a solid composite AN/HTPB-based propellant, was prepared by conventional processing in a mechanical mixer and by applying an advanced processing technique relying on RAM. The experimental results showed that the oxidiser particles, the homogeneity of the propellants, the density and the burning rate are very similar, independent of the processing method used. As a result of the intense mixing capability of RAM, only a fraction of the processing time is required compared to conventional mechanical mixing, implying considerable savings in production costs [21]. It can be concluded that for the propellant studied in the aforementioned research, RAM is a very promising, advanced processing technique that can be applied as an alternative to the conventional mechanical mixing of this high-solid-load propellant composition.

For the RAM mixing process, the impact stimulus of materials is affected by various factors such as material state and acceleration, so it is necessary to test the impact stimulus and compare it with the impact sensitivity data directly, so as to evaluate the process safety more accurately. The mixture temperature evaluation shows that the temperature can be controlled by the heat transfer effectively, even using the designed standard jacketed vessel mixing container for LabRAM, to provide both cooling and heating to mixed materials in the container during RAM operations [30–32]. However, the impact stimuli have not been tested and fully evaluated, although the impact safety has been analysed and calculated [33]. In conclusion, temperature and static electricity

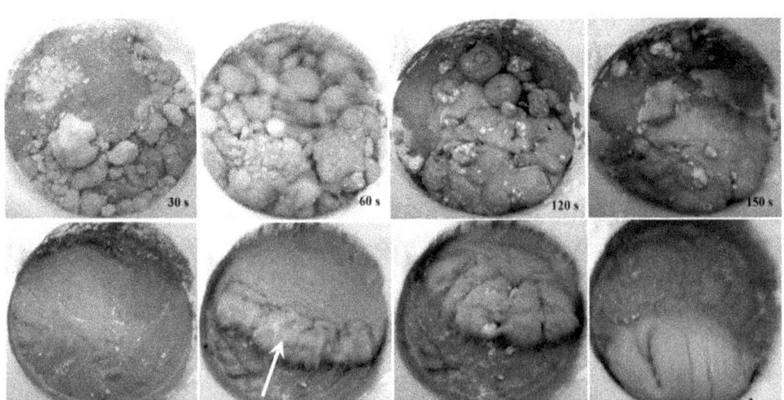

Figure 5.5. Mixing with different times and amplitudes. Top: times; bottom: amplitudes. Reprinted from [29] with permission.

stimuli have proved that the hazard effects are controllable, which indicates that the impact energy obtained from the Bam Fall Hammer Impact Sensitivity Tester (BAM) impact sensitivity meter for 1,3,5-trinitro-1,3,5-tetraazacyclohexane (RDX) is more than 18,000–200,000× the energy delivered.

From the studies considered previously, RAM is able to coat or mix higher concentrations of nanoparticles with liquid binders than conventional techniques. However, this research area is still in its infancy, and further research is needed to determine the solids-loading limitations of RAM for propellant mixtures. RAM has been demonstrated to mix nanoparticles and higher solids loads than conventional techniques, indicating that RAM will likely be able to produce previously unmixable solid propellants and explosives. In addition, the ability of RAM to produce reliable and highly homogeneous mixes makes RAM an attractive option for research or manufacture; though none of these studies specifically consider mixing of HTPB/AP/Al, experiments will need to involve energetics mixtures with oxidisers. For instance, a comparative study on the production of a granular pyrotechnic with magnesium/teflon/viton through RAM and conventional methods was carried out [34]. A 95% w/w was pourable as granules immediately after mixing, making this a low-waste process. Further, the process has been scaled from 25 g to 100 g with near-identical results. It showed a 35% saving per batch and was completed in 5 h compared to 10 h using conventional methods. Overall, RAM was safer, produced less waste, had less environmental impact, was 35% cheaper per batch, and halved the time of production. Another example, PBX shaped-charge warheads, mixed and cast with a bladed mixer (17 made), compared to a batch mixed *in situ* with RAM (18 made), were produced [35]. It was determined that the global standard deviation of the solids content was ±0.217% w/w and ±0.09% w/w for the two test formulations. The mix *in situ* took 0.77 man hours per warhead compared to 1.53 for the mix and cast method. The mix *in situ* required 79% less solvent and produced 63% less hazardous waste, making it more cost effective and environmentally friendly. One of the other benefits of mixing *in situ* is that the building's explosive licence only needs to be for the amount being mixed at that time, which is smaller in each run with RAM compared to larger batch mixers. For this reason, the quantity distance regulation will be lower and have less of a safety impact on other personnel and buildings.

Combining RAM with energetic materials requires a comprehensive understanding of the hazards associated with the equipment on the energetic materials and vice versa. In this chapter, force sensors are used to directly test the impact forces of materials in RAM, and an innovative device and method are proposed which can make the impact force in RAM equivalent to traditional impact sensitivity data, so as to intuitively evaluate the impact safety of RAM.

5.3 Experimental

5.3.1 Raw materials and formulations

The materials used in the experiments reported here are the following: anhydrous sodium sulphate (300 μm; Sichuan Tongqing Nanfeng Co., Ltd, China), spherical aluminium powder (Al, 13 μm and 29 μm; Angang Group Aluminum Powder Co., Ltd, China), hexogen (RDX, $d_{50} = 40$ μm; Gansu Yinguang Chemical Industry Group Co.,

Ltd, China), hydroxyl-terminated polybutadiene (HTPB, type III; Wuxi Ruidkai Chemical Technology Co., Ltd, China), dioctyl adipate (DOA; Aekyung Industry Co., Ltd, South Korea). The main ingredients' mass fractions are listed in table 5.2.

5.3.2 Instruments and equipments

The following instruments and equipment were utilised in the experiments: a Resonant Acoustic Mixer (5 kg: Xi'an Modern Chemistry Research Institute, China), a force sensor (CL-YD-301; Yangzhou Lianzhi Electronic Technology Co., Ltd, China), an acceleration sensor (3097A1; American DYTRAN, USA), and data-acquisition equipment (VibRunner; M+P international, Ltd, Germany).

5.3.3 Impact test devices

The experimental mixed vessel is made of stainless steel with an inner diameter of 130 mm and an inner height of 150 mm. The No. 1 sensor, as shown in figure 5.6(a), is installed on the bottom of the container to test the force which the container

Table 5.2. The mass fractions of the main ingredients in the solid propellants and explosives.

Sample	Solid content/%	Na_2SO_4	RDX	Al (13 μm)	Al (29 μm)	HTPB	DOA
#1	88	58	0	0	30	6	6
#2	84	0	54	30	0	8	8
#3	88	0	58	30	0	6	6
#4	92	0	62	30	0	4	4

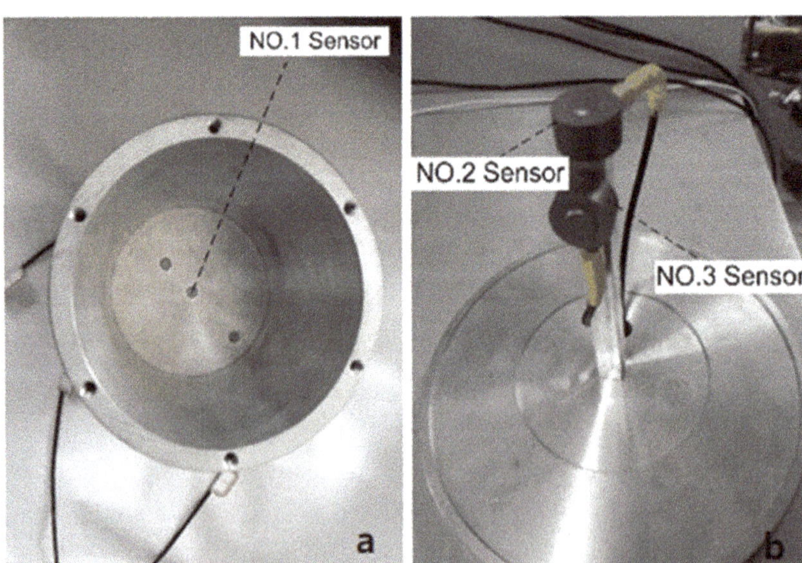

Figure 5.6. The testing devices (a) and sensors for impact forces (b).

bottom exerts on the material; the No. 2 sensor is installed on the straight pillar (perpendicular to the container top) with the sensor surface parallel to the container bottom surface, which is used to test the force which the container top exerts on the material; the No. 3 sensor is installed on the straight pillar with the sensor surface perpendicular to the container bottom surface, which is used to test the force which the container sidewall exerts on the material, as shown in figure 5.6(b).

5.3.4 Impact sensitivity and equivalent impact force tests

In general, the way to evaluate the impact safety of solid propellants and explosives is via the impact sensitivity. The equipment used to measure impact sensitivity includes the BAM impact apparatus and 12-B-type impact apparatuses. The evaluation of impact sensitivity is figured as the explosion percentage of a drop hammer at a certain falling height, where the unit is given as a percentage (%); or it is figured as the falling height corresponding to a certain explosion percentage, where the unit is given in centimetres (cm). The most commonly expressed form is the falling height corresponding to an explosion probability of 50% in the literature (H_{50}).

However, such an explosion would lead to a catastrophic accident, and is not allowed to happen in the practical process. Therefore, the impact sensitivity test method cannot be applied to the safety evaluation of the RAM process, and it is necessary to establish a method that can generate an impact force under an equivalent condition for RAM as during the impact sensitivity test. In other words, where the impact sensitivity of solid propellants and explosives is characterised by the height (cm), the impact force in the RAM process is characterised by the impact force (MPa) and impact force rate (MPa μs^{-1}), respectively.

The equivalent impact force test device is shown in figure 5.7. During the test, the measured materials are loaded between the force sensor and the bearing. To protect the device, a substitute material such as anhydrous sodium sulphate is used to replace the explosives.

5.3.5 Safety evaluation

Step 1: Object selection. Composite solid propellants or cast PBX explosives with a solid content ranging from 84% to 92% were used in RAM; the impact forces of

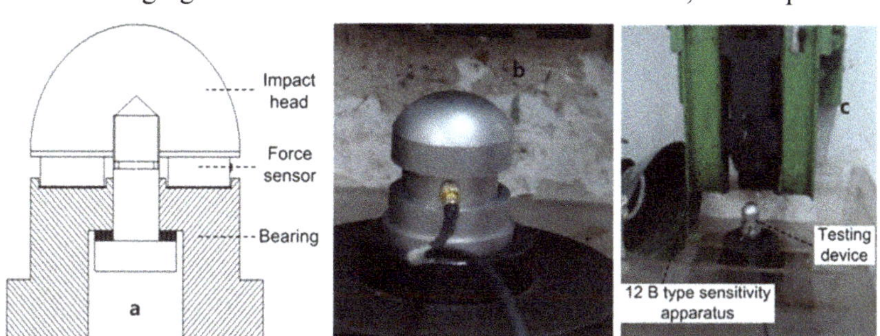

Figure 5.7. The testing device for equivalent impact. (a) Schematic diagram, (b) physical drawing, (c) installation drawing.

propellants or explosives with different formulations (samples #1, #2, #3, and #4) were tested in RAM.

Step 2: Impact test. Considering the maximum acceleration can reach 100 g or more in some extreme cases, the acceleration range during the impact test was 0–60 g. Impact forces corresponding to 100 g and 120 g were obtained by fitting the experimental data.

Step 3: Impact sensitivity analysis. The impact sensitivities of explosives with various formulations were collected, and the formulation with the highest sensitivity was selected.

Step 4: Equivalent impact test. Under the condition of a 3 cm falling height for the drop hammer, the 12-B-type impact apparatus was used to test the impact forces of the materials. At the same time, impact forces corresponding to falling heights of 2 cm and 1 cm were also tested.

Step 5: Safety evaluation. The samples in the RAM test refer to impact forces tests corresponding to 3 cm (2 cm, 1 cm) falling height during the impact sensitivity test. If the sample that had undergone RAM is less than the equivalent impact force, the RAM process can be considered safe in terms of impact force.

5.4 Results and discussion

5.4.1 Impact force of different formulations

The impact force includes force F_{bot}, which the bottom of the container exerts on the materials; force F_{top}, which the top of the container exerts on the materials; and force F_{sid}, which the sidewall of the container exerts on the materials. Figure 5.8 shows the impact forces of $F_{\text{bot-1}}$, $F_{\text{top-1}}$, and $F_{\text{sid-1}}$ for sample #1 changing with acceleration. It can be seen that $F_{\text{sid-1}}$ is 90 kPa, while $F_{\text{bot-1}}$ and $F_{\text{top-1}}$ are 350 kPa at the same acceleration (figure 5.8(a)). The $F_{\text{bot-1}}$ and $F_{\text{top-1}}$ show similar values and trend, which may be attributed to the acceleration, velocity, and other parameters being symmetrical in the positive and negative directions of the equilibrium position in RAM, resulting in the impact forces exerted on the test samples by the bottom and top of the container being almost the same. However, the $F_{\text{sid-1}}$ is much smaller, which is similar to the inertial force (F_{iner}), which may be because the sidewall of the container is perpendicular to the vibration direction of the container and the

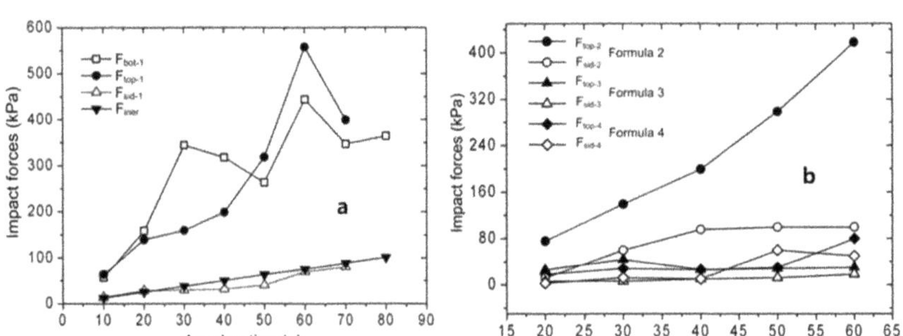

Figure 5.8. The impact forces vs. acceleration of samples. (a) Sample #1, (b) samples #2, #3, and #4.

high-speed moving materials cannot directly collide with the sidewall. $F_{sid\text{-}1}$ is mainly caused by the transfer of inertial force inside the material, which is much smaller than $F_{bot\text{-}1}$ and $F_{top\text{-}1}$.

The inertial force is calculated using equation (5.1):

$$F_{iner} = \rho\, \alpha\, h \tag{5.1}$$

where F_{iner} is the inertial force, ρ is the density of the mixed materials, α is the mixing acceleration, and h is the distance from the sensor to the surface of the mixed materials.

From the above analysis, it can be seen that in the RAM process the most dangerous parts of the impact forces being exerted on the test samples are those of the bottom and top container, while the force exerted by the sidewall of the container is relatively small, and which can be approximately considered as equivalent to the inertial force.

Since the magnitudes and trends of $F_{bot\text{-}1}$ and $F_{top\text{-}1}$ are similar, only the forces acting from the top of the container are tested for samples #2, #3, and #4. The impact forces exerted by the top of the container ($F_{top\text{-}2}$, $F_{top\text{-}3}$, $F_{top\text{-}4}$) and sidewall of the container ($F_{sid\text{-}2}$, $F_{sid\text{-}3}$, $F_{sid\text{-}4}$) are plotted as a function of acceleration in figure 5.8(b). Almost all impact forces do not change significantly with the increase of acceleration except $F_{top\text{-}2}$, corresponding to sample #2, which increases significantly with an increase in acceleration, while the impact forces corresponding to samples #3 and #4, $F_{top\text{-}3}/F_{top\text{-}4}$ and $F_{sid\text{-}3}/F_{sid\text{-}4}$, do not change much with the increase of acceleration, maintaining below 100 kPa. Although the solid content (88%) of sample #3 is the same as that of sample #1, the $F_{top\text{-}3}$ is 30 kPa at 60 g, which is far lower than that of $F_{top\text{-}1}$ (550 kPa) corresponding to sample #1. The magnitude of the impact force of sample #2 exhibits the same trend as that of formula #1, even though the solid content (84%) of sample #2 is smaller than that of sample #1 (88%).

5.4.2 Impact force of sodium sulphate powder

The impact forces of sodium sulphate powder and samples #3 and #4 are plotted in figure 5.9. It can be seen that the impact forces of sodium sulphate powder are lower

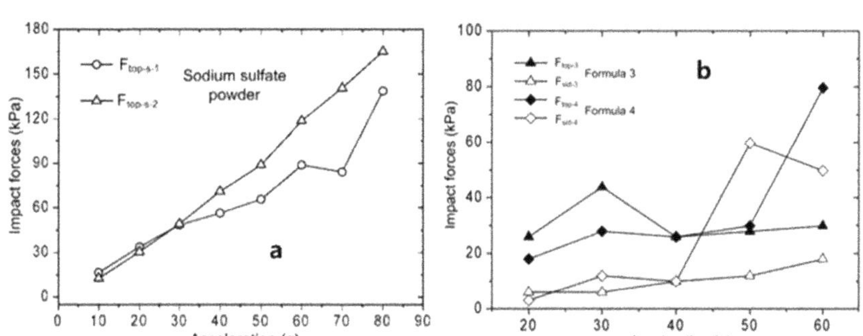

Figure 5.9. Change of impact forces acting on samples with acceleration. (a) Sodium sulphate powder, (b) samples #3 and #4.

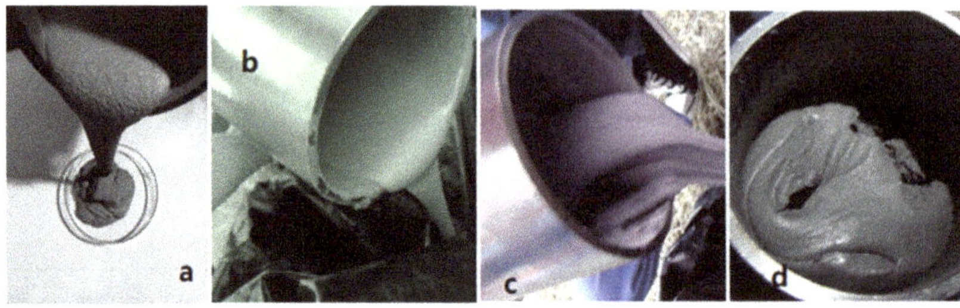

Figure 5.10. Images of the flowing states of the different samples: (a) #1, (b) #2, (c) #3, (d) #4.

than those of samples #3 and sample #4 (100 kPa within 60 g), which may be due to the worse fluidity resulting in a smaller impact force. Samples #1 and #2 with good fluidity lead to an increase of mixing intensity with an increase in acceleration, and then the movement speed and spatter height increase, while samples #3 and #4 have poor fluidity at a near semi-solid state. Good fluidity is related to the rheological properties, solid propellants with good rheological properties (fluidity) can produce propellants with stable and complete structure. Meanwhile, with the increase of acceleration, the increase of movement speed and spatter height were restricted, leading to the impact forces being not as evident as samples #1 and #2. Moreover, due to the smaller particle size of the solid components in sample #3, the viscosity of mixed sample #3 is larger and the fluidity is worse than that of sample #1 under the same solid mass fraction.

Figure 5.10 presents images of the flowing states of the different samples. The materials can flow freely when dumped after sample #1 and sample #2 are mixed, while sample #3 is peristaltic and can be induced to flow with the aid of tools; what is more, sample #4 has the worst fluidity, which is similar to semi-cured concrete and which cannot be dumped.

5.4.3 Equivalence of 'impact sensitivity' and 'process impact force'

5.4.3.1 Lower limit of sensitivity and influencing factors

Generally, the impact sensitivity is expressed by the height of the drop hammer corresponding to the specific probability of explosion, which is mainly expressed as the 50% probability of explosion (called critical drop height, H_{50}) and the 0% probability of explosion (called the lower limit of sensitivity, H_0). The critical drop height H_{50} is used to evaluate the relative safety, and the lower limit of sensitivity H_0 is used to evaluate the absolute safety. The H_0 data of typical single-compound explosives and composite explosives are shown in table 5.3.

As can be seen from table 5.3, for a 2 kg drop hammer, the lower limit of sensitivity H_0 of typical explosives is generally higher than 10 cm, and the corresponding H_0 of PETN is 17 cm, which represents the most dangerous material among them. For a 10 kg drop hammer, the lower limit of sensitivity H_0 of typical explosives is generally 10–50 cm, though the composite explosives containing nitroglycerine, with a H_0 of 3 cm, is the most dangerous one. Furthermore, the lower limit of sensitivity (H_0) and the critical height (H_{50}) are positively correlated with the weight of the drop hammer, but there is no definite functional relationship between them.

Table 5.3. Lower limit of sensitivity H_0 of typical explosives.

2 kg Drop hammer		10 kg Drop hammer	
Explosives	H_0/cm	Explosives	H_0/cm
Tetryl	26	Tetryl	15
2,4,6-trinitrotoluene (TNT)	100	Lamellar TNT	50
Picric acid	82	Ammonium nitrate/aluminium powder/nitrate/calcium stearate 78/11/10/1 (by weight, the same below)	7
RDX	18	Nitroglycerine/nitrocellulose/ sodium nitrate/wood flour 62/3.5/32/2.5	3
1,3,5,7-tetranitro-1,3,5,7-tetracyclooctane (HMX)	32	Ammonium nitrate/TNT/salt 64/16/20	25
Pentaerythritol tetranitrate (PETN)	17	Ammonium nitrate/TNT/ nitrate/wood flour 65.5/12/9/ 1.5	10
Nitroguanidine	47	Ammonium nitrate/TNT/ aluminium powder/RDX 66/ 5/5/24	10
Ammonium picrate	>100	Ammonium nitrate/TNT powder 79/21	50

The H_0 data for samples of RDX, HMX, tetryl, PBX-9010, PBX-9007, Comp B-3, PBX-9407, and Comp B obtained by different drop hammers (2.5 kg and 5.0 kg) are shown in figure 5.11. It can be seen that H_0, like H_{50}, decreases with an increase in the drop hammer weight.

Due to limited experimental conditions, the drop hammer used in our experiment weighed 5 kg, which is only half of the weight of the abovementioned 10 kg drop hammer. We tested the impact forces according to the H_0 values which correspond to the 10 kg drop hammer in table 5.3, which means that the impact forces corresponding to a 5 kg hammer on the material are amplified, which is a stricter method to evaluate process impact safety.

5.4.3.2 Equivalent test of 'impact sensitivity' and 'process impact force'

The impact forces and force rates act as a 'bridge' for impact safety evaluation of RAM, a bridge which connects the impact sensitivity of explosive materials and the impact forces acting on explosives with RAM. As is known to us, the lower limit of sensitivity H_0 of the most dangerous ammonium nitrate explosive with a 10 kg drop hammer is 3 cm [36]. The relationship between impact and time was measured, and the results are shown in figure 5.12. Impact forces are given corresponding to falling heights of 1 cm, 2 cm, and 3 cm, respectively. The corresponding stress peak data are

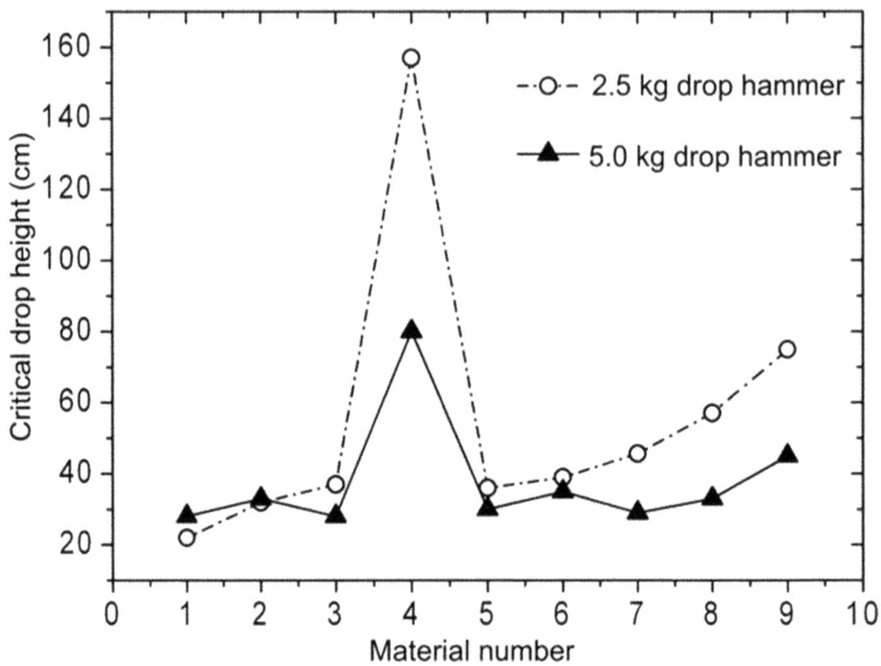

Figure 5.11. The critical height H_0 for different explosives and propellants.

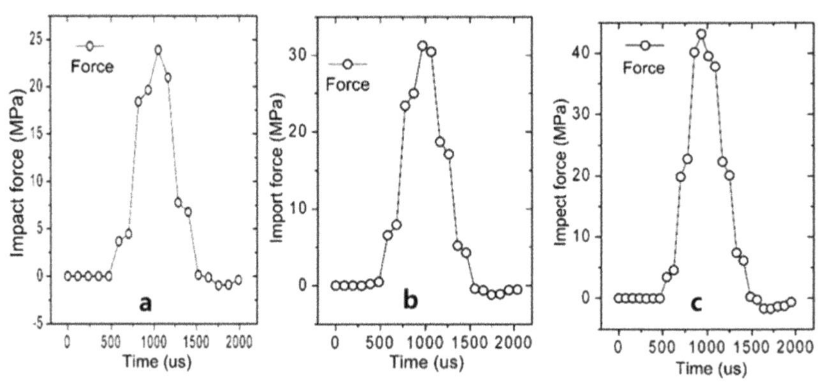

Figure 5.12. Impact forces vs. time with different falling heights: (a) 1 cm, (b) 2 cm, (c) 3 cm.

24 MPa, 31 MPa, and 43 MPa, and the corresponding maximum stress rates are 0.12 MPa μs^{-1}, 0.16 MPa μs^{-1}, and 0.23 MPa μs^{-1}, respectively.

5.4.4 Safety evaluation of impact

The impact of $F_{bot\text{-}1}$, $F_{top\text{-}1}$, and $F_{bot\text{-}2}$ are plotted as the dotted–dashed line shown in figure 5.13. According to the fitting equation, the maximum impact forces are 0.368 MPa, 0.595 MPa, and 0.708 MPa, respectively, when the accelerations are 60 g, 100 g, and 120 g. In order to evaluate the process safety performance, the

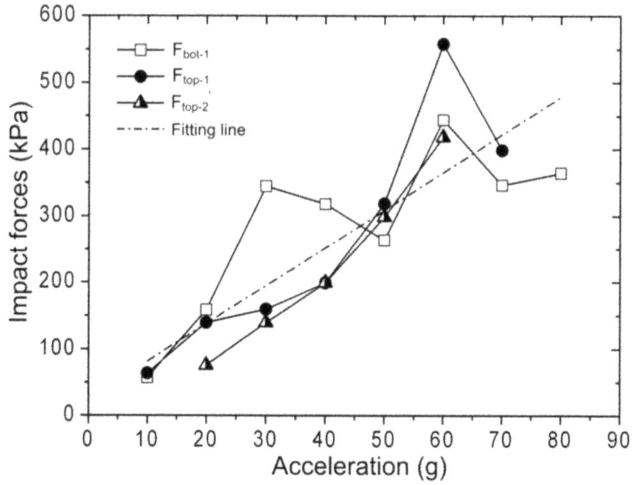

Figure 5.13. Linear fitting to forces as a function of acceleration.

Table 5.4. Safety evaluation of impact forces and equivalent impact forces.

	Falling height			Acceleration of RAM		Safety evaluation
Evaluation of impact force	1 cm	2 cm	3 cm	60 g	120 g	conclusion
The max forces/MPa	24	31	43	0.368	0.708	Safe
The max force rates/MPa μs^{-1}	0.12	0.16	0.23	0.000131	0.000221	Safe

acceleration is fitted and restricted to 100 g and 120 g according to equation (5.2). The fitting equation is given by

$$y = 27.91 + 5.67x \tag{5.2}$$

where y is the maximum impact force acting on the material in RAM, and x is the mixing acceleration. The units of y and x are given as MPa and g (10 m s^{-2}), respectively.

The maximum impact forces acting on the samples in the RAM experiments were compared with those under conditions of 1 cm, 2 cm, and 3 cm falling height, as shown in table 5.4. The data show that the maximum impact forces in RAM are much less than those under impact sensitivity testing. For example, the maximum force corresponding to a 1 cm falling height in impact sensitivity testing is 24 MPa, while the maximum force corresponding to 120 g in RAM is 0.708 MPa; the former force is 33× higher than that of the latter, indicating that the impact force in RAM is much lower and safer than that of the traditional process [37, 38]. The impact force of samples in RAM is 0.0001× of that in the BAM impact sensitivity test, which is perhaps attributable to the single RDX particle being different from composite samples containing RDX (or HMX) in the RAM process.

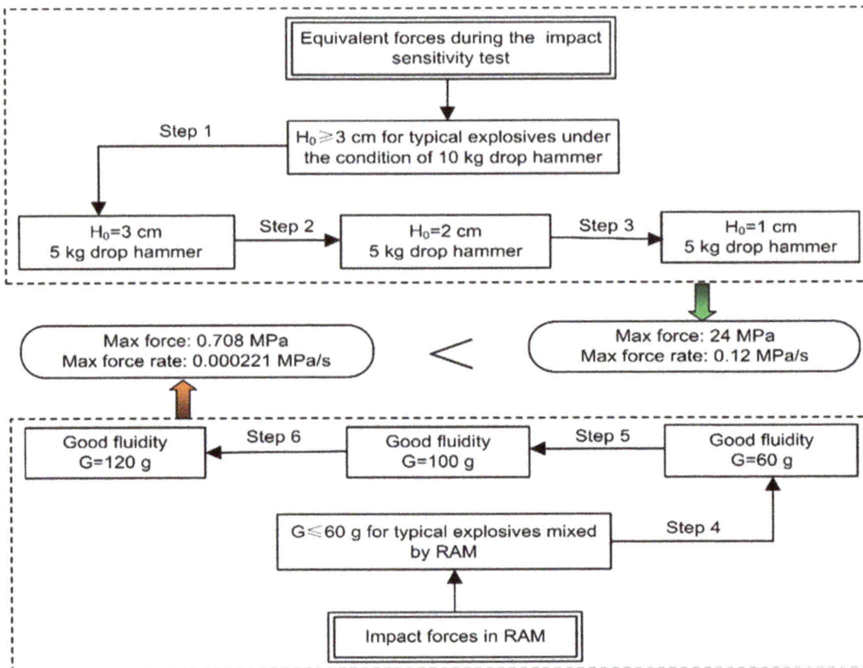

Figure 5.14. Flow chart of the steps for safety evaluation.

In summary, six stricter steps are used in the course of testing and analysis to make the safety evaluation results more conservative and reliable, including stricter hammer mass (step 1), stricter falling height (step 2 and step 3), stricter material type (step 4), and stricter acceleration (step 5 and step 6), as shown in figure 5.14. The safety evaluation methods of RAM technology were outlined, which can be applied to other innovative processing technologies to manufacture energetic materials.

5.5 Prospects and drawbacks of RAM

While improved performance and reliability of a product are highly desirable for any new manufacturing technique, there will be other benefits/drawbacks that need to be considered in judging whether a given technique should be implemented. Safety and environmental factors will be of high interest, but the leading factor will be cost.

There are four consistently reported benefits of RAM that impact all phases of product development. These are (i) shorter mixing time, (ii) a more homogeneous mix, (iii) gentler or low shear mixing, and (iv) less waste. The last of the benefits across all areas of product development is the reduced waste produced by RAM compared to conventional mixing techniques. This will have the greatest benefit in large-scale manufacturing, but also can assist in smaller batches. When a highly viscous formulation is mixed in a bladed mixer there will be a considerable amount of residue on the sides of the container and on the mixing blades [39]. Therefore, more material is lost in the conventional process and, further to this, more cleaning solvents will be needed to clean the vessel for future use [40, 41]. The clean-up phase will incur costs in cleaning materials, protective equipment, operational down time, and the time and cost of the cleaners themselves as well as their associated health risks [42]. The environmental impact of the explosive residue and solvents will also be reduced using RAM, which is an important consideration in modern industry. However, a drawback is that the lower shear might not break down the agglomerates within a mixture. This can be managed in different ways including pre-mixing, pre-wetting, or the intensity of mixing, which is related to the application of vacuum in RAM.

5.6 Conclusions

Overall, RAM offers numerous benefits to mixing existing and new materials with large savings in time and cost, as well as being safer and more environmentally friendly. With that said, RAM is relatively novel in its use in the materials industry and there is still a considerable amount of research required for its benefits to be widely utilised. Based on the present investigation, the following conclusions can be drawn:

(1) The impact stimuli acting on the mixed material are closely related to the fluidity of the propellants and explosives. The higher the RAM impact stimulus, the better the fluidity of the mixed materials.
(2) The maximum impact forces acting on PBX explosives with 84% mass fraction increase with increasing acceleration, with the maximum impact force and impact force rate being 0.368 MPa and 0.000131 MPa s^{-1}, respectively, at 60 g. However, the impact force of PBX explosives with 92% mass fraction is just 0.0796 MPa corresponding to 60 g, which is almost unchanged with acceleration.
(3) For most typical explosives, the lower limit of sensitivity H_0 is higher than 3 cm with a 5 kg drop hammer. Through an equivalent test for a 3 cm falling height, the impact force is 43 MPa, which is nearly 100× of that in RAM, and the impact force rate is 0.23 MPa μs^{-1}, which is nearly 1500× of that in RAM.
(4) RAM is able to mix the current range of solids-loaded propellants and explosives with numerous benefits including shorter mixing time. With RAM, new propellant formulations will be able to be tested at the R&D stage and scaled up more quickly than conventional methods. As such, the rate of new formulations undergoing development is expected to increase significantly.

Acknowledgements

We gratefully acknowledge the National Natural Science Foundation of China (Grant No. 21673178).

References

[1] ResonantAcoustic® Mixing, Resodyne Acoustic Mixers, Butte, Montana, 2013, pp 1–8
[2] Nance D V 2013 *An examination of the RAM's flow field.* 96th ABW/PA Approval and Clearance #96ABW-2013-0384
[3] Nance D V 2013 *Simulating the Resonant Acoustic Mixer* AFRL-RW-EG- 2013-79 DTIC Technical Report

[4] Howe H W, Warriner J J, Cook A M, Coguill S L, Scott L and Farrar L C 2007 *Apparatus and Method for Resonant-Vibratory Mixing,* US Patent 7,188,993 (Resodyn Corp, Franklin St., Butte, MT, USA)
[5] Andrews M R, Collet C and A Wolff C H 2020 Resonant acoustic mixing: processing and safety *Propellants Explos. Pyrotech* **45** 77–86
[6] Ondrey G 2007 Resodyn unveils 'entirely new' mixing technology *Chem. Eng.* **114** 13
[7] Nance D V 2013 *An examination of the resonant acoustic mixer's flow field* AFRL-RW-EG-TR-2013-108 DTIC Technical Report
[8] Wright C J, Wilkinson P J, Gaulter S E, Fossey D, Burn A O and Gill P P 2021 Is resonant acoustic mixing® (RAM) a game changer for manufacturing solid composite rocket propellants? *Propellants Explos. Pyrotech.* **46** e202100146
[9] Sutton G (ed) 2017 *Rocket Propulsion Elements* 9th edn (Hoboken, NJ: Wiley)
[10] Resodyn Acoustic Mixers, Inc., How RAM Mixes – Faraday, available at: https://resodynmixers.com/how-ram-mixes/#faraday (accessed: 9 June 2021)
[11] Coguill S and Farrar L 2014 Resonant acoustic mixing of PBX *40th Int. Pyrotechnics Seminar (Butte, MN, July 13–18)*
[12] Wolff A, Andrews A and Collet C 2019 Additive manufacturing processes applied to energetic materials *16th Rapid Design Prototyping & Manufacturing Conf. (London, UK, April 4–5)*
[13] Ma N, Chen S, Jiang H L, Zhang Z and Qin N 2017 Research progress and prospect on application of resonance acoustic mixing technology in energetic materials field *Ordnance Industry Automation* **36** 20–4
[14] Davey R J, Wilgeroth J M and Burn A O 2018 New age of PBX manufacturing: optimization of RAM *49th Int. Annual Conf. of the Fraunhofer ICT (BAE Systems Land, Glascoed Usk, GB, June 26–29)* 123
[15] Nellums R, Terry B, Tappan B, Son S and Groven L 2013 Effect of solids loading on resonant mixed Al-Bi_2O_3 nanothermite powders *Propellants Explos. Pyrotech.* **38** 605–10
[16] am Ende D J, Anderson S and Salan J 2014 Development and scale-up of co-crystals using resonant acoustic mixing *Org. Process Res. Dev.* **18** 331–41
[17] Hope K, Lloyd H, Konar S, Bull C and Pulham C 2017 Putting the squeeze on energetic co-crystals: high-pressure studies of 2 (CL-20):HMX and CL-20:TNT *New Trends Res. Energ. Mater., Proc. Semin., 20th (Pardubice, CZE, April 26–28)* 184–92
[18] Nelson A 2015 The advantages of resonant acoustic mixing (RAM) for making novel high-energy composite materials *RAM Technical InterChange* August 3–6 (MT, USA: Butte)
[19] Yamamoto C, Miklaszewski E, Dunham J and Shaw A July 8–13, 2018 Issues and observations during scale up of pyrotechnic flare compositions in resonant acoustic mixers *43rd Int. Pyrotechnic Society Seminar* (Denver, CO)
[20] Zebregs M, Mayer A E H J and Antoine E D M van der H 2020 Comparison of propellant processing by cast-cure and resonant acoustic mixing *Propellants, Explos., Pyrotech.* **45** 87–91
[21] Pourpoint T, Wood T, Pfeil M, Tsohas J and Son S 2012 Feasibility study and demonstration of an aluminum and ice solid propellant *Int. Journal of Aerospace Engineering* **2012** 1–11
[22] Rumeau N, Threlfall D and Wilmet A 2015 ResonantAcoustic® mixing–processing and formulation challenges for cost effective manufacturing *Insensitive Munitions Energ. Mater. Technol. Symp.* **47** e202100146
[23] Park J H, Bae K T, Kim K J, Joh D W, Kim D, ha Myung J and Lee K T 2019 Ultra-fast fabrication of tape-cast anode supports for solid oxide fuel cells via resonant acoustic mixing technology *Ceram. Int.* **45** 12154–61

[24] Chen S, Ma N, Yang F, Qin N, Chen C Y, Zhang Z, Sun X P and Xie Z Y 2019 The superiority assessment of resonance acoustic mixing used for laboratory preparation of propellant and explosive *Explosive Materials* **48** 23–6
[25] Davey R J, Wilgeroth J M and Burn A O 2018 New age of PBX manufacturing: optimization of RAM *49th Int. Annual Conf. of the Fraunhofer ICT (BAE Systems Land, Glascoed Usk, GB, June 26–29)* 123
[26] Anderson S R, am Ende D J, Salan J S and Samuels P 2014 Preparation of an energetic-energetic cocrystal using resonant acoustic mixing *Propellants Explos. Pyrotech.* **39** 637–40
[27] Coguill S and Farrar L 2014 Resonant acoustic mixing of PBX *40th Int. Pyrotechnics Seminar. (Butte, MN)*
[28] Cross T A, Nelson A P and Ferguson B P *et al* 2012 Processing benefits of resonance acoustic mixing on high performance propellants and explosives *Proc. 37th Meeting: WPN Group Conventional Weapons, Technical Panel, WPN TP-4, Energetic Materials and Propulsion Technology* 2 *(DTSO Edinburgh)* 190–206
[29] Ma N, Chen S, Jiang H L, Chen J and Shang J B 2015 Experimental study on resonance acoustic mixing of highly viscous explosives and propellants simulant *Guangzhou Chemical Industry* **43** 46–9
[30] Coguill S and Farrar L 2014 Resonant acoustic mixing of PBX *40th Int. Pyrotechnics Seminar* July 13–18 (Montana, USA: Butte)
[31] Wilgeroth J M, Davey R J and Burn A O 2018 Resonant, in-case mixing of a polymer-bonded explosive sharped charge *49th Int. Annual Conf. of the Fraunhofer ICT (BAE Systems Land, Glascoed Usk, GB, June 26–29)*
[32] Lucon P A, Coguill S L and Martineau Z R 2012 Electrostatic generation and methods to mitigate electrostatic discharge, while using the Resonant Acoustic® mixer *38th Int. Pyrotechnics Seminar. (Denver, CO)*
[33] Birkhoff G, MacDougall D P, Pugh E M and Taylor G 2020 Resonant acoustic mixing: processing and safety *Propellants, Explos., Pyrotech.* **45** 1–11
[34] Wolff A 2019 Resonant Acoustic Mixing Performance and Optimization for Energetic Materials. Limited Report. MSIAC, Brussels, Belgium, 2019.
[35] Nelson A and Miller M 2018 Resonant acoustic mixing of high-energy composite materials,
[36] Caicedo J G O 2014 *PhD Thesis* The State University of New Jersey, NJ
[37] Daniel M 2015 Dry polymer coating of energetic materials: feasible or futile *5th Resodyn Technical Interchange (Butte, MT, August 4–6)*
[38] Gruhne M S, Klapötke T M, Lommel M, Wurzenberger M H H, Szimhardt N and Stierstorfer J 2019 Impact sensitivities of explosives explored by the OZM ball drop tester (BIT-132) *New Trends Res. Energ. Mater., Proc. Semin., 22nd (Pardubice, CZE, April 10–12)* 410–17
[39] Leung D H, Lamberto D J, Liu L, Kwong E, Nelson T, Rhodes T and Bak A 2014 A new and improved method for the preparation of drug nanosuspension formulations using acoustic mixing technology *Int. J. Pharm.* **473** 10–9
[40] Zebregs M, Mayer A E H J and van der Heijden A E D M 2020 Comparison of propellant processing by cast-cure and resonant acoustic mixing *Propellants Explos. Pyrotech.* **45** 87–91
[41] Jubb D 2017 Resonant acoustic mixing (RAM) of energetic materials *Off. J. Inst. Explos. Eng.* 28–31
[42] Chavan R B, Thipparaboina R, Yadav B and Shastri N R 2018 Continuous manufacturing of co-crystals: challenges and prospects *Drug Deliv. Transl. Res.* **8** 1726–39
[43] Chen S *et al* 2020 *Chin. J. Explos. Propellants* **43** 383–7

Chapter 6

Reducing vulnerability and insensitive munitions

Adam S Cumming

6.1 Introduction

Insensitive munitions now play an important part of the programmes within all NATO nations and beyond. The intention is to improve safety and reduce accidents through managing the vulnerability of munitions. Proper operation requires a systematic approach based on the understanding of vulnerability and reaction. This field has been developing in the last 40 years, and these studies have produced a workable approach.

NATO ratified STANAG 4439 (IM) [1] and this now forms part of the policy of most NATO member nations. It is recognised as an important approach to meeting the requirement worldwide and has therefore become a normal consideration in munition design.

What do we mean by insensitive munitions (IMs)?

STANAG 4439 defines IMs as follows:

'… munitions which reliably fulfil their performance, readiness and operational requirements on demand but which minimize the probability of inadvertent initiation and severity of subsequent collateral damage to weapon platforms, logistic systems and personnel when subjected to unplanned stimuli …'

History

There have always been accidents with munitions. For example:

James II, King of Scots, being killed by an exploding cannon at the siege of Roxburgh Castle in 1460.

The sinking of UK battle cruisers at Jutland in WWI.

Ammunition depots exploding in UK and elsewhere.

Accidents and violent response are still an accepted risk in many places.

doi:10.1088/978-0-7503-3943-8ch6

However, the increased effectiveness and power of many weapons has changed the balance of risk. Damage can therefore be more catastrophic to the munition owner and prevent successful prosecution of any mission as well as putting people at risk. At the same time, the understanding of risk has improved. This has, of course, not eliminated the hazards.

However, IM policies are the result of experience in the last 50 years and have largely been led by the US, particularly the US Navy. The reasons for this include experience in the Vietnam War and later experiences in the Gulf and elsewhere [2].

It became accepted that there was both a major risk to users and a reduction in the military capability through the risks of accident.

The development of IM assessments complements the standard hazard tests as listed in the UN Recommendations on the Transport of Dangerous Goods [3]. A series of tests were defined and listed in the STANAG and associated documents. They were aimed at assessing the response to a range of likely threats. These are shown in table 6.1 with the acceptable response.

The responses are defined as follows:

- Type 1/2: Detonation or partial detonation.
- Type 3: Explosion.
- Type 4: Deflagration or propulsive reaction.
- Type 5: Burning reaction.

Development

- Vulnerability affects many aspects
    - Manufacture.
    - Transport.
    - Storage.
    - Logistics.
    - Use.

Vulnerability is affected by and applies to the whole system; it is not merely the energetic materials, which are the main reactive component, but also the interaction of these with other components. It is therefore possible to introduce vulnerability into any system through poor design including poor design choices. However, this can and should be managed by proper understanding of the risks and careful design.

Table 6.1. IM test response thresholds.

Threat	Passing Criterion
Fast cook-off	Type 5 Reaction
Slow cook-off	Type 5 Reaction
Bullet impact	Type 5 Reaction
Fragment impact	Type 5 Reaction
Sympathetic reaction	Type 3 Reaction

6.2 Energetic materials and IMs

As energetic materials are the source of most of the hazards it is worth considering them separately.

- Ingredients and compositions
  - Performance and safety are related: greater performance means greater risk, but only for a given set of ingredients. Changing ingredients/formulation constituents can modify this balance.
  - Several developments have taken place in the last 20 years. These have offered new options to assist in IM development [4, 5].
  - The formulators' aim is to provide well-designed and well-understood compositions to meet IM and other service requirements.
  - Existing ingredients can be modified and used/processed in different ways.
  - Composite rocket propellant technology gave rise to polymer-bonded eXplosives (PBXs), particularly in cast or extruded forms.
  - PBXs contain a solid filler in a rubbery matrix; the energetic loading meeting most performance goals (>80%), less in propellants. The proportion can be 95% in pressed compositions.
  - High solids loading means that the most desirable mechanical properties are not fully obtained.
  - However, systematic testing results have shown that cast PBXs are generally less vulnerable overall than melt cast alternatives.

  Newer materials are offering new options for approaching this balance and presenting new solutions. This includes all the components of compositions.

- Hazard assessment and modelling
  - Modelling and mitigation are significant parts of the approach to IM systems.
  - Existing tests replicate accidents and are hard to transfer into models.
  - Research is aimed at diagnostic small-scale tests providing data for validated models.
  - These results can be used to assist with assessment and can produce revisions of IM characteristics and requirements [6].
  - The aim should be to provide *sufficient* information for a systems assessment. There will always be uncertainties, but these should be understood.
- Initiation systems
  - A critical part of the system.
  - Effective, low-vulnerability systems are essential.
  - Electrical, mechanical, and laser systems are being investigated. Effective laser systems are available and are being applied.
- Systems integration
  - Proper use must be made of energetic materials.
  - Systems design and incorporation is crucial if the best results are to be obtained.

- Compositions should be designed to meet system requirements, and the systems should be designed to make best use of available technology.
- Periodic review is needed to ensure that new options are assessed to give better overall behaviour; both in terms of performance and vulnerability.
- Performance and mitigation methods need consideration to ensure a cost-effective solution.

IMs have been produced and fielded. The concept has become standard requirement for new systems and for retrofitting older systems during a mid-life update.

They may now be better described as reduced-vulnerability munitions as they are still energetic and their insensitivity is based on the understanding of their behaviour, the threats to which they will be exposed, and the environment in which they are manufactured, stored, transported, and used. When reports and papers are published it is clear that there is always more work proposed. That is the nature of research and, indeed, a response to changes in understanding of both the technology and the environment as well as of the changing threats.

The example of unknown to detonation transition (XDT) is an indication of this, since removing the possibility of immediate initiation only revealed another mechanism for reaction; this has been shown by delayed detonation in a damaged material where the damage introduced the possibility of reaction.

It must be remembered that munitions are by their nature risky: they are designed to function and release energy in a useful way. However, there will always be the risk of improper functioning. The aim is to manage the risks as efficiently as possible.

There are several aspects of this, but the most important one is to understand the risks. That understanding must cover all the factors that affect vulnerability throughout the lifetime of the munition. Without that understanding, management is very much more difficult, though the level of understanding need only be sufficient to manage risk properly. However, this always needs review in the light of new understanding and an appreciation of new threats. This understanding can stimulate the development of new techniques for risk management. It is equally important to be sufficiently aware of the interdependencies of the various aspects so that the various components can be combined in a suitable manner.

Thus, new ingredients and new formulations need to be considered together with their processing and performance, as well as their vulnerability. Ageing also needs to be considered since reactions continue within the chemical energetic system and this may change the vulnerability signature. All need to be related to the systems in which the components are combined.

This requires an assessment of developing tools; provision of accurate data; empirical testing and development of new tools. It cannot all be done at once, and technical areas develop at different rates. Any theoretical approach must be validated, and, in some cases, a more empirical approach may be all that is possible; indeed, this can be all that is needed at the moment of assessment provided the assessment is as realistic as possible.

Therefore, research must be a mixture of empirical and theoretical work. That balance will change, as noted above, and must be subject to review.

6.3 Research priorities

If an iterative programme of improved understanding is to be obtained, then the user requirements must be understood, and the science and technology directed to meeting those needs. They may not immediately answer all those needs but should offer improvement over the existing or previous position, so that the risk is managed and understood in ways relevant to the user. Therefore, the aims should be as follows:

- To be able to provide advice on vulnerability throughout the life cycle of a munition.
- To be able to predict the characteristics of a munition system sufficiently well to allow proper fielding and management.
- To be able to design reduced-vulnerability systems meeting performance, life, and environmental requirements.
- To understand ignition well enough to design an optimised system,
- To be able to predict the effect of ageing on vulnerability,

Different areas of the user community will have different needs. The balance of the priorities will therefore change depending on circumstances. One issue has been, and continues to be, the need to understand and manage the vulnerability of munitions in urban terrain and in conflicts such as Iraq and Afghanistan where new risks have appeared.

One of these more recent problem areas is outlined below. While this may seem to be peripheral to the main aspects of IM technology, it is an area of increasing importance.

The use of IM systems has also produced problems for disposal in the field, as existing tools for disposal are less adequate for the purpose; that is, such tools were designed for disrupting more traditional explosives but are much less effective with reduced-vulnerability materials. Often unreacted explosives and propellants are recovered from vulnerability trials on next-generation insensitive explosives and propellants. Trials were aimed at demonstrating reduced vulnerability but equally demonstrated the risks of unreacted or partially reacted material being left [7, 8].

This applies to most forms of energetic material, both propellants and explosives, and could apply to reduced-vulnerability ignition systems.

This area is one possible example that is growing in importance as a priority.

6.4 Areas of active work

There are several apparently discrete areas where research is in progress, with different emphases in different countries, based in part on their local priorities. In some cases, they contribute to international collaboration. However, these should be seen as the building blocks for an integrated approach to providing validated IM technology and a clear understanding of the direction of future research.

It can be hard to get these disparate pieces to fit together, and hard, too, to incorporate new options either from within the academic world or from technology developments in non-munitions areas. These newer options may challenge the status quo, but it is not impossible to bring them into use. The development of resonant acoustic mixing (RAM) processing technology, discussed in chapter 5, is one example of new technology having an impact. A further UK-based example of this is the use of materials technology to provide improved modelling tools using the Porter–Gould model, described in several papers at the International Ballistics Symposium and the Detonation Symposium.

In short, integration is a present issue and there will be a continuing need to seek novel solutions to enable the whole technology area to meet future needs.

It is important to examine the various strands of activity in overview before looking in more detail at the research in more detail. Many of these are already joined in some way, and one of the roles of the Munitions Safety Information and Advisory Centre (MSIAC) in NATO is to foster this integration amongst firstly their members and secondly more broadly.

The areas of research can be summarised according to the following subsections.

6.4.1 Detonics

Detonics includes connecting microscale detonation properties to macroscale via studies of the meso-scale. The area is discussed in chapters 7 and 8. The ability to understand and predict behaviour remains an issue for there are more complications than simple theory suggests. The challenge remains to provide enough information to support a satisfactory analysis for use in predicting behaviour.

6.4.2 Testing

Testing can be best understood as providing data for both empirical assessments of risk and vulnerability and now, in addition, to sustain and direct modelling. The original NATO STANAG tests for IM are now seen to have had limitations, and they are being examined to find updates or replacements. It is essential to understand what is being tested and why a test is performed. It is important to avoid a routine approach to testing where the result is disconnected from the risk and merely a process to be followed. Part of the general aim is to move from reliance on large-scale tests to intelligent use of small-scale diagnostic tests supported by models but with validation if required by full-scale testing. It is likely that full scale will often be required for the foreseeable future. The crossover between IM testing and more normal hazard testing has been recognised and future testing needs to provide useful information for a range of applications.

What do we test? Many tests simulate accidents and may therefore give false confidence in the result. The use of PBX explosives appeared to minimise the appearance of shock to detonation (SDT), for instance, but studies on rocket motors indicated that another mechanism could be in operation. Its initial appearance was unexpected and hence it was named 'XDT'. This has been identified as often due to damage produced by the first shock, which damage can sensitise the material to other stimuli, including self-heating. The result can be a delayed but catastrophic response.

How do we refine the STANAG tests to provide more useful information? The STANAG tests were produced in an attempt to use the best scientific information available at the time to produce a series of tests to provide the best possible advice. The understanding has since moved on and so the tests need to change and evolve, too. MSIAC, the NATO agency, is promoting studies of this, as are groups in several nations.

One area of concern is the 'slow cook-off' test where the temperature regime is unrealistic. Another is the 'shaped charge attack', where a Type 3 response is allowed but that would do considerable damage to the surroundings. This needs a detailed understanding of the environment and the behaviour of energetics in that environment. This is far from easy and is one of the possible uses of the studies of the behaviour of energetic ingredients at high pressures. Even knowing the equation of state in those conditions would be helpful. That understanding should also help to design either crystals or compositions with a more benign response. This is partially discussed in chapter 3.

It will be very hard to produce something energetic capable of a benign response in such a harsh environment and using a shaped charge attack, but mitigation and minimisation may well be possible.

Can we provide data to populate models and refine them?

At present a series of models are being used to predict behaviour; this is briefly discussed in chapters 4 and 7. With the increase in computing power, it becomes ever easier to deal with the complexities of detonation/deflagration and related processes. However, burning to violent reaction remains an active area, and one not yet resolved though progress is being made. Much of the real work is being undertaken in the US National Laboratories, but further research is being carried out in France and the UK as well as elsewhere. Any developments need to be validated and sustained with real data. The acquisition of that data needs, and is getting, active collaboration between experimentalists and theoreticians. Laboratory-scale tests of carefully characterised materials in places such a Cambridge, in the UK, are backed with large-scale instrumented tests to sustain and change models to sufficiently validated predictive tools. There is still work to be done and this is an active area, with papers presented at the Detonation Symposium amongst other venues.

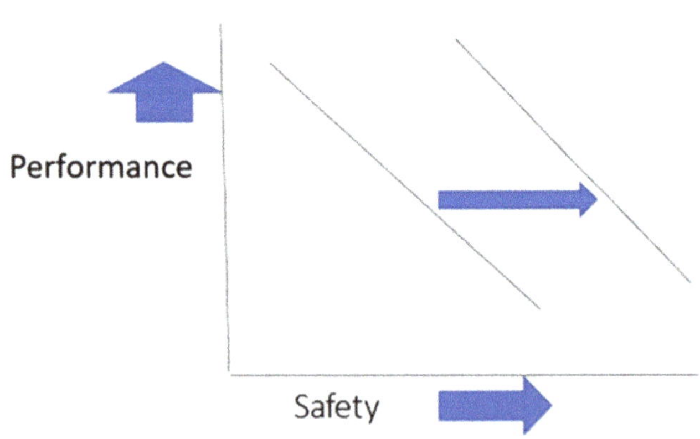

Figure 6.1. Schematic of performance vs. safety.

Can we use small-scale tests to predict full-scale responses?
An example of this recent work can again be found in QinetiQ for the UK Ministry of Defence, looking at understanding small-scale tests.

A recent paper [9], presented in 2014, describes experiments and modelling aimed at understanding the behaviour of highly loaded (90%–95%) pressed HMX-based PBX compositions, when subjected to shock compression and ignition. A range of tubes were designed to examine the contribution of the tube material's properties (i.e. steel versus aluminium, 218.5 MPa) and to examine the effect of reduced confinement (120 MPa). In particular, tests carried out in polycarbonate tubes gave a good insight into the occurring processes.

One result is the proof that such tests are not as straightforward as might have been thought, but that if properly designed with an understanding of their limitations they can provide sufficient information for initial assessment of the likely responses.

Can we connect detonic-scale events to munitions?
One of the areas identified for active research is crystal detonics, where the study of the effects of crystal morphology and size is linked to molecular studies of ignition and output. This is part of the missing link between that fundamental understanding and the larger one of SDT, DDT, and, indeed, XDT. It is not essential to have a complete understanding but such work will provide more confidence in the predictions of behaviour. That confidence could allow this form of analysis to be more rapid and also more cost effective. It will still require expertise and intelligent use to enable it to be useful and not mislead. Recent work by Morrison *et al* [10] has investigated the importance of crystal structure on sensitivity. The assessment and modelling from first principles give a possible route to managing risk through greater understanding.

The integrated approach outlined in chapter 1 would incorporate and integrate this improved understanding.

6.4.3 Hazard analysis

Linked with detonics and testing, hazard analysis now includes examination of the toxicity of the materials being considered and the lifetime and compatibility requirements. Much of this will be covered by testing and detonic studies. Much of this is discussed in chapters 7, 8, 9, and 11.

Threat analyses. These must include new and changing threats and also be capable of modelling and predicting likely scenarios. The data provided will support the direction of research to support risk management both present and future. It is iterative since new materials and new concepts may change the threat priorities. For example, the likelihood of significant quantities of unreacted reduced-sensitivity explosives becoming available will need to be considered. A second example is the impact on the environment, which needs assessment.

Modelling. This involves predicting effects and properties. While the development of models and software can be considered in isolation, it is more usually linked with detonics and testing, as the model provides a means for taking the theory and using it to predict the results. Often applying the theory and the modelling approach will define the data needs and therefore the experiments that are needed to provide that data, so that the models more closely approach reality. The aim should be to move to predictive models and away from curve-fitting approaches, which often have limited value and can lose their validity with new understanding or, indeed, with new materials. A recent programme linked modelling with testing in an attempt to produce a diagnostic prediction capability [11]. There is increasing interest in the modelling of other relevant properties including those associated with life assessments; see chapter 10.

6.4.4 Ingredients

Ingredients are a critical part of the current research programme in most nations. It is clear that these are the source of the performance, be it detonation, propulsion, ignition, or any other form of work. The majority of the ingredients currently in use have been around for many years. This is one of the reasons for calling energetics a mature technology. However, new ingredients are becoming available and, perhaps more importantly, there are also newer versions of existing materials. Much of this is discussed in chapters 2 and 3.

It should also be possible to predict new species with desirable properties, and this is being attempted by the pharmaceutical industry. Ongoing work in the US is applying such tools to energetics, including the machine learning methods discussed in chapter 4. It is necessary to understand the properties and characteristics that are wished for if this to be successful. An example of this is the prediction of the desirability of nitrogen species, as illustrated below. These seem to offer significant performance benefits, but there are no tools to predict their sensitivity and therefore what contribution they might make to an IM solution. At present, making them appears to be the best option, though the two ionic species N_5^+ and N_5^- appear to be very sensitive [12, 13].

One of the factors affecting the existing response of materials is that the technology used in their manufacture is now obsolescent. It is possible to produce ingredients whose size and shape and size distribution is much more carefully controlled and therefore give better characteristics.

Nanomaterials, both those of existing materials such as aluminium and RDX, HMX, etc, and also of new constructs such as loaded nanotubes, are being investigated as promising options. These materials are hard to handle or process, with poor handling producing porosity, which can lead to ignition points through hot spot formation. Here there is another opportunity for the merging of different subareas to work on a unified approach. New processing methods may make these useable [14].

Solid ingredients are only part of the solution, for the binders and other ingredients need consideration, too. The origin of PBXs and composite propellants illustrate the early stages of that process, with the use of rubbery material to give more ideal mechanical properties enabling the delocalisation of the shock. Such inert materials could not contribute to the energy of the system, though the common binder hydroxyl-terminated polybutadiene (HTPB) acted as a fuel in propellants. The solid loading was therefore too high to optimise the mechanical properties. This led to the preparation of energetic binders. The best known of these is gycidyl azide polymer, though others have been produced and tested.

This approach seemed promising but has not been adopted significantly due to cost and changes in priorities for operations. If there is a need they may be employed more extensively. There is, however, a problem in their curing chemistry in that they use isocyanates to produce the polymer, as does HTPB. Isocyanates are highly toxic and so research work is ongoing to produce alternatives.

In summary, the ingredients used must provide improved options for formulations and should include both existing and new options. This area comprises not just new materials—ingredients such as 1,1-diamino-2,2-dinitroethylene (or FOX-7), 3-nitro-1,2,4-triazole-5-one (NTO), and 2,4-dinitroanisole (DNAN) have been and are being investigated—but also improved versions of existing materials. The best known of these is reduced-sensitivity RDX. This is discussed in chapter 3.

6.4.5 Formulation

Well-understood insensitive formulations are needed for all systems. These include the following:

- Warheads.
- Gun systems.
- Rocket systems.
- Ignition.
- Pyrotechnics.

In general terms, there is a relation between safety and performance as in general the higher the performance the lower the safety.

The energy always exists and must be available for the proper use of the materials. All that can be done is to raise the threshold for accidental ignition above that which is likely to be encountered. This can be attempted using less-sensitive energetics. The line therefore moves but the gradient remains.

An alternative approach is mitigation, where the energetic requirements prevent the material choice meeting the needed sensitivity reduction. However, one issue with such systems is that mitigation methods need to be researched and proved to operate in the right way and at the right time. For example, shape-memory alloys and strip laminate cases have been researched to provide ways of relieving pressure and temperature and allow venting in the right timescale, but they have had only partial success.

Research has focussed on providing new options for IM systems. As mentioned in chapter 1, one of the first routes to reduced vulnerability was using rubbery binders to produce improved mechanical properties, which reduce the vulnerability of the compositions to impact. This had been used for composite rocket propellants, and one of the features in the last 20 years has been the drawing together and the increasing synergy of that process of integration. The use of energetic binders was seen as a way of maintaining performance but has so far failed to make a significant mark. New options in melt cast compositions such as DNAN or Guanylureadinitramide (GUDN, or

FOX-12) appear to offer some more traditional routes, but these newer materials must also meet increasingly stringent environmental constraints that affect their potential for introduction. Their invention has produced problems peculiar to these materials involving their toxicity and material properties [14]. Some of this is discussed in chapters 3 and 11.

It is also necessary to sufficiently understand the threats to enable them to be managed. Therefore, here too there is a need for interaction with those working on fundamental science and testing so that all become aware of the problems and the possible solutions.

Formulations need to meet performance, lifetime, environmental, and other constraints and be cost effective in their creation and production. Tools for predicting this performance exist (CHEETAH, EXPLO5, amongst others) and help reduce the time to develop new candidates. This has been an active area for some time, with several examples in the literature.

Such programmes and related ones pertaining to composite rocket propellants and gun propellants indicate that it is possible to provide solutions. However, the solutions have tended to be expensive, and while some have been taken up, this is still an area with possibilities and that continues to be active.

The area that poses the most problems is possibly the incorporation of these compositions and their analogues as missile propellants in a system.

Similar approaches are being taken in the study of missile propellants with the addition of alternative approaches. These include hybrid systems and completely liquid propellant systems. Some of this is discussed in chapters 15 and 16.

Solid propellants were required to become more energetic, and this usually meant the addition of high-performance materials such as nitramines. Unfortunately, this changed the critical diameter and led to materials with a UN Hazrad Classification 1.1 designation and not the 1.3 that was normally considered. The UN Transportation Classifications are a normally used measure for assigning a hazard level and 1.1 is the highest risk [3]. Reduced-vulnerability motors are now a priority and the use of mitigation and better understood approaches is providing answers [15].

While explosives have the highest energy and may appear to offer the highest risk, the rocket motor is larger and, if ignited, can incur significant damage, even leading to detonation in some cases. Therefore, the same approaches are being utilised and the same problems investigated.

6.4.6 Processing

Any new thing needs to be physically made! This is often the constraint that prevents new materials and new approaches being adopted.

Novel approaches and twin-screw extrusion offer some advantages, but recent research on RAM offers promise for developing new routes for the manufacture of suitable compositions. This is perhaps the research topic of greatest interest at present. This has been discussed elsewhere in other chapters, particularly chapter 5. RAM may also allow proper and effective handling of nanomaterials.

In short, processing and manufacture need to be considered. New materials and more tightly specified requirements cannot be applied if the physical means of producing them is unavailable. Therefore, research is required to explore new options for manufacturing which ensures that they can operate cost effectively.

6.4.7 Components

System construction and integration is relevant for IMs; energetics are not the only piece to consider. Their interaction and compatibility with other materials need assessment, and those other materials can add to the solution, or, if misused or wrongly chosen, can open a route to vulnerability. Modern materials are providing the munition designer with choices that would not have been possible 10 years ago.

6.4.8 Performance

It is vital to ensure that we can meet the performance requirements of the users; whatever is produced and offered needs to meet the performance requirements for the system. This has often been quoted as a problem with IMs, in that it is deemed necessary to require a performance reduction. It must be demonstrated that the system fully meets the requirements as well as the IM's need. This also requires that a realistic performance requirement is identified. Other constraints will factor, too.

Cost. To ensure we can afford the proposed solution, whatever options are considered, the cost needs to be one factor in that consideration. Often, however, true costs only become apparent after research and development investment. New options in development will cost more and the cost can only be reduced if those materials and systems are developed to a manufacturing level. Cost awareness is necessary but should not be the only factor.

Developments in all these areas are reported regularly, both at conferences or in the literature. *Propellants, Explosives and Pyrotechnics* is one journal worth following, as is the *Journal of Energetic Materials*. Work is presented at the annual ICT Conference in Germany and at the NDIA IM/Energetic Materials Conference, held at around 18 month intervals. There are other meetings, such as the Detonation Symposium and New Trends in Energetic Materials, held annually in Pardubice and aimed at younger scientists, where information is shared and discussed.

It is important to continually review the range of activities in progress and published in the open literature to understand how the options are changing.

6.4.9 Fundamental science

It is worth recalling the very basic and fundamental aspects drawing on our understanding in the light of formulation and processing research and that on shock physics, modelling, and testing. The question remains: Can we understand the basic properties of energetics well enough to

- Design new molecules and formulations?
- Have enough information and component control to manufacture inexpensively and reproducibly?
- Be able to specify the component requirements, energetic or otherwise, to meet the requirement in a cost-effective way?

- Predict the behaviour of a material or a component, now and after ageing?
- Operate a virtual munition design and testing laboratory with at least 80% effectiveness?
- Be flexible enough to respond rapidly and cost effectively to new demands and new threats?
- Integrate them effectively with other components, making the best use of the properties of each to produce an optimal system or subsystem with the materials available?

6.4.10 System issues

IMs are part of munitions design. The interactions with other components of the munition need to be optimised, and research on composite materials, metals, and mitigation techniques are in progress, though often linked to specific systems.

There is a need for a systems-integration approach that treats all subcomponents as part of a system and works to optimise the behaviour and performance of the whole. Often a piecemeal approach is taken, which reduces effectiveness. Development times for subcomponents differ, which makes this harder, but integration research needs to be developed and sustained with the aim of producing a practical solution for future systems.

In addition, there is a need to understand how the properties of any such system will change with time. Energetic materials are reactive and will continue to change chemically with time and depending on the environment experienced. There needs to be an appreciation of this and an assessment so that any risk can be managed. This too is being assessed [16].

Finally, all aspects of research and the introduction and pursuit of IM goals is subject to cost constraints. The UK uses the 'ALARP' principle: As Low As Reasonably Practical, which should not mean as cheaply as possible but as inexpensively as possible whilst meeting requirements [17].

This means that the requirements also need to be realistic and matched to meeting the threat now as well as over the lifetime of the system, with potential routes to improvement, including technology updates, to meet future IM needs.

6.5 Conclusions

Insensitive and reduced-vulnerability munitions are now a normal requirement, and one that will be maintained or even further developed in the future. We have enough understanding to meet this need and are developing the tools and understanding to improve the response to that requirement. However, the technology base will continue to develop, and improved fundamental understanding will be transferred into component and system applications. It is therefore not a mature area but one that is in continual motion. The options and risks will continue to change, driving a need for better responses to them.

There will continue to be questions of vulnerability and accidents which will drive additional work and research. That research will affect the whole field of energetics, including the civil field, where safety remains critical.

It may be argued that separating IMs and safety is no longer sensible and that it is better considered as a continuum from safety tests through to finished systems. The results at a component level do affect the aggregated system, so all aspects need consideration and the understanding needs to be applied across the technology.

Here, again, we have a reason for an integrated approach with the sharing of knowledge, experience, and expertise.

References

[1] STANAG 4439, Edition X, Policy For Introduction and Assessment of Insensitive Munitions, March 2010.; AOP-39, Edition 3, Guidance on the Assessment and Development of Insensitive Munitions, March 2010

[2] Till M and Kosar K 2006 Formation of the US navy's munition reaction evaluation board (MREB) *Proc. 2006 Insensitive Munitions & Energetic Materials Technology Symp. (Bristol, UK)*

[3] UN 2019 *Recommendations on the Transport of Dangerous Goods, Manual of Tests and Criteria* 6th edn (New York: United Nations)

[4] Powell I J 2016 Insensitive munitions—design principles and technology developments; propellants *Explosives and Pyrotechnics* **41** 409–13

[5] Bordachar S and Lamy P 2012 A new impulse for the French MURAT (IM) Policy *Proc. 2012 Insensitive Munitions & Energetic Materials Technology Symp. (Las Vegas, NV)*

[6] Jacq C, Tomasello K, Baker E and Sharp M 2019 Review and update of insensitive munitions test procedures *Insensitive Munitions & Energetic Materials Technology Symp. (Seville, Spain)*

[7] Lamy P, Leiber C-O, Cumming A S and Zimmer M 1996 ASNR LTTP on insensitive high explosives *Proc. 27th ICT Conf. (Karlsruhe, 25–28 June)*

[8] Cumming A S *et al* 2007 Performance tests of next generation solid missile propellants *NDIA Insensitive Munitions and Energetic Materials Technology Symp. (Miami, Florida)*

[9] Cook M D, Wood A D, Ottley P R and Cheese P J 2014 Observation of sub-detonation response in confined high density HMX based PBXs *18th APS-SCCM and 24th AIRAPT*

[10] Michalchuk A A L, Trestman M, Rudić S, Portius P, Fincham P T, Pulham C R and Morrison C A 2019 Predicting the reactivity of energetic materials: an *ab initio* multi-phonon approach *J. Mater. Chem.* A **7** 33

[11] Thomas Swierk, Steven and Collignon I M 2019 Hazards modelling and simulation—an international collaboration *NDIA Insensitive Munitions and Energetic Materials Technology (Sevilla, Spain)*

[12] Dixon D A, Feller D and Christe K O 2004 Enthalpies of formation of gas-phase N_3, N_3^-, N_5^+, and N_5^- from ab initio molecular orbital theory, stability predictions for $N_5^+N_3^-$ and $N_5^+N_5^-$, and experimental evidence for the instability of $N_5^+N_3^-$ *J. Am. Chem. Soc.* **126** 834–43

[13] Östmark H, Wallin S, Brinck T, Carlqvist P, Claridge R, Hedlund E and Yudina L 2003 Detection of pentazolate anion (cyclo-N5−) from laser ionization and decomposition of solid p-dimethylaminophenylpentazole *Chem. Phys. Lett.* **379** 539–46

[14] Coster P, Henderson C, Hunter S and Pulham C *et al* 2014 Explosives at extreme conditions: polymorphism of 2,4-dinitroanisole *Acta Crystall.* A **70** C896–896

[15] Smit B 2010 Mitigation techniques for reduced rocket motorvulnerability against external thermal stimuli *NDIA Insensitive Munitions and Energetic Materials Symp. (Munich, Germany)*

[16] Paul D *et al* 2013 The effect of energetic materials ageing on system IM response *NDIA Insensitive Munitions and Energetic Materials Technology, (San Diego, USA)*

[17] UK MoD 2018 *Defence ordnance, munitions and explosives safety and environmental management policy* JSP 520 Ministry of Defence

IOP Publishing

Energetics Science and Technology
An integrated approach
Adam S Cumming

Chapter 7

Ignition and detonation in energetic materials: an introduction

William G Proud

Theories on the ignition and growth of reaction in energetic materials have developed significantly over the past two decades with increased experimental knowledge. There is an extensive and growing literature in this field. This chapter deals with the fundamentals of the ignition and detonation process, highlighting the fundamental physics and chemistry of the situation. It is important to recognise that the overall processes may appear to be complex because they represent an interplay between thermodynamics and kinetics; the result is a dependence on pressure, temperature, and time. In specific scenarios, high shock wave levels, for instance, one process may dominate but in general it is a combination of all three. The latter sections of this chapter discuss research relevant to this subject in greater detail.

7.1 Energetic materials

The term 'energetic materials' (EMs) often seems to be a euphemism for 'explosives' but actually describes a wide variety of materials. EMs can be divided into three broad categories: explosives, propellants, and pyrotechnics. Explosives are defined as materials which react to produce a violent expansion of hot gas, an explosion, which rapidly delivers energy to its surroundings. The rate of transformation from solid to hot gas takes place on timescales of microseconds. Propellants are less violent in reaction and, as the name implies, are used to accelerate objects such as missiles and bullets; the time taken to go from solid to gas is of the order of milliseconds. Historically, some materials such as gunpowder can be used either as an explosive charge or as a propellant, the difference in behaviour being dependent on the amount of containment/confinement around the gunpowder: the more confined, the more violent the reaction. Pyrotechnics are systems which react to produce an effect such as smoke, light, or noise with reaction times from milliseconds to many minutes.

doi:10.1088/978-0-7503-3943-8ch7

Comparing EMs with other materials such as petrol, 1 kg of petrol contains six times more available energy as the same mass of trinitrotoluene (TNT), *but* the TNT releases its energy a hundred million times faster. It is this rapid kinetics of transformation that forms the basis of an explosion.

The simplest way to envisage the chemistry of an explosive is as a mix of an oxidiser with a fuel. Historically, the most important explosive was gunpowder, classically having the ingredients potassium nitrate (the oxidiser), charcoal (the fuel), and sulphur (an initiatory compound making it easier for the mix to catch fire and stay lit). Burning between the grains of the different components, called deflagration, controls the rate of gas output.

In contrast, modern military and commercial explosives are based around materials which have the fuel and the oxidiser as separate chemical groupings on the same molecule, effecting a degree of mixing which is impossible to achieve with simple powders. This intimacy means that the reaction, instead of moving as a flame, from grain surface to grain surface, now moves as a wave from molecule to molecule inside the explosive. The chemical reactions take place in a zone which is much less than 1 mm thick, fast moving at velocities of the order of 8000 m s^{-1} and at high pressures of about 250,000 atmospheres (atm). This zone is called a detonation wave. The energy release rate is up to 1000 times higher than that for gunpowder. This high output rate results in these materials being called 'high-order explosives', while gunpowder, which is deflagration controlled, is a 'low-order explosive'.

The output of an explosive can be defined in terms of two factors: brisance and heave. A shock wave can be transmitted from detonating high explosive into materials that cause them to shatter. This shattering is called brisance. In general this only happens when the explosive and target material are in physical contact, e.g. the rock around an explosive-filled borehole in a quarry or the metal casing of a shell. The second effect is produced by the expansion of the hot, gaseous products which lift, separate, and throw materials; this is called 'heave'. For example, 1 cm^3 of explosive can produce approximately 10 litres of hot gas, an expansion of 10,000 times. Most explosive devices produce both brisance and heave. Military explosives tend to have more brisance and less heave. In this case the aim is to first shatter the target directly or to fragment a shell casing, afterwards accelerating those fragments.

7.2 Hot-spot formation

In general it is thought that the impact mechanisms by which explosives ignite and ultimately detonate are thermal in origin. The mechanical energy is envisaged as being converted into heat in localised regions by rapid or large deformations with the formation of 'hot spots'. Although there is a wide agreement that hot spots are the process by which EMs ignite, the precise mechanisms have been open to discussion [1, 2]. The following mechanisms are widely accepted: adiabatic compression of trapped gas spaces, viscous heating of material extruded between impacting surfaces, friction between grains of material, and localised adiabatic shear of material. To these mechanisms other nonmechanical processes such as electrical discharge or laser-heating can be added.

In all cases the difference between critical and subcritical hot spots needs to be drawn. In the case of a critical hot spot the reaction site is of sufficient size that the heat loss via thermal conduction is lower than the energy produced by the chemical reactions. Thus the reaction zones become larger and extend into the bulk material. Subcritical hot spots are those for which, while chemical reaction may break out, the heat loss from the region is higher than the production rate from the reactions. Thus the reaction is quenched.

In liquids, the conditions for critical hot-spot formation have been extensively studied [3–8] and the size, duration, and temperature required have been reasonably well quantified. One advantage of a liquid or a single crystal of explosive is the inherent homogeneity of the system and relative ease of visualisation of the reaction front. However, many explosive systems are heterogeneous and energy concentration relies on an interplay between the components.

Calculations [9] and experiments [10] on single cracks growing in individual energetic crystals show that the energy released from the crack tip is not sufficient on its own to cause sustained reaction. However, friction caused by the rubbing of the fracture surfaces against one another after crack growth is an important mechanism in the production of hot spots.

7.3 Detonation

A detonation is best defined as a reactive shock wave. A shock wave is a high-intensity pressure pulse which moves supersonically, with respect to the sound speed, in the uncompressed medium. The speed of sound in water is around 1500 m s^{-1}; a shock wave would move faster than this, e.g. 1600 m s^{-1}. The 'reactive' element of the description refers to the sense that the energy from the detonation is sustaining the pressure in the detonation wave and moving it forward. There are many ways of representing the process of a detonation, but a relatively straightforward way is to consider the pressure–volume relationship for an explosive.

In figure 7.1(a) there is one line and two curves. The line is the Rayleigh line and represents the locus on which mass and momentum are conserved for a detonation wave. The slope of the line is related to the detonation wave speed, the density of the explosive, and the sound speed in the explosive. The lower curve is the unreacted Hugoniot and represents the conservation of energy for the unreacted material

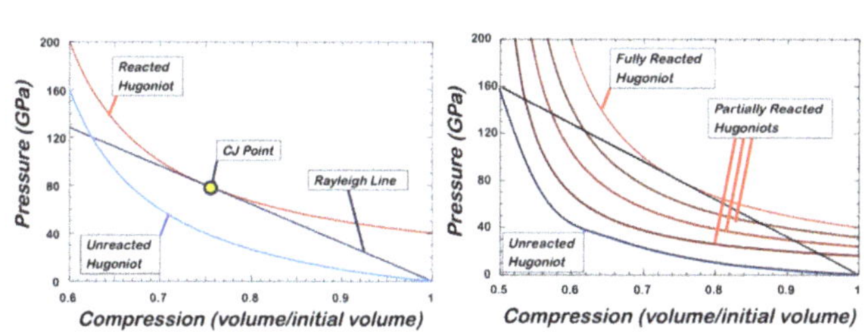

Figure 7.1. (a) Pressure–volume graph of detonation explosive, instantaneous reaction. (b) Pressure–volume graph of detonation with kinetics.

under shock loading. It can be seen that there are two points where the line crosses the unreacted Hugoniot: these are the only two places where mass, momentum, and energy are conserved and so represent the shocked and unshocked states of the material. Any other place in this graph would imply that one of the conservation laws is being broken and so would be ‘unphysical’ (i.e. not possible for a single shock wave).

When a high-explosive detonates, the molecular bonds are rearranged, producing a large energy release, giving a hot, high-pressure gas: the detonation products. As the products are hot and, for a gas, very compressed, the curve representing the conservation of energy here, the reacted Hugoniot, occupies a region in the pressure–volume curve above that of the unreacted material. However, mass and momentum must be conserved so the Rayleigh line and the reacted Hugoniot must meet. This meeting point is called the Chapman–Jouguet point (CJ point) and is the simplest case possible where mass, momentum, and energy are conserved; this is the basis of CJ detonation theory.

It is possible to draw reacted Hugoniots that cross the Rayleigh line at two points. This is possible, does not violate any conservation laws, and is seen in practical situations such as overdriven detonation waves, corresponding to pressures above the CJ point, and low-velocity detonation, which occurs at pressures below the CJ point. While possible, both these situations are, to some extent, unstable, and often systems will revert to the detonation velocity associated with the CJ point.

A more realistic situation is where the detonation wave initiates a reaction that takes place over a finite amount of time. This was independently developed by Zeldovich, von Neumann, and Deoring in the mid-1940s and is called ZND theory. This results in the pressure profile having a structure over the early stages. At the early stage the material is basically unreacted; however, it rapidly moves across a series of states to reach the fully reacted state (figure 7.1(b)). This results in a pressure spike, the so-called von Neumann spike. Figure 7.2(a) shows such a spike measured in high explosive by Bauer in 2001 [11]. This measurement required the use of in-material stress sensors and metal-coated explosives to eliminate electromagnetic pick-up in the sensors from the detonation.

The above arguments apply to a one-dimensional system. However, all practical systems need to be considered as two- or three-dimensional. Explosives have a critical diameter; consider a cylinder of explosive below a certain diameter below the

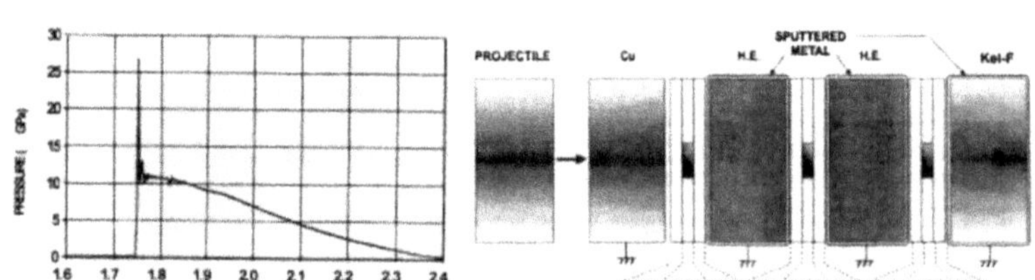

(a) (b)

Figure 7.2. (a) ZND spike seen in detonating high explosive. (b) The experimental arrangement to obtain the ZND measurement. Taken from [11]. Reprinted with permission from AIP.

detonation will die out due to release waves coming from the long edge of the cylinder, effective 'energy leakage', reducing the stress and energy in the detonation wave, snuffing it out. For most military and commercial systems the aim of explosive engineers is to keep the explosive well above this critical limit.

7.4 Deflagration-to-detonation transition

Having recounted the physical processes of hot-spot formation, leading to burning (deflagration) in the material, and a brief review of detonation, there is an obvious need for a linkage between the two. This linkage is extremely relevant to explosives research and engineering, determining, once the material has been struck, heated, or subjected to a shock wave, how the hot spots transition to a detonation wave. Some early research concentrated the buildup of burning, with increasing velocity, to full detonation. However, this faced the limitation shown in figure 7.3. The slope of the line of conservation of mass and momentum must have a negative slope in the pressure–volume space. However, if the line is followed from the start condition 'o' to the reacted Hugoniot this needs a line with a positive slope. Such a gradient implies an imaginary density or velocity, i.e. a value multiplied by root of -1, and is nonphysical.

However, this issue is based on the assumption that the reaction occurs as a single moving wave. It is necessary to consider the physical and chemical processes happening in the EM.

7.5 Process of deflagration-to-detonation transition

One widely accepted mechanism for transition between deflagration and detonation (DDT) was described by McAfee *et al* [12–15]. Here, the critical hot spots lead to deflagration; this is associated with conductive burning at a rate of order 0.01 m s^{-1} though the EM. There is also a faster convective burning, caused by product gases

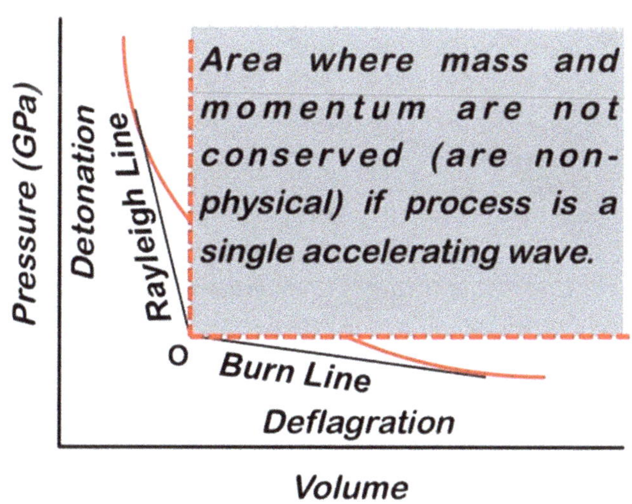

Figure 7.3. The problem of deflagration-to-detonation transition if the process is considered as a single wave.

forcing open channels either pre-existing in powder compacts or produced by impact; it has also been seen that slow cook-off can allow such channels to develop over timescales of hours. This convective burning happens at velocities of order 10 m s^{-1}. The pressure associated with the convection wave also causes the material immediately in front of it to become compacted, represented by the compaction waves shown in figure 7.4(a). This slows the rate of gas percolation, but allows pressure and temperature to increase rapidly. This confined burning presses against the surrounding material forming a dense plug of material that is pushed into unreacted material. This plug forms a considerable number of compressed hot spots in the unreacted material, which cause a rapid pressure buildup and a shock-to-detonation transition (SDT). SDT is discussed in more depth in the next section. A schematic of the process, from a column of granular material, is shown in figure 7.4 with more detail to be found in [16] and in section 7.7 (below).

The transit time between DDT depends on the density of the explosive, degree of confinement, and the nature of the explosive, so is highly variable and subject to precise local conditions. The timescales for DDT are of the order of milliseconds.

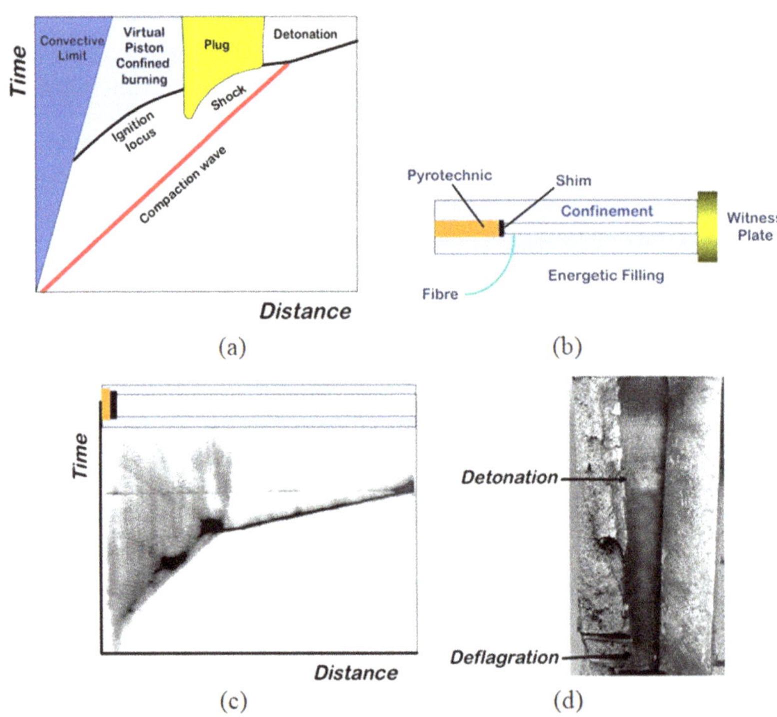

Figure 7.4. (a) Schematic of overall process proposed by McAfee *et al* along the centre line of a stick of explosive. 'Distance' is the distance along the column from the ignition point; vertical axis is time. (b) Schematic of experimental arrangement used by Leubcke *et al.* (c) Streak image showing deflagration-to-detonation transition, showing similarity with the schematic in (a). (d) Recovered confinement showing smooth ring associated with the transition between deflagration and detonation. Reproduced from [16] with permission of The Royal Society Publishing.

A faster process that occurs for high-velocity impacts and stress is a shock-to-detonation transition, which occurs in timescales microseconds.

7.6 Process of shock-to-detonation transition

A shock-to-detonation transition (SDT) occurs in high-stress, shock wave environments. The process is effectively a thermal heating process. Shock waves are associated with rapid compression, high pressures, and temperature rise. After a shock wave has passed over a material due to, for example, an impact, the region around the impact zone is subjected to a temperature rise which allows a reaction to start via shock heating. Again, this can be considered as the formation of a considerable number of hot spots that each send out pressure waves which can coalesce to produce a large pressure pulse. Effectively, thermal explosion occurs after an induction time depending on local conditions. When the detonation wave breaks out it runs through material already compressed and heated by the initial shock, thus causing super-detonation, detonation at a higher velocity than that normally associated with the material. Eventually the reactive shock overtakes the original shock to produce overdriven detonation in the lower-density material. Ultimately, this settles to the 'normal' detonation velocity over some relaxation time and distance (figure 7.5).

7.7 Deflagration-to-detonation studies

Amongst the simplest experiments are those which use metal cylinders, as shown in figure 7.6. The channel along the centre of the cylinder had a diameter of 5 mm and varied in length between 70 and 100 mm (total length 100–130 mm). Initial analysis of what occurred while the reaction took place in the column was performed by a system of post-mortem examinations. Test shots were carried out using cylinders made of copper, steel, and brass. It was found that copper cylinders gave the best

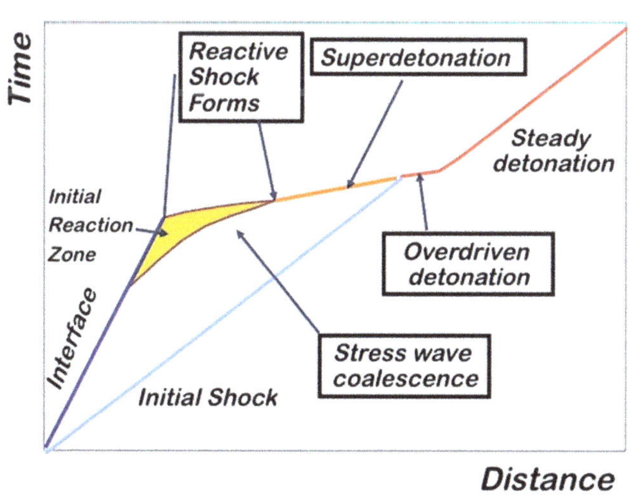

Figure 7.5. The process of shock-to-detonation transition.

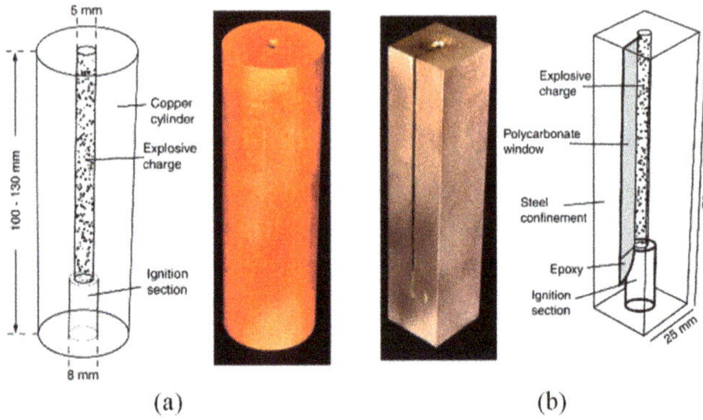

Figure 7.6. Cylinders for use in (a) post-mortem-type experiments; (b) photographic study of DDT. Reproduced from [16] with permission of The Royal Society Publishing.

record of the event that had taken place. Brass proved too brittle to survive a detonation in one piece and steel, although providing some record of the event, did not distort in the ductile manner of copper and as such provided a less clear picture of the processes that had occurred.

Plastics such as polycarbonate and polymethylmethacrylate (PMMA) were not tried as it had been shown previously that the level of confinement offered by these materials was insufficient to withstand the sustained pressures required for a DDT event in PETN or RDX. In experiments where a photographic record of the events was required, it was necessary to use a confinement that afforded direct optical access to the charge whilst retaining the pressure levels of the simple metal confinement.

The confinement shown in figure 7.6(b) was developed by Luebcke [16] based on previous designs by researchers such as Korotkov [17] and Griffiths and Groocock [18]. The confinements used consisted of sections of square steel bar with a 5 mm channel drilled along the centre for the charge and a 1 mm wide polycarbonate window laid into the steel. The window afforded an optical viewing point whilst being thin enough that the level of confinement afforded by the steel is not overly compromised. The length of the charges used varied from 45 to 80 mm depending on the nature of the event that was being investigated.

The explosive charges were pressed incrementally using a hydraulic press. Care was taken to ensure that the height of each individual increment never exceeded more than half of the diameter of the column. Charges below densities of 50% of the theoretical maximum density (TMD) were also incrementally pressed. A system whereby static weights were placed on top of the pressing rod was used in these cases. A pyrotechnic was used to ignite the columns of explosive. The pyrotechnic was ignited using a nichrome wire through which was passed an electrical current. The mixture consisted of 80% potassium dichromate and 20% boron, as developed by Dickson [19]. It had the advantage of producing few gaseous products, so little pre-pressurisation of the charge occurred. The burning temperature is far higher than the ignition temperature of the materials used in the charges. The pyrotechnic

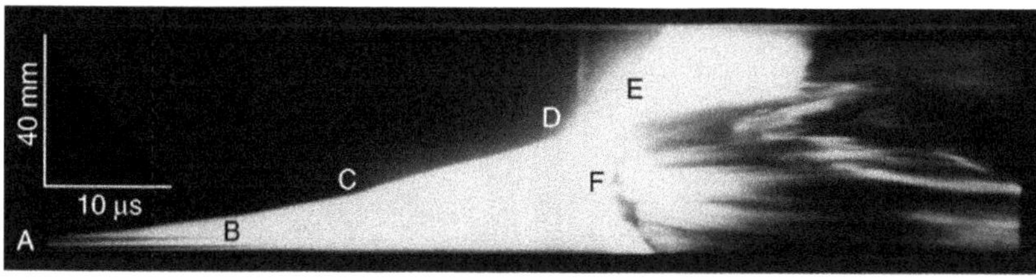

Figure 7.7. 'Normal' or type I DDT event in ultrafine PETN. Taken from [21]. Reprinted with permission from AIP.

was tamped in order to ensure as few gas pockets as possible so as to reduce rapid expansion during reaction. The pyrotechnic was added after the main charge had been pressed into the column. Finally, a small conical aluminium (Al) piece was placed around the electrical wires and pushed into the confinement to reduce rearward venting of the charge to a minimum. As a result pressure generated during the early stages of reaction could be sustained.

Often, a thin piece of copper foil was placed between the pyrotechnic charge and the explosive charge. This was done in order to prevent light emitted during the slow burning of the pyrotechnic charge from being transmitted down the column to the optical fibres and, as a result, prematurely triggering the experimental diagnostics. The brass tube into which the pyrotechnic charge was placed also acted as a light stop as well as an easy check on the amount of pyrotechnic that was used.

The type of data recorded from this experiment is shown in figure 7.7. A full description of the processes can be found in the papers of Gifford [20–22]. However, the streak image shows convective burning (B), the step from deflagration to detonation (D), and the movement of the waves including a retonation wave (F). Addition of high-speed thermocouples [21] to the system allowed the internal states at points along the column to be known prior to detonation.

7.8 Drop-weight studies

A standard technique for sensitivity studies, the drop weight is a system that allows long, low-level pressure pulses to be applied to small samples. In combination with high-speed photography, photodiodes, and stress transducers it is a powerful system for both qualitative and quantitative analysis of material response.

The standard way of analysing the output of a drop weight assumes the weight behaves as a rigid body and hence that one can simply apply Newton's laws of motion. This is used in determining the calibration of a drop-weight force transducer dynamically, by directly relating the measured force both to the deceleration of the weight and also to the position.

In practice, the output signal from a drop-weight machine often has oscillations comparable in size to the signal produced by the mechanical resistance of the specimen. This is particularly true if the drop weight itself is instrumented, e.g. with accelerometers. The reason is that the impact excites the weight below its resonance frequency [23]. Elastic waves, therefore, reverberate around inside until the momenta

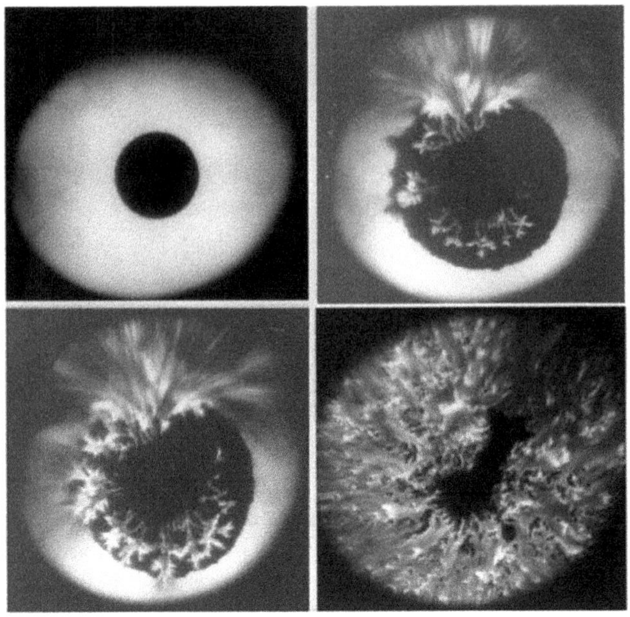

Figure 7.8. The ignition under impact of a thermite composition. Reproduced from [31] with permission of The Royal Society Publishing.

of the constituent parts of the weight have been reversed. Rebound then occurs and the specimen is unloaded. Recent work [24] has demonstrated that it is possible to obtain high-quality data from such machines either by the use of a momentum trap in the weight, if the weight itself has to be instrumented [25], or by careful design of a separate force transducer placed below the specimen, e.g. in [26, 27].

One modification to the drop-weight apparatus that has proved invaluable in the elucidation of explosives' ignition mechanisms is to machine a light path through the weight and to perform the deformation between transparent glass anvils [14, 28–31]. This allows the event to be captured using high-speed photography. An example of a 'classic' high-speed photographic sequence obtained using such an apparatus is given in figure 7.8.

7.9 Second harmonic generation

A recent paper used one of the more unusual optical properties of cyclotetramethylene tetranitramine (HMX, or octogen) to probe the ignition process [32]. The relatively new technique of second harmonic generation (SHG) was used to study the β–δ phase transition in HMX by many authors in the late 1990s [33–36] A second harmonic is generated when intense radiation is incident on molecular crystals with appropriate symmetry properties. β-HMX, the 'chair' configuration, is a centrosymmetric molecule and, as such, is forbidden by its symmetry from generating second harmonics. However, δ-HMX, a 'boat' configuration, has no centre of inversion and generates second harmonics very efficiently. For these second-order processes to be observable the incident radiation needs to be very intense and so a laser is required. In order to produce second harmonics near the peak sensitivity of the high-speed cameras, a Nd:YAG (YAG = yttrium aluminium garnet) laser operating at 1064 nm was used to irradiate the sample allowing the detection of the second harmonic at 532 nm (green). This technique can be applied in a transparent anvil drop-weight apparatus. The second harmonic generation is a near-instantaneous process and so a 9 ns laser pulse can probe for δ-HMX during impact.

The apparatus allowed separate light paths for the laser light to travel to the sample and for the visible light (both light from reaction and SHG) to travel from the sample to the cameras. A diagram of the drop-weight setup is shown in figure 7.9.

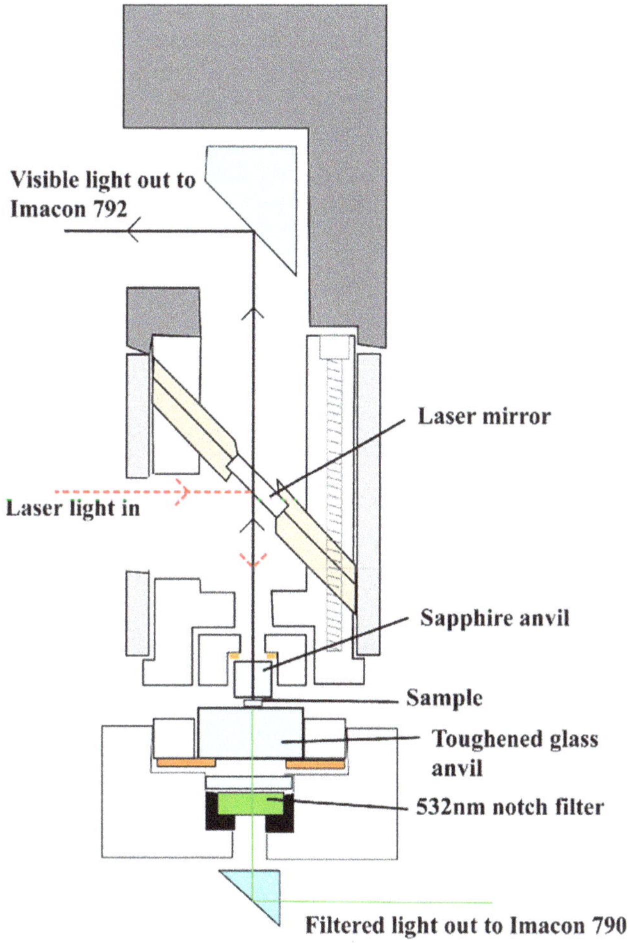

Figure 7.9. Drop weight modified to detect SHG in HMX crystals during impact. Taken from [32]. Reprinted with permission from AIP.

The cameras used were Imacons 790 and 792; a notch filter in front of one of them was used to isolate SHG, while the other followed the visible light from the reaction. A drop height of 39 cm was used, giving an impact velocity of 2.75 m s^{-1}. The height was chosen as the lowest at which the sample would reliably ignite on impact. The stress-time output was a half sinusoid with a peak pressure of 10^7 Pa and duration of the order of 10^{-4} s. A light gate broken by the falling drop weight just before impact was used to trigger the diagnostics, and a photodiode monitored the time of ignition relative to that time reference. While precise impact times are not known for individual experiments, the photodiode gave a good reference as to when the reaction began. The system was such that one frame on the filtered camera coincided with the laser pulse. The visible light camera was set to record a sequence starting just before the laser pulse. Each pulse deposited 0.8 J of energy over an 8 mm diameter spot.

The sensitivity of the system to the presence of δ-HMX, in the sample bulk, was shown by using β-HMX pellets converted to δ-HMX through selective heating. SHG was clearly visible whether the δ-phase regions were on the top or the bottom of the pellet, and so δ-HMX at any depth in the sample could be detected with excellent spatial resolution in pellets up to 1 mm thick.

The time delay between the trigger provided by a light gate broken by the falling drop weight and the laser pulse was varied, in order to probe for δ-HMX throughout the impact. Selected images showing SHG are presented in figure 7.10. The images

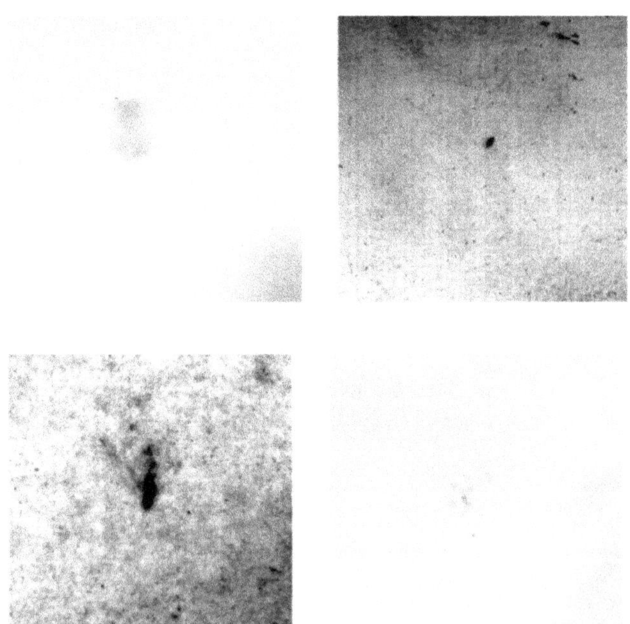

Figure 7.10. Four images showing SHG. The field of view of each image is 10 mm wide. All of the images have enhanced and reversed contrast (black indicates the presence of SHG). Taken from [32]. Reprinted with permission from AIP.

show the second harmonic in narrow regions of the sample. In the experiments all the impacted pellets reacted but no δ-HMX was observed any earlier than 13 μs before ignition. This implies that any transition to the δ phase occurs late in the reaction sequence. It could be the case that either the δ-HMX region grows rapidly before ignition or that after ignition the higher temperature of the burning material converts the rest of the sample before it is completely consumed.

In earlier research, Field [37] saw localised heating along lines and at their junctions and attributed this to shear band formation. It appears that the heating on impact is sufficient to cause conversion to δ phase under these stress conditions before ignition; the patches of δ-HMX observed were small (1 mm across) and very localised. After ignition, larger patches (3 mm across) of δ-HMX were seen before the sample was completely consumed.

7.10 Impact ignition

In another study [38] ammonium nitrate (AN) was placed in an impact cell, shown in figure 7.11. This contained a bed of AN 0.2 mm thick. The front plate was a copper plate 2 mm thick and the rear glass window was 25 mm thick. A mirror, at 45° to the rear of the cell, directed light into a high-speed camera.

A cell was mounted at the end of a 19 mm bore gas gun. The flier plate was a 2 mm thick copper plate mounted on a nylon sabot. The pressure in the impact chamber can be left at atmospheric pressure or pumped to a vacuum of 10^{-3} bar. The flier plate was aligned to the cell to a tolerance better than 0.05°. The width of the stress pulse imparted by the impactor was of the order of 600 ns, this being sufficient to produce a sustained reaction in the system. It was decided to reduce the bed thickness to 0.4 mm, corresponding to a bed thickness of 2–3 grains. This

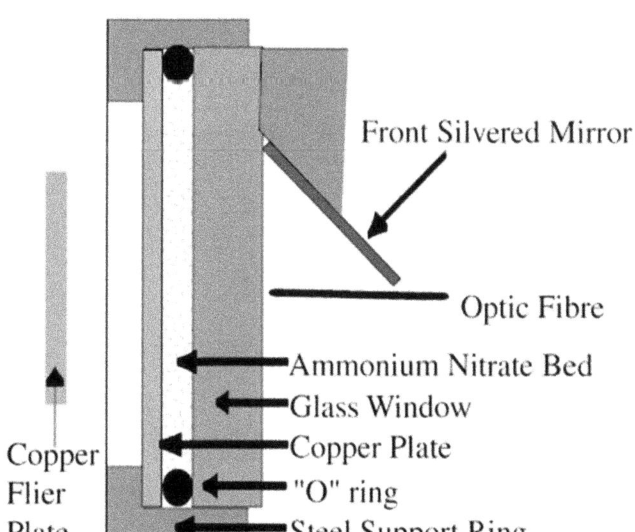

Figure 7.11. Impact cell used for AN studies.

allowed the bed to be observed in its full depth. Tests revealed that the reaction buildup in the thin bed was the same as in previous experiments.

In figure 7.12 the result of an impact on a thin bed at 700 m s^{-1} is shown. The exposure time of each image was 30 ns and the interframe time was zero, as allowed by the Ultra-8 camera. Six frames are shown. Figure 7.13 shows the first two frames in detail; the first frame shows a number of regions of intense light output, while in the second frame some of these regions in the top of the image have joined together. Other regions where the hot spots are more widely spaced, towards the centre of the image, have died out. This appears to represent a distinction between critical and subcritical hot spots. Those hot spots which are in proximity to one another link together possibly as a result of pressure pulses travelling between the hot spots as a

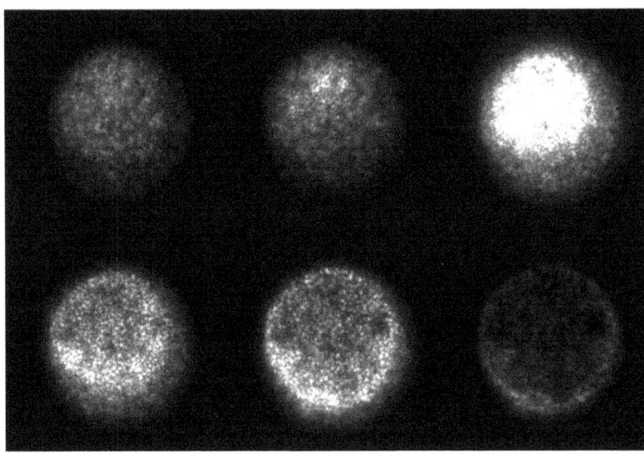

Figure 7.12. High-speed sequence of reaction in a 0.4 mm bed of AN grains, 150–210 μm in size, struck by a copper impactor at 700 m s^{-1}. The exposure time of each frame is 30 ns. Frames are in order left to right, top to bottom. The diameter of the reaction zone viewed is approximately 8 mm. Reproduced from [39] with permission of SPIE.

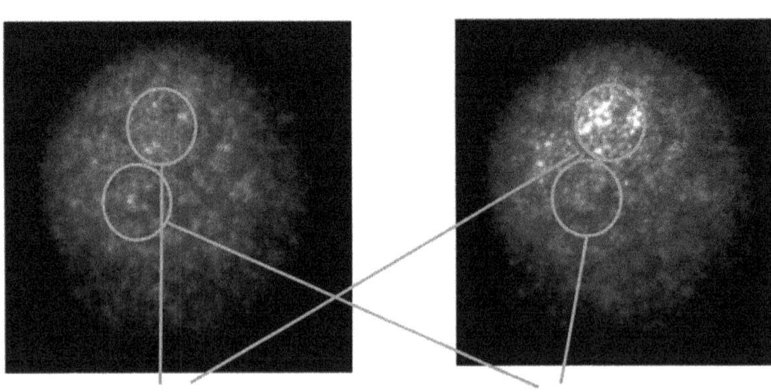

Figure 7.13. Detail of the first two frames of the sequence in figure 7.12. Areas where hot spots are growing and those where hot spots are quenching are highlighted. Reproduced from [35] with permission of SPIE.

result of the production of product gases. Hot spots which are more widely spaced tend to be quenched. In the third frame the reaction front has spread to cover about 55% of the impact area. In the later frames the reaction moves away from the central zone where the AN has been burnt.

This framing sequence provides strong evidence for the formation of both critical and subcritical hot spots. It is possible to use these images and count the hot-spot density and also measure the degree of interaction between individual hot spots.

7.11 Small-scale gap test

In the last two examples, shock waves were considered; a detonation is by definition a reactive shock wave. These examples are complementary. In a small-scale gap test a shock impinges onto a column of EM, while in a cylinder test an EM pushes on an inert surround cylinder of inert material.

The gap test is a shock sensitivity technique in which a sample of the material of interest is placed in contact with a barrier. On the other side of the barrier an explosive charge is detonated. The shock from the charge moves through the barrier, losing energy as it goes, and ultimately passes into the test material. The thicker the barrier, the more the pulse from the charge is reduced and the lower the shock felt by the sample. Experiments are repeated with different barrier or gap thicknesses until a 50:50, or other probability, go:no-go threshold is established.

In the study in question small samples were used [36]. The test samples were pressed into PMMA holders. The cylindrical holders were 25 mm long, 25 mm in diameter, with a central bore of 5 mm. The columns were incrementally pressed with a step height of 0.5 mm to give a very homogeneous column. A variety of pressing densities were studied from 60% to 90%. The holders were then placed in the arrangement as shown in figure 7.14. The holder had a PMMA gap placed above it and a C8 detonator placed on top. The assembly was placed in an armoured box fitted with an observation

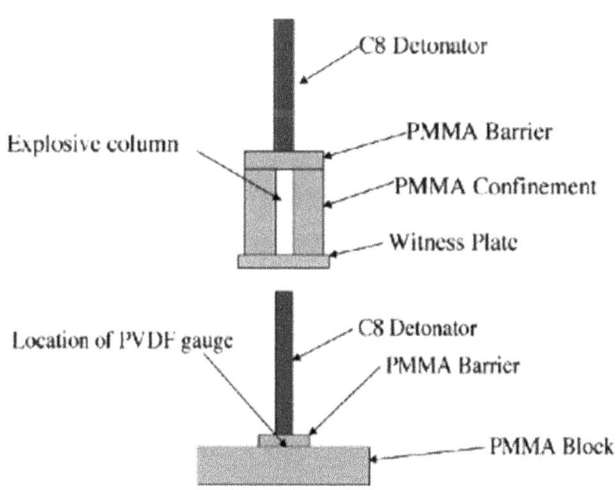

Figure 7.14. Experimental arrangements. Top: for gap test; bottom: for PVDF gauge study. Reprinted with permission from AIP.

window. The camera was triggered by a fibre optic that was fitted near the end of the detonator on top of the gap plate. When light was detected from this source, the camera was triggered. At the lower end of the gap arrangement was a brass witness plate. All contacting surfaces were coated in a thin layer of silicone grease in order to allow good acoustic transmission between the layers.

The streak records from a series of experiments on 90% TMD PETN are presented in figure 7.15. The PETN had an ultrafine grain size. The gaps were

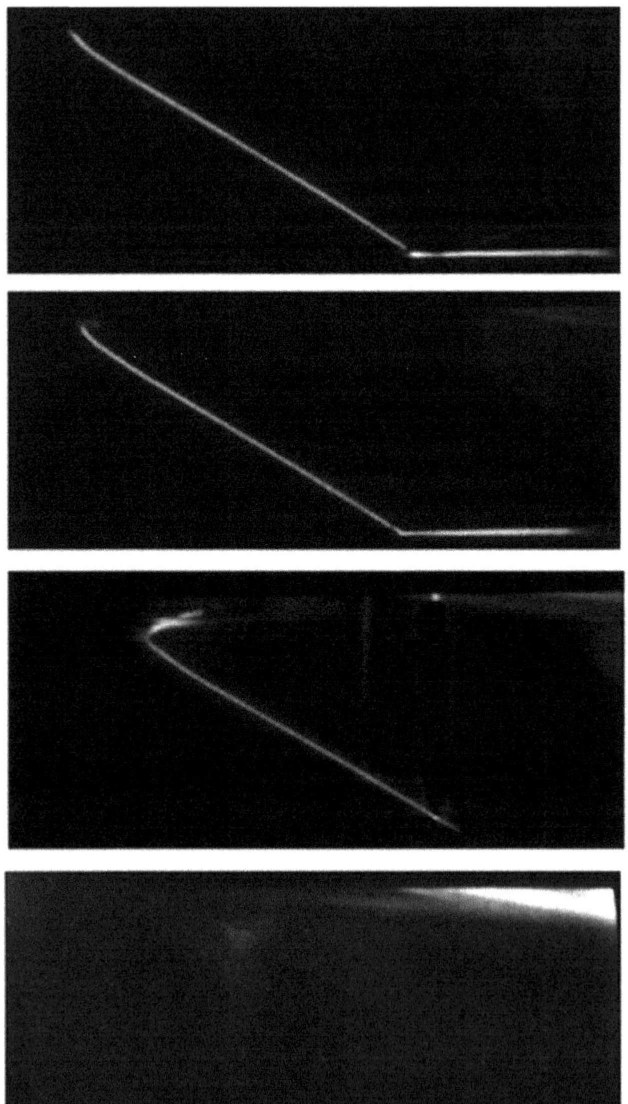

Figure 7.15. Streak records of fine-grained PETN 90% TMD with variation in gap thickness. Top: 3.53 mm; second: 3.63 mm; third: 3.67 mm; bottom: 3.71 mm. Reprinted with permission from AIP.

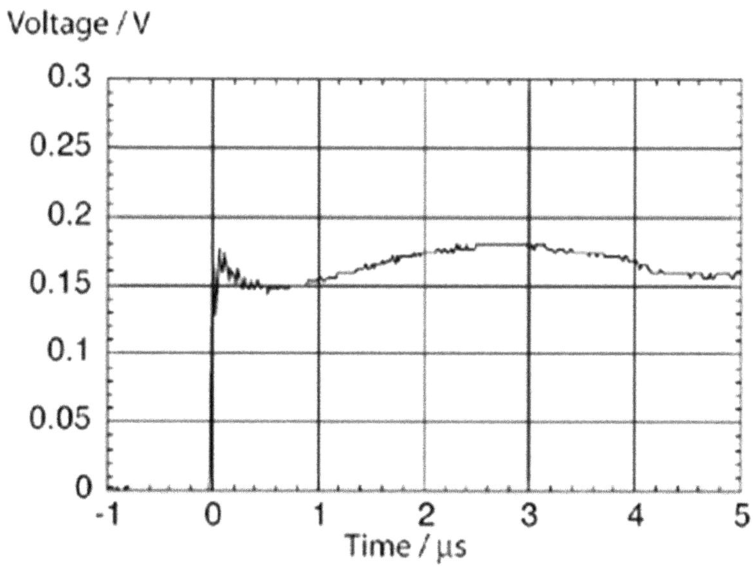

Figure 7.15a. PVDF output from a gauge separated by 5.54 mm PMMA gap from a C8 detonator. Reproduced with permission from CEA.

3.51, 3.63, 3.67, and 3.71 mm. It can be seen that in the top image the detonation is slightly overdriven and was prompt, i.e. at the top of the column. The detonation speed then settled down within 5 mm in the column length. In the second image there is a slight hooking at the top of the column. This is due to the detonation not starting at the top of the column but rather a short distance down it. At 3.67 mm the hooking is severe and the detonation wave breaks out about 8 mm down the column. Finally, at 3.71 mm there is no light associated with the detonation process. The brass witness plate dented in the area exposed to the explosive column, indicating a detonation pressure in all cases except 3.71 mm. For 3.71 mm, only a small dent was seen, indicating a more modest pressure associated with deflagration.

A similar series of results was obtained for conventional grain size material where the gap thickness for failure was shown to be 5.54 mm of PMMA. In a final series of experiments, Dynasen polyvinylidene fluoride (PVDF) gauges [37] were placed between a gap plate and a block of PMMA 100 mm square and 25 mm thick (figure 7.15a). The output from a PVDF gauge is shown in figure 7.20. The output voltage can be converted to stress. For the 5.54 mm gap used in the experiment here this corresponded to a stress level of 2.1 GPa. For a PMMA thickness of 3.67 mm, the stress level was of the order 4.1 GPa. It should be pointed out that the PVDF gauge is recording the stress for several microseconds, much longer than it takes detonation to start.

7.12 The cylinder test

The cylinder test is used to measure detonation performance. In concept it has a very simple geometry: a tube filled with explosive. The material is detonated and, if possible, the deformation of the tube is measured and then related to the detonation pressure and velocity. High-speed recording systems, such as streak photography, allow the rate of expansion of the cylinder to be measured relatively straightforwardly.

However, if the aim is to track gradual changes in the detonation behaviour of a system, due to the presence of additives, especially in the width of the detonation front, this requires a more time-resolved technique. In one case streak photography was supplemented by the use of a velocity interferometer (VISAR) [38]. In another study it was used to look at the effect of Al added to sensitised nitromethane (NM) [39].

In these experiments the streak camera recorded the radial expansion of the cylinder while the particle velocity-time history of the outer surface of the cylinder during detonation was determined using VISAR (figure 7.16). The initial acceleration can be seen to take the form of a series of steps. These are due to the reflection of waves within the copper cylinder. This can be explained with reference to figure 7.17.

The detonation of the NM/Al mix causes the cylinder to expand. A shock wave is sent from the inner wall to the outer wall of the cylinder. This reflects at the outer wall as a release, which moves back towards the centre of the tube. When it reaches the centre it is reflected as compression due to the sustained pressure of the detonation products. Overall, the outer wall of the cylinder accelerates as a series of steps. The cylinder walls become thinner, ultimately shredding into fragments. This effect tends to obscure the step-like structure at later times. Hence the system with the most rapid expansion shows the least resolution of the steps while the highest loading of Al caused the most easily resolved steps.

It is important that the surface is not polished to an optical finish because such a finish is unlikely to be preserved at the high stress and strains associated with the cylinder test. Barker *et al* developed this technique in the late '60s and early '70s [40].

Figure 7.16. The experimental setup. Reproduced with permission from CEA.

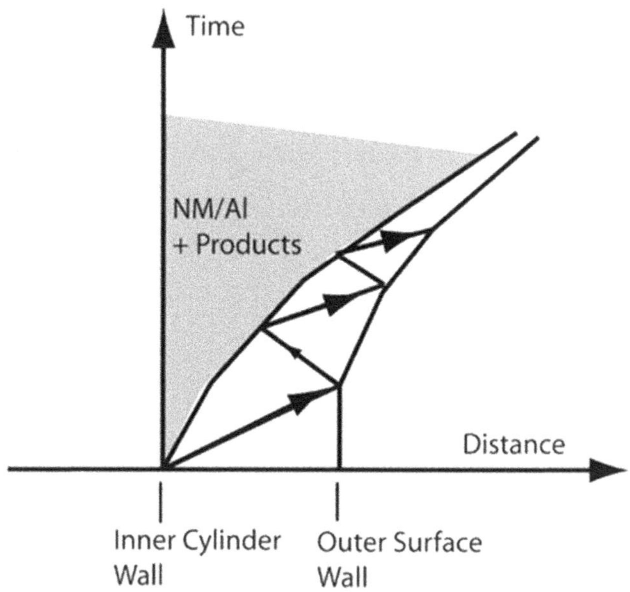

Figure 7.17. Schematic of deformation of the copper cylinder. Reproduced with permission from CEA.

The reflected laser light is captured and split into two beams, one of which is passed through a glass cylinder known as an 'etalon'. Because the refractive index of glass is larger than that of air, the light in this beam is slowed down and hence delayed with respect to the other beam, which passes through air. If the target is stationary, the interference pattern will not change with time when the two beams are recombined. For an accelerating target, however, the reflected beam will be Doppler shifted. The time resolution of the VISAR data is 2 ns and the noise on the photomultiplier produces an error of ±5%.

Copper cylinder expansion tests were carried out on NM/Al compositions containing 20%, 30%, 40%, 50%, and 60% by weight Al. The NM used was Analar grade and the Al particles were spherical, with a mean size of 10.5 μm. The mixtures were prepared by adding 5% by weight of polyethylene oxide to the NM to thicken it, thus helping suspend the Al, then adding the appropriate percentage of Al. The copper cylinders (304 mm long, 25.4 mm internal diameter with a wall thickness of 2.6 mm) were mounted vertically, sealed at the bottom end, and initiated by a small booster pellet at the top.

The overall expansion was viewed with a Cordin-132 streak camera writing at 5.999 mm μs^{-1}. An argon flash bomb and diffusing screen were used to provide backlighting. There was a decrease in expansion speed as the particle loading density was increased with a marked change between 40% and 50%, as shown in figure 7.18.

The accuracy of measurement of the initial expansion phase is limited by the resolution of the film. Analysis of the VISAR traces gave the velocity histories shown in figure 7.19. The traces are offset for clarity. The velocity history in these

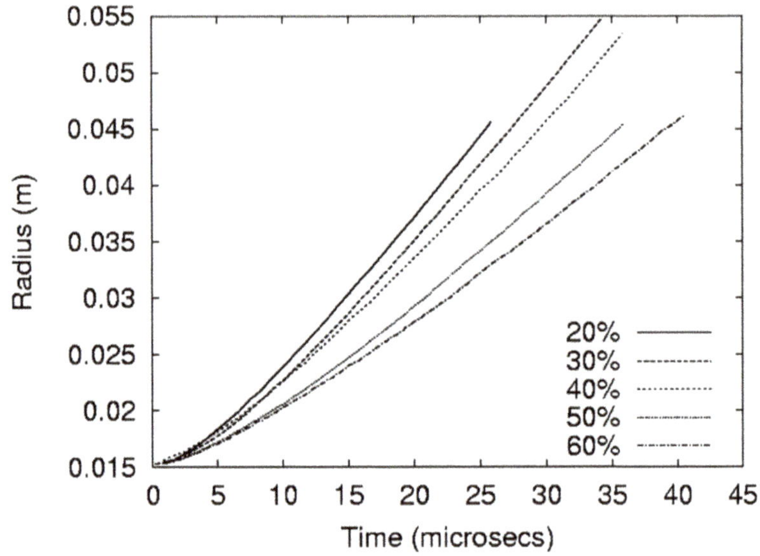

Figure 7.18. Streak camera records. Cylinder expansion radius time plots for various percentage mass fractions of 10.5 μm Al in NM. Reproduced with permission from CEA.

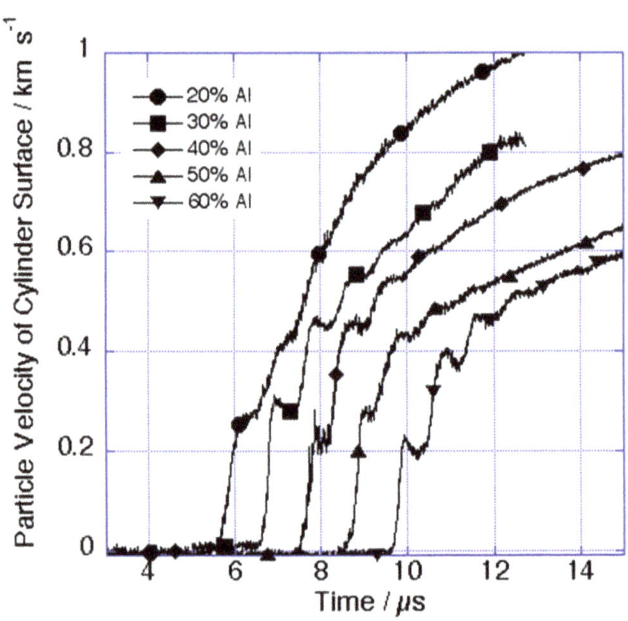

Figure 7.19. Velocity histories from VISAR for the loading densities studied. Reproduced with permission from CEA.

traces tends to flatten after 10 μs but this is due to the loss in sensitivity when the fibre is close to the expanding copper surface. The initial acceleration can be seen to take the form of a series of steps. These are due to the reflection of waves within the copper cylinder. This can be explained with reference to figure 7.17.

The detonation of the NM/Al mix causes the cylinder to expand. A shock wave is sent from the inner wall to the outer wall of the cylinder. This reflects at the outer wall as a release, which moves back towards the centre of the tube. When it reaches the centre it is reflected as compression due to the sustained pressure of the detonation products. Overall, the outer wall of the cylinder accelerates as a series of steps. The cylinder walls become thinner, ultimately shredding into fragments. This effect tends to obscure the step-like structure at later times. Hence the system with the most rapid expansion shows the least resolution of the steps while the highest loading of Al caused the most easily resolved steps.

While the VISAR data measure the rapid acceleration of the outer surface of the copper cylinder in the first few microseconds, the steps are not seen in the resulting streak displacement history. Part of the reason for this is that the stepped region corresponds to the first 2 μs of the process, where the displacements are very small.

From this simple analysis it can be seen that, while the streak camera system is excellent for large displacements, it would be difficult to capture the detail of the early acceleration using such a system. The displacements measured using the streak and VISAR can be combined, as shown in figure 7.20.

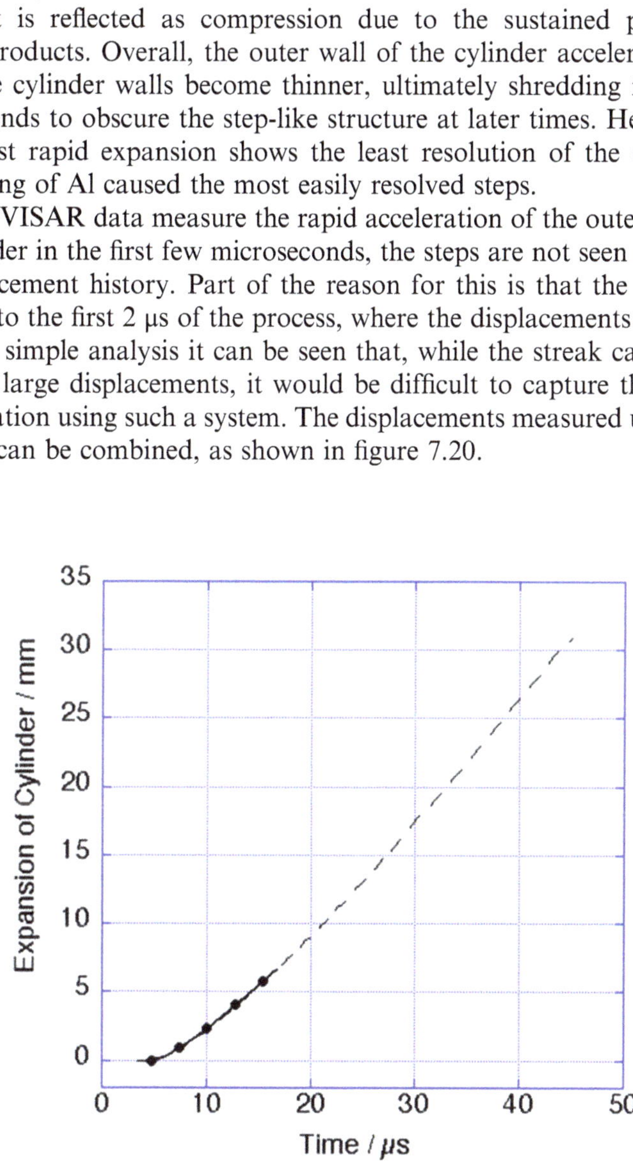

Figure 7.20. Combined VISAR (solid) and streak (dashed) records for 60% Al loading.

7.13 Conclusions

It is hoped that this article has introduced the field of DDT studies. The first part concentrated on the fundamentals of this field while the second outlines specific research areas.

While the above series of examples shows a wide range of techniques applied to energetic systems, ranging from low- to high-velocity impacts, it is far from exhaustive. However, it is possible to draw some simple conclusions. Keeping experimental geometries and interfaces as simple as possible is useful as it allows data to be extracted with limited recourse to complex analysis. In EMs, it has been known for over 50 years that ignition and initiation are multivariable processes and so experimental design is of paramount importance. In some cases, such as HMX, the specific crystal properties allow SHG to be used to look at the ignition process.

While many of the techniques listed can involve complex equipment, some advances, like the use of line lasers to monitor material expansion, can be done with modest resources.

Finally, the desired output from the study is important. Here the emphasis has been on experimental arrangement to understand fundamental processes, so allowing a deeper understanding of how factors like composition affect the ignition and growth of reaction.

Acknowledgements

The author acknowledges the large number of colleagues, students, and collaborators, who have supported and developed the techniques in this review, mainly at the Cavendish Laboratory, University of Cambridge. Principal inspiration and guidance was provided by Prof. J E Field. Other colleagues included Drs and Profs S M Walley, P M Dickson, N K Bourne, P J Rae, J E Blazer, S G Grantham, C R Siviour, H T Goldrein, D J Chapman, M J Gifford, P E Luebcke, H Czerski, D M Williamson, and E J W Crossland, all of whom contributed significantly. The late Dr Avik Chakravarty made a singularly effective contribution to the small-scale gap test. Indispensable technical assistance was provided by D Johnson, R Marrah, R Flaxman, K Fagan, and others. EPSRC, QinetiQ, [dstl], Orica, AWE, and the MoD provided support for much of this research. Drs and Profs I G Cullis, A S Cumming, P Gould, P D Church, D Mullenger, P Haskins, M Cook, and M Braithwaite all provided programmatic guidance. This list is not exhaustive and the author apologises for any omissions. The Institute of Shock Physics acknowledges the support of the Atomic Weapons Establishment, Aldermaston, UK and Imperial College London. Readers are asked to consult the original papers to find greater detail and attribution.

References

[1] Bowden F P and Yoffe A D 1952 *Initiation and Growth of Explosion in Liquids and Solids* (Cambridge: Cambridge University Press)

[2] Field J E, Swallowe G M and Heavens S N 1982 Ignition mechanisms of explosives during mechanical deformation *Proc. R. Soc. Lond.* A **382** 231–44

[3] Bowden F P and McOnie M P 1967 Formation of cavities and microjets in liquids and their role in initiation and growth of explosion *Proc. R. Soc. Lond.* A **298** 38–50

[4] Coley G D and Field J E 1973 The role of cavities in the initiation and growth of explosion in liquids *Proc. R. Soc. Lond.* A **335** 67–86

[5] Bourne N K and Field J E 1991 Bubble collapse and the initiation of explosion *Proc. R. Soc. Lond.* A **435** 423–35

[6] Wlodarczyk E 1992 Analysis of the efficiency of initiation of detonation by 'hot spots' generated by shock compression of gas bubbles included in the explosive. I: analysis of the experimental data *J. Tech. Phys.* **33** 35–61

[7] Wlodarczyk E 1992 Analysis of the efficiency of initiation of a detonation by hot spots generated by shock compression of gas bubbles included in the explosive. II: theoretical analysis *J. Tech. Phys.* **33** 133–66

[8] Bourne N K and Field J E 1999 Explosive ignition by the collapse of cavities *Proc. R. Soc. Lond.* A **455** 2411–26

[9] Holmes W, Francis R S and Fayer M D 1999 Crack propagation induced heating in crystalline energetic materials *J. Chem. Phys.* **110** 3576–83

[10] Chaudhri M M, Field J E, Heavens S N and Coley M 1972 *Studies of the initiation of explosives using high-speed photography (10th Int. Conf. on High-speed Photography)*

[11] Bauer F 2000 Advances in PVDF shock sensors: applications to polar materials and high explosives ed M D Furnish, L C Chhabildas and R S Hixson *Shock Compression of Condensed Matter – 1999* (Melville, NY: American Institute of Physics) pp 1023–28

[12] McAfee M *et al* 1989 Deflagration to detonation in granular HMX *Proc. Ninth Symp. (Int.) on Detonation. Arlington* (VA: Office of the Chief of Naval Research) 265–79

[13] McAfee J M *et al* 1991 Deflagration to detonation in granular HMX: structure and kinetics in the predetonation region *Proc. 1991 JANNAF Propulsion Systems Hazards Subcommittee Meeting* vol 562 (Columbia, MD: CPIA) 153–62

[14] McAfee J M *et al* 1995 Deflagration-to-detonation in granular HMX: ignition, kinetics, and shock formation *Proc. 10th Int. Detonation Symp.* J M Short and D G Tasker (Arlington, VA: Office of Naval Research) 716–23

[15] McAfee J M 2010 The deflagration to detonation transition *Shock Wave Science and Technology Reference Library. 5: Non-Shock Initiation of Explosives* ed B W Asay (Berlin: Springer) pp 483–535

[16] Luebcke P E and Dickson P M *et al* 1995 An experimental study of the deflagration to detonation transition in granular secondary explosives *Proc. R. Soc. Lond.* A **448** 439–48

[17] Korotkov A I and Sulimov A A *et al* 1969 Transition from combustion to detonation in porous explosives *Combust. Explos. Shock Waves* **5** 216–22

[18] Griffiths N and Groocock J M 1960 The burning to detonation of solid explosives *J. Chem. Soc.* **11** 4154–62

[19] Dickson P M and Field J E 1993 Initiation and propagation in primary explosives *Proc. R. Soc. Lond.* A **441** 359–75

[20] Gifford M J, Proud W G and Field J E 2002 Development of a method for quantification of hot-spots *Thermochim. Acta* **384** 285–90

[21] Gifford M J, Proud W G and Field J E 2002 Observations on type II deflagration-to-detonation transitions *AIP Conf. Proc.* **620** 878–81

[22] Gifford M J and Luebcke P E *et al* 1999 A new mechanism for deflagration-to-detonation in porous granular explosives *J. Appl. Phys.* **86** 1749–53

[23] Pope P H 1985 *PhD Thesis* University of Cambridge

[24] Follansbee P S, Regazzoni G and Kocks U F 1984 The transition to drag controlled deformation in copper at high strain rates *Inst. Phys. Conf. Ser.* **70** 71–80

[25] Heavens S N and Field J E 1974 The ignition of a thin layer of explosive by impact Proc. R. Soc. Lond. A **338** 77–93

[26] Field J E, Bourne N K, Palmer S J P and Walley S M 1992 Hot-spot ignition mechanisms for explosives and propellants *Philos. Trans. R. Soc. Lond.* A **339** 269–83

[27] Walley S M, Field J E and Palmer S J P 1992 Impact sensitivity of propellants *Proc. R. Soc. Lond.* A **438** 571–83

[28] Walley S M, Balzer J E, Proud W G and Field J E 2000 Response of thermites to dynamic high pressure and shear *Proc. R. Soc. Lond.* A **456** 1483–503

[29] Czerski H, Greenaway M W, Proud W G and Field J E 2004 β–δ Phase transition during drop weight impact on HMX *J. Appl. Phys.* **96** 4131–4

[30] Henson B F and Asay B W *et al* 1999 Dynamic measurement of the HMX β–δ phase transition by second harmonic generation *Phys. Rev. Lett.* **82** 1213–16

[31] Saw C K 2002 Kinetics of HMX and phase transitions: effects of particle size at elevated temperature *Proc. 12th Int. Detonation Symp.* (Arlington, VA: Office of Naval Research) 70–6

[32] Smilowitz L, Henson B F, Asay B W and Dickson P M 2002 Kinetics of the β–δ phase transition in PBX 9501 *AIP Conf. Proc.* **620** 1077–80

[33] Smilowitz L and Henson B F *et al* 2002 The β–δ phase transition in the energetic nitramine ocathydro-1,3,5,7-tetranitro-1,3,5,7-tetrazocine: kinetics *J. Chem. Phys.* **117** 3789–98

[34] Field J E, Bourne N K, Palmer S J P and Walley S M (unpublished)

[35] Proud W G, Crossland E J W and Field J E 2003 High-speed photography and spectroscopy on determining the nature, number and evolution of hot-spots in energetic materials *Proc. SPIE* **4948** 510–8

[36] Chakravarty A, Gifford M J, Greenaway M W, Proud W G and Field J E 2002 Factors affecting shock sensitivity of energetic materials *12th APS Topical Meeting on Shock Compression of Condensed Matter 2001* M D Furnish, N N Thadhani and Y Horie (Melville, NY: American Institute of Physics) 1007–10

[37] Bauer F 1999 Advances in piezoelectric PVDF shock compression sensors: applications to HE studies *Proc. 25th Int. Pyrotechnics Seminar* (Saint Aubin: Association Française de Pyrotechnie) pp 28–37

[38] Barker L M and Hollenbach R E 1972 Laser interferometer for measuring high velocities of any reflecting surface *J. Appl. Phys.* **43** 4669–75

[39] Proud W G, Field J E, Milne A, Longbottom A W, Haskins P J, Briggs R I and Cook M D 2003 The detonation of NM/Al mixtures *Proc. Fifth Int. Symp. on High Dynamic Pressures* vol 1 (Paris: Commissariat à l'Energie Atomique) pp 135–41

[40] Barker L M and Hollenbach R E 1972 Laser interferometer for measuring high velocities of any reflecting surface *J. Appl. Phys.* **43** 4669–75

IOP Publishing

Energetics Science and Technology
An integrated approach
Adam S Cumming

Chapter 8

Submillimetre spatially resolved observation of detonation phenomena

R Mendes, L Rodrigues, J Pimenta, J Quaresma and J Ribeiro

8.1 Introduction

In this chapter, an experimental methodology used to characterise the detonation behaviour of explosive compositions from small-scale to mesoscale level is presented and discussed. The nonintrusive, high-resolution optical methodology is based on a multi-fibre optical system developed at LEDAP and provided relevant pioneering results in testing energetic materials (EMs; polymer-bonded explosives [1–4], propellants [5], and emulsion explosives [6, 7]) with mini-samples (m-samples) and micro-samples (μ-samples) formed by the collection of few coarse crystals surrounded by a binder [3]. It permits the simultaneous registration of two/three parameters such as shock and detonation velocity (U_S, D) and pressure (P), detonation front (DF) curvature, sizes of the dark zones in the DF corner turning, shock-to-detonation transition (SDT) phenomena, divergent detonation wave (DW) [8] and Mach wave formation, and its attenuation processes. The enhanced information obtained with it allows a better understanding of the processes of the formation and propagation of a DW. This procedure can be applied to the study of new advanced energetic materials. The information gathered with the use of the developed methodology not only allows the identification of the pulsing detonation phenomena in polymer-bonded explosives (PBXs), but also the two main regimes of the unstable DW propagation, in long explosive charges, namely the quasi-periodical longitudinal oscillations (of the continuous DF) and the longitudinal/transversal oscillations (of the cellular DF), as well as the limits of the existence of the two outlined DW regimes. The obtained results seem to show strong evidence of the establishment of self-organisation in the detonation of PBXs, behind the DF, where there is a reaction instability (in shocked energetic particles). This includes the small fluctuations (through the initial particle heterogeneities) which generate a new DW propagation regime, rising to a larger-scale instability.

doi:10.1088/978-0-7503-3943-8ch8

Despite the application of this experimental optical methodology to characterise samples with a mass higher than 10 g, some characteristic tests of m-samples and μ-samples with typical masses (m_0) and volumes (V_0) less than 2000 mg and 2 cm^3 have been developed using the multi-fibre optical technique. These characteristic tests are as follows: the long channel test [1, 3] (V_0 = 0.5–1 cm^3) with crystal scale resolution; the detonation failure cone test [4] (V_0 = 0.3–1 cm^3); the corner turning test [1, 2] (V_0 = 1.5 cm^3); colliding tests [4] (V_0 = 1.5 cm^3); the mini-wedge test (V_0 = 1.0 cm^3); the mini-gap test [3] (V_0 = 0.5 cm^3); and μ-gap tests of crystals, particles, and their clusters ($V_0 < 2$ mm^3, $m_0 < 4$ mg) [3]. Details of some setups and the information obtained from each one are given below.

8.2 Experimental equipment components and arrangements

Experimental setups were instrumented with a multi-channel optical analyser (MCOA) [9, 10]. A basic optic element of a MCOA is a multi-fibre optical probe (MFOP), which is based on a highly precise linear matrix of 64–96 polymethyl methacrylate (PMMA) optical fibres with 250 μm diameter. The application of the MCOA permits simultaneous registration of the reaction radiation transmitted through the explosive bulk, and the stress field induced by the reaction zone in the optical monitor. An optical fibre strip is connected to the recording system, an electronic streak camera (ESC; Thomson TSN 506 N), without any intermediate optics. Spatially resolved measurements of the time-histories of the reaction light intensity, $I(t)$, are performed in a spectral range from 450 nm to 850 nm. Localisations in the stress field, $P(x,Z,t)$, were distinguished with the 250 μm spatial and 1 ns temporal resolution, respectively. The schematic shown in figure 8.1 allows for the simultaneous recording of both the reaction radiance and the shock field caused by the reaction spot in the stacked monitor (SM).

Instrumented setups with MCOA are mainly divided into four configurations, as depicted in figure 8.2.

In the configuration presented in figure 8.2(a), the optical fibre strip is placed on the charge surface and aligned with the direction of the DW propagation. This configuration records the radiation associated with the DW propagation with a spatial resolution of 250 μm, which allows the determination of the detonation velocity in a quasi-continuous

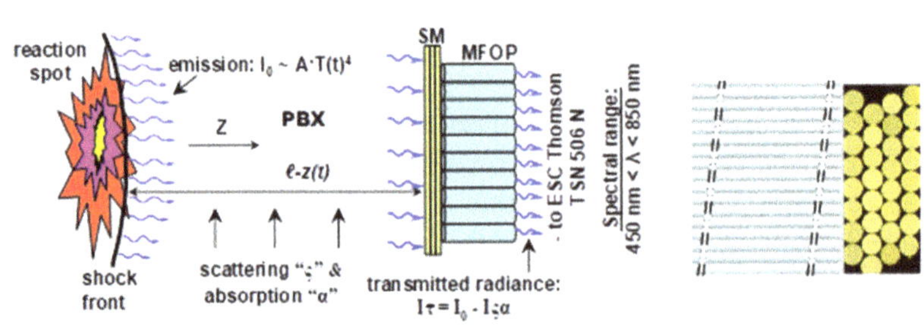

Figure 8.1. Schematics of the spatially resolved measurements of the reaction radiance and the shock field in the stacked monitor. The right image illustrates the fibres' bundle and the 32-channel MFOP matrix assembled by four optic bundles.

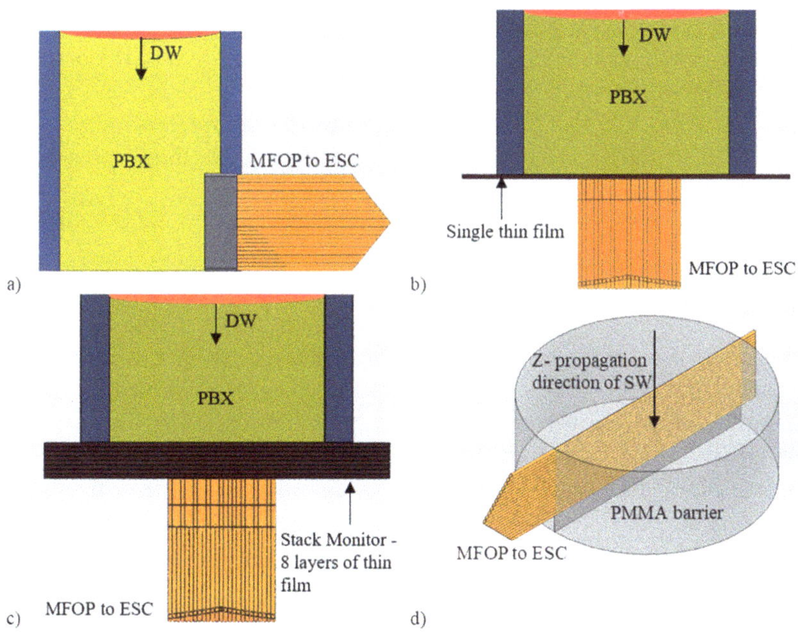

Figure 8.2. The four basic MCOA-instrumented test configurations.

process. This configuration, when applied in a cylindrical explosive charge, is useful to characterise the initiation phenomena and detonation velocity in the steady-state phase. When applied to a conic charge, it enables the characterisation of the detonation extinction process, and the determination of the local detonation velocity as a function of the local diameter.

In the nonintrusive configuration shown in figure 8.2(b), the optical fibre strip is placed outside the charge, at the bottom and along its diameter. Normally, between the top of the MFOP and the explosive charge, a film of transparent or nontransparent material is placed. This arrangement allows the recording of the breakout light from the moment the DW arrives at the end of the charge and, consequently, to determine the DW front curvature as a function of the charge diameter. This configuration, instrumented with MFOP and a transparent film, enables the visualisation of the radiation emitted from the chemical reaction zone (CRZ), some millimetres before the DW attains the MFOP and destroys it. The radiation collected, along the charge diameter and time, shows different intensity levels, providing us with an intensity field, $I(x,Z,t)$, which is related to the different reaction kinetics in space and time.

Configuration (c) is an upgrade of configuration (b), presented in figure 8.2. While configuration (b) contains one transparent thin film, configuration (c) contains a stack monitor formed by a set of thin layers of a transparent material with known Hugoniot characteristics. Commonly, six to eight layers of polyester or polyimide, with thicknesses of 250 μm and 125 μm, respectively, are used. The arrival of the DW at the stack monitor is denoted by a flash of light. After that, a shock wave is transmitted to the stack monitor, emitting a flash at each gap between the layers.

The recorded sets of flashes allow the identification of the shock wave transit time through each layer of the stack monitor. Knowing the thickness of the layers and the transit time of the shock wave, it is possible to calculate the shock wave velocity, *Us,* in the stack monitor for the 64 optical fibres which, coupled with the Hugoniot properties of the material, allows the determination of the shock pressure history induced in the stack monitor. Beyond the information recorded, as in configuration (b), the present configuration (c) instrumented with a stack monitor will permit the registration of the pressure field, $P(x,Z,t)$, generated by the DW.

In the last configuration, represented in figure 8.2(d), the stack monitor is replaced by a polymer 'barrier' (in accordance with the description in the figure 8.2) with known Hugoniot properties, generally PMMA. The optical fibre strip is fixed in the PMMA barrier (between two PMMA plates), and oriented along the propagation direction of the shock wave through the PMMA barrier. This is shock wave can either be generated by a DW of an explosive charge, or by a flyer plate impact. This configuration allows the registration of the shock wave position along the PMMA barrier as a function of time. As a result of the experimental data recovered through this configuration, the shock wave velocity, *Us,* in the PMMA can be calculated which, when combined with its Hugoniot properties, is used to determine the shock pressure history induced in the PMMA.

8.3 Measurement of steady-state detonation velocity and pressure

The measurement of the DW velocity, at the submillimetre scale, was demonstrated with the application of MFOP to characterise the DW propagation in an emulsion explosive (EX) [7, 11]. The EX was based on an ammonium nitrate aqueous solution mixed with oil and emulsifiers in a mass ratio of 84/10/6 and had an effective density of 1.42 g cm^{-3}. Usually, the EX is sensitised with hollow glass or polymer microballoons. A cylindrical charge, encased in a polyvinyl chloride (PVC) tube with 25 mm diameter and 220 mm

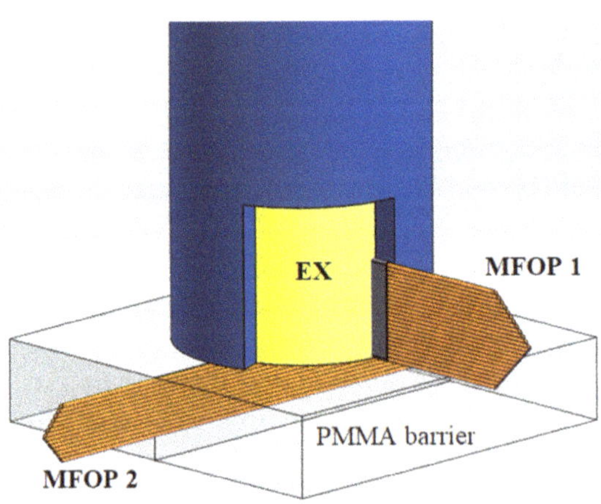

Figure 8.3. Cylindrical charge instrumented with MFOP probes, both connected to the electronic streak camera, to measure the local detonation velocity and induced pressure on the PMMA barrier.

length was used. The fibre strip, with 64 independent fibres, was separated into two parts: a first part, where 52 optical fibres (MFOP-1) were introduced in the EX, as illustrated in figure 8.3; and a second part, where the remaining 12 optical fibres (MFOP-2) were fixed between two PMMA plates. When the DW reaches the end of the charge, a shock wave is generated in the PMMA barrier. The propagation velocity of such shock wave allows the characterisation of the detonation pressure of the explosive composition. Therefore, the MFOP-2 allows the identification of the shock wave velocity in the PMMA barrier in a quasi-continuous registration and, consequently, the determination of the induced pressure history in the PMMA barrier. A similar process to identify the (Z, t) shock wave propagation in an inert barrier was presented by other authors [12, 13], using a discrete registration but in a more extended distance, as opposed to the quasi-continuous optical contact acquisition method herein presented.

To perform the simultaneous measurement of steady-state detonation velocity and pressure, a combination of the configurations depicted in figures 8.2(a) and (d) was used. This combined setup is presented in figure 8.3.

Figure 8.4 shows a typical $(Z–t)$ obtained photochronogram. The first signal refers to the DW propagation in the EX (MFOP-1), while the second part refers to the shock wave propagation in the PMMA barrier (MFOP-2).

Figure 8.5(a) shows the linear trend line, obtained by applying the least squares method to the experimental $Z–t$ points. The slope of the trend line provides us with the mean detonation velocity of the DF propagation, as well as, the respective standard error of the regression lines. Local detonation velocity was calculated

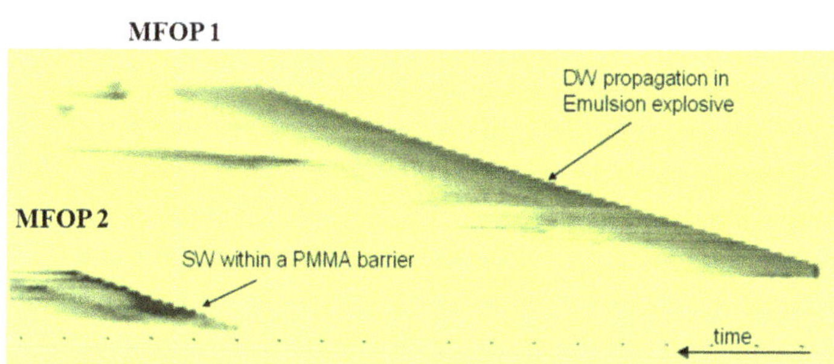

Figure 8.4. Typical streak camera record for evaluation of the steady-state detonation regime at the mesoscale level.

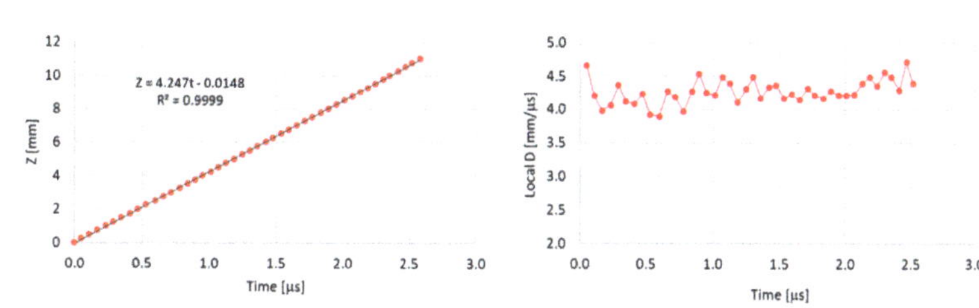

Figure 8.5. Experimental data obtained from MFOP-1: (a) $Z–t$ diagram for DW propagation. (b) Local DW velocity as a function of time.

according to equation (8.1), where the local velocity is obtained as a central derivative. Figure 8.5(b) shows the local detonation velocity as a function of time. Oscillations in detonation velocity are higher than the associated error estimated in 3% of the measurement. Despite the constant value of the mean detonation velocity, it still presents some local oscillations, which have been discussed in several works in the scope of PBX [14–16]. These results show evidence of the different reaction kinetics intensity of the explosive components, or clusters of energetic material.

$$D_{\mathrm{i}} = \frac{Z_{\mathrm{i+1}} - Z_{\mathrm{i-1}}}{t_{\mathrm{i+1}} - t_{\mathrm{i-1}}} \tag{8.1}$$

The arrival times of the shock wave propagation inside a PMMA barrier were read from the light signals of MFOP-2 and plotted against its corresponding distances. A quadratic equation was fitted to the experimental data. The shock velocity in the PMMA barrier, at the interface for $t = 0$ (U_{int}), was obtained from the first derivative of the fitted equation $Z(t)$ for $t = 0$ [12, 13].

An example of the distance–time data obtained in the PMMA barrier is presented in figure 8.6. The particle velocity at the interface, u_{int}, was determined from the Hugoniot relation between shock velocity (U_{int}) and particle velocity (u_{int}) according to equation (8.2):

$$u_{\mathrm{int}} = \frac{U_{\mathrm{int}} - C_0}{s}, \tag{8.2}$$

where C_0 and s are constants describing states obtainable by shock loading of a specific inert material. The Hugoniot parameters for PMMA material were taken from [17] and are summarised in table 8.1.

The initial pressure, P_{int}, in the inert material, at its interface, was determined according to equation (8.3):

$$P_{\mathrm{int}} = \rho_{\mathrm{PMMA}} U_{\mathrm{int}} u_{\mathrm{int}}. \tag{8.3}$$

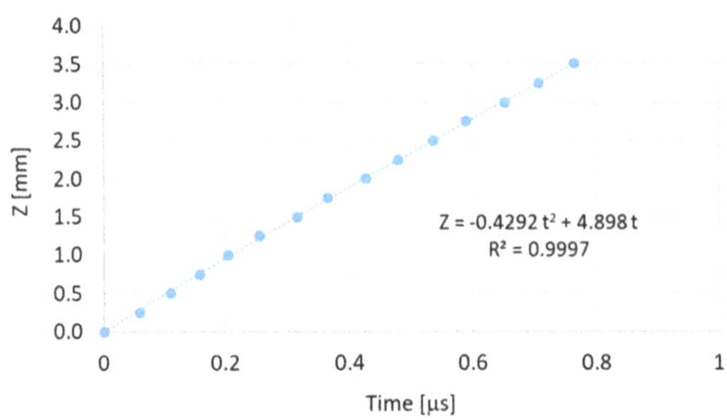

Figure 8.6. Experimental data obtained from MFOP-2: Z–t diagram for the shock wave propagation in PMMA barrier.

Table 8.1. Hugoniot properties.

Material	ρ_0 (kg m^{-3})	C_0 (mm μs^{-1})	s
PMMA	1186	2.598	1.516

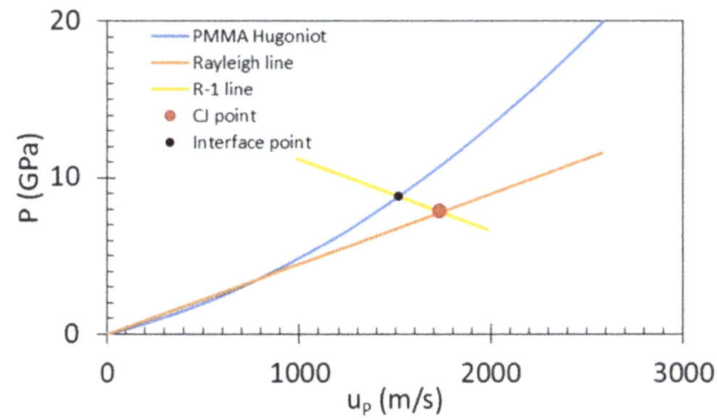

Figure 8.7. Determination of detonation pressure by the impedance-matching technique.

The Rayleigh line of the explosive composition was calculated using equation (8.4):

$$P = \rho_{\text{EX}} Du. \tag{8.4}$$

Impedance-matching techniques were used to determine the detonation pressure. The intersection of the explosive Rayleigh line with its inverse line (R-1), starting at the interface point (u_{int}, P_{int}), allows the determination of the detonation pressure value. Figure 8.7 depicts the referred calculations, and a detonation pressure of 7.8 GPa was determined for an EX with an initial density of 1.06 g cm^{-3}.

8.4 Detonation reaction zone performance tests

A long charge test (see figure 8.8) is a setup that combines the schematics depicted in figure 8.2(a) and figure 8.2(c). They are used for spatially resolved measurements of the detonation reaction zone (DRZ) structure of a DW, which allow the simultaneous assessment of detonation velocity, pressure-stress fields, and DF curvature [18]. The explosive charge is contained in a 20 mm diameter steel tube with 5 mm wall thickness and initiated by a 100 mm booster charge of PE-4A. A set of 6–10 optical fibres (MFOP-1) are fixed on the internal tube surface, and another 54 optical fibres (MFOP-2) are assembled on the charge end. Between the explosive charge and the top of the optical fibres, there is a *Kapton stack monitor* (KSM), consisting of a set of 5–8 Kapton sheets with argon-filled interfaces. Figure 8.8 (right) presents a typical photochronogram recorded with a streak camera. The application of a 64-channel MCOA provided mesoscale probing of the DRZ through simultaneous measurements of the detonation velocity D (including the local oscillations of the DF motion), and the time-resolved localisations in the stress field $\{P(t, X)\}$, induced by the detonation in the KSM and the reaction radiance emitted from the DRZ surface.

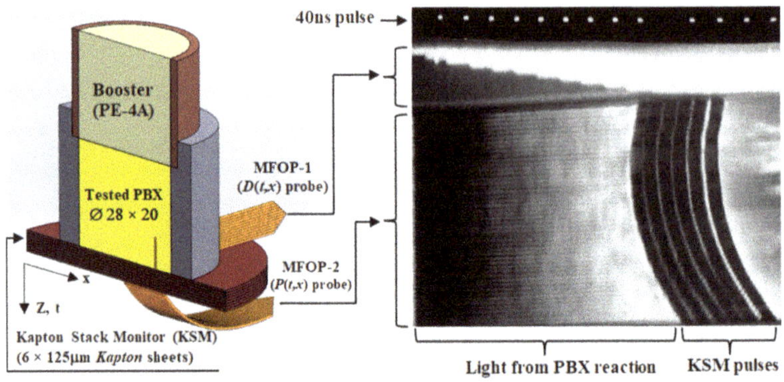

Figure 8.8. Experimental setup of the long charge test (left) and a typical photochronogram (right).

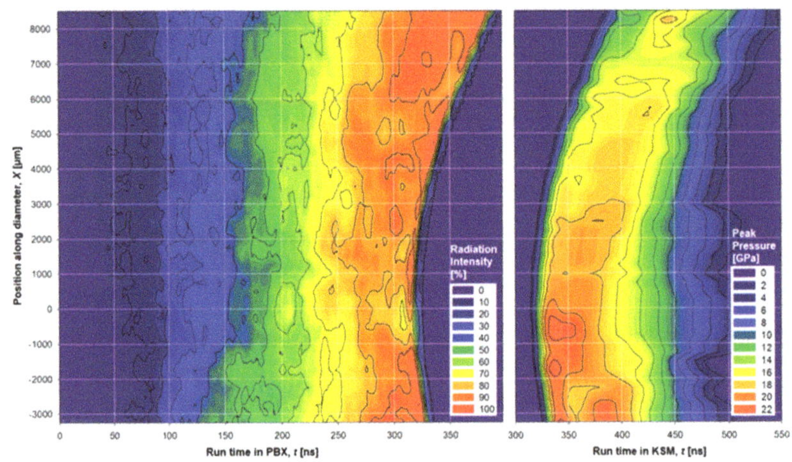

Figure 8.9. Temporally and spatially resolved radiation from the PBX reaction (left) and pressure field $\{P(t, X)\}$ in the KSM (right).

The photochronogram displayed in figure 8.8 resolves the pulsing irradiation from the DF, evidencing the shock front curvature at the charge end and its propagation through the KSM. The sequence of light pulses, that are generated at the argon-filled interstitial gaps between the Kapton layers, allows the reconstruction of the pressure field in the KSM. Localised light emissions can be observed just before the DW arrives at the charge end, indicating that the PBX composition is sufficiently translucent. Pulsating irradiation in the DRZ is related to the effect of kinetics nonequilibrium between the coarse high explosive particles, and the 'dirty binder' on the reaction localisations in the DRZ structure. The effect of the reaction localisations and ejecta on copper confinement performance are described in [14–16, 19, 20].

Local perturbations were spatially resolved in the DRZ structures of PBX composition 'RDX 93/7 wt% HTPB' having an initial density ρ_0 of 1.39 g cm^{-3} and presenting a porosity of 17%. Figure 8.9 shows the temporally and spatially resolved reaction radiation, in arbitrary units, from the tested PBX and the calculated pressure field, $\{P(t, X)\}$, in the KSM, resulting from the treatment of the photochronogram shown in figure 8.8.

The radiation history during the last 330 ns of the DW reaction (until its termination upon reaching the KSM) was acquired by the MFOP-2, which observed the reaction through the transparent layers of the KSM (due to the translucent nature of the composition). This method allowed the observation of submillimetre reaction perturbations, resulting from the inherent inhomogeneities of the composition under test. This technique enables a qualitative assessment and comparison between different PBX formulations, particularly with different particle size distributions (PSDs; and, hence, different final densities and fine-structure properties).

The local shock wave propagation velocity in the KSM can be calculated from the time intervals between the recorded light pulses, emitted by the ionisation of the argon gas that fills the interstitial gaps between the KSM layers. The local pressure can then be calculated as a function of the particle velocity, by applying a polynomial fitting to the empirically obtained shock Hugoniot data for Kapton. In this particular test, significant pressure perturbation amplitudes were measured along the observed axis. The high-pressure localisations are indicative of the cellular structure of the DRZ of PBX. These millimetre- and submillimetre-scale perturbations are detrimental in many applications where a smooth shock wave surface is desirable, e.g. acceleration of liners in shaped charges and explosive welding applications.

8.5 Detonation failure cone test

The detonation failure diameter represents an important parameter of shock sensitivity [21]. To provide spatially resolved measurements of the detonation failure process for compositions with reduced critical diameter, a detonation failure cone test was instrumented with the MFOP and an electronic streak camera (Thomson TSN 506 N) [19]. The setup configuration is presented in figure 8.10. The test PBX sample is encased in a polypropylene shell with 0.6 mm wall thickness, having a cylindrical section of 9.0 mm inner diameter, 40 mm length, and a 9.0° conical

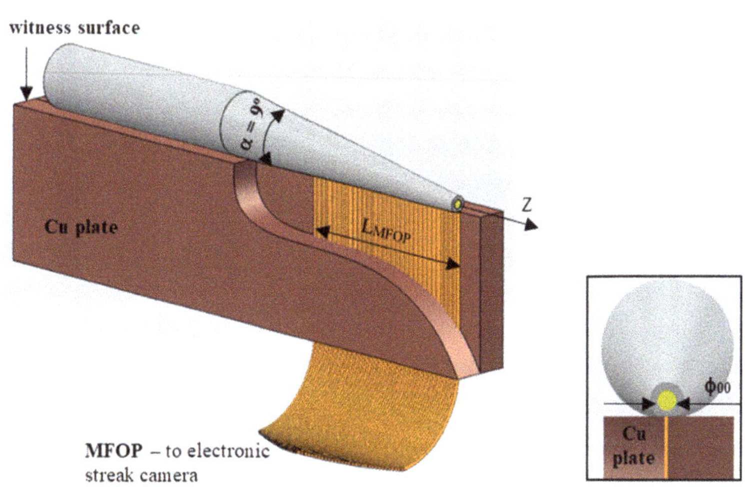

Figure 8.10. Detonation failure cone test instrumented with the MFOP.

section of 48 mm length, in which end the inner diameter is ϕ_{00} of 1.50 mm. A 40 mm booster charge of 9 mm diameter (PBX PE-4A, having 1.55 g cm^{-3} density and 7.80 mm μs^{-1} detonation velocity) was applied for transferring detonation to a sample under test. The MFOP was fixed between two copper plates with 5 mm thickness, leaving the tops of the optical fibres in contact with the cone surface. The surface of the copper plate in contact with the explosive charge is used as a witness surface. Before assembly with the MFOP (see figure 8.10), the cone shell was thinned in its joining zone up to ~50 μm. The shock front transmitted from the DRZ to the shell is recorded through the tops of the optical fibres. Figure 8.11 presents a typical photochronogram for the extinction of detonation in a conical test instrumented with the MFOP. Detonation failure is identified by a break of the slope in the recorded (Z,t) diagram, coincident with the abrupt fall of transmitted light. Using this method, it is possible to identify and distinguish the zone where the shock front dissipates, consequently leading to detonation failure. This is supported by the fact that an optical fibre, when submitted to a shock front, loses its ability to transmit radiation, as can be seen in the 'dark part' under the (Z,t) photochronogram depicted in figure 8.11. Whereas, when the shock front extinguishes (detonation failure point), the radiation is still transmitted through the optical fibre, as seen in figure 8.11.

Equation (8.5) describes the relationship between the Z_i-coordinate of the DF location, and the local diameter of cone ϕ_i, through which the front is moving at instant t_i. Equation (8.6) allows the determination of the local DF velocity, which is obtained from the run time through the space between two adjacent optical fibres. Measurements of D_i values were conducted with an error less than 3%:

$$\varnothing(z) = \varnothing_{00} + 2(L_{\mathrm{MFOP}} - Z_{\mathrm{i}}).\quad tg\left(\frac{\alpha}{2}\right), \tag{8.5}$$

$$D_{\mathrm{i}} = \frac{250\quad \mu m}{\Delta t_{\mathrm{l}}}. \tag{8.6}$$

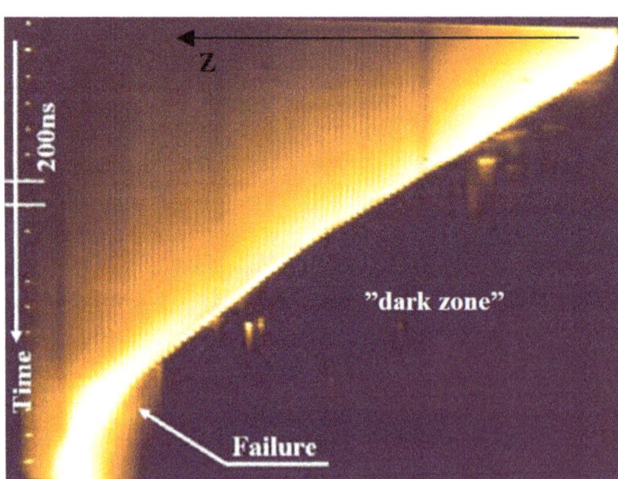

Figure 8.11. Typical photochronogram for detonation failure in a conical test instrumented with the MFOP.

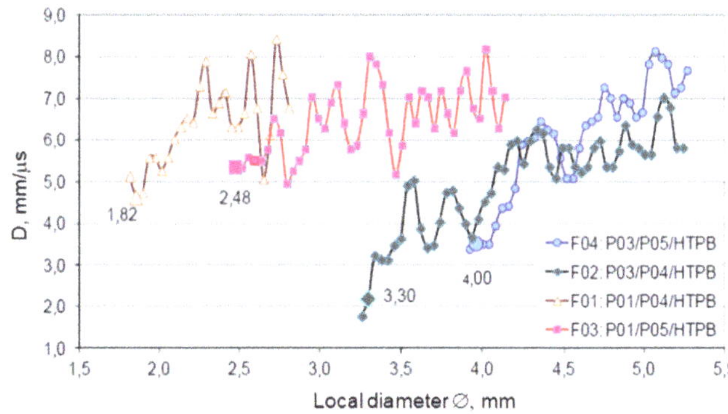

Figure 8.12. Detonation velocity as a function of local diameter. Reproduced from [22] with permission.

Table 8.2. Detonation failure diameters, $\varnothing_f$.

PBX	F01	F02	F03	F04
$\varnothing_f$ (mm)	1.82	3.30	2.48	4.00
ρ_O (g cm^{-3})	1.562	1.573	1.537	1.576

Despite all PBX tested samples being based on HMX 82/18 hydroxyl-terminated polybutadiene (HTPB), the compositions of F01, F02, F03, and F04 were prepared with different HMX particles sizes and crystal quality. The HMX particles used in the compositions are described and characterised in [22]. The $D(\varnothing)$ diagrams of the DF velocity as a function of the local charge diameter are presented in figure 8.12.

Table 8.2 summarises values of the initial density and detonation failure diameter $\varnothing_f$ for several tested compositions [22]. This information can be used to determine the impact of mean particle size and crystal surface quality on the detonability of the compositions. Despite having the same composition (HMX 82/18 HTPB) and a similar initial density, the compositions prepared with a fraction of ultrafine particles, F03 and F04, presented a higher failure diameter in comparison with the compositions prepared with fine particles, F01 and F02, respectively. Compositions where the coarse fraction are recrystallised HMX particles (F04 and F02) also presented higher critical diameter, when compared with the same compositions prepared with coarse crystal from industrial-grade HMX (F01 and F03). This demonstrates the effect that the recrystallised HMX particles have on the composition detonability, which corresponds to a lower composition sensitivity.

8.6 Shock-to-detonation transition: wedge test

The shock-to-detonation transition (SDT) wedge test allows the assessment of the shock sensitivity of explosive materials, based on a measured run time t^* and run distance Z^* to high-order detonation as a function of the initiation pressure, P_0 [23].

The test samples were four pore-free PBX compositions, 'HMX 82/18 wt% HTPB', having a bimodal PSD mixture of coarse particles (P01 or P03) with fine and ultrafine HMX constituents (particles P04 or P05). The formulation F01 was a reference PBX, 'P01/P04/HTPB 65.6/16.4/18 wt%'. Formulations F02, F03, and F04 were composed of formulas of P03/P04/HTPB, P01/P05/HTPB, and P03/P05/HTPB, respectively, keeping the same mass ratio: 65.6/16.4/18 [22]. The density and theoretical maximum density (TMD) values are presented below in table 8.3. The shock sensitivity measurements were conducted in a series of wedge tests instrumented [22, 24] with a calibrated booster, a PVC stacked shock attenuator (SA), a PVC stacked optic monitor (SM), and MFOPs connected to the ESC. Figure 8.13 shows the test schematic and a typical streak camera record obtained at simultaneous testing of PBXs F02 and F04.

In the wedge test setup represented in figure 8.13, the HMX particles of test samples were subjected to a 22.5 GPa shock. The shock front motion speed in the SA and SM was controlled in each experiment; pressure was obtained with the use of a second-order polynomial *P*(*Us*) approximation of shock Hugoniot data for PVC (1.376 g cm^{-3}) [11]. The run time t^* and run distance Z^* to high-order detonation were determined from the analysis of the (Z,t) and $(dZ/dt,t)$ diagrams of front motion at the SDT [1] and, reciprocally, the shock front local velocity inside the optic monitor [7]. Parameters t^* and Z^* were obtained with an accuracy better than 10 ns and 50 μm. Measured t^* and Z^* values of the tested compositions are presented in table 8.3 [22], and compared with the SDT data obtained for PBX-9502 and 100 wt% purified triaminotrinitrobenzene (TATB) [23].

Table 8.3. Run distance and run time detonation initiation.

PBX	F01	F02	F03	F04	Purified TATB [23]	PBX-9502 [23]
Z^* (mm)	0.55	1.25	1.08	1.61	1.45^+	1.45^+
t^* (μs)	0.09	0.21	0.14	0.23	0.20^+	0.20^+

$^+$ Extrapolation.

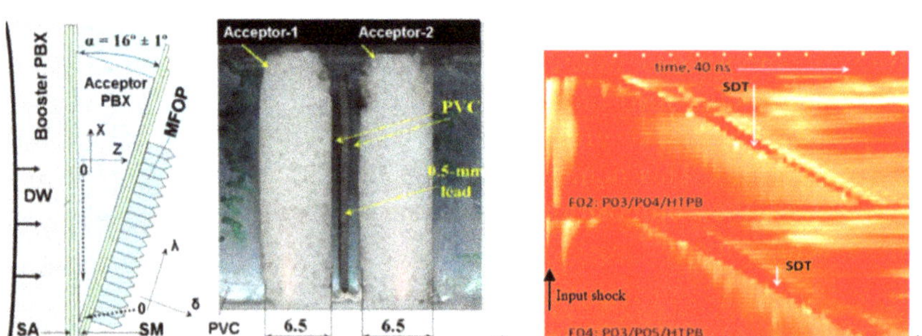

Figure 8.13. Wedge test instrumented with the PVC stacked shock attenuator, PVC optical stack monitor, and MFOPs (a) (reproduced from [25], with the permission of AIP Publishing), and a typical streak camera record (b) [22].

8.7 Shock-to-detonation transition: flyer plate impact test

When a high-velocity flyer plate impacts an explosive sample, a shock wave is formed and propagated in the sample. If the pressure pulse has enough intensity to initiate the explosive, a chemical decomposition reaction starts immediately as the pressure pulse is formed. At the same time, the energy release will support the shock wave propagation and convert it into a DW.

Generally, the distance and the delay time needed for a shock wave to be converted into a DW depends on the amplitude and time duration of the pressure pulse. Consequently, for high-velocity flyer plate impact, efforts are focused on measuring the SDT, run distance Z^* and run time t^*, at submillimetre and microsecond scale, respectively.

A thin flyer plate of Mylar (polyester film) with 25 mm diameter, launched by a mega ampere pulse in a velocity range up to 8 mm μs^{-1}, was used to initiate the explosive samples [10, 26].

To measure the run time t^* and run distance Z^* inside the explosive sample, the MFOP-1 was fixed in the normal direction to the movement of the flyer plate, as represented in figure 8.14.

A cylindrical explosive container with 25 mm internal diameter and 15 mm length was used to characterise the shock initiation of a PBX by a flyer plate impact. The explosive charge was covered by a PMMA barrier equipped with the MFOP-2, which permitted the characterisation of the shock wave propagation through the PMMA barrier and, consequently, the identification of the detonation pressure. As the front of the shock wave, or DW, is propagating in the sample, a light pulse is generated on the MFOP-1 and recorded on the ESC. A typical recorded photochronogram is presented

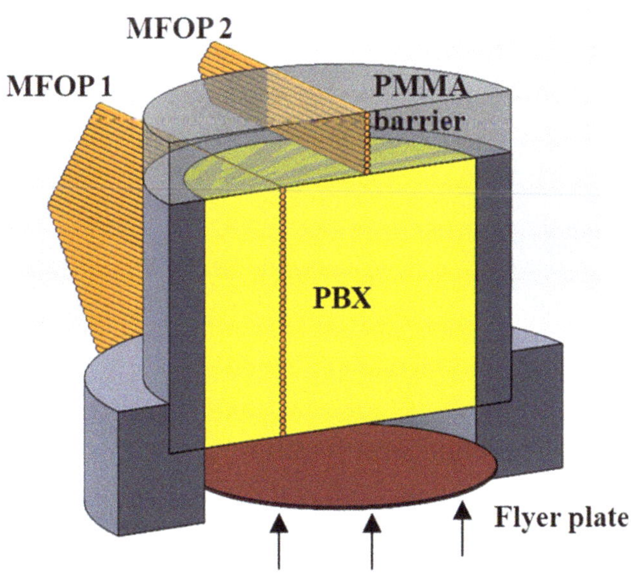

Figure 8.14. Experimental setup to evaluate the run distance and run time in flyer plate impact tests.

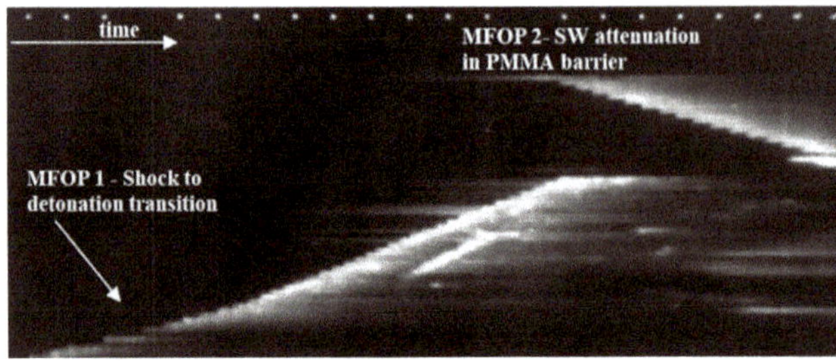

Figure 8.15. Typical photochronogram for SDT experiments initiated by flyer plate impact and shock wave attenuation in PMMA plate [26].

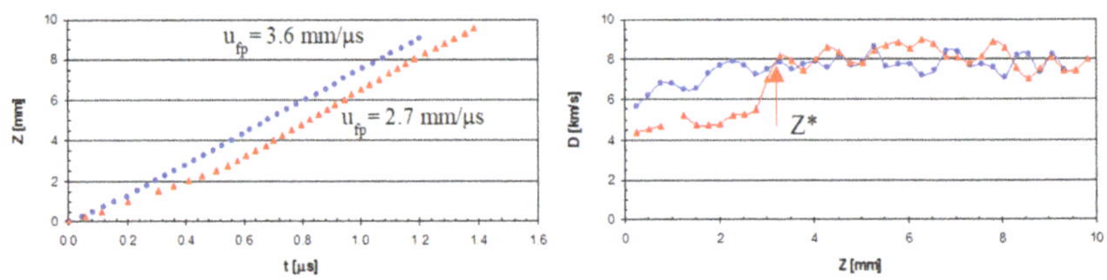

Figure 8.16. *(Z,t)* diagram and local detonation velocity versus distance of SDT experiments.

in figure 8.15. This figure shows two light traces. The first one can be identified as the characteristic slope of the initial shock wave velocity in the explosive. This is followed by an acceleration zone that produces a second slope, characteristic of the detonation velocity, as shown by MFOP 1 in figure 8.15. The second light trace, MFOP-2, shows the shock wave propagation in the PMMA barrier.

This SDT setup, instrumented with the MFOPs, makes it possible to 'continuously' record the evolution process of the shock detonation transition in space and time [26]. Figure 8.16 shows the (Z,t) diagram for two experiments where the PBX samples were shocked with different flyer plate velocities, u_{fp}, and the corresponding effect in terms of local velocity on the initiation process of the PBX sample. As the flyer plate velocity increases from 2.7 mm μs^{-1} to 3.6 mm μs^{-1} the initiation distance Z^* decreases from 3.25 mm to 1.75 mm. Local shock or detonation velocity was calculated according to equation (8.1).

The tested samples were a PBX composition 'RDX 85/15 wt% HTPB' having a bimodal PSD mixture of coarse particles (d_{50} = 96 µm) with fine particles (d_{50} = 22 µm) constituents. This composition had a density of 1.574 g cm^{-3} and a mean detonation velocity of 7.90 mm μs^{-1}.

8.8 Single crystal reaction observation

Submillimetre kinetic localisations and ejecta formation on the detonation conversion of a single β-HMX crystal surrounded by a 'dirty binder' were studied. The application of matrix-type MFOPs allows a panoramic observation [15, 16] of the reaction localisations within and around a coarse HMX particle, and simultaneously the spatially resolved shock field measurements with a complex of 3D

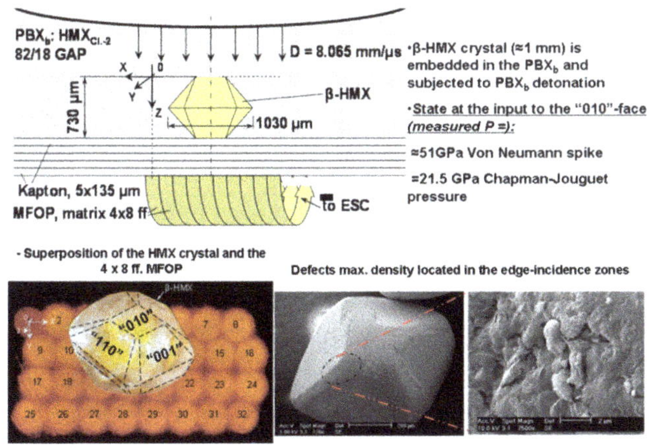

Figure 8.17. Disclosing the reaction localisations and ejecta formation in 1 mm β-HMX crystal surrounded by fine-grained PBX$_b$: test configuration, superposition crystal vs. MFOP, and the SEM image showing the surface microstructure of the typical HMX crystal selected from the coarse fraction 707 μm < d < 1100 μm. Zones of micro-defects maximum density are located at the edges of the crystal.

geometry. Figure 8.17 illustrates experiments covering spatially resolved measurements of the kinetic localisations occurring in the meso-cell of a PBX. This is represented by a β-HMX crystal of size ≈1 mm (DYNO Nobel) surrounded by fine-grained PBX$_b$ as a 'dirty binder' simulant. The PBX$_b$ was a cast-cured material, 'HMX Class-2 82/18 wt% GAP', having an initial density ρ_0 of 1.759 g cm^{-3} (0.998 TMD). A 20 mm PBX$_b$ charge of 28 mm diameter was encased in a 10 mm steel confinement and initiated by a 100 mm booster charge of PE-4A (ρ_0 = 1.56 g cm^{-3}, 7.80 mm μs^{-1} detonation velocity) [15].

Figure 8.17 shows a test configuration depicting, in detail, a superposition of crystal vs. MFOP. The crystal was subjected to the PBX$_b$ detonation from the '010' facet. Scanning electronic microscopy (SEM) of typical HMX crystals (selected from the same as the tested crystal, with a coarse fraction of 707 μm <d<1100 μm) reveals important features of the surface microstructure. Zones of a higher density of micro-defects, such as clusters and encrustations, are located on the edges of the crystal.

The surface micro-defects have a mosaic-like structure; they manifest themselves as a continuum of clusters of 1.5 × 0.8 μm cross-section, spaced from each other by dislocation pits, micro-cracks, and fissures.

Figure 8.18 depicts the streak camera record of the radiation collected by the MFOPs positioned under the β crystal [16]. The first part represents the radiation emitted by the β crystal reaction during the detonation of the fine-grained PBX$_b$ that surrounded the crystal. The second fragment shows a set of flashes that represents the shock wave propagation inside the Kapton stack monitor.

Taking the radiation intensity in the 32 optical fibres it is possible to determine the intensity isochrones every 5 ns, which allows one to construct a set of 5 ns interval frames, depicting spatially resolved reaction localisation sites inside and around the β-HMX crystal, presented in figure 8.19.

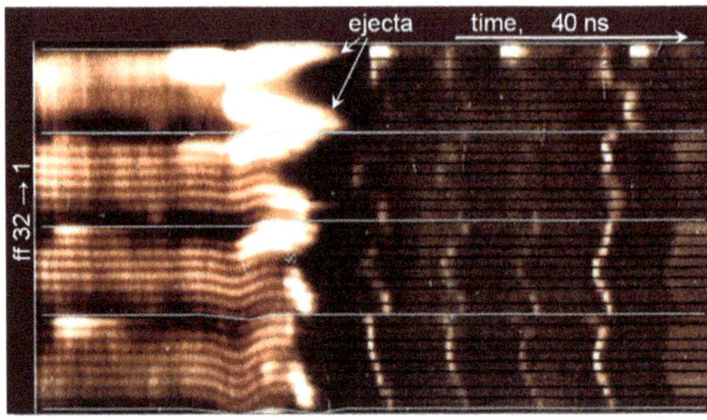

Figure 8.18. Streak camera record for the reaction of 1 mm β-HMX crystal under detonation of the fine-grained PBX_b.

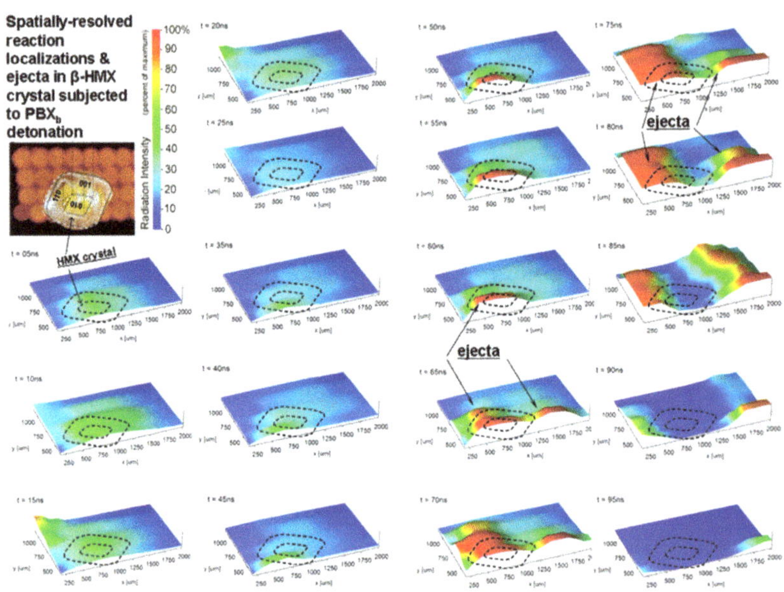

Figure 8.19. Reaction field evolution in the elementary detonation cell, 1 mm β-HMX crystal within a 'dirty binder', subjected to PBX_b detonation followed by ejecta formation at lower slant facets. Time '$t \equiv 0$' is taken when the PBX_b detonation reaches the '010' facet.

As it follows from the reaction field distribution represented in figure 8.19, the reaction appears first in the zone of the '010' facet where the quasi 1D planar collision occurs with the incident DF. In the projection of the '010' facet, the reaction front moves through the crystal's body for 96 ns, having a mean velocity, *D*, of 7.6 μm ns^{-1}, which is 14% and 6% less than the normal detonation velocity in a β-HMX crystal with a 1.91 g cm^{-3} density [23], and the DF velocity in the surrounding 'dirty binder', respectively. However, the reaction is not dominated by the '010' facet: during the 45 ns run (run distance corresponding to approximately from two-fifths to one-half of the crystal's body) the reaction front is emitting a relatively homogeneous light which slowly changes in intensity. Major reactions arise (see frame '$t = 50$ ns') in the edge incidence zones of the neighbouring facets in which, according to the SEM data, the micro-defects are highly pronounced [16]. Due to the specific crystal's topography, the collision geometry between the incident DF and the crystal's slant facet gives an indication of the major role that the shear-driven deformation plays in triggering decomposition chemistry in the surface clusters (SEM picture in figure 8.17), located in the high-defect edge zones [16]. As a result, a major reaction progresses in the edge zones of the crystal's surface being dragged by the PBX_b detonation. The reaction localisations are followed by the formation of the longitudinal/transversal ejecta (see frames '65 ns', '70 ns', and '75 ns'). The overdriven ejecta (~20–25 μm ns^{-1} velocity) are emerging from the lower edge zones, and then reaching the periphery zones of the optical monitor 22 ns later than the reaction front achieves in the '010' projection [16].

8.9 Conclusions

This chapter presents a set of experimental approaches used to characterise the detonation process of explosive compositions from small-scale to mesoscale levels. Experimental approaches are based on nonintrusive, high-resolution optical metrology procedures, wherein a multi-optical fibre strip forms the main element. Tests instrumented with several configurations of such MFOPs permit the simultaneous registration of parameters including the local detonation velocity; the pressure field; the geometrical shape of the DF; the light emitted from the CRZ; the structure of the SDT zone; the detonation failure diameter; and the crystal reaction within a dirty binder. In terms of scale, it allows the testing of PBX micro-samples made up of a few crystals surrounded by a binder. The simultaneously recorded parameters provide a better understanding regarding the processes of formation and propagation of a DW, at the submillimeter scale, evidencing the behaviour of the explosive components (particles and binder) on detonation.

Acknowledgement

A substantial part of the work presented in this chapter stems from the pioneering works of Dr Igor Plaksin at the University of Coimbra, particularly regarding the development of innovative submillimetre- and microsecond-resolved experimental techniques for the study of the onset, propagation, and extinction of the detonation process. These innovations have widened the characterisation and deepened the understanding of the detonation phenomena. Beyond the acknowledgements due to those key contributions, the authors want to remember Igor for his friendship and for being an extraordinary human being.

References

[1] Plaksin I, Campos J, Mendonça M, Mendes R and Gois J C 1998 Interaction of double corner turning effect in PBX *AIP Conf. Proc.* **429** 1002

[2] Plaksin J, Campos I, Mendes J, Ribeiro R and Gois J 1998 Pulsing behavior and corner turning effect of PBX *11th Int. Detonation Symp.* 679–85

[3] Plaksin I, Campos J, Mendes R, Ribeiro J and Gois J Feb. 2000 Mechanism of detonation wave propagation in PBX with energetic binder *AIP Conf. Proc.* **505** 817

[4] Mendes R, Plaksin I, Campos J and Ribeiro J 2000 Double slapper initiation of the PBX *AIP Conf. Proc.* **505** 915

[5] Simões P, Pedroso L, Portugal A, Plaksin I and Campos J 2001 New propellant component, part II. Study of a PSAN/DNAM/HTPB based formulation *Propellants Explos. Pyrotech.* **26** 278–83

[6] Plaksin I *et al* 2008 Novelties in physics of explosive welding and powder compaction *Mater. Sci. Forum* **566** 135–40

[7] Mendes R, Ribeiro J B, Plaksin I and Campos J 2012 Non ideal detonation of emulsion explosives mixed with metal particles *AIP Conf. Proc.* **1426** 267–70

[8] Mendes R, Campos J, Plaksin I and Ribeiro J 2004 Divergent detonation wave processes in PBX based on RDX *Propellants Explos. Pyrotech.* **29** 146–9

[9] Plaksin I, Campos J, Ribeiro J, Mendes R, Gois J, Portugal A, Simões P and Pedroso L 2002 Detonation meso-scale tests for energetic materials *AIP Conf. Proc.* **620** 922–5

[10] Mendes R, Plaksin I and Campos J 1998 Single and two initiation points of PBX *AIP Conf. Proc.* **429** 715–8

[11] Mendes R, Ribeiro J, Plaksin I, Campos J and Tavares B 2014 Differences between the detonation behavior of emulsion explosives sensitized with glass or with polymeric micro-balloons *J. Phys. Conf. Ser.* **500** 052030

[12] Quaresma J, Deimling L, Campos J and Mendes R 2021 Characterization of the detonation pressure of a PETN based PBX with the optical active method *Propellants Explos. Pyrotech.* **46** 875–82

[13] Künzel M, Anastacio A C, Kučera J and Pachman J 2017 OPTIMEX: Detonation pressure determination using passive optical system *Proc. 17th NTREM* 726–30

[14] Plaksin I *et al* 2006 Meso-scale probing of CRZ structure in PBX: DW oscillations from ignition up to failure *AIP Conf. Proc.* **845** 1002

[15] Plaksin I, Coffey C S, Mendes R, Ribeiro J, Campos J and Direito J 2006 Formation of CRZ 3D structure at SDT and at shear initiation of PBX: Effects of front irradiation and reaction localization in HMX crystals *13th Int. Detonation Symp.*

[16] Plaksin I *et al* 2007 Shear induced reaction localization and mechanisms of energy dissipation in PBXs subjected to strong shock *AIP Conf. Proc.* **955** 1427–32

[17] Marsh S P 1980 *LASL Shock Hugoniot Data* (Berkeley, CA: University of California Press)

[18] Plaksin I, Mendes R, Rodrigues L, Plaksin S and Campos J 2016 Characterization of detonation performance and shock reactivity of PBX-materials with use of multi-channel optical analyzer *Proc. 19th Seminar on New Trends in Research of Energetic Materials* 253–64

[19] Plaksin I, Campos J, Direito J, Ribeiro J, Mendes R and Gois J 2004 Birth and death of the PBX' detonation in a Mezzo-scale. effect of the grain sizes and Binder's nature in ignition and failure of detonation *Proc. 35th Int. Conf. ICT*

[20] Plaksin I *et al* 2005 Micro-ejecta from detonation front as an ignored negative factor in performance of PBX detonation *36th Int. Conf. ICT*

[21] Dremin A N 1997 On correlation of shock sensitivity and detonation failure diameter of condensed explosives *Proc. 28th Int. Conf. of ICT – Combustion and Detonation*

[22] Plaksin I *et al* 2012 Exploring the HMX-based PBXs with the TATB shock sensitivity *Insensitive Munitions and Energetic Materials Technology Symp.*

[23] Gibbs T R and Popolato A 1980 *LASL Explosive Property Data* (Berkeley, CA: University of California Press)

[24] Plaksin I, Rodrigues L, Plaksin S, Campos J, Mendes R and Ribeiro J 2010 Effect of the reaction light absorption on the formation of the detonation reaction zone 3D-structure in PBXS *Proc. 14th Int. Detonation Symp.*

[25] Plaksin I, Rodrigues L, Plaksin S, Campos J, Mendes R, Ribeiro J and Gois J 2009 Radiation-induced precursors in crystalline energetic composites *AIP Conf. Proc.* **1195** 137–40

[26] Mendes R, Campos J, Plaksin I and Ribeiro J 2001 Shock to detonation transition in a PBX based on RDX *Proc. 32nd Int. Conf. ICT* no. Energetic Materials 19/1–9/11

IOP Publishing

Energetics Science and Technology
An integrated approach
Adam S Cumming

Chapter 9

A traditional approach to munition life management

Mark Ashcroft

All materials degrade in some way over time and at an increased rate when exposed to harsh environments. In order to demonstrate that munitions[1] remain safe and suitable for service (S3) throughout their life, in accordance with Allied Ordnance Publication AOP-15 [1], nations typically adhere to engineering management systems to provide robust levels of technical assurance. The conduct of life-assessment programs forms part of that engineering system. These are carried out at a number of different stages across the munition life cycle, such as qualification testing to support their entry into service. Once introduced, nations classically apply in-service surveillance principles and methodologies to provide ongoing assurance, as proposed in NATO Standardization Agreement STANAG 4675 [2] and its suite of Allied Ordnance Publications (AOPs): AOP-62, AOP-63 and AOP-64 [3–5]. Often over the service life of munitions a life extension or an expansion of the operational envelope is required, which was not anticipated at introduction into service and encapsulated within the original qualification test program. In these cases, a further evaluation of in-service munitions is required to ensure safety and suitability over any given extended life duration or environmental exposure. In these examples, guidance is provided in Allied Environmental Conditions and Test Publication AECTP-600 [6] to extend munition life. This chapter will discuss the traditional approaches to life management and test programs conducted across three areas of a munition's life: introduction into service, in-service surveillance, and life extension. Although a number of documents provide useful guidance on the structure of test programs, nations adhere to the principles detailed therein differently and they do not prescribe a single protocol. As a result, each nation tailors its program to its own requirements,

[1] In this chapter 'munitions' will be the generic term used to describe military weapons, ammunition, explosive ordnance and demolition stores, etc.

depending on a range of considerations including the size of inventory, the approach to long-term storage, expected usage rates, and so on. Accordingly, this chapter includes discussion on a range of factors that might be taken into consideration in the design and tailoring of munition life-management systems.

9.1 Principles and life-management phases

The aim of life-assessment programs is to promote more effective through-life management strategies and inform decision-making in order to optimize the lives of munitions. Successfully delivered life-assessment programs lead to a number of benefits, such as follows:

- S3 assurance of military capability through improved knowledge of life-limiting failure modes.
- Reduction in disposal efforts and the associated environmental impact.
- More efficient logistics with reduced equipment buys and greater end-of-life visibility.
- Improved stockpile management to utilize munitions with effect ahead of life expiry.

To realize these benefits it is critical that the environmental conditions which munitions experience and the degradation caused to their energetic materials and other components by temperature, humidity, shock, vibration, and pressure are well understood.

According to AOP-46 [7], a life-assessment procedure has five phases:

- *Phase 1: Design assessment and environmental exposure.* Carry out a design assessment in order to identify potential failure modes of the system configuration and constituent materials. This is done in parallel with a review of the environmental exposures identified in the manufacture to target or disposal sequence (MTDS).
- *Phase 2: Life-limiting failure mechanisms.* All potential life-limiting failure mechanisms are identified.
- *Phase 3: Assessment methods.* A decision process where techniques to be used for the life assessment of the munition are identified.
- *Phase 4: Trials.* Trials performed in the form of sequential environmental tests (SETs) to provide objective and quantifiable evidence.
- *Phase 5: Life assessment.* Utilize all evidence accumulated in the above phases and make an assessment of the life span for which the munition remains S3.

These phases will be discussed in turn to provide a logical flow to examples of traditional life-management activities and programs.

9.1.1 Phase 1: design assessment and environmental exposure

Every aspect of munition design, from ingredients within energetic materials and the materials themselves through to complex weapon systems, is manufactured in accordance with accepted engineering specifications. The overall design and its suitability for the conditions identified in the MTDS is assessed in the context of failure mode identification. The components listed for a given system will be used to identify the life-limiting elements in Phase 2 of an assessment.

As packaging can play a critical role in the hazard mitigation of munitions, it is also necessary to fully evaluate the approved packaging of the item when it is in a storage and transport (S&T) configuration. The design assessment must therefore identify the packaging used by the original equipment manufacturer (OEM) and any authorized alternate packaging.

The ability to obtain detailed design information is key in order to enable life-assessment decisions. In the current commercial world, where the need to protect intellectual property dominates, the ability to obtain design information is challenging, even for large military nations. This challenge is amplified for nations with less control over, and limited insight into, an industrial base. Nations with a more limited munitions industrial base and less financial buying power encounter increased issues around the release of sensitive technical information, which hampers munition life management in the first phase of the assessment procedure.

In order to establish the predicted life based on failure modes it is necessary to determine the worst-case environmental conditions to which the item will be exposed. As items have multiple life cycle phases, the worst-case environment for each circumstance is often used to generate the life-limiting environment. This evaluation is based on the munition usage and environments identified in the MTDS. A typical MTDS in terms of munition usage is depicted in figure 9.1.

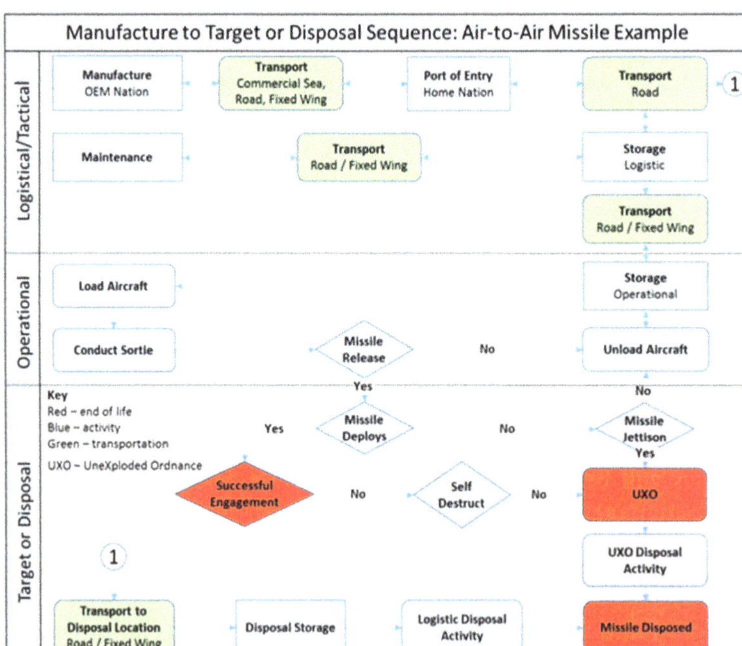

Figure 9.1. Example MTDS usage flow diagram. Reprinted with permission of QinetiQ Ltd, 2022.

The environments identified in the MTDS can be broken down across usage phases to more accurately capture the different environmental exposures at each. An example of the environmental considerations at the logistical and tactical phase is shown in figure 9.2.

There are a number of facets to an MTDS in terms of environmental exposure, such as temperature, humidity, vibration, and shocks. Each aspect needs to be considered in the context of the munition design, both separately and cumulatively.

The climatic regions detailed within the environment section of the example MTDS (figure 9.2) are taken from environmental guidance documents, such as DEF-STAN 00–35 [8]. The Climatic Region A1, for example, represents extreme hot dry environments. These are areas which experience very high temperatures accompanied by high levels of solar radiation, namely the hot, dry deserts of North Africa, parts of the Middle East, Northern India, and South Western USA.

The diurnal temperature cycle for each region is representative of conditions on days when near-extreme temperatures occur. These values are those temperatures which exceed the hotter/colder locations in the category, on average, for a total of 1% of 1 month during the hottest/coldest period of the year. The associated relative humidity is taken from dew point data over the same period. Transit and storage diurnal cycles are representative of the additional effects on induced temperatures of direct solar radiation on days when near-maximum ambient air temperatures occur.

An example of hourly diurnal cycles for Category A environmental conditions is provided in figure 9.3, which is reproduced from [8].

Whilst there are many interrelated aspects of environmental exposure, the example of temperature is used here. The S&T environment detailed in the

LIFE CYCLE PHASE	ENVIRONMENT
LOGISTICAL/TACTICAL PHASE	
Storage: Storage in this phase consists of storage at the manufacturer, port of entry and logistic storage. Items in this phase are being delivered, returned for repair, or stored before going to an operational phase. Items will be in their container at all times and stored in purpose built facilities.	**Environment:** A1, A2, A3, B2, C0. **Duration:** 240 months (Service life). **Packaging:** All S&T will occur in approved packaged configurations.
Transport: Transport in this phase will include delivery from the OEM to storage depots, and return for repairs. All environments and threats were accounted for and the worst case transport scenario for duration.	**Environment:** A1, A2, A3, B2, C0, C1. **Duration:** • Road: 45,500 km (and) • Sea: 300 days or • Road: 7,500 km (and) • Prop Aircraft: 125 hours (or) • Jet Aircraft: 100 hours **Packaging:** All S&T will occur in approved packaged configurations.

Figure 9.2. MTDS environment. Reprinted with permission of QinetiQ Ltd, 2022.

	METEOROLOGICAL CONDITIONS									TRANSIT AND STORAGE CONDITIONS		
Category	A1			A2			A3			A1	A2	A3
Local Time Hours	Ambient Air Temperature °C	Relative Humidity %	Solar Radiation W/m²	Ambient Air Temperature °C	Relative Humidity %	Solar Radiation W/m²	Ambient Air Temperature °C	Relative Humidity %	Solar Radiation W/m²	Induced Air Temperature °C	Induced Air Temperature °C	Induced Air Temperature °C
0100	35	6	0	33	36	0	30	69	0	35	33	31
0200	34	7	0	32	38	0	29	72	0	34	32	29
0300	34	7	0	32	43	0	29	74	0	34	32	29
0400	33	8	0	31	44	0	28	76	0	33	31	28
0500	33	8	0	30	44	0	28	78	0	33	30	28
0600	32	8	55	30	44	55	28	78	45	33	31	29
0700	33	6	270	31	41	270	29	74	170	36	34	31
0800	35	6	505	34	34	505	30	67	500	40	38	35
0900	38	6	730	37	29	730	31	29	800	44	42	40
1000	41	5	915	39	24	915	34	51	960	51	46	44
1100	43	4	1040	41	21	1040	36	47	1020	56	51	50
1200	44	4	1120	42	18	1120	37	45	1060	63	57	54
1300	47	3	1120	43	16	1120	38	44	1020	69	61	56
1400	48	3	1040	44	15	1040	38	43	915	70	63	58
1500	48	3	915	44	14	915	39	43	660	71	63	58
1600	49	3	730	44	14	730	39	44	250	70	62	56
1700	48	3	505	43	14	505	38	46	70	67	60	53
1800	48	3	270	42	15	270	37	48	15	63	57	50
1900	46	3	55	40	17	55	35	50	0	55	50	46
2000	42	4	0	38	20	0	34	53	0	48	44	41
2100	41	5	0	36	22	0	34	56	0	41	38	37
2200	39	6	0	35	25	0	32	59	0	39	35	34
2300	38	6	0	34	28	0	32	63	0	37	34	33
2400	37	6	0	33	33	0	31	66	0	35	33	32

Figure 9.3. Diurnal cycles for Category A environmental conditions. Data reproduced from reference [8].

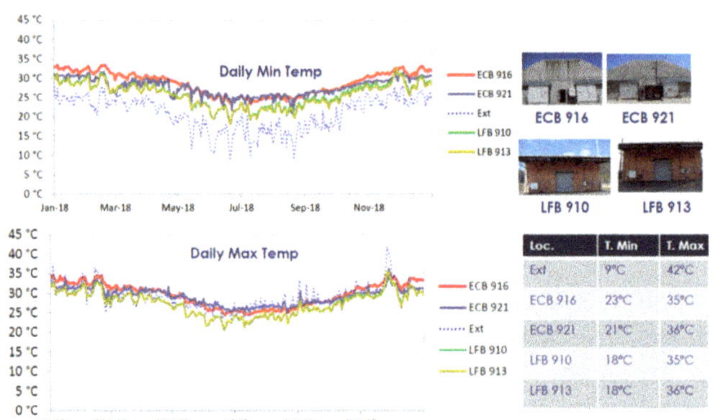

Figure 9.4. Environmental monitoring temperature. Data reproduced from reference [9].

MTDS often gives a range of temperatures, from Climatic Region A1 (non-induced extreme hot dry, range of 32 °C to 49 °C) to C1 (induced intermediate cold, with minimum of −33 °C), as detailed within DEF-STAN 00-35 [8]. As it is intended to assess the maximum S3 life, any determination will focus on the highest realistic storage temperature for that climatic region. The munition configuration, role, and environment are all important here. For example, the maximum temperature for the non-induced Climatic Region A1 environment is considered appropriate in the assessment of logistical storage conditions as munitions are located under cover and not exposed to radiant heating effects. However, the induced Climatic Region A1 environment needs to be taken into account within other areas of the MTDS, such as the transportation or operational phase, where the induced temperature range can reach a maximum of 71 °C. As the prescribed climatic regions often reflect a worst-case perspective of maximum temperatures, the benefits of environmental monitoring can be used to provide further technical assurance through the accurate knowledge of exposure and application of safety margins. An example of temperature data recorded by Thales for the Australian Department of Defence is provided in figure 9.4 [9].

As can be seen in figure 9.3, storage conditions are rarely isothermal. Furthermore, the table in the bottom-right-hand corner demonstrates how locations and buildings around the storage facility experience different extremes of temperature.

The assessed environment is based on the maximum temperature since the accelerated degradation of energetics is typically assessed as a possible life-limiting factor. The life-limiting failure mode that results from low temperatures is not the acceleration of energetic degradation modes, but, more often, structural failures due to embrittlement of materials or thermomechanical stress. Such failure modes are typically realized upon onset of the cold temperature extreme, if at all, and hence long-duration cold exposures are not the focus of test programs or chemical (thermal) ageing calculations.

As natural environmental conditions are not isothermal, especially over large countries, a graphical representation of the acceleration factors associated with individual storage locations around Australia is produced in figure 9.5 as an example of how location can affect munition life. In figure 9.5 the acceleration factor is used to give an approximate equivalency to the period of time the item would have aged if it were stored at a constant 25 °C. The acceleration factor was calculated with an activation energy of 80 kJ mol^{-1}. As the activation energy is an exponential factor in the Arrhenius equation, the value selected can significantly affect predictions. The value of 80 kJ mol^{-1} is often used at temperatures below 60 °C in accordance with both STANAG 4582 [10] and AOP-48 [11]. This value is often seen as a conservative figure for nitrate ester degradation, and it is noted within documentation that most items will have decomposed significantly less after 10 years of storage than what the calculated result has predicted. Nonetheless, combined with temperature data, chemical rate kinetic data can be used effectively to gauge relative life at different storage locations. The application of other activation energies can be applied to determine the life of different energetic materials and degradation mechanisms. Any selection needs to focus on the worst-case scenario.

It can be seen that the storage conditions in Edinburgh, Sterling, and Myambat are more benign than storage at a constant 25 °C, as opposed to storage in Townsville and Darwin, whose climates result in accelerated ageing. As the MTDS

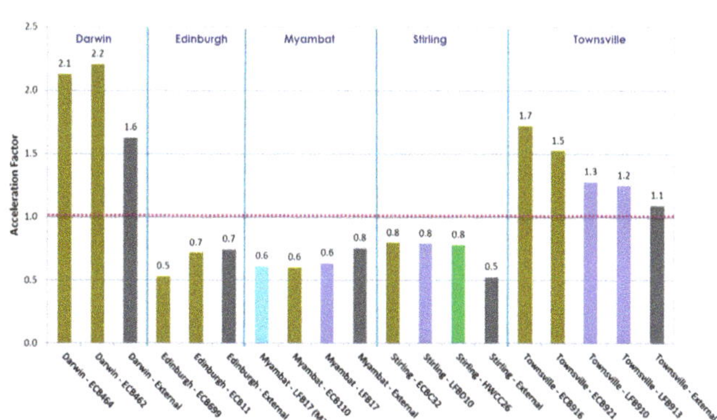

Figure 9.5. Environmental acceleration factors. Data reproduced from reference [9].

for any given munition will unlikely specify a single storage depot the conditions are taken from the worst-case scenario, which in this instance is Darwin. These conditions, and those of other environmental ageing influencers, are factored into the munition design assessment and used to identify failure modes in Phase 2.

9.1.2 Phase 2: life-limiting failure mechanisms

In this phase, all potential life-limiting failure mechanisms are identified given the known or expected components and the level of environmental exposure. The identification of failure modes, and their mechanisms by which they are realized, is often carried out by a subject-matter expert assessment, where the expert has knowledge of munition technologies and ageing.

As energetic materials are used across numerous munition types and by their nature are susceptible to decomposition, energetic subcomponents will be the focus here. Figure 9.6 shows the phase where material ageing is considered, in particular how this can effect a change in material properties and the consequence on the overall military capability [12]. This should be made clear in the figure. 'Insensitive Munition Properties' relates to system-level design characteristics whereas 'Small Scale Hazard Properties' are associated with an energetic material feature.

Some examples of life-limiting failure modes and mechanisms are given below, in a non-exhaustive list, against example munition types of small arms ammunition (SAA), composite rocket motors, and pyrotechnic systems. It should be noted that each system design will have a unique set of probable failure modes in addition to the more generic ones that all system designs will have. The focus below is on those that are common.

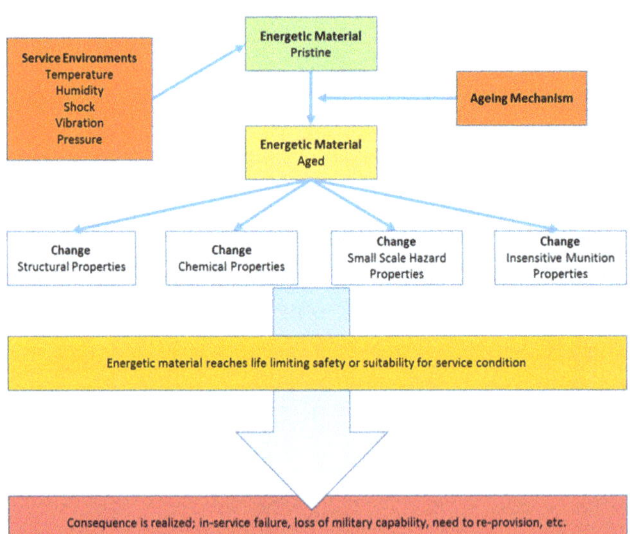

Figure 9.6. Energetic material ageing. Reprinted with permission of QinetiQ Ltd, 2022.

Small arms ammunition

Small arms ammunition typically contains gun propellants that made up from formulations based on the energetic materials nitrocellulose and/or nitroglycerine. An example of propellant extracted out of small arms is provided in figure 9.7.

The ageing of the gun propellants used within SAA can lead to a number of failure modes, some of which are as follows:

- Degradation of nitrate ester ingredients by thermolytic or hydrolytic chemical reactions. These reactions are exothermic, in that they evolve heat, and the lower-temperature dominant reaction of hydrolysis is catalyzed under both acidic and basic conditions. Given the products of nitrate ester degradation are acidic, the process is auto-catalytic and can lead to self-heating above the propellant ignition temperature. There is a clear safety risk of unintended ignition in significantly progressed cases of material degradation. In lesser extremes of ageing mechanisms, the performance of SAA can be reduced and become more variable due to loss of propellant combustion effectiveness.
- Deterioration of nitrate esters by hydrolysis of the nitrocellulose glycosidic bond leads to chain scission and a softer material. The propellant grains can deform if too soft, which alters the burn characteristics and affects the overall performance of the SAA.
- Migration of propellant deterrents, otherwise known as burn-rate reducers, away from the surface of propellant grains and towards the bulk can occur with time as ingredients attempt to reach a state of equilibrium. This allows for a faster propellant burn and higher initial peak pressures, the SAA and weapon effects being reduced performance, increased barrel wear, and overpressurization of gun barrels in the extreme.
- Plasticizer can be lost from the propellant grains due to migration into adjacent materials. Similarly, some plasticizers have a low vapor pressure and if exposed

Figure 9.7. Small arms ammunition. Picture reproduced from reference [12].

to the atmosphere can be lost by evaporation. A reduction in propellant plasticizer content will cause the propellant to become harder and more prone to cracking. Any increase in burn surface area can be the reason for greater performance variation and overpressurization of gun barrels in the extreme.

Composite rocket motors

Composite propellants used within solid rocket motor designs typically consist of filler particles, such as oxidants, and fuel sources for example aluminum, as well as a polymeric binder fuel system, which also incorporates a plasticizer to tailor the mechanical properties of the propellant. Such propellants are made via a cast-cure process to produce a propellant grain structure of the required geometry. An example of a solid rocket motor is provided in figure 9.8.

The ageing of the composite propellants that are used within solid rocket motor designs is complex. Some of the mechanisms are described below:

- De-bonding at the particle–binder interface can occur due to deterioration of the bonding agent, the influence of moisture, and mechanical stress. The microscopic de-bonds increase porosity, which can alter the propellant burn characteristics, that in turn affects performance and increases the risk of overpressurization. A reduction in strength at these interfaces can also affect the mechanical properties of the propellant. This leads to a softer material and risks performance deviations and overpressurization from charge deformation, which affects the burn surface area or leads to nozzle obstruction.
- Oxidation of the polymeric binder within composites produces an increase in cross-linked density and a stiffer material that is more prone to crack formation. In turn, cracks cause an increase in the propellant burn surface area and greater maximum pressures on combustion. This can result in rupture of the pressure vessel, with performance and safety consequences.
- Deterioration of the polymer through hydrolytic chain scission mechanisms results in a softer material and increases the potential of charge deformation and slump. The reduction in mechanical strength risks performance deviations and safety due to overpressurization.

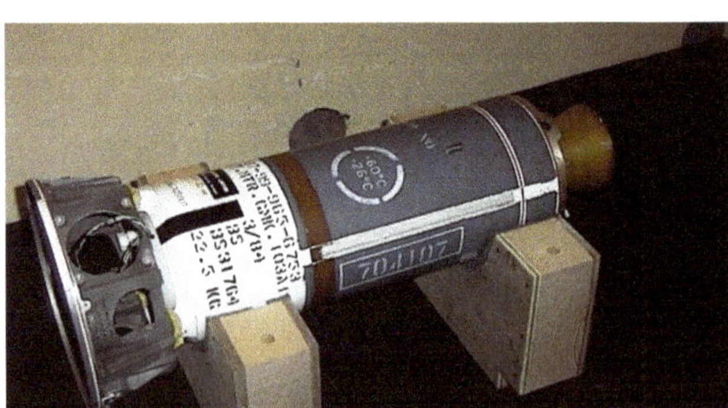

Figure 9.8. Solid rocket motor. Picture reproduced from reference [12].

- Diffusion of plasticizer can affect the mechanical properties of the propellant grain. When present at levels greater than specified this can lead to a softer material that is more prone to deformation. A less than tolerable value enables the propellant to become more brittle and crack. Both mechanisms can cause overpressurization and catastrophic failure, resultant in lost performance and safety consequences.
- Oxidant dissolution and redistribution due to the ingress of moisture can affect the propellant burn due to variation in the oxidant content. A significantly higher presence of oxidant on the surface of the propellant grain may lead to high-pressure spikes on ignition.

Pyrotechnic devices

Pyrotechnic devices typically consist of energetic mixtures that contain a combination of metal fuel and oxidizer. Once ignited these energetic materials are designed to produce heat, light, sound, gas, or smoke through self-sustained exothermic chemical reactions. An example of a pyrotechnic device is provided in figure 9.9.

The ageing of pyrotechnics used within munitions can lead to safety and suitability failure modes. The more common ones are provided below:

- Oxidant materials contained within pyrotechnic formulations are often hydroscopic and will absorb moisture if exposed to the atmosphere. The increase in moisture content can lead to delays in ignition or low heats of combustion, which can cause performance failures.
- Agglomeration of powders can affect the burn surface and lead to a reduction in performance and system failures.
- Size reduction of pyrotechnic pellets or granules on exposure to vibration or shock can lead to a faster burn of powders and a more energetic output than specified. This can cause safety and performance issues, especially if the pyrotechnic is to have an ejection function at a system level.

A primary failure mode is often related to the deterioration of energetic materials and their subsystems. However, the assessment of munitions should not be limited to the energetics alone. Other failure modes that relate to non-energetic components or

Figure 9.9. Pyrotechnic device. Picture reproduced from reference [12].

packaging should also be considered in the design review and assessment of potential failure modes. The failure of printed circuit boards due to tin whisker shorting, the corrosion of steel weakening structures, or the bridgewires in detonators breaking should be given equal consideration.

Once failure modes have been identified there is a need to assess each failure mode to determine its effect and criticality at a system level. This topic is the focus of discussion in Phase 3 of this life-assessment approach.

9.1.3 Phase 3: assessment methods

Whilst there are a number of ways to assess life based on the mode and likelihood of system failure, the Failure Mode Effects and Criticality Analysis (FMECA) method provides one bottom-up approach to the assessment of failure mechanisms. The application of a FMECA method is routinely applied to highlight potential causes of failures in munitions. The FMECA method, in its development, uses a combination of likelihood and consequence to determine the criticality of the failure. One of the criticality methods used throughout the conduct of a FMECA is to assign an overall risk rating or level.

Figure 9.10 is an example of an overall risk rating. This approach combines a likelihood rating with a consequence rating of a particular outcome and allocates each outcome to a specific risk level. In addition to the scale of risk, from Very Low to Very High, the risk level is assigned an alphanumerical to summarize the make-up of likelihood and consequence. To aid this assignment, a series of descriptions are used to define the likelihood that a mishap will escalate to a specific consequence and similarly express the consequence with respect to safety, performance, and the environment. When combined to produce a risk level this view of failure mode criticality aids assessment of remediation efforts and determines the strength of need to evaluate root causes through trials that will be conducted in a Phase 4 activity. Essentially, when practicability and cost are considered, the FMECA is used to focus trials activities on high risk levels where the greatest risk reduction and benefit is offered.

In the context of munitions the consequence is often catastrophic, which places the emphasis on an assessment of the perception of likelihood. A rare example of a lower consequence could be considered when authorized packaging is deemed to have a Hazard Classification Code (HCC) of 1.4S. For reference, the descriptions

Likelihood	Consequence				
	Minor (A)	Moderate (B)	Major (C)	Critical (D)	Catastrophic (E)
Almost Certain (5)	Low (A5)	Medium (B5)	High (C5)	Very High (D5)	Very High (E5)
Probable (4)	Low (A4)	Medium (B4)	High (C4)	High (D4)	Very High (E4)
Occasional (3)	Very Low (A3)	Low (B3)	Medium (C3)	High (D3)	High (E3)
Improbable (2)	Very Low (A2)	Very Low (B2)	Low (C2)	Medium (D2)	Medium (E2)
Rare (1)	Very Low (A1)	Very Low (B1)	Very Low (C1)	Low (D1)	Low (E1)

Figure 9.10. Levels of risk. Reprinted with permission of QinetiQ Ltd, 2022.

behind the UN hazard classification system in terms of Class, Division, and Compatibility Group are provided in figures 9.11 and 9.12 [13]. This means that in a worst-case scenario (unintended initiation) the harmful effects generated by the item should be almost completely contained by the packaging. This, coupled with an assessment that the energetics will have a depreciation in performance as they age, indicates that the item remains compliant with the requirements of HCC 1.4S for an indefinite time period, or for as long as the integrity of the authorized packaging remains. In such an example the criticality of an unintended ignition event over the capability life cycle is never greater than Low (A5). This has been identified by a Minor (A) consequence and an Almost Certain (5) worst-case likelihood. The need to test unintended ignition mechanisms in this case, in order to reduce the likelihood through increased knowledge of the environmental impact, is assessed according to this low criticality of risk.

Further efforts to assess failure modes can utilize previous test data collected from the same design specification, or similar item designs that have been assessed where an element of read-across is deemed appropriate. A trend analysis of ageing mechanisms, or even computational models that have been developed to predict life, can aid in-service life prediction of different designs or inform failure mechanisms in other environments. As confidence in models, or a trend analysis, can be somewhat limited, this information can be used to plan Phase 4 trial-based assessments and to focus in on features of concern as a minimum. The predictive assessments can also be used in support of amelioration efforts and the development of risk mitigation plans, all of which is most useful in lieu of test data that could take some time to conduct and report.

There are times when an assessment of munition life is based only upon the design, the historical data of like energetic compounds exposed to comparable in-service environments, and similarities to other in-service munitions. In some cases this assessment-only approach is all that can be conducted, such as where access to items and/or data is limited or the capability of the acquisition nation is restricted.

Class	Division	Description	Example
1: Explosives	1.1	Substances and articles which have a mass explosion hazard	High explosives
	1.2	Substances and articles which have a fire hazard and either a minor blast hazard or a minor projection hazard or both, but not a mass explosion hazard	Flares
	1.3	Substances and articles which have a projection hazard but not a mass explosion hazard	Display fireworks
	1.4	Substances and articles which present no significant hazard	Ammunition
	1.5	Very insensitive substances which have a mass explosion hazard	Blasting agents
	1.6	Extremely insensitive substances which do not have a mass explosion hazard	Explosive devices

Figure 9.11. UN hazard classification system for Explosives—hazard division. From 'Globally Harmonized System of Classification and Labelling of Chemicals©' (2021) United Nations. Reprinted with the permission of the United Nations [13].

Description of substance or article to be Classified	Compatibility Group	Classification Code
Primary explosive substance	A	1.1A
Article containing a primary explosive substance and not containing two or more effective protective features. Some articles, such as detonators for blasting, detonator assemblies for blasting and primers, cap-type, are included, even though they do not contain primary explosives	B	1.1B 1.2B 1.4B
Propellant explosive substance or other deflagrating explosive substance or article containing such explosive substance	C	1.1C 1.2C 1.3C 1.4C
Secondary detonating explosive substance or black powder or article containing a secondary detonating explosive substance, in each case without means of initiation and without a propelling charge, or article containing a primary explosive substance and containing two or more effective protective features	D	1.1D 1.2D 1.4D 1.5D
Article containing a secondary detonating explosive substance, without means of initiation, with a propelling charge (other than one containing a flammable liquid or gel or hypergolic liquids)	E	1.1E 1.2E 1.4E
Article containing a secondary detonating explosive substance with its own means of initiation, with a propelling charge (other than one containing a flammable liquid or gel or hypergolic liquids) or without a propelling charge	F	1.1F 1.2F 1.3F 1.4F
Pyrotechnic substance, or article containing a pyrotechnic substance, or article containing both an explosive substance and an illuminating, incendiary, tear or smoke-producing substance (other than a water-activated article or one containing white phosphorus, phosphides, a pyrophoric substance, a flammable liquid or gel, or hypergolic liquids)	G	1.1G 1.2G 1.3G 1.4G
Article containing both an explosive substance and white phosphorus	H	1.2H 1.3H
Article containing both an explosive substance and a flammable liquid or gel	J	1.1J 1.2J 1.3J
Article containing both an explosive substance and a toxic chemical agent	K	1.2K 1.3K
Explosive substance or article containing an explosive substance and presenting a special risk (e.g. due to water-activation or presence of hypergolic liquids, phosphides or a pyrophoric substance) and needing isolation of each type	L	1.1L 1.2L 1.3L
Articles containing only extremely insensitive detonating substances	N	1.6N
Substance or article so packed or designed that any hazardous effects arising from accidental functioning are confined within the package unless the package has been degraded by fire, in which case all blast or projection effects are limited to the extent that they do not significantly hinder or prohibit fire-fighting or other emergency response efforts in the immediate vicinity of the package	S	1.4S

Figure 9.12. UN hazard classification system for Explosives—compatibility groups. From 'Globally Harmonized System of Classification and Labelling of Chemicals©' (2021) United Nations. Reprinted with the permission of the United Nations [13].

In these cases, there is a greater emphasis and level of robustness required within a FMECA. Even with a greater focus, there will always be an increased risk without the foundation of technical data to assure S3. The ability to inform technical S3 assurance through the delivery of objective and quantifiable evidence is the focus of a trial or test program, and is the objective of Phase 4 in a typical life-assessment procedure.

9.1.4 Phase 4: trials

As part of an S3 assessment a number of trials are performed in the form of a combination of SETs, the aim of which is to ascertain the safe and suitable life of munitions, to understand their limitations, and inform their use. This takes place through the application of a test program that replicates a worst-case credible threat to the munition of interest. The primary focus in this phase is to obtain objective and quantifiable evidence in support of decision-making. The approach taken by most nations in the conduct of qualification testing is to conform to the principles outlined in AOP-15 [14]. The 'how to test' is addressed within documents produced by a number

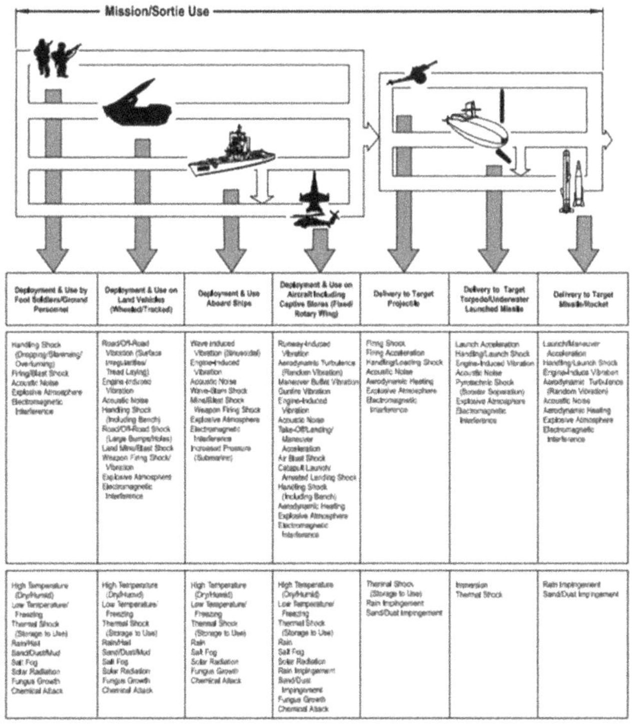

Figure 9.13. Example life cycle histories for munitions. Data reproduced from reference [15].

of other military nations, such as MIL-STD-810 [15] or DEF-STAN-00–35 [16], respectively.

Some examples of life cycle histories associated with different munitions are provided in figure 9.13. The intent is to give an indication of the type of tests that form a SET program for a given MTDS.

The approach to testing differs between nations and often depends on a number of variables, such as the size of the test nation's defense budget, capability of test facilities, system knowledge of munitions obtained, amount of munitions acquired, the intended usage rate, and the availability supplementary qualification test data supplied as part of an overseas acquisition program, to name but a few.

To gain an appreciation of how the approach to S3 trials can differ between nations, an overview based on characteristic S3 trials approach is provided below.

Comprehensive

Often delivered by nations that consist of a large industrial base, with a number of OEMs, and are responsible for the development of munitions and the system integration onto platforms. As a result of domestic manufacture the responsible national authorities tend to request a complete qualification test program in accordance with procedures detailed within comprehensive documents, such as MIL-STD-810 or DEF-STAN-00-35. The SET program is created to represent the full MTDS and is reflective of that nation's munition usage. The qualification data packs that form the output of an SET program can sometimes form part of an overseas munition sale, but not always. The conduct of an SET program and the delivery of results, which constitute objective and quantifiable evidence that are representative of the MTDS, provides the greatest level of technical assurance and risk reduction.

Focused

Typically nations that deliver this type of trials program have some industrial base and test and evaluation (T&E) capabilities, but not those required to fully develop a wide range of munitions to meet capability demand. More often such nations need to accept qualification test data from other nations or OEMs from other nations. In these cases, where the ability to test is permitted, the qualification data set supplied with the acquisition is supplemented with a more limited SET program and is often focused in design to provide technical assurance of S3 around the differences between the MTDS of the OEM nation and that of the acquisition nation. In these instances there is a need to accept and have confidence in test data from other nations to inform safety assessments. This can be in the context of differences in safety culture, which can present procedural and political challenges to rationalize. Where the ability to test is removed due to technical sensitivities, this places greater emphasis on test data acceptance. When tests are contractually limited to, for example, nondestructive techniques only, there is a greater need to obtain more information and technical assurance from these techniques. Hence, these nations can often have more developed, but isolated, capabilities to maximize the benefits of the permitted T&E to assure S3.

Minimal

Nations that deliver these programs usually have little in the way of domestic manufacture or T&E capability. They rely solely on the test data provided by other nations to inform safety cases. For such nations there is an inherent need to acknowledge gaps between test nation MTDSs and their own, to accept a greater level of risk within their munition life assessments and the use of the acquired capability.

In addition to capability, capacity, and information release constraints, the need for nations to perform trials on acquired items is dependent on the number of items that the nation will receive. If this number is insufficient to produce statistically viable results then the benefit of a test program is reduced due to the inability to provide confidence in any evidence gathered. Similarly, the acquisition could be in support of a short-duration life cycle, which will not necessitate a test program as the consumption rate is high or enable the timescales to deliver. In such cases there is a reliance on technical engineering assessments, the basis of which is a risk-assessed FMECA approach.

The Phase 4 trials activity deals solely with testing to inform and assure the decision to introduce into service new munitions to a given country. The assessment of life within Phase 5 of a program then needs to utilize all evidence accumulated in this process to make an assessment of the life span for which the munition remains S3.

9.1.5 Phase 5: life assessment

A Phase 5 life assessment is the assignment of initial service life, given the whole body of evidence provided by the test program delivered in Phase 4. This allocation is in the context of an acceptable level of assurance and the appetite of nations to tolerate residual risk. Accordingly, the approach to how munitions are life assessed differs between national authorities. Some nations apply a philosophy in the assignment of an indefinite initial life that is complimented by mandatory in-service surveillance periods of 3 to 5 years, at which time the trends are reviewed and life adjusted. At this time, if no trend has been assessed to demonstrate a reduction in reliability, the indefinite life is maintained until such a time as is demonstrated otherwise. Other nations apply a policy in the assignment of a more conservative initial life that is incrementally extended at each surveillance period based on expert advice.

The estimation of initial service life and the in-service management of munitions will tend to be determined on a design-based life-assessment approach, as detailed in previous phases of this process. However, actual in-service management and service life extensions in many cases will dictate that a munition-lot-based assessment is required. This requirement is driven from a number of perspectives, including, but not limited, to phased acquisition buys, the establishment of life from the date of manufacture, environmental exposure due to munition storage of lots at different in-country locations, and munition lot variation due to in-specification manufacture that introduces specific sets of properties not matching the general population and their influence on failure mode development. Variations between munition lots increases the need to life assess batches of the same design separately, which increases the workload within a whole life-assessment program. Ideally munition designs are tested to evaluate properties at the extremes of the specification tolerances to determine the impact prior to introduction into service. However, this is often not practicable across all variables and so munition-lot-based properties are typically detected upon in-service surveillance testing.

The assessment of munition life does not end at introduction into service. The predictions made through worst-case T&E approaches need to be validated, and this is often achieved through an in-service surveillance activity. The periodicity of in-service surveillance testing can be determined by how the age-related changes materialize. This is dependent upon the outcome of the failure mode analysis conducted for each munition and any identified predisposition to fail at a particular point in its capability life cycle. The frequency of surveillance model shown in figure 9.14 illustrates key factors in the determination of realistic intervals. In the absence of any formal assignment of periodicity this model recommends points, as depicted on the blue time line, at which safety and reliability activities are conducted based on a percentage of the manufacturer's design life at 20, 55, and 85 percent points. The first percentile is reflective of the need to gather data early on to establish a future test baseline, validate lot acceptance test (LAT) results or provide evidence in support of munition

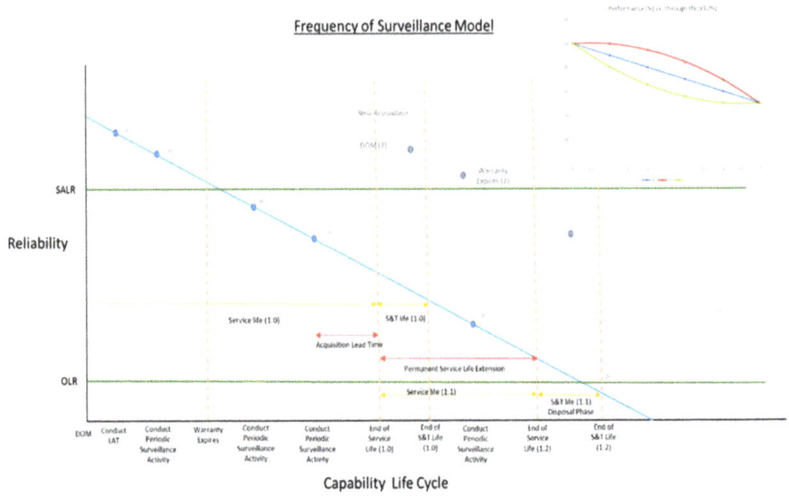

Figure 9.14. Frequency of surveillance. Reproduced from reference [17].

warranties. At this stage the test program is focused on conformance above the service acceptance limit of reliability (SALR), which is typically based on manufacture acceptance criteria. The other 55 and 85 percentiles are placed to establish at least three evenly distributed points on the trend line and enable data to be collected before end of life is reached. The end of service life is the point in figure 9.14 where the operational limit of reliability (OLR) is met and is that minimum level of reliability accepted in service. The principle of these approximate in-service surveillance periods can be adopted in lieu of through-life knowledge on failure modes and material trends. However, if it is known or suspected through testing that there is a more significant reduction in reliability later on in life, the periodicity of the in-service surveillance test point can be placed later on in the life cycle to offer more statistical significance to these changes. Such an approach to in-service surveillance is demonstrated in figure 9.14, whereby the periodicity is refined to pick up on the times throughout the capability life cycle that reliability is most likely affected [17].

The supply of munitions by other nations or an overseas OEM can often be accompanied by a stockpile reliability program and is delivered through the release of periodic ordnance assessments. If assessments are not complimented by testing in the acquisition country this approach places a reliance on the test data produced by the overseas nation or OEM. As the delivery of ordnance assessment data is used to underpin S3 decisions, timely delivery is essential. The level of dependency on this information and delivery based on external timescales can present further issues in the context of different life-assessment philosophies, such as indefinite versus defined life assessment, for example if the OEM nation does not publish a specific munition life within such limited ordnance assessment reports and it is necessary for the acquisition nation to establish a definitive S3 life. In order to do this the scientific understanding of munition ageing is used to inform risk assessments (in accordance with AOP-46 [7]) and there is a greater need for expert judgment in this case.

To gain an appreciation of how the approach to the conduct of in-service surveillance trials can differ, an overview based on surveillance styles is provided below.

Frequent

Frequent and often in-service surveillance programs are often delivered by nations with larger defense budgets and a comprehensive T&E capability. The characteristics of many age-related trends within different design components are tracked, such as energetic materials, subsystem structures, and overall performance. Analysis of data is often refined with statistical trend assessments and predictive models. Information generated can be supplied to other nations as part of the acquisition package. However, these ordnance assessment reports may be provided with limited technical information or statements in terms of munition life assessment. There are no raw data for other nations to interrogate to gain increased confidence in the assessment.

Focused

A more moderate or focused surveillance program is typically conducted by nations with medium-sized defense budgets and T&E capability. A subset of tests that are focused on primary failure modes of concern, or specific munitions with more accessible materials and components to enable tests, is often the focus. There is an emphasis on nondestructive techniques and/or performance for more complex munitions, especially if destructive evaluation is capability or contractually constrained, as with some overseas acquisitions.

Limited

For nations that have smaller defense budgets and restricted T&E capability there is a necessity to fully accept the life-assessment policy of the OEM nation, with no test. Any assessments would typically be unit report based or noncomplex performance evaluation.

If executed effectively, in-service surveillance programs provide ongoing technical assurance of munitions throughout the capability life cycle. As is shown in figure 9.14 this often results in a portfolio of objective and quantifiable evidence that can support life extension decisions. The incremental life extension of munitions can often be accomplished through extrapolation of trend lines with statistical significance and confidence levels. A simple example is given in figure 9.15 with an assessment of stabilizer content. The blue trend line is a line of best fit for the data points, the solid orange and green lines are the 95% confidence intervals, and the dashed lines 95% prediction intervals. In the case of nitrate-ester-based propellants the value of a 0.2% effective stabilizer is taken as a safe storage limit in accordance with AOP-48 [11] and can be used as one metric to gauge the operational limit of reliability.

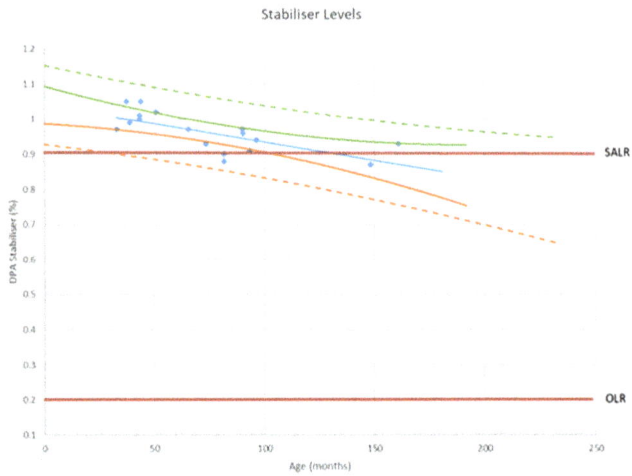

Figure 9.15. Stabilizer assessment. Reprinted with permission of QinetiQ Ltd, 2022.

The prediction of service life in support of life extension exercises relies upon the continued projection of confidence to the remainder of the stockpile through the accurate selection of high-life fleet leaders and their incorporation within a surveillance test program. Fleet leaders are those items that represent munitions that have been exposed to the highest levels of life cycle environments. As a result these fleet leader munitions should display the greatest change in energetic material properties, for example, and potentially highlight failure modes of concern. The selection of fleet leaders is often subjective and dependent on a subject-matter expert evaluation of the exposure levels and consequence to the munition. This is particularly the case where groups of munitions have been exposed to different parts of the MTDS, such as an increase in road transportation distances versus time aboard ship. In these cases an evaluation of which environment is likely to have the largest impact is required. The environmental monitoring and tracking of limits defined within the MTDS, and tested through trials, is key information that can be used to support the selection of fleet leaders.

The approach to service life extension programs is another aspect that differs between nations. Large military nations, for example, have access to comprehensive introduction into service and in-service surveillance test data to incrementally extend life from a trend assessment of failure modes. These nations also have detailed platform knowledge from the integration program and of usage. There may also be an effective munition environmental monitoring program, which has tracked the exposure of the munitions through life and gives added confidence in the selection of fleet leaders.

The last example of life assessment provided here is when there is a need to expand the operational or in-service environment. In such cases an incremental life extension is not feasible due to the more complex nature of environmental exposure and the effect on munitions. In these examples the selection of fleet leaders is once again utilized, and nations would conventionally follow the guidance provided in

AECTP-600 for methodology on how to evaluate these munitions to meet extended life requirements [6]. In scenarios where there is an expansion of the service environment, munitions are selected for participation within the limited SET program that is focused on understanding the life cycle environment that was not anticipated at introduction into service. These programs can present a unique challenge for T&E capabilities as the test conditions can be bespoke to new and novel environments or emerging threats. The capability within T&E establishments is therefore imperative to deliver against nonroutine requirements.

9.2 Conclusions

A successful objective and quantifiable, evidence-based munition life-assessment program delivers a wide range of benefits. At the fore of these is an increased knowledge of safety margins and the ability to inform safe service use, which is gained through an enhanced system knowledge of munition ageing. Through the conduct of trials a more assured capability is delivered with greater confidence that the munition will perform within the given service environment. Any extension to munition life reduces the need for sustainment buys that are costly and can be time constrained. In itself the ability to extend life improves flexibility within equipment plans and helps to mitigate against future capability gaps. The reduction in disposal activities, with munitions staying in service for longer, limits their environmental impact through the prevention of premature disposal and the need to acquire more stock. Lastly, the conduct of supportive T&E activities exercises key enablers to develop domestic defense capability. These prime benefits are all reasons why munition life-management programs have been and will continue to be a requirement for defense forces globally.

The strength in applying a traditional life-assessment approach lies in how programs are then devised to systematically run through the process. This systematic approach provides assurance that all important considerations are captured and increases the likelihood that the end outcome will deliver the benefits articulated above.

There are, however, significant challenges faced by nations in delivering the required robust program needed to realize these benefits. The common challenges are often lengthy delivery times, high costs in facility operation and equipment expenditure, and limited national resources and experience. All the aforementioned needs to be recognized and rationalized to determine the end value of a life-assessment program. An example where benefits are eroded in the traditional system would be a munition that has a short in-service life requirement against a backdrop of lengthy delivery times within the program. In this example, within the context of traditional approaches detailed here, a focused program style is often adopted. This would entail upfront environmental test requirements being stripped back, a short lease of life assigned to the munition, and the assignment being heavily supplemented by in-service surveillance evaluations. These compromises drive the need to make efficiencies in this space, through optimization of processes without adding cost, to deliver all the benefits in a shorter period of time.

A traditional approach to munition life management has been summarized in this chapter, with the processes detailed here having been applied for many decades by a number of nations. Under an increased demand for efficiencies there is a need to optimize each area of a life-management process and undertake an even smarter approach. This look to the future with more modern technologies and methodologies applied to the life-management of munitions will be the focus of chapter 10.

Acknowledgments

The author would like to acknowledge colleagues in QinetiQ, both in Australia and UK, who have supported the development of this chapter; specifically, Mr Craig Fallshaw (QinetiQ AUS), Mr Stewart Dawson (QinetiQ AUS), Dr David Tod (QinetiQ UK), and Mr Michael Firth (QinetiQ UK). A special thanks is also required to the Australian Department of Defence; in particular, Mr Bernard Smith-Roberts (Explosive Materiel Branch), Mr David Seccombe (Thales AUS), and SQNLDR Timothy White (Explosive Materiel Branch) in enabling the use of images and referenced material.

References

[1] NATO CNAD 2009 AC/326, SG/B, AOP-15 Edition 3 – Guidance on the assessment of the safety and suitability for service of non-nuclear munitions for NATO armed forces, NATO Standardization Agency (NSA), Brussels, Belgium, April

[2] NATO CNAD 2017 AC/326 CASG, SG/B, STANAG 4675 – Edition 1 – In-service surveillance (ISS) of munitions, NATO Standardization Office (NSO), Brussels, Belgium

[3] NATO CNAD 2017 AC/326 CASG, SG/B, AOP-62 – Edition A Version 1 – In-service surveillance of munitions general guidance, NATO Standardization Office (NSO), Brussels, Belgium, 2017

[4] NATO CNAD 2017 AC/326 CASG, SG/B, AOP-63 – Edition A Version 1 – In-service surveillance of munitions sampling and test procedures, NATO Standardization Office (NSO), Brussels, Belgium, 2017

[5] NATO CNAD 2017 AC/326 CASG, SG/A, AOP-64 – Edition A Version 1 E – In-service surveillance of munitions condition monitoring of energetic materials, NATO Standardization Office (NSO), Brussels, Belgium, February

[6] NATO, CNAD 2007 AC/327, AECTP-600 – Edition 2 – The ten step method for evaluating the ability of materiel to meet extended life requirements, NATO Standardization Agency (NSA), Brussels, Belgium, April

[7] NATO CNAD 2006 AC/326, SG/B, AOP-46 – Edition 1 – The scientific basis for whole life assessment of munitions, NATO Standardization Agency (NSA), Brussels, Belgium

[8] Defence Standard 00-35(PART 4) 1999 Issue 3, Environmental Handbook for Defence Materiel Part 4: Natural Environments, Ministry of Defence, 7 May

[9] SQNLDR T 2019 White and D. Seccombe, Prognostic EO Condition Monitoring, Parari

[10] NATO CNAD 2007 AC/326, SG/A, STANAG 4582 JAS (Edition 1) – Explosives, Nitrocellulose-based Propellants, Stability Test Procedure and Requirements Using Heat Flow Calorimetry, NATO Standardization Agency (NSA), Brussels, Belgium, March

[11] NATO CNAD 2008 AC/326, SG/A, AOP-48 – Edition 2 – Explosives, Nitrocellulose Based Propellants, Stability Test Procedures and Requirements Using Stabilizer Depletion, NATO Standardization Agency (NSA), Brussels, Belgium, October
[12] Ashcroft M and Dodds J 2017 QINETIQ/17/04098, Explosive Ordnance Surveillance, Parari
[13] United Nations 2021 ST/SG/AC.10/30/Rev.9, Globally Harmonized System of Classification and Labelling of Chemicals
[14] AOP-15 Edition 3 2009 Guidance on the Assessment of the Safety and Suitability for Service of Non-Nuclear Munitions for NATO Armed Forces, April
[15] MIL-STD-810G 2008 Environmental Engineering Considerations and Laboratory Tests, United States Department of Defense Test Method Standard, 31 October
[16] Defence Standard 00-35 Issue 3, Parts 1 to 4, Environmental Handbook for Defence Materiel, Ministry of Defence, 07 May 1999
[17] Australian Government – DoD – Capability Acquisition and Sustainment Group, Explosive Ordnance Safety and Reliability Program Plan – Version 1.0, (ENG-EMB) 140, September 2019

IOP Publishing

Energetics Science and Technology
An integrated approach
Adam S Cumming

Chapter 10

Improved systematic life management of munitions

Ruth Pettifer, Claire Hobman, Jenna Hulme, Ian Tinsley and Dave Tod

One of the major uses of energetic materials is in munitions and weapon systems. For energetic materials to be successfully used for such purposes, there must be effective integration of these in the system and a sufficient understanding of their properties across the life cycle. This is demonstrated in chapter 9, which discusses the traditional approach to life management from a worldwide perspective.

However, there remains areas for improvement with regards to the optimisation of the through-life management for weapons and munitions. There have been wide-ranging developments in this field in recent times, including spiral development, which is a concept that originates from software engineering. The importance of the better availability, and therefore application, of data through the future use of digital threads is also discussed in this chapter.

Systems engineering principles have been integral to the whole-life management of weapon systems for decades. As a basic premise, time and money should be spent early on in a project, to really understand the problem, in order to avoid costly errors later. In addition, appropriate life-cycle management approaches are adopted.

With regards to a systems engineering methodology for the safety and suitability for service aspects of life management, this means that understanding the failure modes of the system will continue to be an important approach. It will also be imperative to understand the specific risks and hazards facing a system, which are inextricably linked to their experienced environments, to enable focused assessment as part of through-life management in those areas.

In this chapter, the types of testing currently used in the through-life management process are described, with a focus on where there have been notable developments in recent years. Areas where there is a clear need for more representative analysis, such as in the mechanical testing of solid rocket motor (SRM) propellants, are discussed. The importance of accurate and reliable data, along with the potential for

doi:10.1088/978-0-7503-3943-8ch10

an increase in the use of modelling are also highlighted to demonstrate the value of an integrated approach to life management.

10.1 Introduction

There is an increasing drive to introduce weapon systems into service in a more rapid manner and to reduce the cost of ownership through life. However, as discussed in chapter 9, the generation of sufficient evidence to ensure the safety and suitability for service (S3) of munitions and weapon systems remains a fundamental and critical part of stockpile ownership.

Clearly, the newer processing techniques for energetic materials, such as resonant acoustic mixing, flow chemistry, and additive manufacturing (see chapter 5) will likely have a disruptive effect on the through-life management process. With the move from war-era 'bucket' manufacture of energetic materials to more modern 'precision' energetics [1], there is an obvious need for the tools and techniques for the analysis of energetic materials to keep pace. In addition, artificial intelligence and machine learning (see chapter 4) will see increasing prominence in order to make best use of data across the through-life management process.

In recent decades, there has been a shift towards a smarter approach to the through-life management of munitions, with a need to adopt a more scientific basis to aspects of the process [2–7]. Smart testing is defined as the assessment of intelligent data from tools and techniques which help to achieve a desired objective [3, 5].

This can be described as the use of a robust systems engineering approach to risk mitigation through the application of appropriate assessment tools. It tailors the process to the specifications of the munitions being tested to ensure the data obtained are valuable and well utilised. Effective and efficient methods are chosen, which could mean replacing physical testing with a wider use of predictive modelling. A smart approach is most successful when a thorough knowledge of material specifications, ageing behaviours, and an environmental history of a munition is available [2, 8].

Smart through-life management should enable more accurate predictions of a munition's safe life, therefore allowing stockpiles to be kept in service for longer and reducing the need for disposal. The ultimate aim of this approach is to be more effective, efficient, tailored, and better validated than previous methods. In some cases, such as with respect to certain laboratory methods of analysis, improvements are starting to be realised. However, with others, for example the development of more representative multi-axis vibration testing, the research is still ongoing.

In this chapter, the tools and techniques which are being used to enable smart through-life management are discussed, along with any challenges that may be present. These include using systems engineering to identify failure modes, risks and hazards, and where spiral development may play an increasing role in the future.

Nations which are deploying a smarter approach to through-life management are described along with the areas in which laboratory-based testing of energetic materials have seen improvements in recent years and those aspects which may see further evolution. In many cases, these methodologies have been developed and used in other industries but their application to energetic materials is not necessarily widespread or the nature of such compositions presents difficulties from an analytical perspective. Identifying and understanding such issues is a necessity if advances in these areas are to be realised.

Finally, the two broad approaches to life-assessment modelling of energetic materials are detailed. The potential for the use of digital threads and twins for ultimate integration purposes is also highlighted.

10.2 Systems engineering and its approach to weapons development

10.2.1 Introduction

One of the cornerstones of enabling smarter through-life management is a systems-engineering-type approach to risk mitigation. Systems engineering has many definitions depending upon the industry and context within which it is used. One of the most concise comes from NASA, who define the discipline as 'a methodical approach for the design, realisation, technical management, operations and retirement of a system' [9].

Systems engineering works across the multidisciplinary arenas of requirements setting, design, solution development, testing, interoperability, technical management, operation and retirement of complex systems that function together to meet a desired outcome. It can include all hardware, software, equipment, facilities, personnel, processes, procedures, and architectures to produce a useful system.

When it comes to complex weapons they are, by their very nature, systems within themselves: they integrate numerous mechanical, electrical, energetic, and control subsystems to produce a multifunctioning product, designed to deliver effect through the prosecution of selected targets. Moreover, the weapon itself may form part of a larger parent system, such as a launch aircraft, ship, or land vehicle, which may orchestrate and coordinate functions such as target acquisition and tracking, launch and manoeuvre. Further still, multiple launch platforms may now be fully integrated through the use of tactical data and satellite links. When considered in this context, the concept of a 'weapon system' is far reaching, with individual weapons themselves forming subsystems within larger systems.

From the foregoing, systems engineering may be viewed as a necessary component in the development and sustainment of complex weapons.

10.2.2 Systems engineering and the life management of munitions

Complex guided weapon systems come in many guises dependent upon their intended target set, means of carriage, and release, i.e. whether they are air, land vehicle, ship, submarine, or personnel launched. In a systems engineering context, the first principle is that of requirements setting. Armed forces have to bridge capability gaps to meet emergent threats in changing and complex operational theatres. Capability developments are far and wide reaching, from anti-tank weapons to defeat ever-developing armour, complex over-the-horizon capabilities to engage and prosecute stealth-equipped enemies, or the ability to protect from increasingly ubiquitous unmanned air vehicles (UAVs).

The low detectability signatures and the relatively low speeds associated with UAVs, in particular, pose significant challenges to weapon designers, where traditional high-velocity, high-lateral-acceleration (Latax) missiles would otherwise be too fast to manoeuvre within the required engagement envelope [10]. In all cases, any weapon system, and therefore the energetic materials within them, must meet the requirements of being safe and reliable as well as performing to an acceptable standard.

The through-life management of weapon systems begins with requirements setting and the resultant solution development. Even at this early stage, selection of the correct materials and components is of paramount importance; therefore, consideration of their properties and reliability, such as mean-time-between-failure (MTBF), is needed. In addition, the materials play a large part in defining potential performance envelopes, and this sometimes dominates discussions over other trade-offs.

As discussed in chapter 9, the likely failure modes play an important part in informing the designer's ability to predict component, subsystem, and system life, through the application of well-proven engineering laws. Some organisations employ a failure modes and effects analysis approach in order to assign the risk priority number of that issue and for ranking failure modes as well [11]. However, whilst mechanical and electronic assemblies are well understood in terms of their tolerance and reaction to environmental stimuli such as vibration, shock, or heat, energetic materials present greater difficulties.

Life prediction of energetic materials is more of a significant challenge due to their complicated behaviour and relative intolerance to being operated outside of strict limitations. The critical energetic components on a complex weapon comprise the propulsion subsystem (ignoring air-breathing propulsion via gas turbines normally found on long-range weapons), the fuzing subsystem, and the lethal package or warhead. Aside from these more obvious candidates, most weapon systems also contain smaller pyro-devices such as gas generators for control actuation, initiators for thermal batteries, or safety-release mechanisms such as fin-unlock devices. In all instances, these critical components must be carefully managed throughout the weapon's manufacture to target or disposal sequence (MTDS) in order to maintain safety, reliability, and performance.

10.2.3 Systems engineering, failure modes, and risk management

As discussed in more detail in chapter 9, post-delivery, the MTDS may be divided into discrete phases including storage, handling and transportation; loading; carriage, release, and, if applicable, jettison; terminal effect; disposal. Each phase attracts hazards posed by the environment; for example, excessive temperatures or humidity in storage may cause significant chemical changes sufficient to alter the properties of energetic materials used in solid fuel rockets, warheads, or pyro-devices.

Low-frequency vibration or shock during logistic transport can cause damage to external and internal components, whilst the carriage environment presents its own uniquely hazardous stimuli such as gunfire shock from the launch platform or high-frequency vibration from coexisting systems such as the host platform's engines, pumps, or generators. Moreover, for air-launched guided weapons in particular, the presence of other stores on adjacent launch pylons may give rise to undesirable aerodynamic effects such as shock waves, buffet or air-spill from engine intakes, all of which have an adverse effect on the overall 'health' of complex weapon systems. These stimuli are challenging enough when considered in isolation, but when acting in concert present an increased risk due to concomitant and cumulative effects.

In order to explain how systems engineering works toward managing through-life risk, safety, reliability, and performance, the following fictional weapon system (figure 10.1) is considered.

10.2.3.1 System overview

A homing infra-red (IR) guided missile, carried on a fourth- or fifth-generation combat aircraft. The missile is designed to engage hostile aircraft and medium-sized UAVs. A description of the main subsystems is given in the following subparagraphs and represented in figure 10.1.

10.2.3.2 Guidance and control subsystem

A staring array IR seeker to track, engage, and prosecute hostile aircraft or medium-to-large UAVs, using lock-on-after-launch. It employs proportional navigation to maintain a direct line of sight between the seeker and the target as the fly-out range closes, using its onboard autopilot and guidance algorithms. Control is achieved through tail fins in a symmetrical cruciform arrangement to minimise cross-coupling between the missile's x, y, and z axes and freedom to achieve Latax manoeuvres. Most electronic components are tested at subsystem level using, for example, uniaxial vibration tests, to comply with MIL-STD 810 [12], before being tested again as part of the all up round (AUR). The control fins are powered by electrical actuators, although they are unlocked using pyrotechnic release mechanisms once the missile leaves its post-launch safe escape field.

10.2.3.3 Lethal package

The warhead is a typical 6 kg blast-fragmentation unit consisting of 2 g tungsten cubes, packed around an RDX explosive fill. Detonation in the vicinity of the target

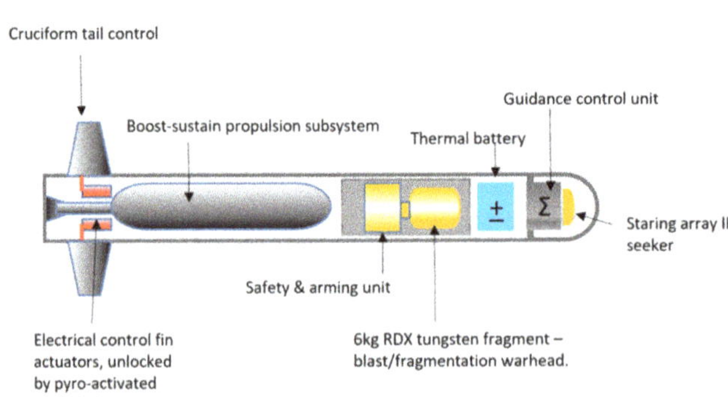

Figure 10.1. Generic air-to-air missile system schematic.

generates fragments with initial velocities of approximately 2000 m s^{-1} in order to couple energy into the target [13].

10.2.3.4 Fuzing subsystem
Typical of other air-to-air missiles, this example uses an optical laser fuze.

10.2.3.5 Propulsion subsystem
The missile is propelled by a solid fuel rocket motor, filled with composite modified double-base (CMDB) propellant. The propellant grain is sized to achieve a boost-sustain thrust profile. A known drawback with CMDB propellants is their propensity for cracking at low temperatures, though the addition of elastomeric polymers to the colloidal matrix gives some low-temperature strain capabilities and assists in overcoming this problem [14]. It should also be noted that such a composition would not be considered as high performance. In terms of this, the most optimum SRM performance is still achieved through the use of ammonium perchlorate (AP) and hydroxy-terminated polybutadiene with aluminium. However, there remains a desire to move away from AP-based systems for environmental and healthcare reasons.

The rocket would have been tested at subsystem and system level in the range −50 °C to +70 °C via sequential environmental tests (SETs). The SET methodology employs preconditioning of the AURs by exposing them to the hot and cold limits of this range, before subjecting them to numerous tests including vibration and shock, which are representative of the transport and operational environments, moisture and dust ingress. After these tests, the AUR is broken into its subsystem parts and analysis is conducted on each component, including energetics, using breakdown, test and critical analysis (BTCA), procedural guidance for which is found in Def Stan 00-035 [15].

10.2.3.6 Electrical power subsystem
The electrical subsystem is powered by a thermal battery, activated by a small pyrotechnic squib which is fired from the host aircraft launch pulse to melt a lithium bromide eutectic electrolyte. The thermal battery has been tested at subsystem level and has an inherently long storage life and good reliability.

A knowledge of the key subsystems greatly assists the systems engineer in identifying the most vulnerable components within the missile. This allows an assessment of potential damage mechanisms that could lead to particular failure modes and, ultimately, present risk to the user or third parties over the course of the weapon's life cycle.

By way of illustration, the structural elements of missiles in general are well understood, easily modelled for the expected load cases, and are subject to inspection and maintenance throughout the MTDS. Electronic components are, for the most part, reliable and MTBFs may be predicted with some accuracy, based upon design, comprehensive and established evaluation such as vibration testing across all axes and in-service experience of components used in similar environments and systems.

By far the most difficult subsystems for which to assess risk are those containing energetics. For example, solid fuel rocket motors, fuzes, and warheads are intolerant of being stored outside of temperature and humidity limits set through manufacturer's design certifications or advice received from national authorities.

Overexposure to heat, cold, or cycles between both can result in changes to the propellant or explosive properties caused by migration, chain scission and cross-linking, to name but a few [13, 14]. During normal storage and use, damage such as cracking or dusting of energetic materials is almost impossible to detect without removing AURs from service and exposing them to costly and time-consuming radiography intervention.

Any damage sustained can almost certainly be exacerbated by subsequent mechanical insult when the weapon is loaded to, and carried on, its host launch platform. For example, a rocket propellant which has hardened due to improper storage can then be susceptible to fracture when carried for long periods; kinetic heating in flight, combined with the weapon's response to vibration, aerodynamic buffet, or gunfire can all contribute to causing damage.

In terms of consequence, a propellant damaged through cracking or case de-bond may burn at a rate which exceeds that of its design due to the now locally increased propellant surface area, resulting in rocket vessel overpressure and potential case breach or burst. Moreover, a damaged rocket propellant grain may also give rise to nozzle damage and asymmetric/irregular thrust due to the potential for irregular gas flows and nonuniform shock waves being generated through the nozzle.

Warheads are also susceptible to damage, with some modes such as cracking resulting in out-of-specification performance or others including the explosive increasing the potential for friction or 'nipping' if allowed to migrate into mating surfaces such as screw threads, etc. Furthermore, and despite the inherent and obvious safety risk to the user, damaged warheads can also present a clear danger to third parties, increasing the risk of collateral damage, or in the case of Non-functioning, contributing to an unexploded ordnance threat.

Smaller pyrotechnic subsystems can also be susceptible to damage from environmental stimuli and vibration. For example, this presents a risk of fins either unlocking too close to the launch platform, presenting an aircraft self-damage risk, or failing to unlock altogether, resulting in a missile which is unable to guide through a loss of directional control.

Despite their relative size in comparison to rocket motors and warheads, pyrotechnic devices are nevertheless vulnerable and critical objects [14]. On modern in-service weapons these small components are sealed within the AUR and their replacement is currently only conducted at manufacturers' specialist facilities during mid-life refurbishment programmes. Nevertheless, line replacement of these relatively simple and inexpensive items could be considered during future designs.

Table 10.1, though not exhaustive, shows some of the common or more likely damage mechanisms and the typical risks they pose to guided weapon systems in general throughout their lives and over the MTDS. Many components are constructed from materials which display highly predictable behaviours. When stressed, metals or composite materials used to construct the missile body tube,

Table 10.1. Example manufacture to target or disposal (MTDS) failure-mode matrix. MTDS summary risks for a generic air-to-air missile system.

Subsystem	Propulsion
Critical components	Propellant grain, rocket initiator, rocket casing, nozzle.
Storage	Early life storage unlikely to cause energetic degradation provided compliant conditions are maintained and data loggers are used to verify conditions. Longer-term ageing can cause thermochemical damage such as chain scission, cross-linking, and plasticiser migration, which can lead to increased risk of mechanical failure, e.g. bulk cracking, surface cracking, and/or dusting.
Handling, transportation, and loading	Logistic handling exposes energetic components to risk of mechanical damage. Where energetics have been subjected to environmental exceedances such as excess temperature or humidity, vibration or shock accumulate and contribute to (or exacerbate) cracking, dusting, or case de-bond. Propulsion subsystem can be vulnerable to moisture/sand/dust ingress during logistic handling.
Carriage	Operational carriage on launch platform induces vibration. Launch platform systems (e.g. aircraft engines) can induce vibration. Gunfire shock and vibration increases risk of damage to energetic components. Kinetic heating may further contribute to thermal cycling and damage. Aerodynamic shock may cause mechanical insult to energetics.
Release-to-target	Cracked, dusted, or voided propellant susceptible to inherently high rocket vessel pressures (typically 30 MPa). High risk of damaged propellant leading to case rupture or burst. Asymmetric propellant burn can disrupt gas flow through nozzle, leading to asymmetric thrust/departure from envelope. Enemy fire risk.
Qualitative risk of subsystem failure	*Early Life Cycle* Likelihood (L) = Medium, Severity (S) = Catastrophic *Later Life Cycle* Likelihood (L) = High, Severity (S) = Catastrophic
Comments	Risks associated with propulsion subsystems increase over time due to ageing and/or mistreatment. In-service surveillance (ISS) programmes are inherently costly and lengthy. Early life-cycle mitigation may be provided through use of well-understood energetics of known provenance, supported by data logger information, diligent storage management, and good husbandry. Enemy fire poses a risk from fragment or bullet attack and fire. Mitigated to some extent by insensitive munitions (IMs) testing and compliance.

Subsystem	**Lethal package and fuzing**
Critical components	Safety and arming device, fuze, warhead.
Storage	Early life storage in compliant and data-logger-monitored environment unlikely to cause significant degradation of explosive, but again ageing becomes key issue over time. Thermochemical and mechanical changes present significant risk.
Handling, transportation, and loading	Logistic handling exposes energetic components to 'mechanical insult'. Where energetics have been subjected to environmental exceedances such as excess temperature or humidity, vibration or shock accumulate and contribute to cracking, dusting, or charge liner separation.
Carriage	Warhead energetics are subjected to vibration and shock during carriage. Integrity of components is critical to safety, reliability, and performance.
Release-to-target	Risk of unintended detonation on launch. De-bonded charge liner may result in loss of warhead performance or failure. Fuze safety, reliability, and performance are affected. Enemy fire risk.
Qualitative risk of subsystem failure	*Early Life Cycle* Likelihood (L) = Medium, Severity (S) = Catastrophic *Later Life Cycle* Likelihood (L) = High, Severity (S) = Catastrophic
Comments	Assurance historically obtained via costly/lengthy ISS. Mitigation may be provided via use of single-source explosives with known provenance. Through-life condition monitoring is challenging. Enemy fire poses a risk from fragment or bullet attack and fire. Mitigated by IM compliance.

Subsystem	Guidance and Control
Critical components	Autopilots, tail control servos, fin unlocking squibs, seeker head.
Storage	Generally there is a low risk to embedded guidance and control components during storage due to their sealed nature. Pyrotechnic fin actuator locks remain of concern and are dependent upon compliant storage.
Handling, transportation, and loading	Subsystem components are exposed to shock/vibration during logistic transportation; may be mitigated by damping and shock-absorbing systems. Weapon system is vulnerable to moisture and sand/dust ingress.

(Continued)

Table 10.1. (*Continued*)

Subsystem	Guidance and Control
Carriage	Subsystem components exposed to wider band vibration during carriage. Vulnerable to moisture and sand/dust ingress.
Release-to-target	Undetected pre-existing guidance failure to launch or guide. Environmental damage to energetics may result in early, late, or no fin unlock, which may increase risk of aircraft self-damage or collateral effects.
Qualitative risk of subsystem failure	*Early Life Cycle* Likelihood (L) = Low, Severity (S) = Catastrophic *Later Life Cycle* Likelihood (L) = Low, Severity (S) = Catastrophic
Comments	Guidance and control systems subject to rigorous testing in all three axes. Mitigation via software-in-the-loop and hardware-in-the-loop testing during manufacture or at change points. Subsystems designed to undergo built-in test on pre-loading, loading, and pre-use. Guidance and control system may be subject to 'go/no-go' pass fail.

Subsystem	Missile structure
Critical components	Body tube section, connector clamps, wing attachment, fin attachment, and bearings.
Storage	Humidity, temperature, and moisture may compound to cause corrosion, rubber/plastic seal degradation, and moisture ingress. Mitigation provided through provision of compliant, monitored environmental storage.
Handling, transportation, and loading	Risk of damaging weapon during logistic handling and transportation increases risk of ageing harming missile structure, wings, fins, and/or sensors.
Carriage	Operational carriage presents environmental hazards such as sand or rain ingress, along with mechanical stimuli through vibration, gunfire shock, etc.
Release-to-target	Degraded structural integrity may result in failure of rocket body tube, wings/wing mounts, fins, fin bearings, or subsection interface mechanisms (e.g. clamping bands, etc).
Qualitative risk of subsystem failure	*Early Life Cycle* Likelihood (L) = Low, Severity (S) = Medium *Later Life Cycle* Likelihood (L) = Low, Severity (S) = Medium
Comments	Weapon is designed with load factors of safety and is subject to quality control on build. Inspection/maintenance provides mitigation for visible defects such as corrosion, mechanical damage, or leaching of explosive matter.

Subsystem	Power system
Critical components	Thermal batteries, distribution.
Storage	Thermal batteries are sensitive to heat, humidity, and moisture ingress.
Handling, transportation, and loading	The power distribution system is subject to vibration. A sealed system is less likely to experience moisture or sand ingress.
Carriage	Operational carriage presents environmental hazards, e.g. sand or rain ingress, mechanical stimuli through vibration, shock, etc.
Release-to-Target	Environmentally degraded thermal batteries may be at risk of overheating or combustion through cumulative effects of poor storage and vibration.
Qualitative risk of subsystem failure	*Early Life Cycle* Likelihood (L) = Low, Severity (S) = Low *Later Life Cycle* Likelihood (L) = Low, Severity (S) = Low
Comments	Mitigation provided by batteries largely being contained in sealed housings. Weapon containers have desiccators inside which are subject to regular inspections. Any weapons which indicate moisture ingress are quarantined.

wings, fins, and attachment points respond in a manner which may be predicted with reasonable accuracy through well-proven techniques such as finite element analysis (FEA) modelling packages.

Outputs from these techniques can be used to predict failure cases such as fatigue cracking and, more importantly, when they are likely to occur in a particular environment. A similar claim can be made for electronic circuits, where the design is verified by computer modelling packages and, again, MTBFs can be used to drive a preventative maintenance policy.

Unfortunately, the behaviour and potential failure of energetic materials is less easily predicted due to their molecular complexity and, at least when compared to metals, more variable response to environmental stimuli. As discussed in chapter 9, the energetic materials and explosive component health are subjected to in-service surveillance (ISS) programmes where sample weapons from the stockpile are withdrawn and subjected to BTCA. This is to investigate how the energetic material has responded over time to stimuli such as hot and cold cycling, vibration, or in-service wear and tear [15].

The results obtained from ISS therefore become pivotal in any decision to award life extensions or envelope expansions of the weapon. Whilst comprehensive, ISS is costly, time consuming, and heavily reliant on appropriately representative samples being drawn from the stockpiles, it nevertheless serves to provide assurance to the end user that the wider stockpile is safe to use within prescribed limits of employment or performance.

Despite the inherent challenges associated with ISS, knowledge of the most critical weapon subsystems is of paramount importance in targeting surveillance efforts to inform risk reduction. The challenge for the future will undoubtedly lie in the development of predictive techniques to accurately and safely predict energetic dilapidation and risk over the course of the MTDS. Table 10.1 shows example qualitative risks of subsystem failure to illustrate the likely level of risk and therefore where mitigations need to be provided.

10.2.4 Systems engineering and spiral development of weapons

The United Kingdom (UK) Ministry of Defence (MOD) procures weapons using a project management framework that follows the concept, assessment, demonstration, manufacture, in-service and disposal (CADMID) cycle [16]. This is illustrated in figure 10.2.

During the CADMID cycle, all requirements are set in the early stages of the project through the construction of a user requirement document (URD), which serves to inform the basis of solution development. Dedicated decision gates are established to act as 'go/no go' checkpoints, based on project cost and feasibility, known as the 'initial gate' and 'main gate'. If the concept withstands scrutiny and the initial gate is passed, the assessment phase sees the generation of a system requirements document (SRD) and capability assessment, ahead of the main gate decision point. If the solution is still seen as viable, prototype capabilities are demonstrated to validate the solution's compliance with the URD and SRD. Following manufacture, the system is brought into service and managed through its MTDS to end-of-life disposal.

In the UK, weapons procured in this manner are subject to incremental life extensions during the in-service phase, due in part to the difficulty in being able to predict the longer-term behaviour of the energetic materials. ISS forms an important part of through-life management and, where either necessary or called for, mid-life

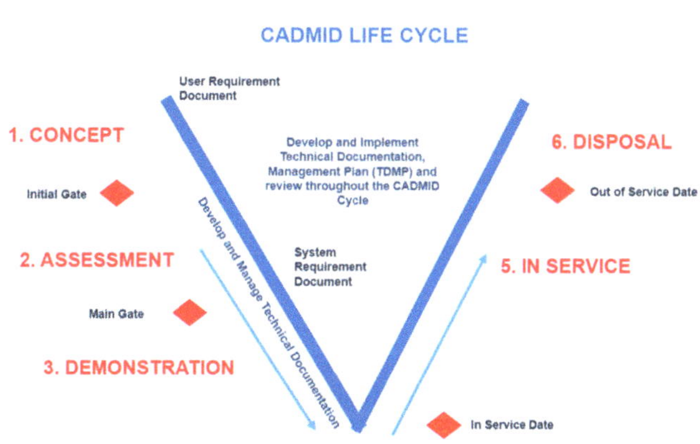

Figure 10.2. CADMID project life cycle.

refurbishment. This approach allows incremental life extension and limitation expansion, whilst providing assurance to the user.

Spiral development is analogous to the Agile methodology used by the software industry [17]. There is currently interest in whether this approach could be adopted for aspects of weapon design and whole-life management [18, 19]. Spiral development differs to that of CADMID in that it aims to provide a more expedient initial capability, followed by incremental developments or updates. Variance in product features is accepted, provided that the initial output meets the minimum viable product criteria, which for weapon systems must be safety, reliability, and performance.

This has proved successful in, amongst other things, software and application development, where an iterative approach to fielded capability is achieved by serial software revisions. One of the key features in Agile or spiral development is the early setting of requirements, which are prioritised as follows:

> *'Must have' requirements.* These are requirements that must be met in order to meet regulatory aspects and achieve acceptable levels of safety, reliability, and performance. They form the minimum useable subset (MUS) of requirements and are captured in a prioritised requirements list (PRL).
>
> *'Should have' requirements.* These are those which would cause a loss of capability or degree of pain if they were not included as part of the MUS, but could be compensated for by some form of workaround. An inability to incorporate a 'should have' requirement must not have an impact upon safety, reliability, or performance.
>
> *'Could have' requirements.* These are those which would not adversely affect performance, certainly in the early tranches, but need to be considered later in the programmeme.
>
> '*Won't have this time' requirements.* These are those which do not make the PRL but will not affect safety, reliability, or performance if left out at a particular stage. However, they should be retained for inclusion during later tranches or if requirements change during development.

Having set the requirements, Agile development concentrates on iterative development, where the initial capability is fielded to be 'safe enough' and 'reliable enough' with performance to an acceptable standard. Capability development is achieved through discrete, rapid work packages or 'sprints', after which the product is reviewed and any areas for redevelopment are allocated a priority in a subsequent iteration or tranche.

Agile development of weapons presents a significant challenge for the systems engineer. Initial, expedient capabilities are sufficiently new enough to have confidence in the state or condition of critical components like the energetics, but this confidence will diminish over time due to the unpredictability of such materials as they age. Therefore, meeting an initial operating capability will most likely come with attendant limitations on system life, air carriage hours, storage life, and the environment within which it can be deployed.

Development of future weapon iterations may prove challenging if energetic components are to be reused, especially when attempting to validate and verify their condition after already being exposed to deployed service. Under such circumstances, ISS would most likely still be required to obtain evidence to verify any claims and validate any arguments that the system remains safe and suitable for service.

Nevertheless, the Agile concept raises an interesting point when considering the use of modular weapon systems: the ability to replace subsystems at user level has the potential to make an Agile approach more attractive in terms of not having to return entire AURs to their manufacturer for costly and time-consuming ISS programmes.

The safety case for weapons exists in a continuum, and if its validity is to be upheld during iterative development then a safety case report must be constructed ahead of fielding any subsequent tranche. Again, this must ensure that appropriately robust safety arguments can be made to support iterative development or life extension for the period of time they cover.

In summary, Agile development of weapons is possible, but to be successful will most likely be subjected to deployment, lifeing, or carriage limitations during the initial or early-fielded capability tranches. Capability extension, though challenging, could be pursued. This would be through the greater use of energetics of known provenance and performance at the design stage, combined with in-service read across from similar systems, supplemented by appropriate employment of subsystem testing on critical components and forensic surveillance of previous tranches. Agile or spiral development will inherently carry some risks, but the critical success factor lies in the ability to produce sufficiently strong arguments to support safety, reliability, and performance, balanced against quantified and appropriately considered risk.

10.3 Smart through-life management

10.3.1 United Kingdom research

The UK MOD has actively pursued research in the field of smart through-life management for decades. Work in this area is also referred to as whole-life assessment. Papers by Barnes *et al* [3–5] review the shortcomings of previous through-life methods and propose new approaches to improve them.

One of the dominant themes explored is the utilisation of more intelligent data to inform the safety and suitability of a weapon through its life. Traditional data-collection processes often produce excess data that do little to inform the ageing behaviour and failure modes of the munitions. The process of accelerated ageing, for example using reaction models such as Arrhenius to project the reaction rate onto the true environmental conditions, are described.

It is noted, however, that failure mechanisms occur in different temperature regimes, meaning accelerated ageing does not always accurately describe ageing behaviour. The method must also assume the total environmental stressors experienced by the weapon over its whole life, which is often overestimated. This is why it is essential to have good baseline data, something that the authors noted had previously been lacking.

It was proposed that the accelerated ageing stage should be limited, and that munitions should be subjected to continued ISS. Destructive testing can be reduced by utilising more modern testing techniques that require less energetic material and nondestructive evaluation (NDE).

Automatic data logging and heat flow calorimetry (HFC) were recommended, although it was discussed that the application of this technique at the time of publication had not been completely agreed upon. It should be noted that HFC is now considered a routine test for the stability of nitrate-ester-based propellants [20] and it also is part of a suite of tests considered acceptable for assessing explosive compatibility [21].

In two papers regarding the simulation of heat flow curves for nitrocellulose (NC)-based propellants, Itkis and Bohn state that measuring heat flow curves near the in-service range have become standard for stability and compatibility assessments. However, their work highlights the need for an approach for understanding the contributions to the total heat flow from the various individual reactions taking place in the NC-based propellant [22, 23].

In reference [5], Barnes and Sharp conclude that greater collaboration between allied nations is needed. Despite the drive for international standards such as North Atlantic Treaty Organization (NATO) Standardization Agreements (STANAGs) there is still a lack of transparency with important munitions data due to intellectual property (IP) rights, which had at the time of writing vastly hindered progress. This situation has not improved in the intervening years.

Finally, further to the limitations around IP rights, perhaps an even more significant issue is that of security. There are many organisations and groups around the world, such as the Defence Scientific Advisory Council (DSAC), NATO AC/326 Sub Group A—Energetic Materials Team, and Joint Army Navy NASA Air Force (JANNAF), which have commissioned studies in this area but, due to security limitations, these publications are not openly available.

Previous research into improving weapons qualifications practices is also explored by Barnes *et al* [6]. This paper envisions a smart testing framework that focuses on intelligent tests with key purposes. It highlights the lack of computer modelling used in current practice, which is often due to a dearth of complete data to make use of available models. The paper recommends taking advantage of opportunities to collect valuable baseline data during the qualification regime so that materials are better understood and models can be used, leading to more accurate safe-life assessments.

Progress made in the exploitation and verification of data collected by environmental data loggers, and how this can be used in life extension decisions, is described by Richards [24]. Data loggers can provide relevant information on the environmental stressors experienced by a weapon, to inform a view of its health. How these data are exploited is key to ensuring that users make the right decisions on munition safe life. Data must therefore be well validated.

It was suggested that logger data should be compared and verified against three individual sources. The sources of verification should be from other identical weapons in the stockpile, where possible. Alternatively, data collected from other equipment in the same environment or from historic databases can be used. Several commonly used degradation models are assessed by Richards [24] including chemical decomposition, high cycle fatigue damage, low cycle fatigue damage, and humidity degradation. The benefits of tracking weapons through their life via dedicated data loggers, to avoid errors in estimating the life history of individual units, is underlined.

In recent years, this concept has been developed in more detail. In particular, a health usage monitoring systems data utilization project was described by QinetiQ [25]. This aimed to explore how environmental data loggers can be used to support qualification and through-life management of complex weapons. The paper details the need for a better understanding of failure modes and ageing mechanisms while using a targeted approach to data collection and testing.

A cost-benefit analysis is proposed as a method of determining the number of failure modes to be tracked. Together, these factors form the foundation of a health and usage monitoring system (HUMS) / munition health monitoring (MHM) system. In agreement with the authors of [3, 4] and [6], a paper by Rix [25] defines the importance of sufficient baseline testing and the need for advanced knowledge of failure modes in all materials to support the development of physical life prediction models. The use of data loggers on both in-service and instrumented munitions to provide the boundary conditions are also discussed.

Rix [25] also describes generic degradation models, for example Arrhenius methods, but the need for bespoke, physical models is highlighted, along with the requirement for the derivation of specific material activation energies. One of the issues with using generic activation energies, such as those defined in STANAG 4170 [26], is that they might not adequately describe the key failure process and lead to the incorrect designs for trials. If, for example, the activation energy driving the main failure mode is thought to be 'high', then conducting a trial will generate data rapidly but the material will also age poorly in high temperature environments. However, if the activation energy for the main failure mode is low, such as in the case of some diffusion processes [27, 28], then it will be difficult to generate data quickly during a trial and data will come from ISS as the material is likely to age well.

In addition to this, the activation energy may change during the life cycle of the asset [29]. If this is the case, any models which use this activation energy need to be refined as this value changes as a consequence of ageing. Physical models would allow for the impact of multiple ageing processes to be explored in unison, something that is not possible with Arrhenius methods, and testing would provide initial validation for these.

One approach for providing some data for such physical models for certain energetic materials, such as polymer-bonded explosives (PBXs), would be through the use of universal validation experiments [30]. This would involve using well-characterised methods such as the compression split Hopkinson pressure bar, the Brazilian test [31], and the Taylor test that come under the high rate or shock regime. These tests are designed to be below the ignition threshold so they are dominated by the mechanical response.

It is noted that while ageing algorithms for weapons exist, they have generally not been fully exploited due to a lack of completeness. Limitations like this are demonstrated when unexpected behaviour is detected during surveillance. Consequently, the paper by Rix [25] suggests that ISS will always be required to support predictive models. Advancing these models is made more challenging by the lack of material data and the inability to validate models directly. The paper suggests that physical tests, such as accelerated ageing, are still of benefit for deriving the rate expressions of the degradation mechanisms. The munition life index approach has been shown to be an extremely promising method of through-life assessment, and has been demonstrated for a small number of fast jet air platforms. It is suggested that there is a need for greater automation of the process, to reduce manual interaction and allow for wider deployment. This paper makes good progress in outlining a smart testing framework; however, it is clear that significant development of modelling techniques and understanding of ageing is still required.

To support the need for wider use of modelling techniques, a predictive modelling framework for ageing munitions has also been developed by QinetiQ [32]. This research aimed to demonstrate that environmental ageing can be predicted through modelling, using algorithms developed for specific ageing processes along with environmental data. A proof-of-concept for a predictive modelling software tool was presented. The tool comprises a database to store data informing the ageing behaviours of a weapon. Response surface methodology [33] is proposed as a technique to execute predictive models using the database, and legacy data were used to verify the concept. While this paper provides a good framework in order for predictive modelling to be realised, it is evident once again that more needs to be done to define the predictive models that will enable this system. Further discussion on predictive modelling techniques can be found in section 10.5.

10.3.2 Global research

The previous section described research conducted within the UK. There have also been significant advances in smarter through-life management principles globally over recent decades. Specifically, there have been many programmes in the area of munitions health management (MHM) including those involving the European Defence Agency (EDA) [34] and the Munitions Safety Information Analysis Center (MSIAC) [35]. This is a vast topic area, with security limitations again restricting the publication of certain aspects of these programmes. A selection of some of the open literature in these areas, from a global perspective, is discussed in this section.

Many papers have tested the validity of using sensors to monitor the environmental stressors that weapons are subjected to during qualification and in service. The process involves tracking instigators of key failure modes and collecting data on parameters such as temperature, pressure, humidity, and vibration. Recent advances in technology have allowed the benefits of wireless, automated data logging to be realised [36–42].

It is evident that a substantial amount of progress has been made in the last 15 years towards smart through-life management in the United States of America.

The Naval Ordnance Safety and Security Activity (NOSSA) command [43] has a programme formed to combine the efforts of their insensitive munitions, explosives safety, and quality evaluation programmes with the purposes of improving scientific validity and the value of data and testing over the munition's life cycle, from concept through to decommissioning. The goal is to further employ the use of sensors, smaller-scale testing, nondestructive testing, and a wider use of modelling.

The United States (US) Air Force has developed a data-analysis system evaluating computed tomography (CT), capable of inspecting the motors of ballistic missiles, as described in the report of Tussiwand *et al* [40]. The system is capable of detecting defects in the propellant grain and de-bonds/delamination in the motor interfaces. The flaws, measured by CT and evaluated by analysis, can be automatically converted to faceted surfaces and imported into a structural/ballistic analysis software, which performs coupled fluid-structural-ballistic-thermal analysis of the motor and determines how the flaws will propagate. Other nondestructive techniques a being explored, including eddy current and ultrasound, which can be used for quality control during manufacturing.

Additionally, environmental, mechanical, and chemical sensors have been developed, along with SRM propellants models, including the effect of ageing and chemistry on the mechanical properties. Alongside this is the development of the US Air Force's 'aging and surveillance and health monitoring' doctrine for motors efforts, detailed in reference [36]. A description of a near-term demonstration which could serve as a prototype is also given, with the intention of proving the validity of health monitoring. Though the paper states the desire for demonstrating this in the 2014–15 timeframe, no literature describing the results of this has yet been found. A ballistic-thermal-burnback platform allows for the acquisition of internal and external temperature, humidity, and acceleration/load data through sensors to use as boundary conditions. It is asserted that the use of sensor data vastly increases the accuracy of ageing models, allowing them to be applied to individual motors. Data warehousing is used to store and utilise data.

The US Army has employed an integrated health monitoring system to monitor health and perform prognostics and diagnostics analysis of missile systems, as described in [44]. This system, known as the remote readiness asset prognostics and diagnostics system, has been field tested, with the purpose of assessing the reliability of weapons. The paper discusses the use of probabilistic engineering methods to analyse data, such as those based on stress and strength, cumulative damage functions, and Weibull analysis.

These models can be used to compute degradation in material properties as a result of exposure to thermal and humidity cycling, shock and vibration, and/or a corrosive environment [37]. Temperature, humidity, and shock sensors that monitor missile storage and transportation environments are used to define the material properties. Chemical kinetics or age-related relationships enable one to determine changes in modulus, strain energy, or other properties of this nature to be determined. Finally, FEA is applied to degraded material properties to evaluate induced internal stresses and predict current and future health. This will provide a realistic assessment of the method's reliability.

Similar activities have been conducted at Rheinmetall Waffe Munition GmbH [45]. Their paper described an external data logger system and corresponding analysis and prediction methodology for evaluating the safe life and health of gun ammunitions. Due to the complexity of the physical and chemical processes involved in the ageing of explosives, it was concluded that there is still a requirement for continuous validation of these models through experimental investigations. This includes both laboratory-scale tests as well as those performed under operational conditions. This research was unable to conclusively validate the prediction models; however, experimental data was found not to be in contradiction with the simulated data.

In the Netherlands [40], environmental monitoring and near real-time response capability during storage, transport, and operations has become routine practise since 2001. Micro-electro-mechanical systems sensors are placed next to munitions, the data of which are transferred via satellite for analysis. The health of the munitions is determined through the use of ageing models, alongside nondestructive testing. Little information is given on the specifics of the ageing algorithms.

Similarly, the effective logistics monitoring system for ammunition (ELISA) is a MHM system demonstrator developed for the Italian Navy by MBDA [40]. It monitors the munition through the detection of temperature, acceleration, shock, and humidity sensors, and uses a thermal ageing model based on Arrhenius to provide diagnostics of munitions and prognostics of their life expectancy.

A systems engineering approach is used by Orbital ATK to develop a diagnostic and prognostic health management (PHM) capability for SRMs [46]. The importance of boundary condition history is noted. This research was able to successfully demonstrate the ability to capture and process relevant data from a rocket motor and to use this to provide a prediction of the asset's health and safe life. This was done using both empirical and chemical ageing surveillance methods, and physical mechanistic models to chemically evolve the materials using their environmental history. This was then used to assess the probability of failure of the motor's components. Comparing the predicted properties to measured characteristics showed that the projection was well within the model's uncertainty, thus proving its validity.

A similar PHM approach is described in a research paper from China [47]. In this, PHM methodology was implemented to assess the suitability and life prediction of missiles. The data-acquisition system and database framework are described, and a storage life prediction method based on data-driven physical models and prediction models is proposed. PHM is defined as a method of fault prediction based on the system history, failure threshold, and life distribution model, and the ability to make health management decisions based on this. The system functions to predict the time and location of system failure as well as its remaining life.

The system is composed of two parts, an ISS sensor system and the models that use this information to predict residual life. In that paper, wireless miniature sensors are described as the sensor technology. The PHM technology analysis models included those for a range of aspects including fault diagnosis, life prediction information feedback, and machine learning and data processing. Data-acquisition hardware was installed both inside and outside the missile.

Together, these are capable of collecting comprehensive information, storing life history data, and using forecasting and decision-making software to predict events and provide action suggestions. The paper promotes the use of independent component analysis rather than analysis of independent measurement parameters, as many of these are not independent of each other. The research makes significant progress in the development of a smart testing system and appears to make large use of modelling and automation. Again, there is a lack of description around specific models and testing techniques used.

It is evident that there is a global interest in improving weapons qualification and testing practices, and that the limitations of previous methods are well established and agreed upon. The majority of qualification procedures described follow general mutual principles: identification of baseline data, identification of dominating failure mechanisms, establishing appropriate ageing algorithms, the application of *in-situ* sensors to collect ISS data, and using physical testing where necessary.

In general, there is clear consensus on the necessity of gaining an in-depth knowledge of failure mechanisms and ageing phenomena as part of the through-life process. However, it has been highlighted that the current level of knowledge is still lacking in this area and so further research is needed into the physics behind munition ageing.

Several papers also noted that thorough baseline data is essential for establishing accurate health and safe-life evaluations. This can be hindered by issues around the sharing of intellectual property. Barnes and Sharp [5] noted that the Swedish MOD vision of 'in safety we share, in technology we compete' had not been realised in 2007. Shared data repositories, to encourage greater collation of testing data and testing and modelling mechanisms between institutions, in order to ensure a complete picture of a weapon's properties and behaviour and to avoid the same research being conducted twice, would be of benefit.

The MSIAC Energetic Materials Compendium (EMC) provides a resource for such requirements [48]. However, this does rely on organisations disclosing data without concerns around IP protection. If a truly smart approach is to be adopted then there needs to be a willingness for the sharing and acceptance of such information for when the material under test is exactly the same. This would lead to the avoidance of unnecessary duplication of analysis.

It is also clear that a smarter approach to through-life management has been advocated for decades. There should be a recalibration against the original sources for such material to ensure the interpretations are in line with modern thinking.

Finally, the widespread use of sensor technologies and data-logging systems is evident. It was found that the practice of logging environmental data is well established, with several papers corroborating the reliability of the resulting data. Sensor technology has advanced significantly in recent years, and the use of wireless, automated data logging has proved beneficial.

10.4 Energetic materials analysis

In a smart approach, efforts are focused on testing to identify failure modes and to provide data from such key areas. However, care must be taken to ensure unexpected issues do not emerge or are not adequately identified. Furthermore, it is imperative that an iterative approach is taken with regards to failure modes. If an emergent issue occurs during the life of a munition, this should be fed back such that lessons are learned for future testing. A failure-mode approach is detailed in chapter 9 and has been well established for many years [2, 49].

It is likely there will always be a requirement to perform some degree of physical testing of energetic materials as part of through-life management. Many methods for this have been well established for decades. In this section, areas where improved analytical methods have been developed, which show promise for life-assessment purposes, are discussed.

10.4.1 Accelerated ageing and data analysis

As part of the life-assessment process, the energetic materials may be subjected to artificial ageing. This may be conducted to provide predictions that the behaviour of the materials in the long term are not expected to give cause for concern. The ways in which this may be conducted, such as the need to consider the atmosphere around the sample, humidity and the surface area, as well as limitations with the methodologies, are discussed in references [50–61].

The foci for many papers on accelerated ageing and the subsequent analyses are on chemical and thermal techniques. Whilst clearly these are important areas for study, many, such as stabiliser depletion in nitrate-ester-containing propellants, are well understood [62–64]. It is the mechanical properties which are key to how many energetic materials, such as SRMs, fail. Whilst chemical decomposition of a particular compound may provide an indication that there may be changes in the mechanical properties, the actual quantitative change can only be measured through tests designed to look specifically at the mechanical characteristics of the material. These are discussed in section 10.4.3.

Essentially, ageing involves heating the material for a defined period of time, and is usually conducted at multiple temperatures to enable the rate of reaction at different temperatures to be calculated. A limitation of this approach is that reactions which take place during the artificial ageing experiments may not occur at lower temperatures, and for some materials there may be a change in the activation energy at lower temperatures [65, 66].

Additionally, before any ageing is carried out, consideration should be given to the likely failure mode [51]. It has been found that some analytical techniques and methods do not always detect the degradation of some high-energy compositions and so alternative ways of following the ageing processes need to be devised [67, 68].

The ways in which the different mathematical treatments can be applied to materials such as solid rocket propellants to determine the rate of reaction at different temperatures are also well established [27, 69–73]. As discussed by Roduit *et al* [74] one of the issues with energetic materials testing is it can be difficult to collect sufficient data points due to problems such as a lack of samples. In addition, oversimplification of the kinetic analyses can lead to incorrect estimation of shelf life. As a result, efforts in recent years have borne research concerned with being able to predict the thermal stability of energetic materials, with advanced kinetic modelling, even if there are limited numbers of data points [74, 75].

Chelouche *et al* [76] concluded that there was a sparsity of literature dedicated to the accurate assessment of the real in-service time. This is of utmost importance to ensure safe storage and use. The correctness of the results from artificial ageing plans is strongly linked to the precision of the equivalent in-service time (EIST) from the time–temperature profiles. As a result, the authors proposed combining principal component analysis with data from a vacuum stability test for a double-base rocket propellant as a new way for monitoring the RIST/EIST times of this material. The approach was validated using two similar double-base rocket propellants, and they proposed a simple model to predict the EIST.

Kumar *et al* [77] also noted that the dominant mechanism of advanced energetic propellants (AEPs) is not predicted during storage, is accelerated by higher temperatures, and is clearly not understood. They carried out ageing studies on two AEPs and then characterised them using mechanical and ballistic testing as well as thermal analysis to determine the activation energies for the propellants that are used in SRMs. They set a failure criterion for the prediction of the shelf life of the SRM based on the tensile test results and combined the thermal and mechanical data to estimate the shelf life of the AEP.

The results from tensile testing were also used by Wang *et al* [78] to make predictions on the in-service lifetime of nitrate ester plasticized polyether (NEPE) propellants. Ageing was performed at four elevated temperatures before the strength limit of the materials was determined. This demonstrated that this parameter varies during ageing but reduces quickly with prolonged ageing. The authors used the Bethelot equation to predict the in-service time of NEPE propellants.

Other integrated methods for making predictions about the safe in-service time of materials such as solid rocket propellants [79, 80] is through the use of computer models such as FEA [53, 81]. This is seeing increased use with respect to energetic materials.

In the example in reference [82], it was demonstrated how data from laboratory testing using high-performance liquid chromatography and HFC could be used in FEA models to understand if there would be heat buildup in the centre of large-calibre munitions, which would be a cause for concern. A similar approach was taken by Roduit *et al* [83], who used FEA methods coupled with data from thermal analysis tests such as differential scanning calorimetry (DSC) to show that accurate kinetic descriptions enable the determination of the effects of scale, geometry, heat transfer, thermal conductivity, and ambient temperature on the heat accumulation condition for a PBX.

The US Army has designed a system to monitor missile storage and transportation environments on a real-time basis [44]. The materials properties data are determined using sensor information [84], which is used in conjunction with chemical kinetics or age-related relationships to determine changes in mechanical properties such as modulus or strain energy. The degraded material properties are then used in FEA to evaluate induced internal stresses and predict current and future factors of safety [44].

Roduit *et al* [74] also discussed the importance of continually monitoring the shelf life of energetic materials through the use of data loggers. This illustrated the benefits of combining data loggers with complex kinetic models to more accurately understand the decomposition status of a NC-based propellant. The importance of validating and refining models using measured data was underlined.

10.4.2 Life assessment testing: considerations and advances

Many of the techniques for characterising energetic materials for life-assessment purposes are well established [51, 85]. A thorough review by Trache and Tarchoun [86] has shown that there are a large number of analytical methods available for the analysis of nitrate-ester-based propellants. The authors concluded that in order to truly characterise such materials, it is important to combine the results from a number of different methods and critically examine and assess them together.

In recent decades, there has been a shift from general materials characterisation to a more scientific approach in ensuring that only the most appropriate tests are applied to energetic materials in order to truly understand their stability and ageing behaviour [3–6, 87]. It is also important for nations which procure their inventories from overseas to understand the testing and evaluation undertaken in the country of manufacture and to maintain in country knowledge and expertise to ensure ordnance, munitions, and environment safety [3, 88].

As a result, there have been collaborative and integrated approaches to updating testing standards for particular energetic materials [89, 90]. The aims of these have included to standardise and improve test procedures and to assess if new methodologies are available or required.

Despite the fact that energetic materials have been widely investigated experimentally and theoretically for decades, there are still unanswered questions regarding their thermal decomposition mechanisms that are critical for understanding the overall reaction processes [85]. As a result, there has been the increasing use of hyphenated analytical thermal analysis techniques to study such mechanisms. These include the use of detection methods such as Fourier transform infra-red (FTIR) or mass spectroscopy (MS) with thermogravimetry (TG) [85, 91, 92] and the ability to measure gas pressure during HFC [93]. An overview of where hyphenated thermal analysis methods fits into the assessment of energetic materials is shown in figure 10.3.

Another area which has seen development in terms of thermal analysis is the use of high-pressure DSC to study energetic materials [94–96]. As energetic materials operate in pressurised environments, such a methodology may enable more representative testing to take place compared with routine DSC, which is a widely available tool [97].

During life-assessment testing, a few of the characteristics, such as stabiliser content or the percentage of a certain ingredient, are absolute. However, many of the characteristics are relative, with the aim being to look for change, for example

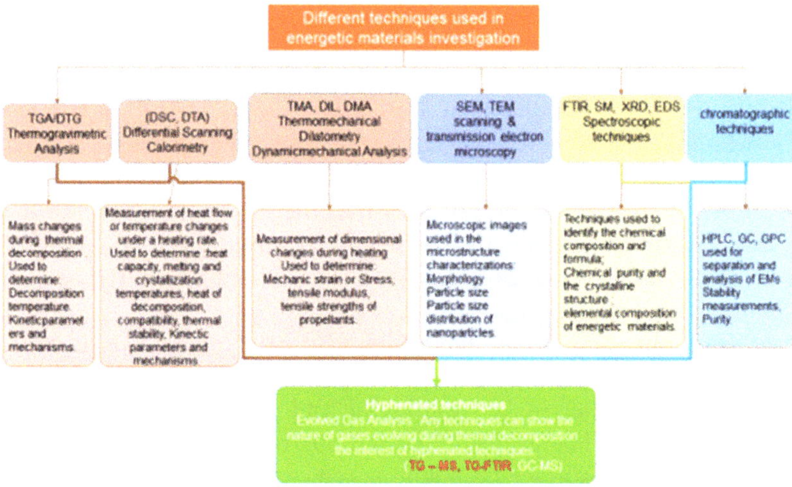

Figure 10.3. Different analytical techniques used to assess energetic materials [85]. Reproduced with permission © Taylor & Francis Ltd.

from previously tested samples or qualification data. Small-scale hazard testing data falls between the absolute and relative properties. It allows determination as to whether there is a gross change in sensitiveness to a specific stimuli. However, the results typically lack the fidelity to allow a conclusion as to whether a small change is part of a trend or improving or worsening hazard properties over time. One of the ways in which this could be improved would be through the use of companion systems, which are discussed in section 10.4.3 [98].

Using additional instrumentation to gain further information about a particular test is also another way in which energetics analysis can be improved to enable a smarter approach. When assessing the impact response of LLM-105 energetic material, the authors used high-speed photography, mass, and optical spectroscopy and optical-radiometry to enable the small-scale impact response of an explosive to be studied in considerably more quantitative detail than previously reported, thus enabling complex behaviour to be described [99].

Morley also demonstrated the benefits of such a methodology to investigate the reactions of HMX [100]. As part of this, improvements were made to the fall hammer sensitivity test and the split Hopkinson pressure bar by coupling to a mass spectrometer to improve quantification and identify the reaction products.

With regards to small-scale hazard testing, the need to understand the variability of the data and what is required to enable meaningful interpretation from the tests have been reported as ways in which this area of life assessment could be enhanced. Further to this, ensuring that the tests are conducted in a standardised way is imperative [98].

The Health and Safety Laboratory (HSL) in the UK carried out a detailed study into the electric spark test, with one of the conclusions being that there were several unexpected results, suggesting that the apparatus was not performing as exactly as predicted. They recommended that future work should be carried out on the measurement and understanding of spark energy deposited into a test sample and the variables involved with the deposition [101].

Figure 10.4 shows the running mean for a Rotter impact test for a UK test house (Energetic Materials Testing Assessment and Policy). This method is a means of assessing the impact sensitivity of energetic materials. The data are provided from December 1990 to 2021, and it can be seen that there is a significant variance over the 31 years. By understanding the cause of the variability and taking steps in reducing or removing it, significant progress towards improving the test would be ensured [98].

However, as figure 10.5 shows, there may also be a trend towards a continued reduction in the yearly test volume available for analysis. It is therefore imperative to ensure that all data collected is meaningful and that the limitations of it are thoroughly understood. The criticality of nations retaining suitably qualified and experienced personnel has never been so marked.

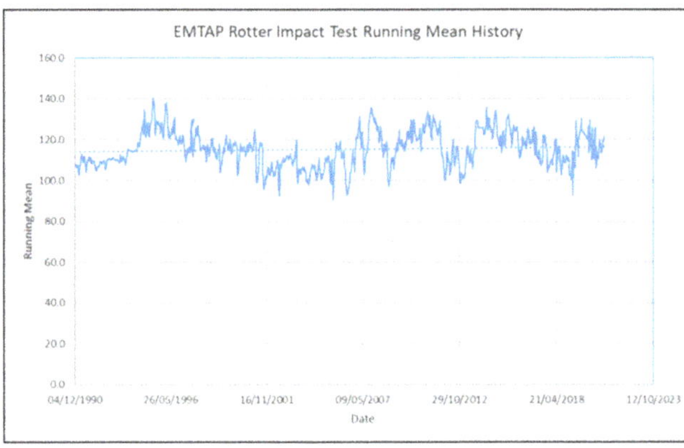

Figure 10.4. Energetic Materials Testing Assessment and Policy Rotter impact running mean data from December 1990 to June 2021. Reproduced from reference [98], with permission of QinetiQ Ltd, 2022.

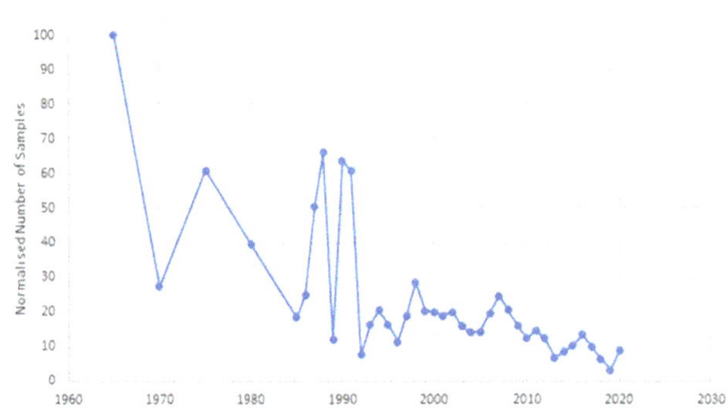

Figure 10.5. Normalised number of small-scale hazard tests conducted from 1960 to mid-2021. Reproduced from reference [98], with permission of QinetiQ Ltd, 2022.

Doherty's detailed review on sensitivity data discussed testing features which are applicable to small-scale hazard testing worldwide. This illustrated that in order for the data and reporting to be meaningful, the results must contain sufficient detail, which depends on the purpose(s) for which the data were collected [102]. The need for better data management in general is vital if the results from life-assessment testing are to be better integrated.

In the same paper [102], Doherty notes that an advantage of the Rotter impact test is that it uses a reference standard to provide a baseline comparison. However, there are at least eight other forms of testing available to understand an explosive's sensitiveness to impact [103]. In recent years, Cambridge University have demonstrated the potential for using a split Hopkinson bar apparatus as an improved and standardised way of performing impact testing [103, 104].

Whilst Green *et al's* [105] paper on the use of DSC for thermal hazard assessment of new chemistry, 'Avoiding explosive mistakes', is not about energetic materials, it makes a valid critique of how testing is conducted and disseminated. It is important that all details are reported so that experiments can be replicated. In addition, consideration should be given to how the testing is conducted as otherwise the results could be misleading and the conclusions dangerous.

10.4.3 Uniaxial, baxial and triaxial mechanical testing

For complex weapons where mechanical failure is deemed critical, such as in a rocket motor, different nations when designing a system may use a range of approaches to failure criteria (strain or stress based, for example) and the resultant margin of safety (MoS) determination [106].

The most common mechanical test to failure that is conducted for the assessment of SRM propellants, as well as for some of the more rubbery PBXs used in warheads, is the uniaxial tensile test [107]. Samples for this are easy to produce as they are relatively small and the test is straightforward to conduct as the deformation is applied in one axis only. Video techniques allow the strain to be measured as the sample is deformed, with the load being measured simultaneously. Although the cross-sectional area of the sample will be reduced during the test, these factors are not normally taken into consideration, though application of the material's Poisson ratio can refine these—bearing in mind that this may be strain dependent in itself.

Video imaging techniques such as digital image correlation (DIC) can also be used to look at the localised strain at points in the sample as it deforms and may also give an estimation of the Poisson ratio [108]. A more thorough determination of the Poisson ratio is possible through the use of dilatometry, and specifically the use of a Farris gas dilatometer (FGD), during the uniaxial tensile test. As well as following the Poisson ratio, the FGD measures the onset of vacuole formation and de-wetting between the binder-filler matrix [109] and thus has the capacity to give important information that would be needed in full constitutive modelling. These factors relate to the damage, including cumulative damage, that arises separately to the binder ageing, which is traditionally taken to be the main ageing mechanism for composite propellants and PBXs.

However, a limitation is that the stress states in a cylindrical asset that is confined within a rocket motor casing are not uniaxial, and therefore multiaxial testing is required to predict failure thresholds. In a rocket motor those stresses at the conduit will be biaxial, whereas those in the bulk will be triaxial. Gathering biaxial and triaxial properties more routinely would provide better information about the potential for cracking to form and would allow more accurate structural assessment models, such as FEA, to be completed.

The uniaxial tensile test is well established and very good at understanding differences in the material properties between assets and also the effects of ageing. In terms of crack prediction in triaxial stress states, it is the best comparator that currently exists, but one cannot predict triaxial cracking from uniaxial testing. The strain capability of a material is overestimated by uniaxial testing, and this needs to be accounted for when uniaxial testing is used as a comparator. Triaxial stress states reduce toughness, and if the ageing mechanism influences the molecular drivers for toughness (loss tangent) then concern is relevant. Instruments that measure the tensile properties of propellants can also be used to conduct testing in compression mode [51, 110].

Specific biaxial and triaxial testing does not form part of the minimum requirements for qualification as listed in AOP7 [111]. In some countries generic knock-down factors are sometimes applied to the uniaxial tensile test results that allow estimation of the biaxial and triaxial properties [106]. However, the only true way to provide accurate multiaxial properties is to use results from samples tested in these stress states as these will be dependent on the propellant formulation. As a result, extensive characterisation is sometimes conducted for the particular design, but this is not universal.

Within the laboratory environment, the biaxial stress properties are relatively easily obtained by using a strip tension test piece. In this, the sample is loaded uniaxially. However, the sample has a large width-to-height ratio, particularly compared to standard uniaxial test pieces, and this drives the stress state into biaxiality. The thickness of the test piece needs to be such that strains measured on the surface can be assumed constant through thickness. Testing of this type is facilitated by DIC. Another method is to apply the load in simple shear, but this requires additional interfaces to be created and causes problems with both sample preparation and the success of the test. Schematics of both are given in figure 10.6.

A more difficult configuration is that needed to drive the stress field into triaxial mode. In simple terms, a large sample will form a triaxial stress field within its bulk so the problems of sample weight and ability to extract from most rocket motors will be magnified. A study into the triaxial failure of polymeric materials from both analytical and theoretical aspects was conducted by Schapery and others in the early 1960s [112]. For experimental analysis they describe the use of the poker chip specimen as being satisfactory for determining fracture under triaxial tensile stress, and this geometry remains as the most well-known experimental test setup for assessing the triaxial stress state, with earlier work using similar test pieces being used to study bond lines [113–115]. It should also be noted that ductile failure in metals is driven by triaxial tension in necked uniaxial tension test pieces, and the methods to understand this process could be extended to rubbers.

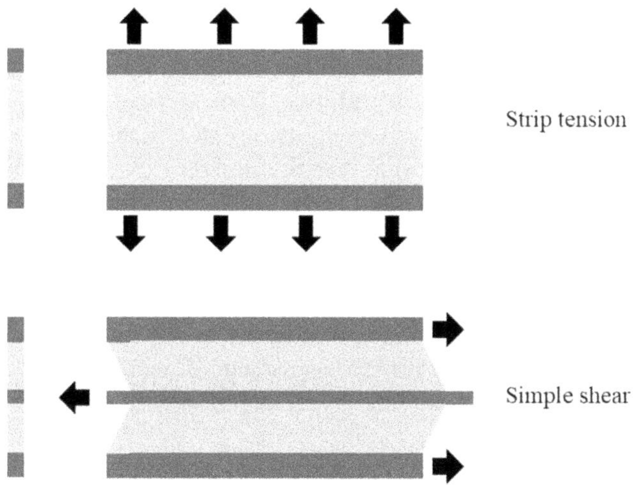

Figure 10.6. Biaxial tensile test pieces.

Other various pieces of equipment can be obtained which try to mimic triaxiality by loading the sample in three dimensions. Such techniques are extremely sophisticated and the rate of application of stress in each of the three directions can be varied to suit the user requirements. However, although all these methods are valid for typical rubber-like materials, they start to deviate from this with highly loaded energetic materials. The key to any result from an energetic material is that the sample tested is representative of the bulk material; if it is too small, the influence of the filler distribution within the sample to be tested will be unrepresentative.

Obviously, from a standard surveillance programme using a typical air-launched rocket motor, it would be difficult to obtain sufficient samples in the biaxial and triaxial form to provide adequate data. One method of overcoming this is to establish the relationship between the uniaxial tensile properties and those of the biaxial and triaxial using carton samples. These are samples of bare propellant which are poured into 'cartons' and cured. They can be produced as required and enable complex samples to be made. If the relationship can be defined then the surveillance can be undertaken using standard uniaxial testing.

However, it should be recognised that any relationship between uniaxial, biaxial and triaxial properties are very likely to change as the propellant ages. Therefore, it is necessary that this issue be covered with the initial carton testing as well as the inherent difference between the properties of a carton sample and one cast as a rocket motor grain. As with other samples needed during initial material qualification, there could be limitations in the amount of material made from any one batch. Once any changes are understood, a material model that can predict the differences in biaxial and triaxial properties could then be used with the appropriate FEA model to assess the change in stress states in a real system. Limited but targeted routine testing on samples from ISS assets could then provide confidence that this approach remains valid through the life cycle.

10.4.4 Crack growth failure

Often rocket motors will exhibit a degree of cracking. Generally, once a crack is discovered the asset is normally withdrawn from service as the potential consequences of the cracks are unknown. Cracks cause extra burning surface and, as such, if sufficient extra pressure is generated then the case may overpressurise. They can be detected by NDE tests, such as X-ray radiography. Detecting the physical existence of a crack at one point during an ISS programme, even if it is small, does not give information about its propensity to propagate, hence the decision to usually withdraw the asset from service. Conversely, the absence of a crack does not give information about how a crack may be initiated in the future. Even then, not all nations conduct NDE as routine for ISS assessment of rocket motors, instead relying on proof firings and testing, including visual inspection, all of which necessitate the destruction of the asset.

Many authors have studied crack growth in propellant with the purpose of determining the integrity and service life of a motor so that unnecessary disposals are reduced. The crack growth mechanism in experimental testing has been found to be through a blunt-growth-blunt-growth mechanism of extension, which is highly nonlinear [116].

Much research has been conducted in trying to establish fracture models that can predict the crack growth in a propellant grain if a crack or defect is present, for example by using a cohesive zone model [117]. This is distinct from the general constitutive models that aim to predict the complex behaviour of defect-free propellant under general loading conditions.

However, the approach still requires the generation of suitable experimental data, such as cohesive fracture energy determined from single-edge notch test samples. Such processes are not routinely applied to the study of motors entering service. Studies are sometimes conducted later in the lifetime of the system when a failure occurs.

Understanding fracture in materials is an essential element of life prediction, so small-scale trials need to be conducted during the qualification stages of a system. For most non-energetic materials, studies of crack growth and resistance to cracking are routine as the fracture is generally linear. Typical trials would involve the rate of crack growth with the application of a repeated stress. No designer would commit to a concept without an understanding of the crack resistance of a material, but with energetic materials, which are structural, this is not the case and fracture testing does not form part of the minimum qualification requirements.

Many theories exist for the way in which filled rubbers fail, and a standard set of experiments provide the data necessary to populate the relevant equations. Therefore, the principles are well defined and need to be applied more generally to energetic materials to understand fully how initiation of cracks can occur and how the crack grows under varying environmental conditions.

What is not clear from the above methods is whether there are any suitable algorithms that allow for changes to be interrogated in an easy manner, for example the monitoring of crack initiation and/or growth over a time period by removing the previous baseline information. In addition, the definition of a critical crack size needs to be defined and then looked for during ISS, and this may be asset and composition dependent. This therefore remains an area for research towards a more integrated smart through-life testing approach.

10.4.5 Data for constitutive material models

Even without considering fracture properties, the nonlinear behaviour of the viscoelastic propellant makes generation of constitutive materials models difficult. A method to conduct biaxial dynamic mechanical analysis (DMA) testing to look at the viscoelastic nonlinearities is described in [118]. It was concluded that the new measurements indicate that constitutive models need to take these into account in order to represent the complete 3D behaviour.

More complete viscoelastic plastic constitutive models have been proposed for energetic materials such as that described in reference [119]. This introduces an irreversible deformation component through a compaction/dilatancy parameter with the viscoelastic part described by a generalised Maxwell model, which is a standard methodology for linear viscoelasticity. The model was calibrated using a large database of experimental data, which included DMA testing, cyclic compression, including some under confinement, and tensile tests with relaxation and recovery stages (hysteresis). The model validation was based on a three-point bending test and a Brazilian test, though with the latter there was some discrepancy in the experimental and simulated results. It was noted that the test results highlighted induced damage anisotropy observed from cyclic tests and that this is not taken into account in the model.

Such a model indicates the large number of tests that are known to give useful information about the strength of propellants but it also indicates that there is currently no fundamental understanding to link those disparate responses with a predictive scheme. Such data are useful to collect on all energetic materials. There needs to be more work to determine the degree to which all of these data need to be collected given minor changes in formulation, as the utility of the type of constitutive model described in reference [114] would be driven by the cost of characterising each separate material.

10.4.6 Bond testing

One of the fundamental failure modes in rocket motors is failure at a bond line. Depending upon the design this could be between the propellant and the rubbery insulating liner or between the insulating liner and the case. Either has the potential to cause catastrophic failure of the motor. These can range from immediate failure on launch, with consequential damage, or loss of the launch platform to failure of the mission.

Bond testing is very difficult as the creation of the test piece from sampled motors is unlikely to represent the stress state of the bond line in the system. Typical configurations, such as a conical bond test piece, use a small section of the bond line, and during tensile testing the failure load is used in the stress analysis of the system to produce the structural design record. In general, constant strain rate bond testing is easier as it is quicker, but conducting testing at constant load more closely represents conditions where low-level biaxial and triaxial strains take place over a longer period of time.

In such test configurations the insulator material can deform as the sample is loaded. However, in the rocket motor configuration the insulator is continuous and cannot deform in the same way as in the test piece. Such differences in failure mode are likely to produce a variance in the bond strength of the system. Very few methods are available to determine the bond strength in these cases. One classic technique is the so-called blister de-bond method, wherein a small hole in the casing is used to apply gas pressure to the bond line and by increasing the loading the gas pressure for failure of the bond growth can be determined [120]. Unfortunately, though, such a system is very difficult to configure and getting several equivalent replicates from one asset is hard.

Bond failures in rocket motors are often determined by a NDE technique, such as X-ray analysis or CT. However many NDE techniques require the presence of an air gap to detect the bond failure. The classic 'kissing bond', i.e. two surfaces in intimate contact but unable to bear tensile load across the plane of the contact, is difficult to detect. A method to try to aid this is to cool the asset down such that the differing coefficients of thermal expansion allow the closed gap to open and thereby be detected by typical NDE techniques. However, too low a temperature and there is a danger that the asset will be stressed in an unrepresentative way, and if the temperature is not low enough the full extent of the de-bond might not be established.

Full understanding of the bond strength of a new system during the concept and pre-production stages is best addressed by using structural test vehicles, which have a simpler stress state but are manufactured using the correct system materials. These can be used to measure bond strength in environmental trials to try to drive bond failure and its detection via an NDE technique. Using this evidence, the bond strength of the production asset can be established without destroying valuable assets.

10.4.7 Nondestructive evaluation

Advances in NDE imaging techniques continue apace. Much of this is not just due to improved and/or new techniques and the sources and sensors that are used, but also in the advancement of computing power and development of more suitable algorithms.

X-ray radiography is still used as the best NDE tool. Historically, this utilised wet film and images were taken in two dimensions. Nowadays, the acquisition of such data is in digital form. This methodology tends to enable the highest resolution images to be collected.

A complementary technique which is seeing increasing use is CT, which allows better characterisation of the location and size of defects because the images are collected in 3D. An increase in source energy allows more types of assets to be interrogated, for example those with thick cased walls.

Ultrasonic inspection is also routinely used for bond-line assessment but further improvements are desired, with reference [121] describing a methodology which uses a time-frequency analysis algorithm. There are potentially other sources that can be used, such as microwave radiation for bond-line assessment [122]. Less proven is the ability of ultrasonic monitoring to consistently detect changes in the elastic modulus of the propellant or PBX fill as it ages, though there is sufficient interest that allows research in this area to continue with the aim to measure the changes externally through any casing so that this remains a true NDE technique.

A new method for *in-situ* characterisation of the dynamic modulus and monitoring of ageing was investigated using an electro-mechanical impedance (EMI) technique [123]. Whilst the process requires only a small number of embedded piezoelectric sensors to be incorporated in order to reflect the mechanical properties of SRMs, the methodology is not quite a true NDE technique. However, the authors claim it could replace the DMA test method, though it is likely that the viscoelastic EMI model will require each material to be characterised to determine the numerical relationship between the dynamic modulus and the measured impedance. Nonetheless, rather than being used for *in-situ* monitoring through the application of sensors it may be possible that the technique could be used as a 'nondestructive' test on samples.

10.5 Modelling

The use of modelling and simulation (M&S) in underpinning smart through-life management is centred on a modelling tool hierarchy integrated with instrumented experiments and materials tests [124]. It is this integration of the two approaches, experimental and computational, that makes the approach smarter. Key to all of this is the creation and use of appropriate, reliable, and accurate data in order to avoid the old adage of 'junk in, junk out' when it comes to M&S. Additionally, there has to be effective understanding of the materials in order to make credible predictions of future behaviour. There will always be a requirement for some degree of ISS in order to check that the energetics are behaving as expected or to provide data for the refinement of the model.

It is necessary to initially define what is meant by M&S. Models are there to provide useful information because most questions need more of a response to that shown in figure 10.7 [124].

This is not a model because a direct comparison with a defined comparator gives the answer. A different question requires a model to give an answer, as shown in figure 10.8. Smart through-life management is this second type of question.

Modelling thus represents a series of different tools of varying complexity linked within a hierarchy. This includes models from full numerical simulation methods through to fast-running engineering models (FREMS). The former are

Figure 10.7. Simple question. Reproduced from [124], with permission of QinetiQ Ltd, 2022.

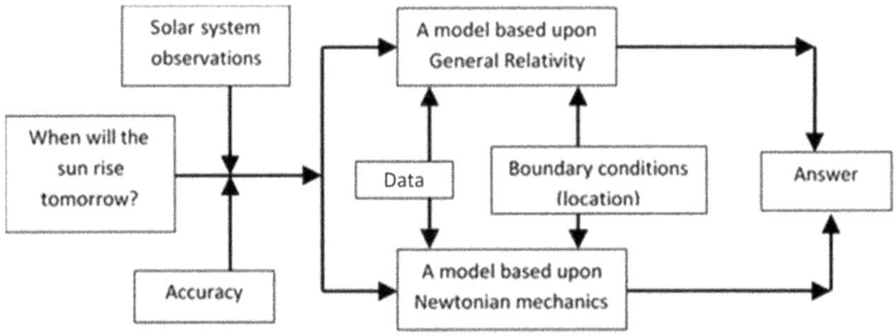

Figure 10.8. Question requiring a model. Reproduced from [124], with permission QinetiQ Ltd, 2022.

termed low-level models and the latter high-level models. The experimental element includes laboratory-based material tests to characterise the material models with instrumented multi-scale experiments providing model verification and validation and design space characterisation. Its success demands a commitment to continued investment in both models and experimentation [124].

A key part of this is the ability to integrate the life-assessment models for energetic materials. This assumes that there is a known mechanism for the energetic material failing in some way [49]. The topic of modelling with respect to energetics is vast but focusing specifically on life assessment there are broadly two approaches. There is the fitting of acquired data, which may be used to make predictions of how the material or system may behave in the future. The other is an *ab initio* or a physics-based model approach.

Evidently, if the same prediction is made from the two different methods this gives confidence in the modelling. However, if the projections are radically different this implies a lack of understanding of the process which requires further investigation and/or there is a probability that an assumption that has been made is invalid.

In order to predict the lifetime of an energetic component, there are four types of factors that need to be known. It may also be necessary to take account of cumulative damage such as those due to thermally and mechanically induced stresses, as well as probabilistic factors due to the variation in properties and history of the assets. The areas where data are required, and therefore may be integrated into a model, are as follows [49]:

- The critical ageing mechanisms.
- The rate expression and so the temperature dependence for these ageing mechanisms from kinetic measurements.
- The service environment of the asset across storage, deployment, and operational use.
- The design limits of the energetic component.

To achieve this, though, is exceptionally challenging. Ideally, predictive testing and modelling would be able to demonstrate the link between the stimulus and the failure and so explore the potential future state of a weapon system under different scenarios. Furthermore, no degradation models, such as those for thermal degradation of energetic materials, currently capture the true complexity of all materials involved [49].

In the future, machine learning, a branch of artificial intelligence pertaining to the creation of models that can effectively learn from past data and situations [125], may assist with the development of accurate models for the life assessment of energetic materials so long as the input data and interpolation functions are good. The use of such approaches, which includes experimental design, can also be informative for ensuring that only the most appropriate experiments are conducted in order to obtain as precise and reliable data as possible [126]. However, obtaining validated data, which is not necessarily available for all required aspects for the life prediction modelling of energetic materials, may remain a challenge.

From an integrated energetic materials perspective, the provision of validated data sets for the development of life-assessment models would be a significant advancement in this field. MSIAC's EMC is an important resource for such activities [48]. However, as discussed previously, this relies on organisations being willing to freely share data without concerns about IP.

Due to the complex nature of energetic materials, including aspects such as the range of components in formulations, and the fact that the materials are not always uniform and that they might contain defects or not behave in an ideal manner, life-assessment modelling has focused on more achievable smaller subset areas. If the goal of increasing the use of modelling is to be reached, then efforts need to concentrate on producing energetic materials to much tighter specifications with less batch-to-batch variations, otherwise the models will never be able to make as accurate predictions as desired.

10.5.1 Service life prediction modelling

As stated in reference [127], the aim of service-life assessment is to provide the most reliable and accurate prediction of service life possible in support of assuring safety, mission performance, and effective cost. This document provides significant detail on the range of activities which can be conducted in the service life prediction field, including modelling. Key papers on this topic include that by Cunliffe *et al* on 'the ageing and life prediction of composite propellant motors' [128] and Faulkner and Tod on 'service life prediction methodologies' [129].

Degradation models can be used to optimise life-assessment estimates. This generally involves the rate of decomposition measured during testing being compared to the rate of degradation estimated for the service requirement, or actually measured in service [49].

An overview of the approach for a predictive service-life model for a composite propellant is shown in figure 10.9. The general methodology can be applied to a range of energetic materials such as other propellant compositions [51].

Using figure 10.9 as an example, there are a number of challenges in developing a predictive service-life model. There may be insufficient data or understanding about how microstructural changes in the grain occur, which may be as a result of the fact

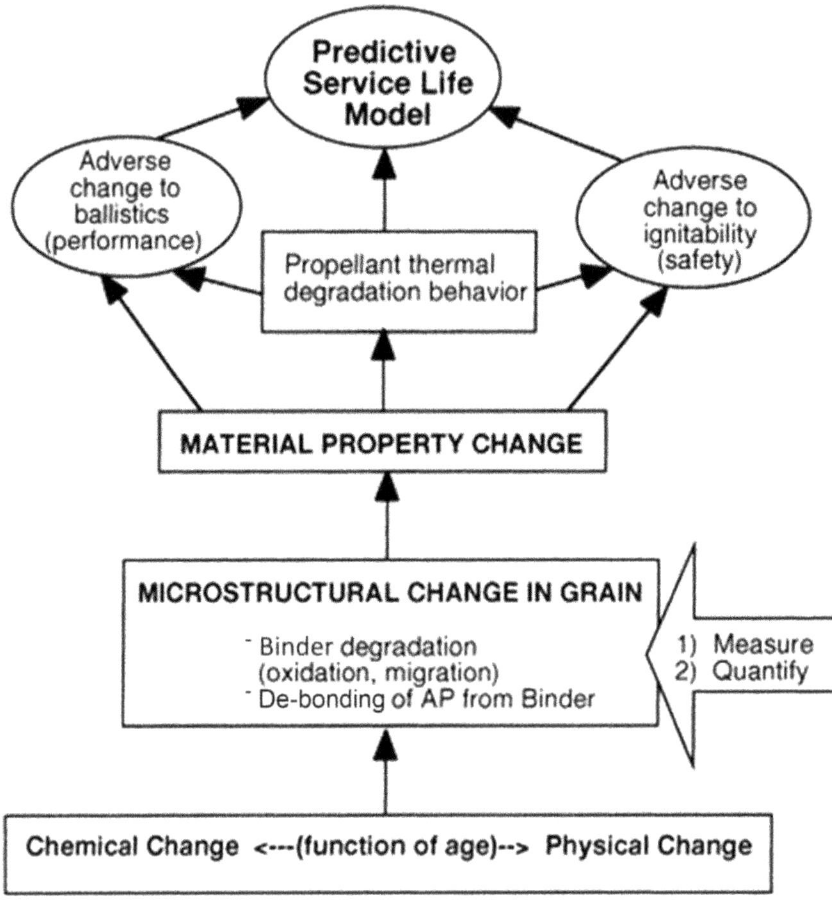

Figure 10.9. Inputs for predicting the service life of a composite propellant Reprinted from [130], copyright (2002) with permission of Elsevier.

that the analytical techniques for the material assessment are insufficient for the purpose. It may be very difficult to directly validate the model once constructed, and the tools that are potentially used as part of the modelling process, such as FEA, may be deficient [49].

A key part of degradation modelling is the development of validated ageing algorithms. In order to achieve these, there needs to be an integration of the fundamental scientific understanding of the energetic materials, their associated ageing mechanisms, and the failure modes of the weapon system [49]. There has been a significant amount of research into ageing algorithms conducted over the decades [131–138] and the use of activation energies for describing decomposition processes, and integrating these with information on the environment to provide a prediction of change [139], is very well established [111, 140].

In most research, the estimation of service life still relies largely on the inverse power law in its basic chemical reaction rate (Arrhenius and Berthelot) and cumulative fatigue damage (Miner–Palmgren) forms [40, 141, 142]. In 2010, Bayern-Chemie [40], used inverse analysis of a sensor's data to demonstrate that naturally aged motors, stored in a bunker for 5 years, had aged as expected by models. This method has been confirmed by several users applying accelerated ageing and reverse-computing the propellant modulus to indicate ageing effects. Unfortunately, testing based on the inverse power law rarely remains reliable when the compression ratio is set greater than 10:1, with failures occurring above this. While these methods show reasonable success, their limitations highlight a clear requirement for more physical models to allow a combined data analysis with better accuracy.

A service-life prediction analysis tool [141, 143, 144] was developed that was suitable for use on rocket motors with cast composite solid propellant grains stored under statistically varying environmental conditions. The Service Life Prediction Programme (SLPP) was a suite of computer programs written in Fortran which brought together different aspects needed to improve capability for predicting the service life of rocket motors. The failure prediction was based on either stress- or strain-based failure models, such as usually used in manufacturer design MoS calculations.

The program is designed to use real temperature data from in-service conditions. Where this data is absent the programme can generate data using probabilistic techniques based on Weibull statistics, which at its basic level describes the seasonal variation with a diurnal variation imposed on top. At its simplest, rudimentary material properties such as mechanical stress and modulus can also be described by a Weibull distribution. What is then needed for a composite propellant is how the material chemically ages and how this transfers to changes in those material properties. This model incorporated the kinetics for the differences in the in cross-link density of the binder by thermally driven oxidative ageing and also a mechanical cumulative damage factor, which is also dependent on the thermal environment. A schematic of this process, which shows the separate Fortran modulus in bold, is given in figure 10.10.

Even this model is very simplistic and does not take into account other failure mechanisms and/or more complex damage models, including fracture mechanics and potential effects of vibration on the mechanical damping properties.

Attempts have also been made to use Bayesian methods to predict the useful life of a munition. Bayesian methods incorporate prior knowledge about the solution to an estimation problem. This involves combining prior knowledge (the prior) about what is expected with knowledge from the actual observations (the likelihood). The final solution is a combination of this prior and likelihood [139].

Such an approach has merit for service life prediction because a prognostic approach to estimate the remaining useful life based on the current state of the systems can give more accurate predictions of the true remaining useful life. If these are used alongside the correct measurement data and analysis, these can optimise stockpile management and thus reduce whole-life costs [139].

Bayesian methods were applied to a representative degradation model from laboratory experimental data for a composite propellant which had been exposed to a range of temperature and humidity conditions [139]. Whilst the research

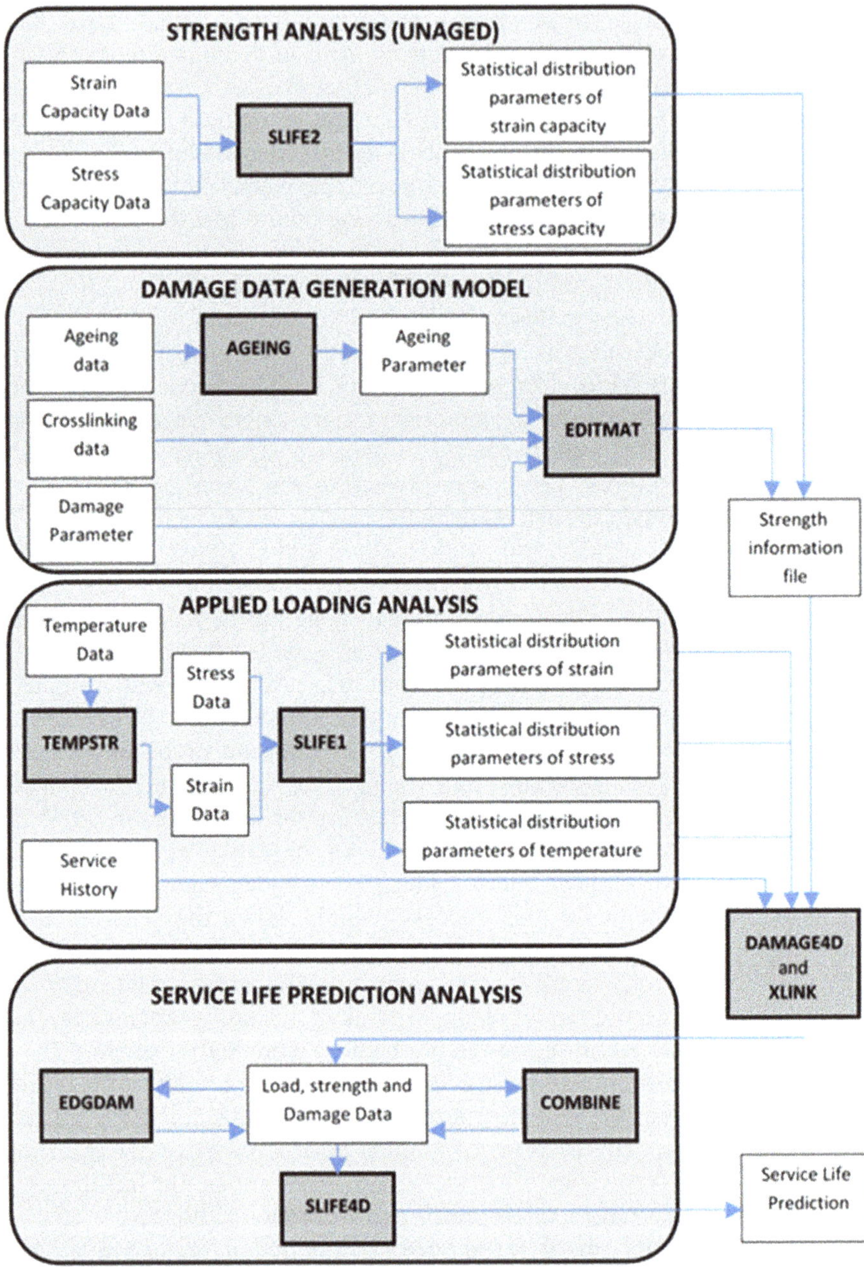

Figure 10.10. Schematic of a service life prediction programme. Reprinted with permission of QinetiQ Ltd, 2022.

demonstrated the principles of the approach, the need to measure sufficiently accurate and reliable data at numerous points of the asset's life was underlined, which remains a significant challenge with respect to energetic materials. The modelling showed that this factor was considered more notable than the measurement of the environment, where an approximate knowledge of the environmental conditions was sufficient for reasonable predictions.

Other authors have noted, however, that the energetic components within assets may have different degradation mechanisms and so activation energies. This means caution should be used in globally describing temperatures and humidities as relatively benign. In addition, whilst low temperatures may be beneficial in terms of not accelerating processes such as stabiliser depletion, they still need to be taken into account as they can cause other issues [145].

10.5.2 *Ab initio* or physics-based modelling

A potential alternative integrated modelling approach is through the use of *ab initio* or physics-based models. These are different to those discussed in the previous section as they are derived from first principles [146].

As an example, group interaction modelling [147] has been very successful at predicting the mechanical properties of many homopolymers and block copolymers that interact via van der Waals forces. Michalchuk *et al* [148, 149] have also used this type of modelling approach for predicting the reactivity of energetic materials, which enables their ranking in terms of impact sensitivity, and as a possible tool in designing energetic compounds with tailored properties. Finally, a physics-based model has been developed in analytic form that enables most of the thermomechanical response of a composite energetic material to be predicted under in-service environmental conditions [138].

This methodology is thought to have application to the field of service life prediction. This is because it has the potential to predict both the consumption of qualified life and the true end of life of an energetic material. As they are based on the physics of the materials, they are likely to be better suited to modelling situations such as inhomogeneous ageing and failure modes which can be a function of multiple factors. Potentially, the models would be component specific, which would enable each asset to have its own life assessment. For even greater integration, there is the potential for these models to be incorporated into the munition itself although this is likely to require the development of FREMS and highly power-efficient hardware [49].

Research conducted to improve propellant constitutive models is presented in reference [36]. Attempts to predict the macroscopic behaviour of propellants from the microscopic behaviour of its constituents were made. It was found that by linking the mathematical descriptions of the structure of the particle pack, the chemical characteristics of the polymer and the particle–binder interaction, it is possible to create a macroscopic constitutive law that can be used in finite element codes. Unfortunately, it was found that the model was too computationally expensive to be of use on any simulation of significant size.

In order to develop physics-based models for service life prediction, a number of inputs would be required, including the following [49]:

- Extensive material characterisation and knowledge of the materials.
- Defined ageing mechanisms and the associated ageing algorithms.

- Component failure modes.
- Data logging to monitor and record the driving environmental parameters.
- Digital thread/twin to exercise models to predict failure/remaining life.

In summary, in order to have integrated models for energetic materials which can be used to explore scenarios and facilitate a smart approach, a number of criteria must be satisfied. These include environmental data, such as temperature and humidity, coupled with material properties that have been accurately and reliably measured and can be validated through the use of data loggers and other real measurements. Any model therefore needs to be able to demonstrate quantitatively where the energetic material currently is in terms of the degradation pathway and allow the accurate prediction of remaining life, given potential future environments and scenarios [49].

10.5.3 Companion assets and trepanning

Another way of improving integration from a through-life perspective is via the use of companion assets for testing. These should be, where possible, assets from similar batches or near-similar batches. This is because of known batch-to-batch variations in properties, which can make analysis of results sometimes difficult.

By using companion assets, at least this batch-to-batch variability can be discounted and better statistical confidence in the effects of ageing can be gained. It should be noted that we are not here concerned about the natural variability in properties at manufacture but instead the effects of ageing. Once these are characterised they should be able to be used to estimate the effect on different batches that have slightly different baseline properties. If the companion assets age badly then it is probable that other assets from different batches will also do so. Conversely, if there is next to no ageing then there is good confidence that the system does not age significantly.

Even with a relatively large number of possible companions there will still be a limit in setting assets aside in benign stores for long lengths of time (typically up to 20 years), which then feed into naturally aged in-service surveillance and life extension activities. In addition, a representative amount of testing should be done on baseline assets to make sure that they have been characterised to a good fidelity—usually more so than any aged asset. If safety and suitability for service use (S3) qualification testing is required, which it usually is, then this requires further 'companions' from the same batch as a suitable 'baseline'.

Even for small-arms munitions these requirements to eliminate or even reduce the possible effects of batch-to-batch variation are problematic. For larger munitions there will not be sufficient numbers from the same batch that can fulfil this ideology—in fact, for some munitions a batch will perhaps only fill five or six assets.

In general, even if there were no limits on the size of each batch the drive must be to reduce the number of complex weapons that are subjected to costly surveillance testing using traditional breakdown and sectioning procedures. As well as being costly to run, these activities are also time consuming. By definition, they also remove the asset from the stockpile.

A more conceptual idea for using companion assets is with periodic sampling of the same asset over a length of time, thus comparing the test results with their own baseline. This has two initial benefits: first, that batch-to-batch variation (manufacturing variability) should be completely eliminated; second, that the same assets can be used to generate test results over a length of time, therefore reducing the number of assets that need to be examined. The methodology of how this sampling fits in with routine in-service surveillance is shown in figure 10.11.

In a sense, a methodology around using the same companion asset may have been previously conducted. It is not inconceivable, for example, that a rocket motor was cut into sections after production and then these lain down for samples to be taken and tested at various time intervals. There are many disadvantages to this in that the sections will not be as representative of the motor as they should be due to differences in properties such as stress states, changes in diffusion profiles of chemical species, varying response to thermal expansion, and many other factors.

As such, the most ideal way to sample companion motors would be to remove material from the same asset at regular intervals and to test this. Radiography and/or other NDE techniques could also be undertaken at the same time to assess for any changes in the bond line and/or charge design, such as due to slump. However, these are likely to needed to be conducted at ambient temperature so as not to disturb the precisely known 'ageing conditions'.

Trepanning is a process by which samples can be extracted from a system without affecting the integrity of the unit. This is not a novel concept but much of the discussions about the process and the types of munitions it has been used on are not available in the open literature.

The advantages of trepanning are that the effect of ageing can be determined using one companion asset over a period of time, and thus there is confidence that the ageing effect is real. The normal trepanning method is to remotely use a trepanning tool to remove a 'core' sample from the rocket motor, for example. The

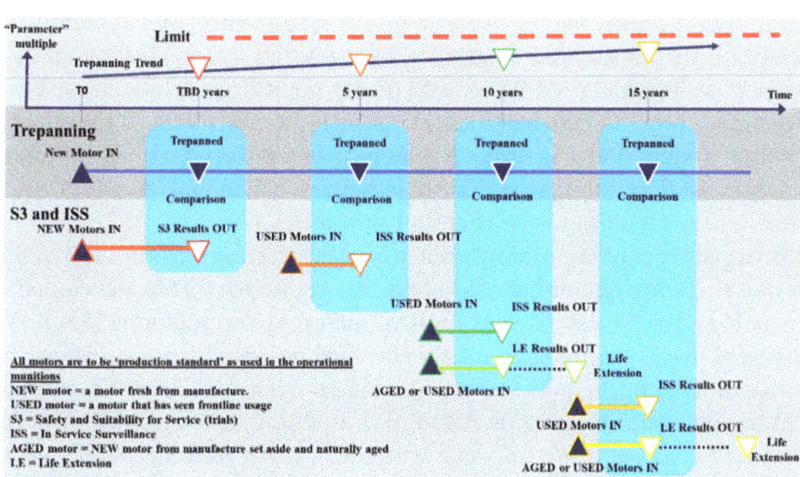

Figure 10.11. Surveillance testing timeline and trepanning. Reprinted with permission of QinetiQ Ltd, 2022.

hole that remains is then filled and sealed against atmospheric influence, including purging with an inert, nitrogen atmosphere if that is the standard for that particular motor.

This method provides a mechanism to provide the maximum number of samples for testing from one asset, rather than to section the motor and then further machine those test pieces.

As the trepanning technique allows removal of a core of material from the rocket motor, this enables suitable chemical and mechanical analysis samples to be prepared for testing on material from different positions, from the conduit to the case. This means that as well as the bulk propellant being tested the effects at the conduit surface and at the propellant/liner/case interfaces are examined, thus enabling a full analysis of how the rocket motor is ageing. The trepanning sampling technique then preserves any future samples from effects due to internal stresses, as well as due to diffusivity and/or permeability of species.

Whilst the trepanning method of sampling does lead itself to those assets that are lain down for a long length of time, there is no reason that companion assets cannot be placed in a slightly less benign thermal environments. The standard S3 qualification and life extension activities tend to use elevated temperatures to reduce the time of the ageing programme, which as more often than not are solely based on generic Arrhenius degradation kinetics. Additionally, there are different potential ageing mechanisms in a subsystem/component, let alone a full complex weapon AUR. On top of this, there is caution about how mechanisms at the typical accelerated ageing temperatures (60 °C static temperature or up to 71 °C when diurnally cycling) really extrapolate back and represent the ageing at more typical storage and/or service temperatures.

Exposing companion rocket motors to perhaps a temperature of 40 °C for a very long length of time may help bridge the gap between what happens in natural ageing and that which occurs in accelerated (forced) ageing and give some element of life extension if it is desired. The concept could be extended to higher temperatures, including diurnal cycling, which will also give a greater mechanical stimulus due to the differentials in the thermal expansion between the materials that make up the companion (case/liner/propellant), as well as driving diffusion processes. The use of the companions allows the environmental inputs to be controlled so that their effects can be better understood. As well as batch-to-batch variation the differences in environments that a munition in service experiences, which may be unknown, have a great effect on the changes to the material properties.

The history of trepanning munitions is not a new concept; traditional methods of bomb disposal using trepanning and steaming techniques were developed during World War II to gain access to and remove melt cast formulations [150]. Whilst it was recognised that trepanning could also be used potentially for sampling the asset for test it does not appear that this was taken up in earnest, and the methodology seemed to be very much focused on disposal. Later on the methodology of executing a similar bomb disposal using abrasive water jet cutting as a method of access was examined. This allowed generation of much larger openings than could be produced using the circular trepanner as well as access to hard-to-reach components. Again, it is not clear whether sampling for laboratory testing had been routinely conducted on assets that had been opened up.

The trepanning in these instances cuts through the casing. Sampling of assets has been conducted using a similar methodology but with the assets partly broken down to gain access to the material, such as through the aft-end plate of a bomb. In these instances, samples are effectively corked bored out using the same methodology rather than use sectioning. The difference is that, like sectioning, this is a one-off activity and no further sampling can be conducted at a later date. This is because, firstly, the availability of material is limited by the depth of the sample you want and therefore only a small volume of the total available is used; secondly, the assets have been fully broken open and would require suitable resealing, including possibly gas purging, to prevent unintended ageing mechanisms taking place.

10.6 Digital threads and twins

A trepanned asset can be considered a real-life digital twin. The ultimate way of integrating energetic material data for smart through-life management is through the use of digital threads and twins. A digital twin is a digitally modelled version of the system used to mirror the real system through development and service. The principles of smart through-life management build towards this idea by promoting the wider use of modelling and more advanced ISS. While the realisation of digital twins for weapons systems has not been widely established, several papers have begun to derive the concept for similar systems [151].

A digital twin is defined as a virtual model of a physical thing [152], which in a service-life scenario for a munition can be used in the following:

- Assessing condition-based service life.
- Supporting service life extensions.
- Establishing initial service life.
- Cartridge-activated devices (CADs)/pressure-activated devices (PADs) health monitoring.
- Failure investigation support.

To explore this further, consider the case of a munition going through ISS [153]. As discussed previously, the purpose of this is to ensure that the energetic components are not degrading faster than expected during operational life. Traditionally, the evidence is gathered via an ISS programme that involves the breakdown and analysis of a sample of operational weapons, which has the consequence of reducing the available stockpile. At the end of the certified life, these weapons are then withdrawn from operational service. A poor understanding of the ageing processes, coupled with the need to manage operational risk, means that in many cases weapons may be withdrawn despite having significant useful life remaining.

An improved and more integrated approach starts with a detailed understanding of the physical properties of the materials and the systems and how these age and change with operating environment. Combining the predictive power of these tools with real-time monitoring of the operational environment experienced by the weapon, on a weapon serial number basis, allows both focused surveillance activities and a much improved ability to dynamically manage the stockpile. This includes facilitating the switching of assets between environments if appropriate. These contrasting approaches are illustrated in figure 10.12 [153].

Key elements of this smart surveillance activity are the concepts of the weapon digital twin and the associated digital thread of evidence. This is an exemplar of embracing the Fourth Industrial Revolution and the breakthrough of digital engineering. It opens up the opportunity to create a digitally enabled test and evaluation enterprise that supports the deeper management, capture, integration, and exploitation of evidence during the capability generation cycle [153].

Defence is rapidly realising the transformative effects of digital technologies on the cost and 'time to market' of increasingly complex platforms. Hence, the next generation of systems will benefit from the use of model-based systems engineering, digital twins, and digital threads [153].

Current in-service weapons could be used as test beds to accommodate these techniques and build on the advances being made, leaning from the civil sector where similarly highly regulated industries such as automotive and civil air have embraced digital engineering technology [153].

This would enable the validity and pace of decision-making to be increased through establishing an evaluation digital thread, a single assured version of the truth, which supports information reuse [153].

To deliver this, a number of key building blocks need to be established. Firstly, the necessary digital infrastructure, tools, and processes to create, share, use, and manage knowledge and information, harnessing the benefits of Big Data capture and analytical techniques to provide near real-time, coherent information. The digital infrastructure and tools could enable federated experimentation across all key stakeholders [153].

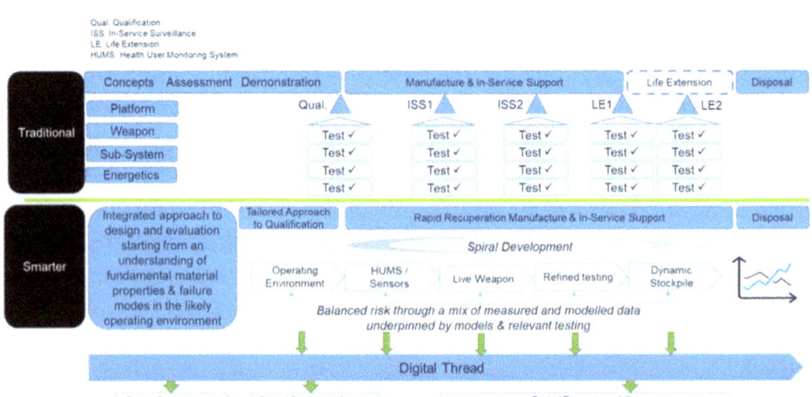

Figure 10.12. Traditional through-life management vs. an enhanced and fully integrated approach incorporating a digital thread [154]. Reprinted with permission of QinetiQ Ltd, 2022.

In addition to this, the following are also important to achieving digital threads and twins [153]:

- Model-based engineering and systems engineering to derive cost and risk balanced testing strategies from evaluation needs, requiring a deep understanding of the evidence required to meet defence standards and how this can be compiled progressively through life and increasingly from models. This needs to be coupled with an ability to consider and recommend alternative methodologies for the capture of evidence.
- Experimental design, data evaluation, and data analytics facilitating the broader use of deeper data-analysis techniques.
- Integration and exploitation of synthetic tools and techniques through understanding the standards and approaches needed to know how to integrate models and simulations both together and into live components.
- Digital and IT engineering to bring pace and agility to projects through deeper digital integration of geographically separate virtual and live components.
- Development and validation of the tools to allow evidence derived at subsystem level or early in the life cycle to be used with confidence for safety and assurance management.

Whilst there are many clear benefits of digital threads/twins, building a fully integrated system to allow the life prediction of a complex weapon system, for example, would be hugely difficult, not least in terms of cost. In 2017, West and Blackburn provided analysis of the challenges that the US Department of Defense would face in creating digital threads and twins for weapon system development. They concluded that if this was to be realised, the effort would require a national commitment on par with the Manhattan Project [154].

Similarly, as described in reference [155], the potential for digital twins to assist with SRM design is detailed along with the key technologies required and the associated challenges. One of these is that data is the basis of the digital twin, which has multiple sources, dimension and attributes. The authors state that there are many problems, which include scattered data, difficult characterisation, unclear evolution mechanism, and high noise. They note the urgent need to establish a standard system of data modelling and management systems and an integration standard including data pretreatment, multivariate data fusion, and digital representation methods.

Despite this, there have been efforts towards developing digital twins in recent years. Kraft and Johnson [152] reported the work by the Naval Sea Systems Command (NAVSEA) on digital twin models for a double-base propellant and composite propellant. For the double-base propellant model they recorded the temperature in the cockpit of the aircraft on which the CAD/PAD would be utilised. They used this to create a temperature prediction model which was validated against data from a data logger that was installed in the static aircraft, and this model was refined based on the collected data. Sampling of the data took place every 20 min for 1 year and the data was downloaded every 3 months.

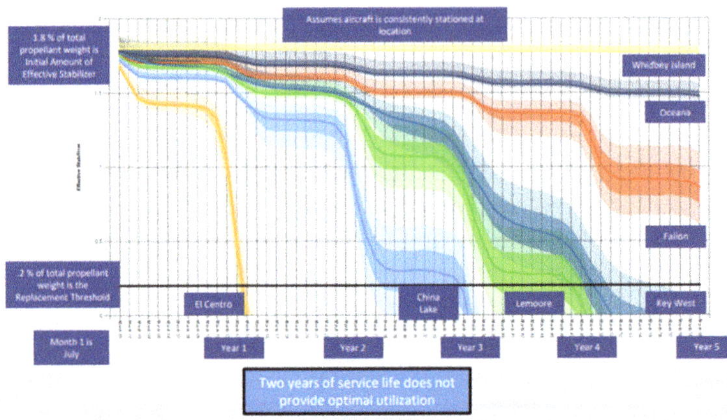

Figure 10.13. Predictions of stabiliser consumption at various locations in US. Reproduced from [152] with permission of US Navy.

The data were combined with models of stabiliser decomposition from accelerated ageing test data on the propellant to provide a prediction of the profile of stabiliser depletion at various locations in the US. This is shown in figure 10.13 and it enables predictions to be made of when the stabiliser depletion is likely to be considered so significant that the propellant has to be removed from service.

10.7 Conclusions

A smarter approach to through-life management of munitions is being demonstrated worldwide and nations continue to strive to make improvements in this field. As part of this, understanding the failure modes, risks, and hazards of the energetic materials to enable focused testing remain important. Laboratory instrumentation continues to evolve, offering enhanced ways of analysing energetic materials, but the need to obtain meaningful data has never been so critical. It is likely that service life prediction models will continue to develop, with physics-based models becoming more important. These will make use of improved sensors and HUMS systems, along with enhanced laboratory data. Data from companion motors may also be inputted into the models to provide a more accurate picture of the true life of the munition. Finally, it has been demonstrated that the ultimate integration of all the data from across the through-life management process can be achieved through digital threads and twins.

Acknowledgements

The authors would like to acknowledge dstl (Pete Fincham, Dr Giles Clark, and Gavin Bishop) and the UK MOD (Phil Cheese).

Andy Rix (QinetiQ), Dr Peter Gould (QinetiQ), Mike Firth (QinetiQ), Dr Paul Steward (QinetiQ), and Kay Rawley (MBDA UK) are thanked for informative discussions. Holly Pennington (UK MOD) is recognised for her contribution in shaping this chapter.

Finally, the QinetiQ Fellowship Scheme is acknowledged for the funding of the writing process.

Glossary

AEP	Advanced energetic propellants
AOP	Allied Ordnance Publication
AP	Ammonium perchlorate
AUR	All up round
BTCA	Breakdown, test, and critical analysis
CAD	Cartridge-activated devices
CADMID	Concept, assessment, demonstration, manufacture, in-service and disposal
CMDB	Composite modified double-base
3D	Three dimensions
DIC	Digital image correlation
DMA	Dynamic mechanical analysis
DSAC	Defence Scientific Advisory Council
DSC	Differential scanning calorimetry
ELISA	Effective logistics monitoring system for ammunition
EMC	Energetic materials compendium
EMI	Electro-mechanical impedance
EIST	Equivalent in-service time
EMTAP	Energetic Materials Testing Assessment and Policy. A UK working group for the small-scale hazard testing of energetic materials
FGD	Farris gas dilatometer
FEA	Finite element analysis
FEM	Finite element model
FREMS	Fast-running engineering models
FT	Fourier transform
HFC	Heat flow calorimetry
HSL	Health and Safety Laboratory
HUMS	Health and usage monitoring system
IP	Intellectual property
IR	Infra-red
IT	Information technology
ISS	In-service surveillance
JANNAF	Joint Army Navy NASA Air Force
L	Likelihood
Latax	High velocity, high lateral acceleration
M	Medium
MHM	Munition health monitoring/management
MOD	Ministry of Defence
MoS	Margin of safety
MSIAC	Munitions Safety Information Analysis Center
MTBF	Mean-time-between-failure
MS	Mass spectroscopy
M&S	Modelling and simulation
MUS	Minimum useable subset

NATO	North Atlantic Treaty Organization
NAVSEA	Naval Sea Systems Command
NDE	Nondestructive evaluation
NEPE	Nitrate ester plasticized polyether
NOSSA	Naval Ordnance Safety and Security Activity
PAD	Pressure-activated devices
PBX	Polymer-bonded explosives
PCA	Principal component analysis
PHM	Prognostic health management
PRL	Prioritised requirements list
RIST	Real in-service-time
S	Severity
S3	Safety and suitability for service
SET	Sequential environmental test
SLPP	Service life prediction programme
SRD	System requirements document
SRM	Solid rocket motors
STANAG	Standardization Agreements
TG	Thermogravimetry
UAVs	Unmanned air vehicles
UK	United Kingdom
URD	User requirement document

References

[1] Defence Science and Technology Laboratory 2021 *Energetics: The Future Is Explosive* DSTL

[2] NATO 2006 *The Scientific Basis for the Whole Life Assessment of Munitions* AOP-46 NATO MSIAC

[3] Barnes P 2002 *Overview of UK Initiatives to Enhance the Whole Life Assessment and Surveillance Policies for Munitions* MSIAC, UK MOD

[4] Barnes P 2003 *Scientific & Technology (S&T) Advances to Improving the Whole Life Assessment of Munitions* MSIAC, UK MOD

[5] Barnes P and Sharp M 2007 Smart assessment & testing of weapons *IMEMTS 2007 (Miami, USA)*

[6] Barnes P, Sharp M, Sloan M, Flower P, Steward P and Mullenger D 2008 *Using enhanced scientific understanding of energetic materials and consequences for their qualification* DDESB33 UK MOD

[7] Hooper G and Bryant R 2011 *Smart Assessment and Testing* FTS3/DOSG/7196, Issue 1 MOD and Cranfield University, UK

[8] Swedish Defence Research Agency 2009 *Insensitive Munitions and Ageing* FOI-R-2874 SDRA

[9] NASA 2020 *Systems Engineering Handbook* 2nd rev (Washington, DC: NASA)

[10] Garnell P and East D J 1977 *Guided Weapon Control Systems* (Oxford: Pergamon)

[11] Gilchrist W 1993 Modelling failure modes and effects analysis *Int. J. Qual. Reliab.* **10** 5

[12] US Department of Defense 2019 *Environmental Test Methods and Engineering Guidelines* MIL STD-810H US DoD

[13] Cranfield University 2004 Department of Environmental and Ordnance Systems, Warheads Handbook
[14] Bailey A and Murray S 1988 *Explosives, Propellants and Pyrotechnics Land Warfare* vol 2 (London: Brassey's)
[15] UK Ministry of Defence 2017 *Defence Standard 00–035 Environmental Handbook for Defence Materiel Part 3—Environmental Test Methods* DSTAN 00-035 UK MOD
[16] Richard F 2020 British Defence Acquisition—A Primer on the Process (28 May 2020) https://overtdefense.com/2020/05/28/british-defence-acquisition-a-primer-on-the-process/
[17] Agile Business Consortium 2017 *Agile Project Management Handbook* 2nd edn (London: APGM International)
[18] Pennington H 2021 *Safety and suitability evaluation for spiral development and rapid capability engineering independent case study* WSRF0011 QinetiQ/21/01148
[19] Pennington H 2021 *Safety and suitability evaluation for spiral development and rapid capability engineering: adapting S3 evaluation approaches* WSRF0011 QinetiQ/21/02087
[20] NATO 2007 *Explosives Nitrocellulose Based Propellants Stability Test Proceedure and Requirements Using HFC* STANAG 4582 NATO
[21] NATO 2001 *Chemical Compatibility of Ammunition Components with Explosives (Non-Nuclear Applications)* STANAG 4147 NATO
[22] Itkis D and Bohn M 2021 Simulation of heat flow curves of NC-based propellants—part 1: determination of reaction enthalpies and other characteristics of the reactions of NC and stabilizer DPA using quantum mechanical methods *Propellants Explos. Pyrotech.* **46** 1188–203
[23] Itkis D and Bohn M 2021 Simulation of heat flow curves of nc-based propellants—part 2: application to DPA stabilized propellants *Propellants Explos. Pyrotech.* **46** 1204–15
[24] Richards D 2016 *Validation and exploitation of munition environmental data* STO-MP-AVT-268 NATO
[25] Rix A 2020 *Health usage monitoring systems—strategy for lifing algorithms for complex munitions* WSTC 0503 QinetiQ/19/03516/3.0
[26] NATO 2008 *Principles and Methodology for the Qualification of Explosive Materials for Military Use* STANAG 4170 NATO
[27] Sell T, Vyazovkin S and Wight C 1999 Thermal decomposition kinetics of PBAN-binder and composite solid rocket propellants, combustion and flame *Combust. Flame* **119** 174–81
[28] Vyazovkin S and Wight C 1999 Kinetics of thermal decomposition of cubic ammonium perchlorate *Chem. Mater.* **11** 3386–93
[29] Tunnell R 2018 *BTCA CE study (phase 2)* DGM 1815, sub task 16/B/17 QinetiQ/18/04089
[30] Church P and Gould P 2021 Universal validation experiments for energetic materials *QinetiQ*, QINETIQ/21/03375
[31] Williamson D 2009 *Fracture and damage assessment in PBXs and propellants: a review* SP1209 Cambridge University
[32] Ingamells V 2013 Predictive modelling: report on the modelling framework QINETIQ/13/00876, *QinetiQ*, September
[33] Ingamells V, Hills D and Lynch N 2010 Response surface modelling development phase 1, *QinetiQ*, March 2010
[34] European Defence Agency 2021 EDA's munition health management project expands https://eda.europa.eu/news-and-events/news/2021/11/25/eda-s-munition-health-management-project-expands

[35] MSIAC 2017 Integrated munitions health management https://msiac.nato.int/workshop/integrated-munitions-health-management
[36] Ruderman G 2011 *Health management and service life for air force missiles—past and future* RTO-MP-AVT-176 Air Force Research Laboratory, NATO
[37] A. N. M. INFO 2007 Sensor application and modelling programme
[38] Schultz E 2016 *Munition health monitoring—feedback from the use of environmental data loggers* L-193 MSIAC
[39] ATK Launch Systems IHPRPT phase 3 sensor application and modelling programme
[40] Tussiwand G, Stiharu I and Buswell J *Data gathering, sensors, and integration of sensors* RTO-TR-AVT-160 NATO
[41] Mueller R Wireless sensing for munitions health management, Raytheon Missile Systems
[42] Crosse B, Adamson E and Back C 2001 Whole life munitions and weapons systems monitoring for security and serviceability *PARARI 2001 5th Australian Explosive Ordnance Symp. (Canberra, Australia, November, 2001)*
[43] Swanson R 2002 *U.S. Navy Weapons and Ordnance Safety Life Cycle* DDESB30 NOSSA
[44] Marotta S, Kudiya A, Ooi T, Toutanji H and Gilbert J 2005 Predictive reliability of tactical missiles using health monitoring data and probabilistic engineering analysis *viation and Missile Research Development and Engineering Center (AMRDEC)*
[45] Kratzsch K *Munition health monitoring—evaluation of ammunition load data gathered under operational conditions* RTO-MP-AVT 176, Paper NBR 28–1 NATO DOI:https://doi.org/10.14339/RTO-MP-AVT-17603/22/2010
[46] DeVries D, Hyde S and Davis I 2017 *A systems engineering approach to evolution of physics-based prognostic health management of ageing solid rocket motor system assets* STO-MP-AVT-26 NATO
[47] Teng K, Han J, Ge F and Zhang G 2018 Research on missile weapons health management and storage life prediction methods based on PHM technology *Prognostics and Systems Health Management Conf.*
[48] Hollands C 2022 MSIAC Energetic Materials Compendium (EMC) https://msiac.nato.int/products-services/energetic-materials-compendium-emc
[49] Rix A 2019 *Health usage monitoring systems—strategy for lifing algorithms for complex munitions* WSTC0503 QinetiQ/19/03516/4.0
[50] de Klerk W 2015 Asssessment of stability of propellants and safe life times *Propellants Explos. Pyrotech.* **40** 388–93
[51] Tunnell R 2017 Overview and appraisal of analytical techniques for ageing of solid rocket propellants *Chemical Rocket Propulsion* (Switzerland: Springer) pp 743–69
[52] Youngberg J and Burstein P 1995 Three dimensional computed tomography for rocket motor inspection *Proc. SPIE* **2455** 291–8
[53] Yildirim H and Özüpek S 2011 Structural assessment of a solid propellant rocket motor: effects of ageing and damage *Aerosp. Sci. Technol.* **15** 635–41
[54] Davis D 2001 Use of dilatation in understanding composite propellant ageing *37th AIAA/ASME/SAE/ASEE Joint Propulsion Conf. and Exhibition (Salt Lake City, UT, 8–11 July)*
[55] Faulkner G, Thompson A and Buswell H 1996 The penetrometer: non-destructive testing of composite propellant rocket motor grains to determine ageing characteristics *AGARD PEP Symp. on Service Life of Propellant Systems* Athens, Greece
[56] Judge M 2003 An investigation of composite propellant accelerated ageing mechanisms and kinetics *Propellants Explos. Pyrotech.* **28** 114–19

[57] Wilker S, Ticmanin U and Stottmeister L 2001 Influence of atmospheric conditions on the ageing of single, double and triple base propellant *12th Symp. on Chemical Problems Connected with the Stability of Explosives* Sweden

[58] Torry S, Earl J, Cunliffe A and Tod D 2006 Effect of plasticiser diffusion on the mechanical properties of double base propellants *Insensitive Munitions and Energetic Materials Technology Symp. (Bristol: UK)*

[59] Asthana S, Divekar C and Singh H 1989 Studies on thermal stability, autoignition and stabilizer depletion for shelf life of CDMB propellants *J. Hazad. Mater.* **21** 25–46

[60] NATO 2008 *Explosives, Nitrocellulose-based Propellants, Stability Test Procedures and Requirements Using Stabilizer Depletion* AOP 48 NATO

[61] Weterings F, Herder G and de Klerk W 2003 Mechanical analysis on rocket propellants *J. Therm. Anal. Calorim.* **72** 921–9

[62] Haberman J 1986 *High performance liquid chromatography of propellants part 1—analysis of M1, M6 and M10 propellants* Technical Report ARAED-TR-86017 US Army Armanent Research and Development Center

[63] Oehrle S 1997 S Analysis of stabiliser degradation products using hplc and photodiode array (PDA) detection *J. Energ. Mater.* **15** 21–3

[64] Fengqi Z, Shuyun H, Rongzu H, Hongxu G and Fang H 2007 A study of kinetic behaviors of the effective centralite/stabilizer consumption reaction of propellants using a multi-temperature artifical accelerated ageing test *J. Hazard. Mater.* **145** 45–50

[65] Bunyan P 2003 Activation energy measurements on a double base propellant by micro-calorimetry over the temperature range 40 °C–90 °C, paper no. D1 *4th Int. Heat Flow Calorimetry Symp. on Energetic Materials (Leeds, UK)*

[66] Bohn M 2018 Characterization of ageing of NC-based double base rocket propellants by not often used measurement properties *8th Nitrocellulose Symp. (Bergerac, France)*

[67] Bunyan P 2001 Accelerating rate calorimetry studies on experimental hybrid compositions *12th Symp. on the Chemical Problems Connected with the Stability of Explosives (Karlsborg, Sweden)* 31–43

[68] Tunnell R *PhD Thesis* Imperial College

[69] Bohn M 1996 Methods and kinetic models for the lifetime assessment of solid propellants *AGARD PEP Symp. Service Life of Solid Propellant Systems AGARD CP-586 (Athens, Greece)*

[70] Cerri S, Bohn M, Menke K and Galtetti L 2009 Ageing behavior of HTPB based rocket propellant formulations *Cent. Eur. J. Energetic Mater.* **2** 149–65

[71] Bohn M 2012 Generic (2012) formulation of performance assessment quantities for stability, compatibility and ageing of energetic materials *43rd Int. Conf. of the ICT (Karlshue, Germany)*

[72] Smirnov L 2004 Chemical physics of decomposition of energetic materials problems and prospects *Russ. Chem. Rev.* **73** 1121–41

[73] Bohn M 1994 Prediction of life times of propellants—improved kinetic description of the stabiliser consumption *Propellants Explos. Pyrotech.* **19** 266–9

[74] Roduit B *et al* 2019 Continuous monitoring of shelf lives of materials by application of data loggers with implemented kinetic parameters *Molecules* **24** 2217–42

[75] Roduit B, Hartmann M, Folly P, Sarbach A and Baltensperger R 2014 Prediction of thermal stability of materials by modified kinetic and model selection approaches based on a limitted about of experimental points *Thermochim. Acta* **579** 31–9

[76] Clelouche S, Trache D, Maamache I, Fouzi Tarchoun A, Khimeche K and Mezroua A 2021 A new experimental way for monitoring of the real/equivalent in-service-time of double base rocket propellant by coupling VST and PCA *Defence Technol.* **17** 440–9
[77] Kumar A, Sadavarte V, Santosh Mada S, Pande S and Chavan P 2021 Studies on the effect of ageing on thermal, ballistic and mechanical properties of advanced energetic propellants *Propellants Explos. Pyrotech.* **46** 626–35
[78] Wang N, Qian W, Sui X and Xiong Y 2014 Life predictions of NEPE *Propellants Explos., Pyrotech.* **39** 102–7
[79] Wang N, Qian W, Sui X and Xiong Y 2014 Life predictions of NEPE propellants *Propellants Explos. Pyrotech.* **39** 102–7
[80] Bunyan P, Cunliffe A, Davis A and Kirby F 1993 The degradation and stabilisation of solid rocket propellants *Polym. Degrad. Stab.* **40** 239–50
[81] Niu R-M, Zhou Q-C, Xiong C, Yu-Tao J, Zhen W and Zheng J 2014 Experimental and numerical analysis of mode II fracture between propellant insulation *Int. J. Adhes. Adhes.* **52** 1–10
[82] Adams S 2017 Integration of heat flow calorimetry in surveillance of marine corps ammunition *10th Int. HFC Symp. (Crane, USA)*
[83] Roduit B, Borgeat C, Berger B, Folly P, Aebischer J, Fierz H and Stoessel F 2008 Advanced tools for thermal ageing and safety analysis *ICTAC (Istanbul)*
[84] Pearson L and Graham R 2004 Sensor application and modeling programme review, Jannaf health management and reliable service life prediction, Hill AFB, Ogden, Utah
[85] Benhammada A and Trache D 2020 Thermal decomposition of energetic materials using TG-FTIR and TG-MS: a state-of-the-art review *Appl. Spectrosc. Rev.* **55** 724–77
[86] Trache D and Tarchoun A 2019 Analytical methods for stability assessment of nitrate esters-based propellants *Crit. Rev. Anal. Chem.* **49** 415–38
[87] Tunnell R, Tod D, Ashcroft M, Sloan M and Henderson E 2018 Defining a reliable testing regime for tetryl *49th Int. Conf. of the ICT (Karlsruhe, Germany)*
[88] Rutter N and Tunnell R 2019 Defining the UK methodology for the stability testing of energetic materials *11th Int. Heat Flow Calorimetry Symp. (Pfinztal, Germany)*
[89] Vogelsanger B, Müller M and Folly P 2016 Scientific/technical workshop on nitrocellulose testing *7th Int. Nitrocellulose Symp. (Montreal, Canada)*
[90] de Klerk W and Bouma R 2008 Characterisation of aged polymer bonded eXplosives development of STANAG 4666 *39th Int. Conf. of the ICT (Karlsruhe, Germany)*
[91] Materazzi S and Vecchio S 2011 Evolved gas analysis by mass spectrometry *Appl. Spectrosc. Rev.* **46** 261–340
[92] Erikson W 2018 Thermal decomposition of energetic materials *US army Materiel Command Coal-Gasification and Waste Munitions to Fuel Working Group Meeting (Middleton, IA)*
[93] Goetz L, Phillips A, O'Brien R and Morris A 2019 Initial assessment of ageing of developmental double base propellants using HFC and stabilizer depletion *11th Int. HFC Symp. (Karlsruhe, Germany)*
[94] Muravyev N, Monogarov K, Bragin A, Fomenkov I and Pivkina A 2016 HP-DSC study of energetic material. Part 1. Overview of pressure influence on thermal behavior *Thermochim. Acta* **631** 1–7
[95] Cervantes O and Montgomery J 2014 *High Pressure DSC on LLM-105 and other High Explosives* (Livermore, CA: Lawrence Livermore National Laboratory)

[96] tHoen C and Klerk W 2018 HP-DSC combined with microscopy applied to explosives and propellants *North American Thermal Analysis Society* Philadelphia, PA
[97] Abd-Elghany M and Klapötke T 2018 A review on differential scanning calorimetry technique and its importance in the field of energetic materials *Phys. Sci. Rev.* **3** 20170103
[98] Steward P 2021 EMTAP strategy discussion document *QinetiQ*, QinetiQ/EMEA/MLW/TS211352
[99] Williamson D, Gymer S, Taylor N, Walley S, Jardine A, Glauser A, French S and and Wortley S 2016 Characterisation of the impact response of energetic materials: Observation of a low-level reaction in 2,6-diamino-3,5-dinitropyrazine-1-oxide (LLM-105). *RSC Adv.* **6** 27896–900
[100] Morley O 2021 *PhD Thesis* University of Cambridge
[101] Wagstaff D and Hooker P 2014 *A Review of Studies into the Use of EMTAP Test 6 (Electric Spark) Apparatus* (UK: HSL)
[102] Doherty R 2018 Making sense of sensitivity data *21st NTREM Conf. (Pardubice, Czech Republic)*
[103] McMaster P, Morley O and Williamson D 2021 A novel method for assessing impact sensitveness of energetic materials *12th ISICP*
[104] Morley O and Williamson D 2020 Pressure and temperature induced red-shift of the sodium D-line during HMX deflagration *Commun. Chem.* **3** 1–6
[105] Green S, Wheelhouse K, Payne A, Hallett J, Miller P and Bull J 2020 On the use of differential scanning calorimetry for thermal hazard assessment of new chemistry: Avoiding explosive mistakes *Angew. Chem. Int. Ed.* **59** 15798–802
[106] NATO 1997 *Structural assessment of solid propellant grains* AGARD-AR-350 NATO
[107] Davis J R 2004 Introduction to tensile testing *Tensile Testing* 2nd edn (Materials Park, OH: ASM) ch 1
[108] Cui H R, Tang G J and Shen Z B 2016 Study on the viscoelastic poisson's ratio of solid propellants using digital image correlation method *Propellant, Explos. Pyrotech.* **41** 835–43
[109] Farris R J 1968 The influence of vacuole formation on the response and failure of filled elastomers *Trans. Soc. Rheol.* **12** 315–34
[110] Sućeska M, Matečić-Mušanić M and Fiamengo I 2010 Study of mechanical properties of naturally aged double base rocket propellants *Cent. Eur. J. Energetic Mater.* **7** 47–60
[111] NATO 2003 *Manual of Data Requirements and Tests for the Qualification of Explosive Materials for Military Use (Allied Ordnance Publication)* AOP-7 NATO
[112] Lindsey G H, Schapery R A, Williams M L and Zak A R 1963 *The Triaxial Tension Failure of Viscoelastic Materials* (Pasadena, CA: Aerospace Research Laboratories)
[113] Yerzley F L 1939 Adhesion of neoprene to metal *Ind. Eng. Chem.* **31** 950
[114] Gent A N and Lindley P B 1959 The compression of bonded rubber blocks *Proc. Inst. Mech. Eng.* **173** 111
[115] Gent A N and Lindley P B 1959 Internal rupture of bonded rubber cylinders in tension *Proc. Royal Soc.* **249** 195–205
[116] Liu C 1997 Crack growth behavior in a solid propellant *Eng. Fract. Mech.* **56** 127–35
[117] Bo H, Yutao J and Zhou C 2012 Simulation of crack propagation in HTPB propellant using cohesive zone model *Eng. Fail. Anal.* **26** 304–217
[118] Jalocha D and Constantinescu A N R 2015 Prestrained biaxial DMA investigation of viscoelastic nonlinearities in highly filled elastomers *Polym. Test.* **42** 37–44

[119] Caliez M, Gratton M, Frachon A, Benelfellah A and Picart D 2014 Viscoelastic plastic model and experimental validation for a granular energetic material *Int. J. Energ. Mater. Chem. Propuls.* 339–71
[120] Burton J and Jones W 1971 Theoretical and experimental treatment of fracture in an adhesive interlayer *Trans. Soc. Rheol.* **15** 39
[121] Liu T, Li J, Cai X and Yan S 2018 A time-frequency analysis algorithm for ultrasonic waves generating from a debonding defect by using empirical wavelet transform *Appl. Acoust.* **131** 16–27
[122] Gokul A, Srinivas K and Dhanasekaran J 2019 Bond quality inspection of bonded surfaces using microwave NDE *Int. J. Adv. Microwave Technol.* **4** 213–17
[123] Zhang S, Xiao L and Qu W 2021 Ageing of solid propellants investigate by the electro-mechanical impedance technique *J. Propul. Power* **37** 68–76
[124] Cullis I and Mills B 2020 *Smart qualification for general munitions* QinetiQ/20/02614 QinetiQ
[125] Andrews M, Baker E, Collet C and Ferran M 2021 Energetic materials selection for performance and safety *AVT-340 Research Workshop on Preparation and Characterization of Energetic Materials*
[126] Demko A, Barnes B, Moody H and Rice B 2021 Burning rate prediction through application of machine learning to curated datasets *AVT-340 Research Workshop on Preparation and Characterization of Energetic Materials*
[127] NATO CP-586-AGARD 1997 AGARD *Conf. Proc. 586, Service Life of Solid Propellant Systems*, Greece
[128] Cunliffe A, Davis A and Tod D 1996 Ageing and life prediction of composite propellant motors *AGARD PEP Symp. Service Life of Solid Propellant Systems (Athens, Greece)*
[129] Faulkner G and Tod D 1996 Service life prediction methodologies *AGARD PEP Symp. Service Life of Solid Propellant Systems AGARD CP-586 (Athens, Greece)*
[130] Celina M, Minier L and Assink R 2002 Development and application of tools to characterise the oxidative degradation of AP/HTPB/Al propellants in a propellant reliability study *Thermochim. Acta* **384** 343–9
[131] Stenson R 1972 *Factors govening the storage life of solid propellant rocket motors* Technical Note No. 47 ERDE
[132] Young G 1973 Factors governing the storage life of solid propellant rocket motors, WAC/223/014
[133] Cunliffe A 1999 A methodology to predict the service life of composite propellants, TP06, Milestone 568T2 1XX4
[134] Torry S 1999 *Humid ageing of composite propellants* WSS/WX4/CR990101 DERA
[135] Hobman C 1999 *Service life prediction techniques for rocket motors* WSS/WS3/CR990740 DERA
[136] Torry S 2007 *Oxygen Consumption of Propellants* (Presentation to RAO)
[137] Rawley K 2008 QinetiQ/08/00433
[138] Dodds J 2017 *WSTC predictive modelling—final report* QinetiQ/17/01032 QinetiQ
[139] Mills A and Kuponu O 2014 Bayesian munitions prognosis, Sheffield University, QinetiQ/14/02032
[140] Sloan M, Salsbury S and Keeton G 1996 Chemical safe life prediction for cast double base rocket propellants *AGARD PEP Symp. on Service Life of Solid Propellant Systems (Athens, Greece, 10–14 May)*

[141] NATO 2016 *Integrated munition health management (IMHM)* STO-L-192 TR-AVT-212 MSIAC
[142] Tse F and Pritchard R 2007 Life assessment methodology for solid propellant rocket motors NSWCIH, NAWCWD
[143] Margetson J and Wong F C 1996 Service life prediction of solid rocket propellant motors stored in a random thermal environment *AGARD PEP Symp. on 'Service Life of Solid Propellant Systems' (Athens, Greece)*
[144] Dale G F 1999 *Computer programme for probabilistic modelling of rocket motor service life* WSS/WX4/CR980878/1.0 DERA
[145] Hobman C 2007 *Environmental monitoring of SeaWolf magazine storage: discussion of data from 1999 to 2006 including ageing effects* QinetiQ/07/00514 QinetiQ
[146] Winstone M 2021 *Technical Evaluation Report* STO-MP-AVT-187 T NATO
[147] Porter D 1995 *Group Interaction Modelling of Polymer Properties* (New York: Marcel Dekker)
[148] Michalchuk A, Trestman M, Rudic S, Portius P, Fincham P, Pulham C and Morrison C 2019 Predicting the reactivity of energetic materials: an ab inito multi-phonon approach *J. Mater. Chem. A.* **7** 19539–53
[149] Michalchuk A A L, Hemingway J and Morrison C A 2021 Predicting the impact sensitivities of energetic materials through zone-center phonon up-pumping *J. Chem. Phys.* **154** 064105
[150] Plumley H J 1963 *Trepanning of cased explosives by etching* US Patent 3,108,918 29
[151] Hulme J 2021 Smart weapons testing: a literature review *QinetiQ* 26 July 2021
[152] Kraft S and Johnson A 2019 CAD/PAD digital twin programme overview *PARARI Australian Explosive Ordnance Safety Symp. (Canberra, Australia)*
[153] Britton A 2021 *Next generation test & evaluation for guided weapons* QinetiQ/PR21/03367 QinetiQ
[154] West T and Blackburn M 2017 Is digital thread/digital twin affordable? A systemic assessment of the cost of DoD's latest Manhattan Project *Proc. Comp. Sci.* **114** 47–56
[155] Fei X, Fengchen Q, Bing S and Yuzhu F 2018 Digital twin of solid rocket motor, problem and challenge *11th Int. Symp. on Computational Intelligence and Design*

IOP Publishing

Energetics Science and Technology
An integrated approach
Adam S Cumming

Chapter 11

Recursive molecular similarity (R.Mo.S): an innovative algorithm for selecting a subset useful for toxicology prediction

C Alliod, R Denis, G Jacob and R Terreux

11.1 Introduction

Military research has the mission to develop the explosives and propellants of tomorrow. But for the last two decades, its role has evolved and is now focused on the performance improvement of energetic molecules, while also taking into account toxicological and environmental concerns. Today's goal is to develop energetic molecules whose hazards for human health and the environment are reduced. Therefore, a research program in collaboration with the ArianeGroup was set up in order to develop optimized tools to predict the toxicity of high energetic materials (HEMs) for the future.

A significant effort has been made to develop new energetic materials for ammunition, explosives, and propellants, hereafter referred to as HEMs. The goal was to replace black powder with new materials whose performance is improved with a decreased sensitivity, while also enhancing the safety of people and goods [1]. Their environmental impact is also taken into account [2].

Energetic materials have many military and industrial applications [3], and therefore have to possess the following properties: high thermal stability, low molecular weight of the gases produced during their combustion, the ability to detonate (for explosives, either by ignition or by employing detonators), and insensitivity to shocks, friction, and mechanical stimuli. HEMs include explosives, propellants, and pyrotechnic devices (i.e. fireworks or light signals).

Over the last few years, there has been an urgent need to develop technologies capable of identifying the toxic effects produced by chemicals on human beings and the environment. Indeed, a large number of *in vitro* test systems have been developed.

doi:10.1088/978-0-7503-3943-8ch11

Genetic toxicity testing, which studies substances that can damage DNA, is often required by government regulatory agencies before an industrial chemical can be produced [4–7]. The most commonly *in vitro* tests routinely used as an initial toxicological screening in chemical development are gene mutation systems in bacteria (also named the Ames test)[8–10]. Because of the number of mutagenicity tests in general use, a variety of harmonized approaches and internationally agreed test protocols have been deployed. These approaches and test protocols should be applied to new tested chemicals and have been described by Registration, Evaluation, Authorization, and Restriction of Chemicals (REACh) [11–13] (Annex I and IV to IX) and the Organization for Economic Cooperation and Development (OECD) [14].

In this study, a database was created using different European databases [15–19], and was harmonized with the approaches set up by REACh and the OECD testing guidelines. These databases contain many different molecules, all of them having information about their experimental results from the Ames test. The prediction of the Ames test was used as a specific example.

However, *in vitro* and *in vivo* tests are long, complicated, and relatively expensive and require the synthesis of the material. Thus, alternative research methods [20–23] such as predictive tools are encouraged. This is why techniques based on the use of computer tools (*in silico* methods) [24, 25] have been developed for this study. It is also important to mention that there is only a very limited toxicological knowledge for HEMs and no clear links with conventional systems.

Improving predictive accuracy has long been the main driving force for the development of new *in silico* prediction techniques. The latter use mathematical approaches to predict toxicity. Several methods can be used such as the quantitative structure activity relationship (QSAR) [26–28] or on-the-fly QSAR (OTF-QSAR) [29].

We highlighted and demonstrated in a previous study [29] that similarity is a critical parameter and plays an essential role in the prediction of toxicity. The search for structural similarity among molecules is one of the most conventional tools used for *in silico* predictions [30, 31]. Based on a virtual selection in a database, the similarity supposes that all the analogous compounds to a compound in question also have similar biological activity. An important point for this methodology is the use of good structural descriptors which are representative of the molecular features responsible for the relevant activity.

Another *in silico* prediction tool is machine learning (ML) [32]. This is a subfield of computer science which explores the construction of algorithms that can learn from data and make predictions on them. The algorithms used in ML are typically classified according to the learning mode they use. They allow a system, to some extent, to adapt its analyses and response, based on the analysis of empirical data from a database (learning data). This analytical method allows scientists to produce reliable, repeatable decisions and results. In this study, we specifically utilise supervised learning methods [33] and decided to test and compare decision trees (DTs) [34], ensemble methods (EMs), artificial neural networks (ANNs) [35] and support vector machines (SVMs) [36].

A DT represents a set of choices in the graphical form of a tree. It models a hierarchy of tests, with each branch representing the result of a test. Thus, it allows one to classify the data into two classes (in our case, mutagenic/non-mutagenic). The learning algorithm used in this method tests all the possible ways and, finally, selects the best tree that suits our data. This process of top-down induction of DTs is by far the most common learning strategy.

The goal of an EM is to create different DTs resulting from the original data and then produce new subsets of data. A majority vote combines all the individual trees to form one final DT (one final prediction). This method offers an increase of the stability and the robustness of the prediction over the methods using a single DT. In general, the more trees that are built, the stronger the forest is, and the more accurate the results are. The learning algorithm depends on which data are modified when the different trees are created. If the modified data belong to the training data, the algorithm is a 'bagging' algorithm (bootstrap aggregating). If the modified data belong to the output data, the algorithm is a 'boosting' algorithm.

There are two types of bagging algorithms: the random forest algorithm (RF) [37] and the extremely randomized trees algorithm (ExtraTrees) [38]. With the RF algorithm, each tree in the ensemble is built with the replacement of the training set at each draw. With ExtraTrees, each tree in the ensemble is created with no substitution of the training set at each draw, and the best separation at a node is done at random.

The most popular boosting algorithm is known as adaptive boosting (AdaBoost) [39], introduced in 1995 by Freund and Schapire. The core principle of AdaBoost is to fit a sequence of weak learners on repeatedly modified versions of the data. The data modifications, at each successive iteration, consist of individually applying adjusted weights to each of the training samples. Then, the learning algorithm is reapplied to the reweighted data to produce the final prediction. As iterations proceed, examples that are difficult to predict receive ever increasing influence.

In this study, an explicit list of 17 molecules (explosives or ingredients for the formulation of energetic materials) was chosen, all of them having well-known experimental results for the mutagenicity test. This explicit list was associated with a toxicological bibliography for the tests. The toxicity of this set of molecules was unknown to the predicted system.

The strategy of this study, which gave rise to a patent [40], uses the search of structural similarity with specific parameters to predict the toxicity of HEM molecules. Thanks to a new, smarter algorithm, named recursive molecular similarity (R.Mo.S), the number of selected molecules is increased. A more complete and meticulous selection of the molecule from the database designed to be used for the prediction is obtained, and the performance and the quality of the predictions are improved. It can identify a suitably diverse subset from a large number of initially presented compounds (whole initial database), which will be used subsequently for mutagenicity predictions. The object of this chapter is to show that our smarter algorithm can be encoded and applied itself to a wide range of different similarity indices for the prediction of the activity of chemical compounds. This will allow us to achieve a rational design of HEMs both in terms of efficacy and weak toxicity level.

In a quest for better and better predictive accuracies, our predictive system will be evaluated and compared with software packages often used in the literature and the scientific community such as Lazar and Advanced Chemistry Development (ACD/Percepta).

11.2 Data and methods

11.2.1 Data sets

Five European databases were chosen: the European Union Reference Laboratory-European Commission for Alternatives to Animal Testing (EURL-ECVAM) database [15], the JRC QSAR Model database, the Carcinogenicity Genotoxicity eXperience (CGX) database [16], the Carcinogenic Potency Database (CPDB) [17], and the Instituto Superiore di Sanità (ISSTOX) database [18]. One inter-agency program, a Health and Human Services inter-agency program named the National Toxicology Program (NTP) [19], was also chosen. We specifically selected the positive and negative (mutagen/non-mutagen) results. All other data (inconclusive, equivocal, or uninterpretable) were eliminated in order to increase the quality of the experimental data and to decrease bias, false responses, or errors.

All the experimental data for each toxicological test were merged into a new considerable and extensive database created with the 2014.09 release of the Chemical Computing Group's Molecular Operating Environment (MOE) software. Thus, after elimination of duplicates, the Ames test database contained a total of 7723 molecules.

11.2.2 Ames test

Five strains of bacteria were used. These five strains of *Salmonella typhimurium* include TA1535; TA1537 or TA97a or TA97; TA98; TA100; and TA102. All information pertaining to this is contained in an OECD Guideline, no. 471, named the Bacterial Reverse Mutation Test [10].

11.2.3 Similarity

11.2.3.1 Similarity using the Tanimoto coefficient

As introduced previously, the *in silico* prediction of the biological activity of a questioned molecule (query) is based on the biological activity of well-known and similar molecules. Such similar molecules are selected by comparing their structure to the query using the MACCS 166 structure descriptors [41, 42].

MACCS 166 are a particular type of descriptors, named a structural key, which characterizes the structure of the molecules. It uses a table of 166 sufficiently complex molecular fragments to discriminate most of the molecules. A vector of 166 values is generated for each molecule, in which each value is a natural integer (positive or zero) representing the occurrences of each of the 166 molecular fragments in the molecule under consideration (figure 11.1). Thus, the structural key MACCS 166 characterizes each molecule based on a table of 166 molecular fragments chosen in a sufficiently complex manner to discriminate between different molecules. Each fragment value is called a 'characteristic' of the molecule. A MACCS characteristic of zero value therefore reflects the absence of the

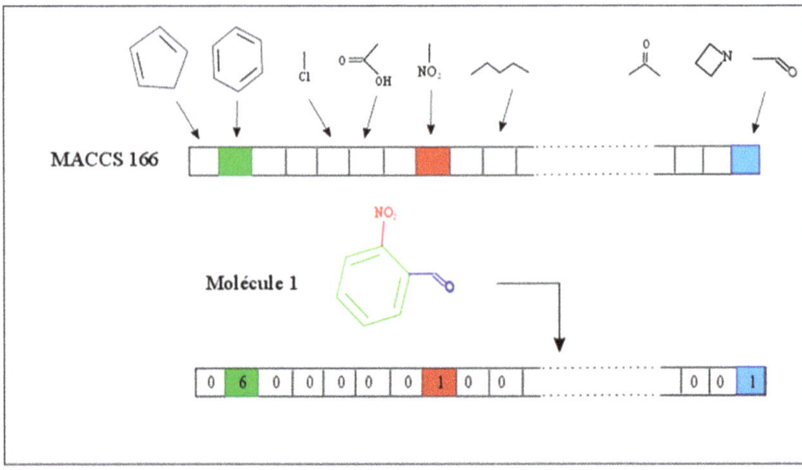

Figure 11.1. The MACCS vector. The MACCS descriptor is a specific descriptor possessing a vector of 166 molecular fragments characterizing the occurrences of each fragment in the tested molecule. The aromatic ring, containing six carbons, corresponds to the 2nd molecular fragment presented in the vector. The presence of this fragment in the tested molecule causes the appearance of a 6 at position two in the vector (6 times the presence of a carbon). The nitro group corresponds to the 8th molecular fragment presented in the vector. The presence of this fragment in the tested molecule causes the appearance of a 1 at the 8th position in the vector. It is the same for the last ketone group in position 166 of the vector.

corresponding fragment in the structure of the molecule. On the contrary, a positive value of a MACCS characteristic indicates the presence of or the number of times that the corresponding fragment is present within the molecule.

A similarity coefficient is calculated from the MACCS 166 vectors of the query and a possibly similar molecule from the database. This is a real value between 0 and 1, so that 0 means that the given two molecules are *very* dissimilar and 1 means that they are *very* similar.

The commonly used similarity formulation (equation (11.1)) is the Tanimoto coefficient (T) [43] based on the MACCS 166 fingerprint.

The MACCS 166 fingerprint represents the indexes of the molecular fragments that are present (i.e. positive occurrence) in the characterized molecule. For example, molecule 1 of figure 11.2 has a fingerprint (2, 3, 5) since only fragments number 2, 3, and 5 are present in this molecule. Fingerprints may also be written as vectors of bits, so that missing fragments are represented by 0 and present fragments by 1, e.g. (0, 1, 1, 0, 1) for the molecule 1.

Given two fingerprints A and B, the Tanimoto coefficient $T(A, B)$ is defined by

$$T(A, B) = \frac{\alpha}{\beta + \gamma - \alpha}, \tag{11.1}$$

where α is the number of fragments *common to both molecules* (i.e. indexes common to both fingerprints), β is the number of fragments *present* in the first molecule (i.e. the length of fingerprint A), and γ is the number of fragments *present* in the second molecule (i.e. the length of fingerprint B).

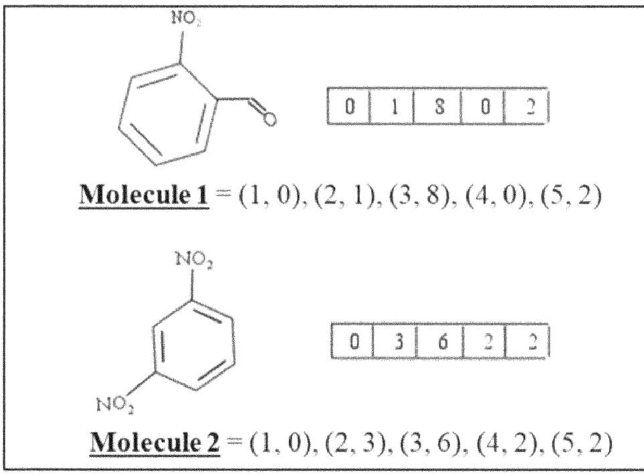

Figure 11.2. Fingerprint of molecules 1 and 2. The example above shows the transformation of a classic MACCS vector constituted here of five characteristics into a new couple of values, allowing one to take into account the non-expression of the same characteristic MACCS.

Representing these two fingerprints by bit vectors a and b, the Tanimoto coefficient equivalently reads as

$$T(a, b) = \frac{\sum_i a_i b_i}{\sum_i a_i^2 + \sum_i b_i^2 - \sum_i a_i b_i}. \tag{11.2}$$

Using this formulation, the similarity coefficient between a fingerprint (2, 3, 5) (molecule 1 of figure 11.2) and (2, 3, 4, 5) (molecule 2 of figure 11.2) is thus equal to 0.75.

However, the Tanimoto coefficient does not take into account the non-expression of the same characteristic (i.e. the number of fragments that are missing in both molecules) and the level of expression of each characteristic (i.e. comparing the possibility of multiple occurrences of a given fragment in both molecules). To address these two issues, we propose a new similarity formulation.

11.2.3.2 A new similarity coefficient

The similarity formulation used in R.Mo.S. is based on several key improvements described in this work.

First of all, the MACCS 166 is represented as the complete vector of the occurrences of each molecule fragment, e.g. (0, 1, 8, 0, 2), for the molecule in figure 11.2.

Then, we use a generalization of the Tanimoto coefficient for ensembles A and B, which is the Jaccard index, $J(A, B)$ [44, 45]:

$$J(A, B) = \frac{|A \cap B|}{|A \cup B|}, \tag{11.3}$$

where (E) denotes the measure of the ensemble E (e.g. the surface in 2D).

Using classical manipulations, it can be conveniently rewritten into

$$J(A, B) = \frac{|A \cap B|}{|A \cap A| + |B \cap B| - |A \cap B|}. \tag{11.4}$$

Given the MACSS 166 vectors A and B of two molecules, we thus need to define the measure of $(A \cap B)$. Considering that each fragment molecule used in MACCS 166 is independent, we can simplify this measure by the sum of the local measurement of two numbers of occurrence of a same molecule fragment:

$$|A \cap B| = \sum_i s(a_i, b_i), \tag{11.5}$$

where s denotes this local measurement that we call the local similarity of the two molecules.

For the sake of comparison, the Tanimoto coefficient can be expressed by this formulation using the particular local similarity defined so that $s(x, y) = 1$ if both x and y are positive, $s(x, y) = 0$ otherwise.

Among the number of possible formulations, we chose to express this local similarity as the composition of a so-called conversion function φ with a so-called signed distance function *dist*:

$$s(a, b) = \varphi(dist\,(\mathrm{a, b})), \tag{11.6}$$

with

$$dist\,(a, b) = \ln\left(\frac{a}{b}\right) \text{ and } \varphi(d) = \exp\left(\frac{-d^2}{2\sigma^2}\right). \tag{11.7}$$

The conversion function φ has the form of a Gaussian function so to transform the unbounded value returned by the distance function to a similarity value that lies between 0 and 1 (figure 11.3). The σ parameter can be used to tune the decrease rate of the conversion. In our experiments, we used $\sigma = 1$.

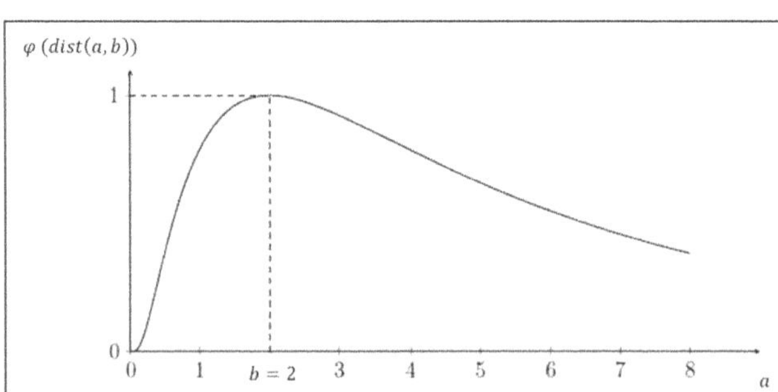

Figure 11.3. Plot of the local similarity measure $\varphi(dist(a, b))$ according to a, with $(dist(a, b) = \ln\frac{a}{b})$ and $\varphi(d) = \exp\left(\frac{-d^2}{2\sigma^2}\right)$ for $b = 2$ and $\sigma = 1$.

The distance function is based on the natural logarithm of the ratio of the two values, extended to extreme cases with the conventions that $dist(0,0) = 0$ (thus $s(0,0) = 1$), $dist(0,\alpha) = +\infty$ (thus $s(0,\alpha) = 0$), and $dist(\alpha,0) = -\infty$ (thus $s(\alpha,0) = 0$) for any positive α.

The magnitude of $d(a, b)$ is related to our intuition of a distance between two numbers of occurrence of the same molecule fragment. For example, two identical numbers (null or positive) means that the two molecules are very close for this particular characteristic, thus leading to a null distance and finally to a local similarity of 1. Conversely, if one molecule is missing a given fragment while the other molecule has one or more of the same fragment, then the local distance is $\pm\infty$, thus leading to a local similarity of 0.

Using these definitions, formula (11.4) can be simplified by noting that both terms $(A \cap A)$ and $(B \cap B)$ are equal to the size of the MACCS descriptor (166 in our case).

To summarize, our new similarity expresses as

$$J(A, B) = \frac{|A \cap B|}{332 - |A \cap B|} \text{ with } |A \cap B| = \sum_{i=1}^{166} \exp\left(-0.5 \ln\left(\frac{a_i}{b_i}\right)^2\right).$$

Comparing the two molecules of the figure 11.2, with vectors of occurrences (0, 1, 8, 0,2) and (0, 3, 6, 2, 2), respectively, our new formulation leads to the following approximated local similarities (1, 0.55, 0.96, 0, 1) and thus to a similarity coefficient of **0.54**. With the same inputs, Tanimoto coefficients lead to the local similarities (0, 1, 1, 0, 1) and the similarity coefficient **0.75**.

11.2.4 Smarter algorithm: recursive molecular similarity

We now describe the use of the Ames database consisting of 7723 molecules, all defined by their MACCS descriptors. These MACCS descriptors also define the unknown molecule we want to test.

At a certain similarity, for example here 80% (figure 11.4), the first step of the R. Mo.S. algorithm aims to recover all the molecules which are similar to the target molecule in a more accurate manner, thanks to the two improvements previously stated. We created a subset containing only molecules similar to the unknown molecule we want to predict.

At the end of this first step, a new (fictitious) target molecule, called a 'signature molecule', will emerge from previously recovered similar molecules. In order to generate this virtual molecule, defined by its MACCS and which will serve as a request for the next iteration of the process, two possibilities have been considered:

1. The associated value for each MACCS characteristics of all the molecules found is defined as the average of the corresponding values for the selected molecules (to the nearest integer).
2. The value associated with each of the MACCS characteristics is defined as the value corresponding to the most common value present among the selected molecules. In case of ambiguity or if two values are present the same number of times, the higher value is preferred.

During the development of the process, the first possibility seemed less consistent since it could generate values not present among all of the molecules selected. On the

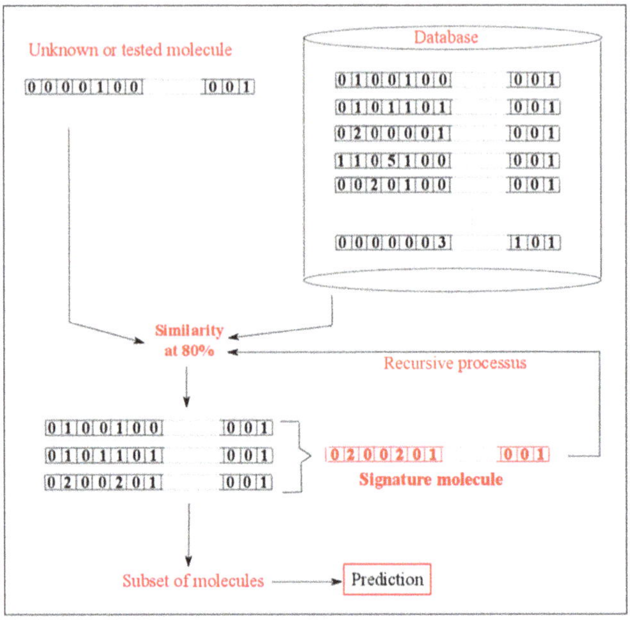

Figure 11.4. R.Mo.S.

other hand, the second possibility makes it possible to obtain a signature of the fictitious target molecule having a higher percentage of similarity with the initial target molecule. It is this second method that will be used by default.

Next, we undertake an iterative process to ensure that the signature molecule, on going through the same method as the previous step, identifies some similar molecules in the initial database (above a set threshold).

Over the iterations, signature molecules are constructed from the MACCS descriptors of the selected similar molecules in the initial database. A new similarity search is performed from these virtual molecules each time.

After a certain number of iterations, all the similar molecules from the different signature molecules are retrieved. Thus a new database is constructed. This latter is our new database used for the prediction of the toxicity. The prediction method is ML.

11.2.5 Machine learning

The python package sci-kit-learn was used to implement ML.

11.2.6 Software packages

ACD/Percepta, version 2012 (http://www.acdlabs.com), aims to provide comprehensive and innovative solutions to accelerate research and enhance competitiveness.

Lazar, version 1.0.2 (http://lazar.in-silico.de/predic), is an open source tool for the prediction of toxic properties of chemical structures. It derives predictions from a database for a query structure with experimental toxicity data.

11.3 Results

We used the Ames test as an example of the mutagenicity category first test. We tested all the ML methods mentioned in the introduction: SVM, ANN, DT, EM, RF, and the ExtraTrees and AdaBoost algorithms.

The first step was to determine which ML method is the best to associate to our smarter algorithm. We identified the zone of occurrence (number of combinations that associates *N* similar molecules selected by R.Mo.S with *x* percent of similarity) for which the ML technique will lead to a correct prediction of the 17 molecules of the explicit list.

Two important parameters can be used to tune the behavior of the R.Mo.S algorithm: the percentage of similarity and the number of molecules selected in the database which are similar to the tested molecule. At this stage of development and prediction, we still did not know which R.Mo.S parameter (similarity and number of molecules) was going to work to make a correct prediction with the ML methods on all the 17 molecules of the explicit list. Indeed, we were looking for similarity parameters and the number of similar molecules selected by R.Mo.S, valid for a specific test, and which would lead, with an appropriate ML technique, to a correct prediction for any molecule.

We therefore tested, both empirically and automatically (i.e. via ML), all the similarity parameters (from 50% to 100% similarity between the tested molecule and all the molecules contained in the database) and the number of possible selected molecules in the database (from 1 to 1000 molecules) that would give us a correct prediction for the entire explicit list of 17 molecules (figure 11.5). We chose not to go below 50% of similarity to keep the system as accurate as possible. Indeed, below 50% similarity, we considered that the R.Mo.S algorithm would no longer distinguish between very different molecules. Also, as we noted in previous calibrations performed with the OTF-QSAR, using over 1000 selected molecules similar to the tested molecule would lead to less and less accurate predictions.

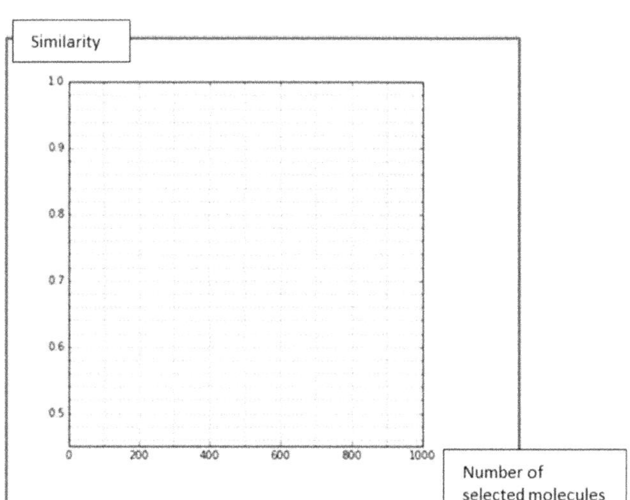

Figure 11.5. Grid of similarity according to the number of similar molecules selected by R.Mo.S.

On a grid parameterized by the percentage of similarity and the maximum number of selected molecules, each dot represents a parameter leading to a correct prediction of the 17 molecules in the explicit list. Therefore, each spot on the grid constitutes an accurate prediction for the 17 molecules. Ideally, we are looking for a broad zone containing points (figure 11.6) that would indicate that the ML technique used is robust and efficient at this percentage of similarity and with this number of similar molecules selected. This area of space corresponds to the validation area of the ML technique.

Isolated points indicate that the prediction of the 17 molecules is valid only for this specific parameter of similarity with the particular number of similar molecules selected (figure 11.7). In this figure, for example, with 800 molecules selected in the database at a similarity of 70% with the tested molecule, the 17 charged molecules will be correctly predicted.

11.3.1 For the machine learning methods using DT algorithm

In order to determine the best cut of the tree, Gini indicators and Shannon's entropy were used. Both characterize the gain in homogeneity of the subsets obtained by divisions of the whole tree: first by the measurement of the impurity and then by the measurement of the information gain. The learning algorithm tests all the possible combinations of the tree and selects the one whose indicator is maximized to provide the best possible prediction. Below are the results we obtained with the DT algorithm (figure 11.8).

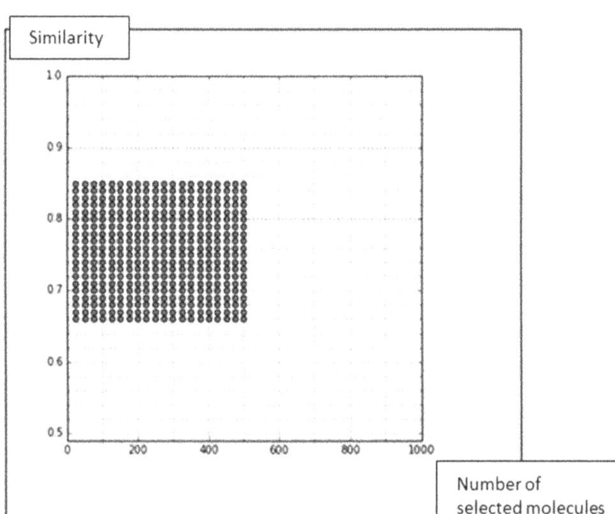

Figure 11.6. Large zone of space containing points. Each red dot in the space is a correct prediction for the 17 molecules of the explicit list. This large area of the space containing points guarantees the robustness and efficiency of the method. Indeed, taking a central point on the grid would provide a correct prediction for any type of molecule.

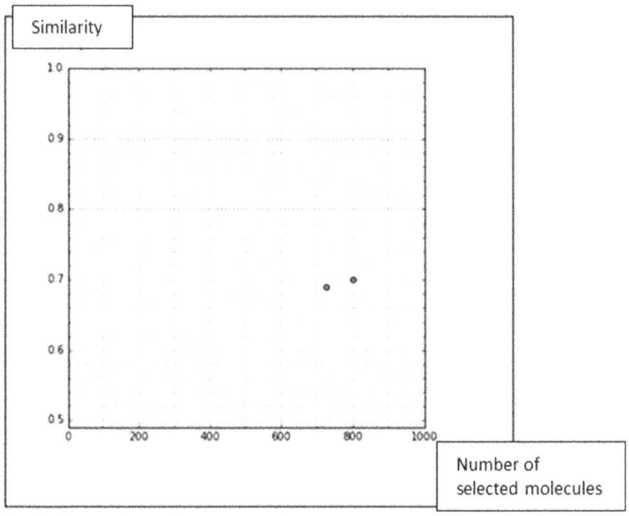

Figure 11.7. Grid containing isolated points.

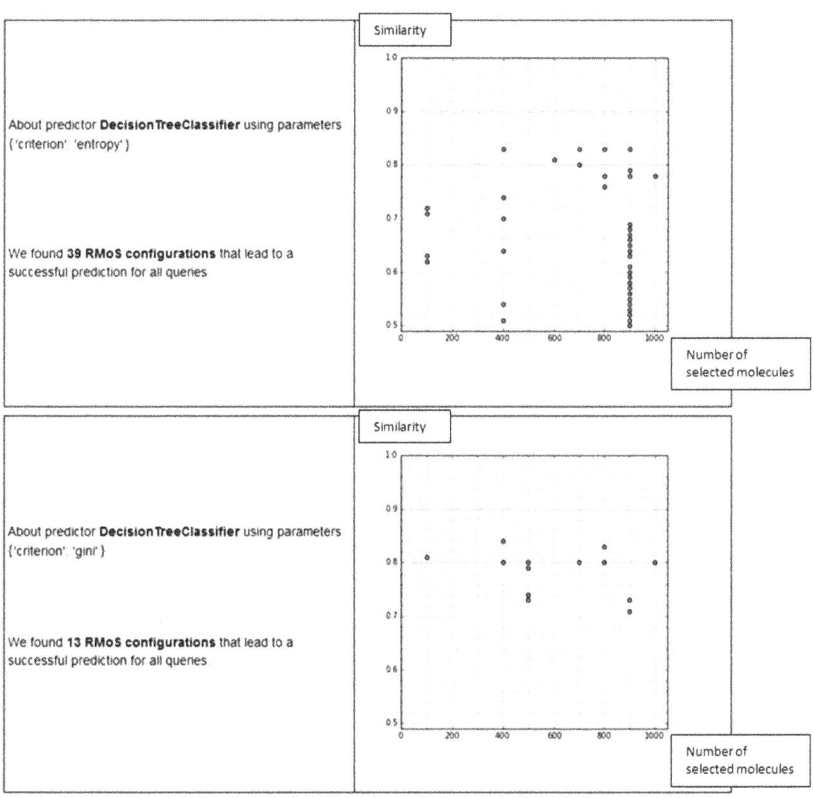

Figure 11.8. Grid results for the ML technique using DTs. The parameters used are the indicator, Gini and Shannon entropy. The indicator is the only variable parameter of this technique. The grid associated with Shannon's entropy finds 39 combinations, leading to a correct prediction of the 17 charged molecules. The grid associated with the Gini index only suggests 13.

We can easily see that the grid associated with Shannon's entropy offers 39 combinations, leading to a correct prediction of the 17 charged molecules, whereas associated the Gini index only offers 13 combinations of correct prediction. Indeed, for example, with the Shannon's entropy indicator, at 70% similarity with 400 similar molecules selected from the total Ames test database, the DT technique leads to a correct prediction of the 17 molecules of the test base.

The best DT is the one preferentially using Shannon's entropy. However, with this method no large area of space is visible, resulting in an optimal R.Mo.S parameter set.

11.3.2 For the machine learning methods using random forest algorithms

The complexity and size of the trees must be controlled to reduce memory consumption. This is made possible using a specific parameter, named the 'estimator'. This parameter defines the number of trees to create and combine to obtain the best possible prediction. In general, the greater the number of subsets of trees are built, the more robust the forest is and the more accurate the results are. We therefore tested the following number of estimators: 10, 50, 100, 200, 300, and 400. This is the results we obtained with the RF algorithm (figure 11.9) in terms of the number of estimators.

By testing a number of estimators of 10, 50, 100, 200, 300, and 400, we realized that beyond a critical number of subsets of trees created, the prediction results cease to improve significantly and the precision decreases. This critical number was determined to be 300 estimators for the EM using the RF algorithm. This critical number of 300 corresponds to the size of the random subsets of characteristics to be taken into account when dividing a node.

Indeed, above and below 300 estimators, only 116 and 110 combinations are obtained, unlike the 126 with 300 estimators.

In addition, we can see that at 300 estimators several large areas of space are present to ensure the robustness of the technique (figure 11.10).

11.3.3 For the machine learning methods using the ExtraTrees algorithm

First, what distinguishes the bagging ExtraTrees algorithm from the RF is that the data set used to generate a subset is not preserved when forming another subset. There is no reuse of the data once it has been used in forming a tree.

We present the results we obtained with the ExtraTrees algorithm using either the Gini index (figure 11.11) or Shannon's entropy (figure 11.12) depending on the number of estimators (number of trees generated for each subset).

First of all, this ML technique shows that it is superior to all other techniques. Indeed, we go from a number of possible combinations of 39 for the DT and 126 for the technique using RF to about 227 with the technique using ExtraTrees.

We see that using both the Gini index or Shannon's entropy, the number of possible configurations increases with the number of estimators. Above 200 estimators, the number of configurations decreases again, showing that, similarly to the RF technique, the prediction results cease to be significantly improved beyond

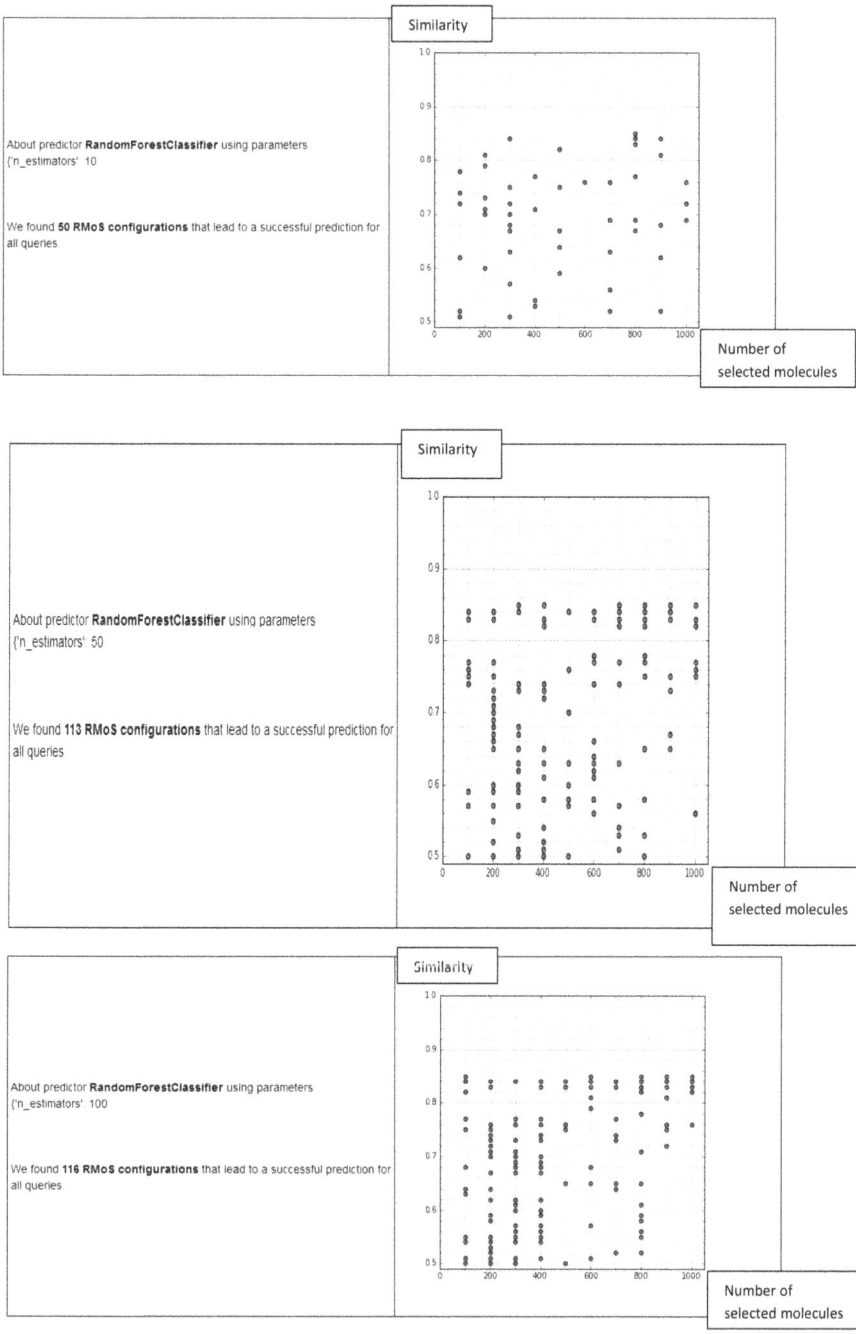

Figure 11.9. Grid results for the ML technique using RF. The parameter used is the number of estimators. The number of possible combinations increases significantly with the number of estimators, leading to a critical number of tree subsets created. Indeed, the number of combinations decreases over 300 estimators.

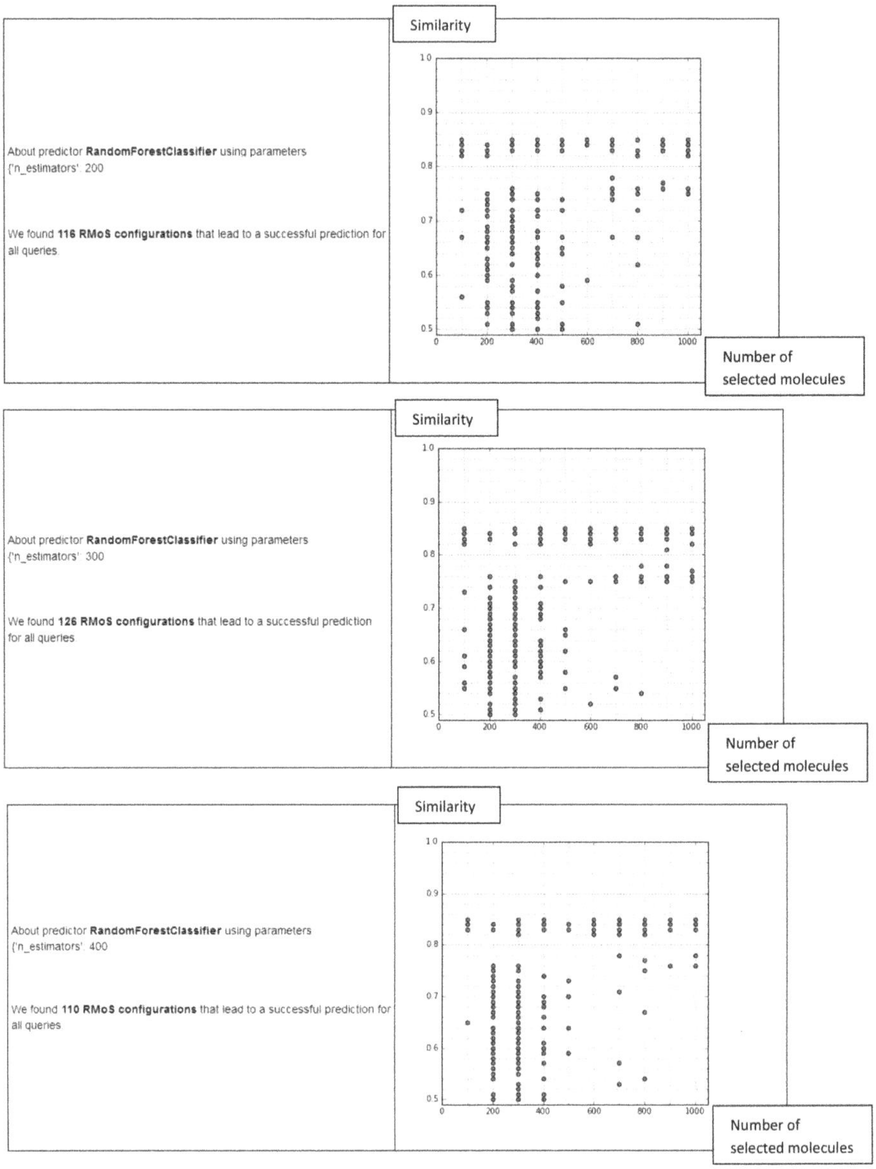

Figure 11.9. (Continued.)

a certain critical number of subsets of trees created. This critical number is 200 estimators for the ML technique using the ExtraTrees algorithm.

In addition, we can see that there is a slightly higher number of possible combinations with the Gini index.

A very large number of large areas of space are presented to ensure the robustness of the technique (figure 11.13).

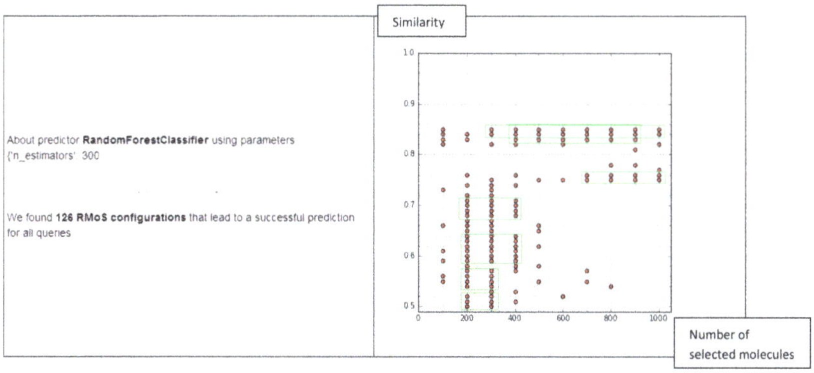

Figure 11.10. Several large areas of space are displayed with 300 estimators for the ML technique using RF.

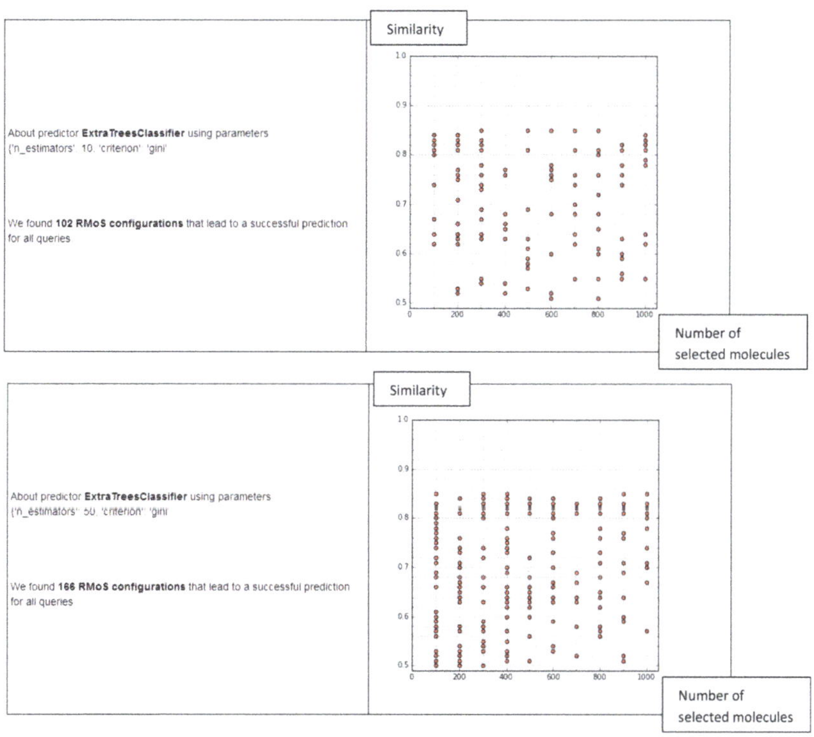

Figure 11.11. ExtraTrees algorithm using the Gini index.

11.3.4 For the machine learning methods using the AdaBoost algorithm

The basic principle of the AdaBoost boosting algorithm is to generate different distributions of the data and then adaptively modify these distributions by adjusting the weights of weak and strong learners. Then, it is necessary to increase the weight of the points which are wrongly classified and slightly reduce

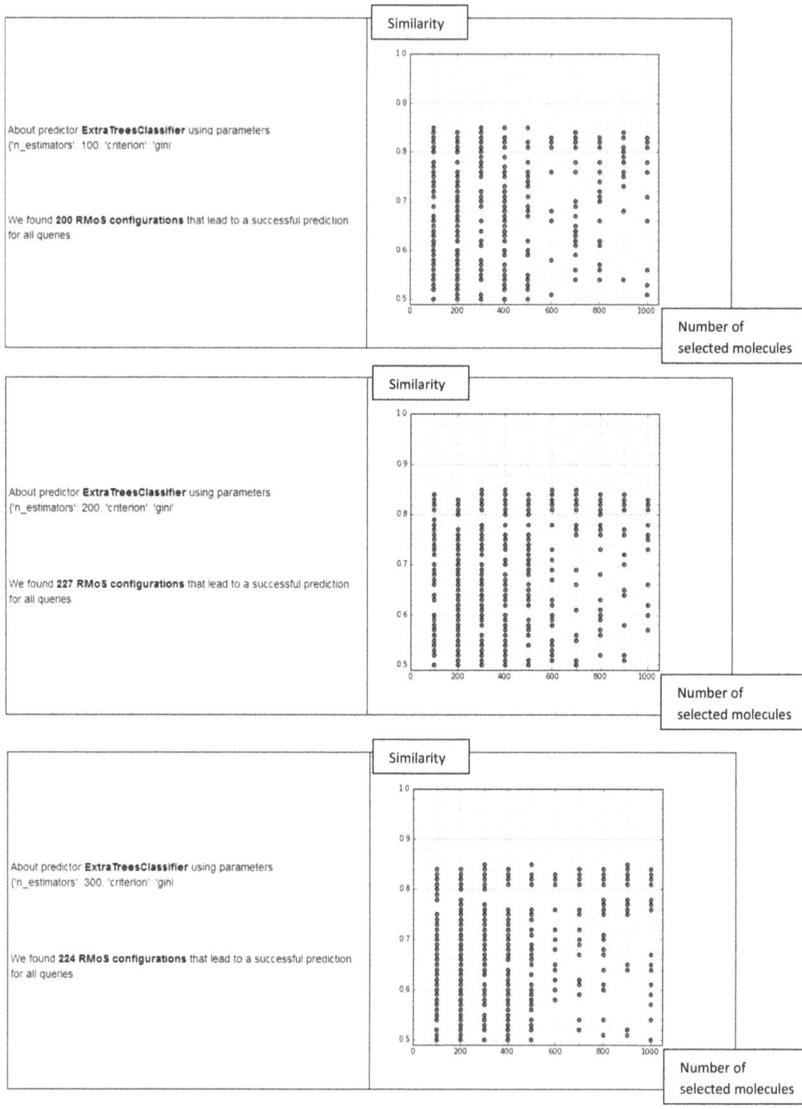

Figure 11.11. (Continued.)

the weight of those which are correctly classified. Thus, over time, examples that are difficult to predict receive an ever greater influence. This converts weak learners into strong learners and, through a majority voting, forms a strong final distribution.

Unlike neural networks and SVMs, the learning process selects only those characteristics that increase predictive power. This dramatically decreases the execution time. As with all other EMs, the AdaBoost algorithm uses estimators in order to be able to perform prediction.

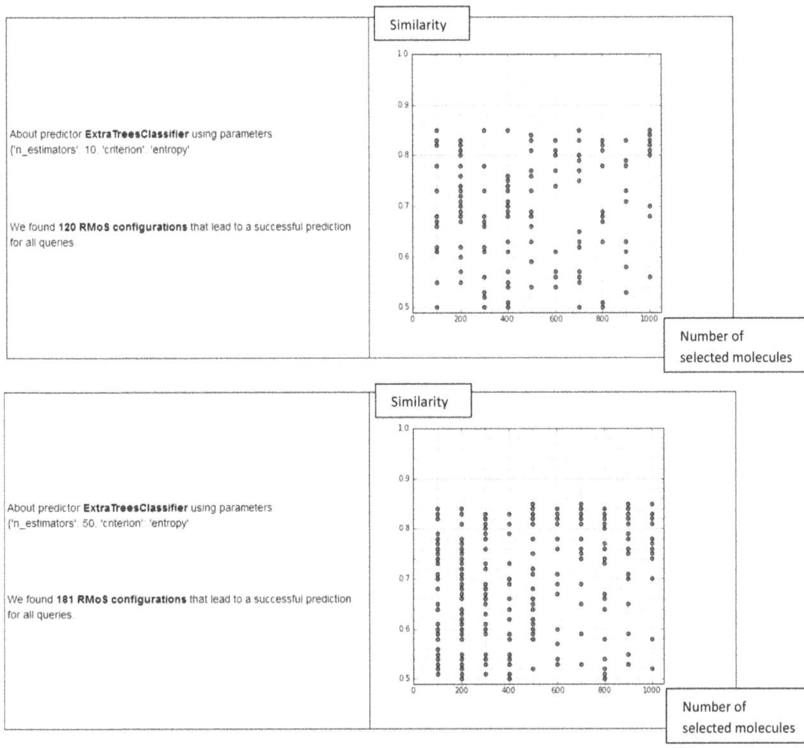

Figure 11.12. ExtraTrees algorithm using Shannon's entropy.

Here we present the results we obtained with the AdaBoost algorithm as a function of the number of estimators (number of distributions generated) (figure 11.14).

Many configurations are possible, the best of which is the one with 100 generated distributions. Indeed, there are 55 possible combinations that would lead to a correct prediction. However, this algorithm does not outperform the technique using ExtraTrees.

In table 11.1 we give the number of possible configurations for the correct prediction of the 17 molecules of our explicit list with the various estimators using the RF, ExtraTrees, and AdaBoost algorithms.

Finally, whilst DT, AdaBoost, and RF give much better results than ANN and SVM, they are still inadequate. The results highlight that the best ML technique is unequivocally the technique using the EM with the ExtraTrees algorithm. It makes it possible to obtain a large number of possible combinations of reliable predictions. Thereby,

Boosting: ExtraTrees > Boosting: RF > Bagging > Simple decision tree.

After having determined the best ML technique thanks to our 17 molecules from the explicit list, the second step was to determine which percentage of similarity was

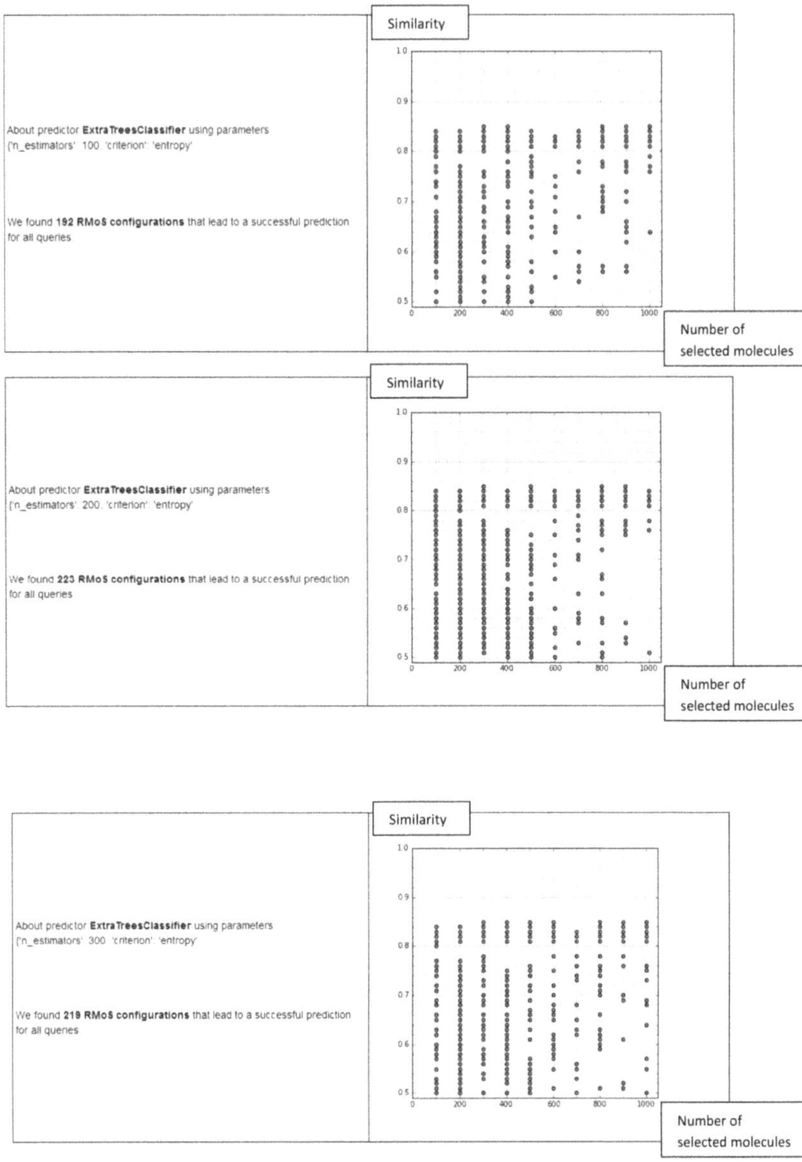

Figure 11.12. (Continued.)

the best to obtain a correct prediction for any type of molecule. Indeed, with the EM ML technique using an ExtraTrees algorithm, we obtained 227 possible configurations leading to a good prediction of the 17 molecules. In order to refine the results and to better calibrate the system so as to find the optimal area of space that would allow a correct prediction for any molecule, we enhanced our explicit list of 17 molecules with other explosives or ingredients for the formulation of explosives

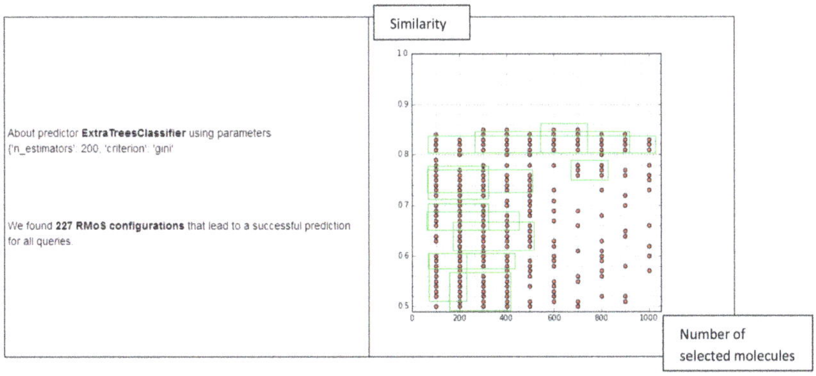

Figure 11.13. Several large areas of space are displayed with 200 estimators using the Gini index with the ExtraTrees algorithm.

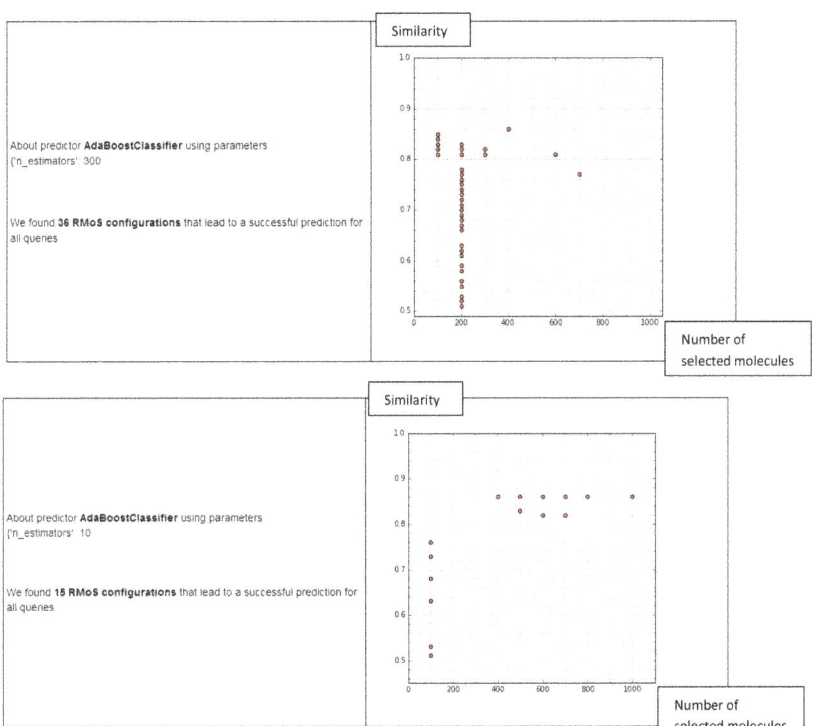

Figure 11.14. Grid results for the ML technique using the AdaBoost algorithm.

like plasticizers, liquids, or pyrotechnic molecules. Finally, we obtained an explicit list of 57 molecules (table 11.2).

After a various number of tests, we determined that the best percentage of similarity was 81% (a very good percentage for prediction) and a minimum of 400

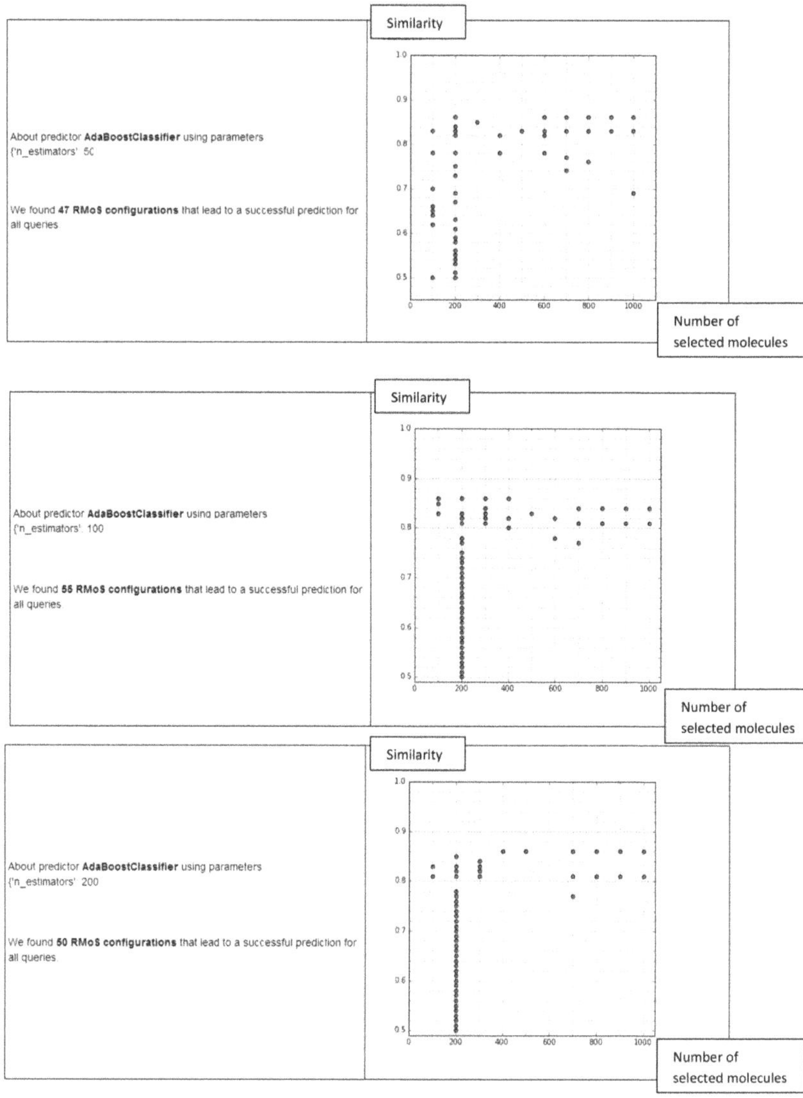

Figure 11.14. (Continued.)

molecules retrieved from the Ames test database with our smarter algorithm is required for a good prediction for the 57 molecules (and for any type of molecules).

The predictions of these 57 molecules are presented in table 11.3.

We also compared the results of our 17 molecules from the explicit list with other *in silico* prediction methods (classic QSAR and OTF-QSAR) and to softwares usually used in the scientific community such as ACD/Percepta and Lazar (table 11.4).

Using the usual software, mutagen and non-mutagen molecules are classified according to a threshold value of 0.5; everything under 0.5 is classed as a non-

Table 11.1. Number of possible configurations for the 17 molecules of the explicit list with the various estimators.

	Number of possible configurations			
		ExtraTrees		
Estimators	RF	Gini index	Shannon's entropy	AdaBoost
10	80	102	120	15
50	113	166	181	47
100	116	200	192	55
200	116	227	223	50
300	126	224	219	36
400	110	195	208	27

Table 11.2. Number of molecules of the enhanced explicit list.

	Number of molecules	Prediction with our predictive system
Initial explicit list	17	17/17
Plasticizers	11	11/11
Oxidizers	6	5/6 (1 doubtful)
Liquids	8	8/8
Stabilizers	6	6/6
Pyrotechnics	9	8/9 (1 doubtful)

Table 11.3. Prediction of the 57 molecules of the enhanced explicit list.

Charges	Experimental data	Accuracy	Plasticizers	Experimental data	Accuracy
Ammonium dinitramide	+	67%	Nitroglycerin	+	88%
Picric acid	+	95%	TMETN	–	92%
Dinitroanisole	+	96%	DANPE	+	82%
NGu	–	60%	Dibutyl sebacate	–	76%
PETN	–	99%	DEHA	–	73%
Tetryl	+	94%	DIBP	–	86%
TNT	+	94%	Diallyl phtalate	–	82%
1,2-Dichloro-4-nitrobenzene	+	98%	DOTP	–	83%
1-Chloro-2,4-dinitrobenzene	+	99%	DOP	–	82%
1-Chloro-4-dinitrobenzene	+	97%	DEGDN	+	66%

1-Methyl-2-nitrobenzene	–	73%	EGDN	+	84%
1-Nitronapthalene	+	98%	Oxidizers	Experimental data	Accuracy
2,3-DNT	+	90%	Nitric acid	+	68%
2,4-DNT	+	93%	Ammonium nitrate	–	78%
2,5-DNT	+	92%	Ammonium perchlorate	–	71%
2,6-DNT	+	90%	Nitrous oxide	+	76%
3,4-DNT	+	92%	HAN	–	60%
3,5-DNT	+	72%			
1,3,5-Trinitrobenzene	+	99%	Liquids	Experimental data	Accuracy
1,3-Dinitrobenzene	+	98%	Perchloric Acid	–	64%
Trinitroglycerol	+	95%	Diethyl ether	–	71%
Stabilizers	Experimental data	Accuracy	DMAZ	+	93%
2-Nitrodephenylamine	–	62%	Hydrazine	+	68%
3-Methyl-1,1-diphenylurea	–	68%	Methylhydrazine	+	77%
Calcium carbonate	–	99%	Nitrogen tetroxide	+	61%
Diphenylamine	–	63%	TMEDA	–	91%
N-Methyl-4-nitroaniline	+	78%	UDMH	+	73%
Triphenylamine	–	74%			
Pyrotechnics	Experimental data	Accuracy	Pyrotechnics	Experimental data	Accuracy
Barium chlorate	–	79%	Strontium chloride	–	93%
HNS	+	91%	Strontium nitrate	–	99%
Paris Green	–	98%	Strontium sulfate	–	58%
Sodium nitrate	–	97%	Styphnic Acid	–	62%
Strontium carbonate	–	100%			

'+' means mutagen molecules and '–' means non-mutagen molecules. The percentage corresponds to the percentage of confidence in the prediction by our predictive system.

mutagen and everything above a mutagen. In order to take into account the variability of the methods and the probability of false positives, we decided to increase this threshold to an interval of 0.4 to 0.6. In this interval, named 'the doubtful zone', the predictive system does not know if the molecule is a mutagen or not and prefers not to classify the molecule. Thus, the molecules are considered non-mutagen between 0 and 0.4 and mutagen between 0.6 and 1.

Finally, we also decided to test these results (R.Mo.S using a percentage of similarity of 81% and 400 molecules retrieved + ML methods) on a validation set of molecules constituted of pharmaceutical molecules. This last set of molecules consists of a large number of molecules, very diversified and unknown to the

Table 11.4. Prediction results of various *in silico* methods and common softwares.

Molecule names	Exp. data	QSAR prediction	OTF-QSAR 85% of similarity	R.Mo.S + machine learning	Lazar	ACD/ Percepta
Picric acid	+	Doubtful	+	+	+	+
Dinitroanisole	+	Doubtful	+	+	+	+
NGu	–	Doubtful	–	–	+	Doubtful
PETN	–	Doubtful	–	–	–	Doubtful
Tetryl	+	Doubtful	+	+	+	+
TNT	+	Doubtful	+	+	+	+
1,2-Dichloro-4-nitrobenzen	+	Doubtful	–	+	+	+
1-Chloro-2,4-dinitrobenzen	+	Doubtful	–	+	+	+
1-Chloro-4-nitrobenzen	+	Doubtful	–	+	+	+
1-Methyl-2-nitrobenzen	–	Doubtful	–	–	–	Doubtful
1-Nitronaphthalen	+	Doubtful	+	+	+	+
2,3-DNT	+	Doubtful	+	+	+	+
2,4-DNT	+	Doubtful	+	+	+	+
2,5-DNT	+	Doubtful	+	+	+	+
2,6-DNT	+	Doubtful	+	+	+	+
3,4-DNT	+	Doubtful	+	+	+	+
3,5-DNT	+	Doubtful	+	+	+	+

'+' means mutagen molecules and '–' means non-mutagen molecules. All the molecules are predicted doubtful with a classic QSAR. There are three errors with the OTF-QSAR, especially situated on chloro-nitrobenzen, and some good responses with the different softwares. There is no error with our technique using our smarter algorithm and a ML using EM with ExtraTrees as prediction *in silico* methods.

system. This set of molecules was compared with the results of the ACD software in order to show the quality of our system (table 11.5).

11.4 Discussion

The improvement of the general process and the prediction steps used a very careful selection of similar molecules. To achieve a high efficiency of similarity, the structures of the molecules are represented by what we call structural keys.

These keys are a particular type of descriptor, making it possible to characterize each molecular structure and especially to discriminate between them. There are many structural keys, but one of the best-known structural keys is the MACCS 166 structural key.

Once a structural key has been derived from a chemical structure, it is possible to compare two molecules. Standard softwares (including MOE) often make reference to the Tanimoto metric for the calculation of similarity and to compare two

Table 11.5. Validation set of molecules.

Molecules	Experimental data	Accuracy	ACD software
1-Methylpyrene	+	87%	+
1-Methyl-4-nitrobenzene	+	70%	–
1,2,3-Benzotriazole	+	60%	+
1,2-Dichloroethane	+	82%	+
1,4-Dithiane	–	D (55%)	+
1,4-Oxanthiane	–	92%	–
1,8-Dinitronaphthalene	+	69%	+
2-Aminopyridine	–	88%	–
2-Methylbenzamide	–	63%	–
2-Nitropropane	+	78%	+
5-Nitro-o-toluidine	+	70%	+
Acetamide	–	68%	–
Aniline	–	96%	–
Aspartame	–	100%	–
Benzothiazole	–	88%	–
Benzyl chloride	+	95%	+
Butanal oxime	+	93%	+
C.I Acid Orange 3	+	79%	+
C.I Allura Red 17	–	99%	–
Caffeine	–	100%	–
Caprulyl chloride	+	97%	+
DEHP	–	76%	–
Dibenz[a,h]anthracene	+	88%	+
Diisobutylphtalate	–	84%	–
Diisodecylphtalate	–	86%	–
Di-n-octylphtalate	–	87%	–
Estradiol	–	72%	–
Ethylene glycol	–	90%	–
HNF	+	/	D (50%)
Hydrazine hydrate	+	61%	/
Hydrazine perchlorate	+	64%	/
Isopropyl methylphosponic acid	–	66%	–
p-Chlorophenyl methyl sulfide	–	84%	–
p-Chlorophenyl methyl sulfone	–	81%	–
p-Chlorophenyl methyl sulfoxide	–	92%	–
Phenylacetonitrile	–	74%	–
Pigment yellow 74	–	93%	–
Pitavastatin	–	100%	+
Propionitrile	–	81%	–
o-Cresol	–	95%	–
Stearic acid	–	95%	–
L-Taurine	–	88%	–

(*Continued*)

Table 11.5. (*Continued*)

Molecules	Experimental data	Accuracy	ACD software
Theophylline	–	91%	–
Thiourea	–	96%	–
Vitamin E	–	97%	–
		39/45	39/45

'+' means mutagen molecules and '–' means non-mutagen molecules. The percentage corresponds to the percentage of confidence in the prediction by our predictive system.

molecules. However, it is important to remember that this metric is only formulated for pairs of Boolean vectors (i.e. two state variables). Indeed, the Tanimoto coefficient does not take into account the non-expression of the same characteristic MACCS in two molecules (zero MACCS value for the two molecules) and the fact that the two compared molecules can have different expression levels. This is why, in this study, we modified and improved the Tanimoto coefficient in two different ways.

First of all, in order to take into account the non-expression of a same characteristic MACCS in two molecules, the molecules are defined by two new values.

Then, in order to take into account the different expression levels, we recall that two identical characteristics have a local similarity of 1 and, inversely, two different characteristics have a local similarity of 0. Thanks to the first improvement, the non-expression of a same characteristic MACCS in two molecules has a local similarity of 1. The second improvement consists in finding an intermediate similarity to define different levels of presence of a MACCS characteristic.

The distance is a particular parameter used to express the difference between two molecules and in particular between two MACCS characteristics. Euclidian distance has been tested but it does not reflect the idea of dissimilarity or levels of expression; it expresses a gap between the values. Among the number of possible distances, we chose to rely on the ratio between the two expression levels of a MACCS characteristic and then applied a natural logarithm function on the difference. This distance is then transformed into a local similarity value thanks to a conversion function with a Gaussian shape. The local similarly applied to each MACCS, which highlights differences, will lead to the selection of molecules.

Thus, a zero distance, and therefore a maximum local similarity of 1, is associated with two identical values of the MACCS characteristics. Conversely, an infinite distance, and therefore a minimum local similarity of 0, is associated with two MACCS characteristic values, considered to be totally dissimilar.

The smarter algorithm, R.Mo.S, can predict the toxicity of an unknown molecule thanks to three steps. The first step aims to recover all the molecules similar to the target molecule thanks to the two improvements previously stated. Thus, at a certain similarity, a subset is created containing only molecules similar to the unknown molecule that we want to test.

The second step is to bring out a signature molecule that will be representative of the similar molecules retrieved from the unknown molecule. This virtual molecule, defined by its MACCS, will serve for the next step of the process. The new signature molecule obtained contains a high percentage of similarity with the initial unknown molecule.

We then proceed with an iterative process to recover similar molecules to this virtual molecule. Over the iterations, signature molecules are constructed and a new similarity search is performed each time with these virtual molecules. After a certain number of iterations, all the similar molecules from the different signature molecules are retrieved and a new database is constructedm which is then used for the prediction of the toxicity.

According to our results, the preferred *in silico* prediction method is ML. Additionally, we tested an ANN, a SVM, a DT and an EM.

The artificial neural network. First of all, using an ANN, a lot of time was spent testing many network compositions based on different publications. Indeed, there is a very large number of types of neural networks depending on the task to be performed, the nature of the data, the complexity of the problem to be solved, the calculation time, and the software used. In the end, we abandoned this method, after trying all these topologies and making sure to avoid over- and underfitting, and in part due to the simulations being extremely long.

The support vector machine technique. This technique is among the best in cases where data are limited. In addition, despite the fact that SVMs are a powerful tool, the storage and computing requirements increase very quickly with the amount of training data. Indeed, the complexity of an SVM is at least quadratic in respect to the size of the learning data. Thus, despite our efforts, the SVM was very complicated to interpret and also to set up for our problem. Indeed, this method is not adapted to our data because our learning set is too big, and the algorithm too slow and much too expensive in terms of calculation time.

The decision tree method. As illustrated in figure 11.8, we can easily see that the grid associated with Shannon's entropy proposes 39 combinations, leading to a correct prediction of the 17 molecules from the explicit list, whereas with the Gini index the grid proposes only 13 possible combinations. The best DT is the one that preferentially uses Shannon's entropy. However, with this method, no large zone of space is visible, resulting in an optimal R.Mo.S set of parameters.

The ensemble method. This method improves the stability and robustness of the prediction compared to the DT method. The complexity and size of the trees must be controlled to reduce memory consumption. This is possible by defining the values of certain parameters, including the number of estimators. This parameter is difficult to choose because it defines the number of trees to create and therefore to combine to obtain the best possible prediction. The number of estimators can be chosen by increasing the number of trees, calibration after calibration, until the accuracy stops showing improvements. However, this can take considerable time. In general, the greater the number of subsets of trees are built, the more robust the forest is and the more accurate the results are. We therefore tested the following number of estimators: 10, 50, 100, 200, 300, and 400.

By testing a number of estimators ranging from 10 to 400, we realized that, beyond a critical number of subsets of trees created, the prediction results cease to improve significantly and the accuracy decreases. This critical number was set to a number of 200 estimators for the EM using any algorithm. This critical number of 200 corresponds to the number of trees combined for the ExtraTrees and AdaBoost algorithms. As illustrated in table 11.1, 227 configurations are possible, leading to a correct prediction of 17 molecules from the explicit list.

We found that the DT, AdaBoost, and RF algorithms gave better results but were still inadequate. After a long period of development and calibration, we found that the best ML method is the one utilising an EM with the ExtraTrees algorithm. These results are based on the explicit list of HEM molecules with known experimental results for the Ames test.

On establishing the best ML method, we determined the percentage of similarity and the number of molecules needed in the database to obtain a correct prediction for any type of molecule. After enhancing the explicit list of molecules with other molecules (explosives and ingredient in the formulation of explosives) we determined the optimal values of 81% similarity and 400 molecules retrieved. This percentage and this number of molecules is the best to obtain a good prediction for 55 molecules out of 57 for the new explicit list of molecules. The two incorrect molecules were predicted as doubtful (tables 11.2 and 11.3).

Then, we compared the results with the other *in silico* prediction methods (classic QSAR and OTF-QSAR) and to softwares usually used in the scientific community such as ACD/Percepta and Lazar (table 11.4). The results of the QSAR and the OTF-QSAR are extracted from our other publication [29].

We can see that with the QSAR technique, all 17 molecules from the explicit list are predicted as doubtful. The prediction is based on the whole database and the general rules of the classic QSAR are not adapted to our HEM molecules.

With the OTF-QSAR there are three errors present with chloro-nitrobenzen. The OTF-QSAR is better than the QSAR but the problem concerns what is learned by the system. We can see with the ACD/Percepta software that all the non-mutagen molecules are predicted as doubtful. Indeed, this software cannot manage to predict non-mutagen molecules.

The Lazar software is good, but there is still one error present with the nitroguanidine molecule.

Finally, all the molecules are correctly predicted with our smarter algorithm, which selects, in a particular manner, similar molecules to a query and performs thereafter the *in silico* prediction technique using ML.

Finally, we also tested these results (R.Mo.S using a percentage of similarity of 81% and 400 molecules retrieved + ML methods) on a validation set of molecules especially constituted of pharmaceutical molecules to show the robustness and the quality of our predictive system. This last set of molecules was unknown to the system and was compared with the results of the ACD software (table 11.5). We can see that,with our system, 39 molecules out of 45 are correctly predicted. It is as good as ACD/Percepta, which is a software specialized in pharmaceutical molecules. Our predictive system is calibrated on HEMs and makes better predictions than any other tested *in silico* technique. Moreover, we demonstrate that our system is also as good as commercial software for the toxicity prediction of pharmaceutical molecules.

11.5 Conclusion

In collaboration with ArianeGroup, in this study a new *in silico* prediction technique has been developed. This latter is based on the prediction of HEM molecules. All the HEM compounds were evaluated for evidence of potential mutagenicity (using the Ames test). The different databases created come from different European databases.

The strategy of this study, giving rise to a patent [40], selectively and iteratively uses the search of structural similarity refined with specific parameters such as fingerprints, metrics, and characteristics. We developed our totally new and smarter algorithm, recursive molecular similarity (R.Mo.S).

The smarter algorithm is totally new in this field. It is able to identify a suitably diverse subset of molecules from a large number of compounds. This thus brings numerous advantages, in particular increasing the number of selected molecules, obtaining a more complete and precise selection of the molecules of the database intended to be used for the prediction and improving the performance as well as the quality of the predictions. This method is also advantageously adaptive because it can be easily used for all the biological tests and all types of molecules (pharmaceutical, cosmetic, explosives, etc). It also can be applied itself to a wide range of different similarity indices for prediction.

References

[1] Klapötke T M 2012 *Chemistry of High-Energy Materials* 2nd edn (Berlin: de Gruyer)

[2] Agrawal J P 2010 *High Energy Materials: Propellants, Explosives and Pyrotechnics* (New York: Wiley)

[3] Nair U, Asthana S, Subhananda Rao A and Gandhe B 2010 Advances in high energy materials *Def. Sci. J.* **60** 137–51

[4] Thybaud V *et al* 2007 Strategy for genotoxicity testing: hazard identification and risk assessment in relation to *in vitro* testing *Mutat. Res. Toxicol. Environ. Mutagen.* **627** 41–58

[5] Müller L *et al* 2003 Strategy for genotoxicity testing and stratification of genotoxicity test results—report on initial activities of the IWGT Expert Group *Mutat. Res. Toxicol. Environ. Mutagen.* **540** 177–81

[6] Blakey D, Galloway S M, Kirkland D J and MacGregor J T 2008 Regulatory aspects of genotoxicity testing: from hazard identification to risk assessment *Mutat. Res. Toxicol. Environ. Mutagen.* **657** 84–90

[7] Kirkland D J *et al* 2005 Testing strategies in mutagenicity and genetic toxicology: an appraisal of the guidelines of the European Scientific Committee for cosmetics and non-food products for the evaluation of hair dyes *Mutat. Res. Toxicol. Environ. Mutagen.* **588** 88–105

[8] Gollapudi B B and Krishna G 2000 Practical aspects of mutagenicity testing strategy: an industrial perspective *Mutat. Res. Mol. Mech. Mutagen.* **455** 21–8

[9] Mortelmans K and Zeiger E 2000 The Ames Salmonella/microsome mutagenicity assay *Mutat. Res.* 29–60

[10] OECD 1997 *OECD Guideline for Testing of Chemicals: Bacterial Reverse Mutation Test* Test No. 471 OECD
[11] Hengstler J G *et al* 2006 The REACH concept and its impact on toxicological sciences *Toxicology* **220** 232–9
[12] EU 2007 *Réglementation REACh.* L 396/46 Journal officiel de l'Union européenne
[13] Fédération Nationale des Travaux Publics 2008 *REACh—Guide Pratique (Paris: Fédération Nationale des Travaux Publics)*
[14] OECD 2015 *Guidance Document on Revision to OECD Genetic Toxicology Test Guidelines*
[15] Kirkland D, Kasper P, Müller L, Corvi R and Speit G 2008 Recommended lists of genotoxic and non-genotoxic chemicals for assessment of the performance of new or improved genotoxicity tests: a follow-up to an ECVAM workshop *Mutat. Res. Toxicol. Environ. Mutagen.* **653** 99–108
[16] Kirkland D, Aardema M, Müller L and Hayashi M 2006 Evaluation of the ability of a battery of three *in vitro* genotoxicity tests to discriminate rodent carcinogens and non-carcinogens: II. Further analysis of mammalian cell results, relative predictivity and tumour profiles *Mutat. Res. Toxicol. Environ. Mutagen.* **608** 29–42
[17] Gold L S, Manley N B, Slone T H, Rohrbach L and Garfinkel G B 2005 Supplement to the Carcinogenic Potency Database (CPDB): results of animal bioassays published in the general literature through 1997 and by the national toxicology program in 1997–1998 *Toxicol. Sci.* **85** 747–808
[18] Benigni R, Battistelli C L, Bossa C, Tcheremenskaia O and Crettaz P 2013 New perspectives in toxicological information management, and the role of ISSTOX databases in assessing chemical mutagenicity and carcinogenicity *Mutagenesis* **28** 401–9
[19] Tennant R W 1991 The genetic toxicity database of the National Toxicology Program: evaluation of the relationships between genetic toxicity and carcinogenicity *Environ. Health Perspect.* **96** 47–51
[20] EU 2010 *Directive 2010/63/UE du Parlement européen et du Conseil du 22 septembre 2010 relative à la protection des animaux utilisés à des fins scientifiques* L 276/33 Journal officiel de l'Union européenne
[21] Tannenbaum J and Bennett B T 2015 Russell and Burch's 3Rs then and now: the need for clarity in definition and purpose *J. Am. Assoc. Lab. Anim. Sci. JAALAS* **54** 120–32
[22] ECHA 2015 *Guidance on Information Requirements and Chemical Safety Assessment Chapter R.7a: Endpoint Specific Guidance*
[23] Pfuhler S *et al* 2009 Reduction of use of animals in regulatory genotoxicity testing: identification and implementation opportunities—Report from an ECVAM workshop *Mutat. Res. Toxicol. Environ. Mutagen.* **680** 31–42
[24] Van de Waterbeemd H and Gifford E 2003 ADMET in silico modelling: towards prediction paradise? *Nat. Rev. Drug Discov.* **2** 192–204
[25] Nendza M *et al* 2013 A comparative survey of chemistry-driven in silico methods to identify hazardous substances under REACH *Regul. Toxicol. Pharmacol.* **66** 301–14
[26] Gramatica P 2011 *A Short History of QSAR Evolution* (QSAR Research Unit in Environmental Chemistry and Ecotoxicology)
[27] Cherkasov A *et al* 2014 QSAR modeling: where have you been? Where are you going to? *J. Med. Chem.* **57** 4977–5010

[28] Cronin M T D *et al* 2003 Use of QSARs in international decision-making frameworks to predict health effects of chemical substances *Environ. Health Perspect.* **111** 1391–401

[29] Alliod C, Chemelle J-A, Jacob G and Terreux R 2017 Ames test prediction on high energy molecules by on-the-fly QSAR (OTF-QSAR) *Propellants, Explosives and Pyrotechnics* **42** 24–35

[30] Bender A 2005 *PhD Thesis* Darwin College

[31] Karakoc E, Cherkasov A and Cenk Sahinalp S 2006 Distance based algorithms for small biomolecule classification and structural similarity search *Bioinformatics* **22** 243–51

[32] Bishop C M 2006 *Pattern Recognition and Machine Learning* (Berlin: Springer)

[33] Hastie T, Tibshirani R and Friedman J 2009 *The Elements of Statistical Learning: Data Mining, Inference, and Prediction* 2nd edn (New York: Springer)

[34] Quinlan J R 1986 Induction of decision trees *Mach. Learn.* **1** 81–106

[35] Sexton R S and Dorsey R E 2000 Reliable classification using neural networks: a genetic algorithm and backpropagation comparison *Decis. Support Syst.* **30** 11–22

[36] Lebrun G, Lezoray O, Charrier C and Cardot H 2007 An EA multi-model selection for SVM multiclass schemes *Lecture Notes in Computer Science* **4507** 260–7

[37] Breiman L 2001 Random forests *Mach. Learn.* **45** 5–32

[38] Geurts P, Ernst D and Whenkel L 2006 Extremely randomized trees *Mach. Learn.* 3–42

[39] Zhu J, Zou H, Rosset S and Hastie T 2009 Multi-class AdaBoost *Statistics and Its Interface* **2** 349–60

[40] Terreux R, Alliod A, Denis R and Jacob G 2018 Method and device for selecting a subassembly of molecules for use in predicting at least one property of a molecular structure *FR Patent WO 2018/234718 Al*

[41] O'Donnell T J 2009 *Design and Use of Relational Databases in Chemistry* (Boca Raton, FL: CRC Press) p 8

[42] Fernández-de Gortari E, García-Jacas C R, Martinez-Mayorga K and Medina-Franc J L 2017 Database fingerprint (DFP): an approach to represent molecular databases *J. Cheminformatics* **9** 9

[43] Bajusz D, R_cz A and H_berger K 2015 Why is Tanimoto index an appropriate choice for fingerprint-based similarity calculations? *J. Cheminformatics* **7** 20

[44] Kosub S 2016 A note on the triangle inequality for the Jaccard distance arXiv:1612.02696

[45] Levandowsky M and Winter D 1971 Distance between sets *Nature* **234** 34–5

IOP Publishing

Energetics Science and Technology
An integrated approach
Adam S Cumming

Chapter 12

Sustainable energetic materials

Carlos Ferreira and José B Ribeiro

Eco-design is a standard regulated approach that integrates environmental criteria into product design in order to prevent, reduce, or mitigate the potential environmental impacts associated with the whole life cycle of a product or system. The application of this approach to energetic materials will allow not only reduction of resource (materials and energy) consumption and indirect emissions during production but also reduce direct and indirect emissions during the use and end-of-life phases. This chapter is also intended to show that eco-design for energetic materials promotes the use of the best available technologies for the defence sector. This includes eco-labelling, which helps to identify products or systems proven to be environmentally preferable to protect both biodiversity and human health, as well as to facilitate recycle of content or remanufacturing and minimise the use of hazardous substances while controlling costs.

12.1 Introduction

Companies that produce energetic materials are usually focused on delivering products with good operational properties and in accord with safety regulations or standards. Other criteria, however, are becoming increasingly important for consumers, and among those are environment issues. The inclusion of environmental criteria in the company's business goals possibly provides a marketing advantage over companies without this type of concern. The companies that incorporate environmental criteria in their production chain also obtain the benefit of reducing costs by, for instance, using less materials and requiring less energy for the production of the same product (Polverini 2021, Ribeiro *et al* 2013).

Over and above any company's motivation to improve their environmental profile, the regulations and legislation that limit environmental impacts are becoming increasingly demanding. Carbon emission limitations (European Commission 2021) and REACH regulations (European Commission 2006) are a good example of the environmental impositions that companies must comply with in order to reduce the impacts on human health and the environment. The pursuit of the reduction of

doi:10.1088/978-0-7503-3943-8ch12

environmental impact and costs associated with the products or services delivered by a company can be obtained by employing an eco-design approach, as the main goal of eco-design is to reduce to a minimum the overall environmental impact of a product or service (European Commission 2009).

The employment of eco-design allows the development of innovative design solutions considering the whole life cycle. Therefore, the environmental concerns of products or services are incorporated from the design phase, taking into consideration the means by which they are produced, their performance in use, and the possibility of repair, recycling, or disposability in appropriate environmental conditions (Chun *et al* 2018). Hence, eco-design is an effective tool for improving the energy efficiency and sustainability of products, by promoting industrial competitiveness and innovation with products demonstrating improved environmental performance throughout the internal market (European Commission 2009).

Eco-design has a legal framework in the European Union (EU) that consists of the Ecodesign Directive (European Commission 2009), which establish rules for improving the environmental performance of products (e.g. household appliances and engineering products). This directive defines mandatory thresholds for the energy efficiency of the products in the EU market (independently of where they are manufactured), and contributes to the prevention of barriers to trade, improvement of product quality, and ensuring environmental protection (Polverini 2021). This directive establishes a consultation forum for stakeholders, which includes formal representatives from EU countries, industry, and civil society who have access to the implementation of the directive and regulation. The Ecodesign Directive is complemented by the Energy Labelling Regulation (European Commission 2017), which ensures that consumers are informed of the product's energy efficiency with a simple instrument in the form of an energy label (Polverini 2021).

In this chapter the advantages of employing eco-design strategies for the whole life cycle of energetic materials is discussed. The main aim is to present the life-cycle assessment (LCA) methodology as a suitable tool to assess quantitatively the potential environmental enhancements associated with the eco-design strategies or measures applied to energetic materials. For that purpose, some examples of LCA applied to energetic materials is presented to help show those advantages.

12.2 Eco-design

Eco-design can be defined as the design of products or services which considers their whole life cycle. Thus, it is a life-cycle thinking approach, with the aim of reducing the environmental impacts associated with those products or services (European Commission 2009, Chun *et al* 2018, Bassani *et al* 2022). The fundamental principle of life-cycle thinking, or circular systems, is that all chain processes of a product are interconnected, so a change in one of those phases (design, development, production, transport, use, and disposal) will have an influence on the rest of the life-cycle phases. Therefore, eco-design appears as a way to develop products that are desirable for stakeholders but, at the same time, environmental criteria are included in all phases of the life cycle to present an enhanced environmental profile (Chun *et al* 2018).

The eco-design approach, according to the Methodology for the Ecodesign of Energy Related Products (MEErP; Kemna 2011), should follow this set of eco-design parameters that refer to the phases of the life cycle of products: raw materials; manufacturing; packaging, transport and distribution; installation and maintenance; use; and, finally, end of life. For each of these parameters some environmental indicators, when important, must be evaluated:

- Predict consumption of resources and energy.
- Predict emissions to air, water, and/or soil.
- Predict pollution through physical effects (e.g. noise, vibration, radiation).
- Predict the waste that will be generated.
- Identify solutions for reuse, recycling, and recovery of materials and/or of energy.

The Sustainability Guide, developed by SVID (a Swedish Industrial Design Foundation) in 2009, also presented a similar approach to achieve sustainable products or services[1]. That approach is divided into seven steps: materials; manufacturing; product; distribution; use; end of life; and sales and marketing. For each one of the abovementioned steps, ways are indicated to improve the environmental profile of products and services using eco-design.

Materials. Selection of materials is an important step in the eco-design of a product, as their associated impacts are influenced by different factors. The materials used need to be durable and recyclable, with lower weight, less life-cycle energy requirement, and to not be hazardous for human health, as well as also be designed in such a way that components or parts are used for facilitating recycling or repairing.

Manufacturing. Sustainable manufacturing aims to optimise the environmental profile of products and services considering the entire life cycle. Therefore, a company which employs a manufacturing with environmental concerns increases their social and environmental benefits by creating products and services, and making the processes themselves, more sustainable. Some examples of employing a sustainable manufacturing are processes that conserve energy and natural resources, improvements associated with assembly, reparability, refurbishment, and disassembly, valorisation of useful parts that can be directly reused in another product; thus, optimising processes in order to reduce to a minimum the environmental impact.

Product. Increasing the life cycle of a product is of great importance in order to get the most out of the materials and energy embedded into that product. For that purpose, it is important to know clearly the material flow of a product. This assists in making sustainable decisions that will influence its life cycle. Therefore, it is key to develop strategies to optimise the useful life of a product and to decide the best strategies to follow. This is in addition to delivering products that meet the requisites of potential buyers, and are decisions that will have a significant influence in the use phase of a product or service.

[1] https://sustainabilityguide.eu/

Distribution. Distribution is an important part of the product or service life cycle in which a company needs to consider a full-circle supply chain. The following are all important: optimisation of the distribution and transport systems, utilisation of reusable packaging or with reduced package waste, and choice of local markets to reduce transport distance. These will aid in decreasing the impact associated with this life-cycle phase. To achieve the optimisation of distribution it is vital to understand and collaborate with both the up- and downstream supply chain. Thus, this involves an understanding of what the vital chains are or if new business models are better for the company.

Use. In the use phase of products or services, it is important to ensure that the consumption of resources or energy is as low as possible, and that less pollution is generated. The products should be designed with the consideration that maintenance or repair is easy to perform, with the aim of increasing its life cycle, and that it is possible to adapt the products for updates or reuse.

End of life. The final stage of the eco-design defines what happens in the end of life of a product or service, which is usually the beginning of a new life for a product or system. Design has to consider waste-management strategies for the products or services and it should be possible to perform disassembly or reassembly, dismantling, upcycling, recovery, or recycling.

Sales and marketing. Green marketing is also important to companies that wish to inform their customers that environmental issues are being considered in the life of their products. It is an opportunity to differentiate their core business, and encourage the customers to buy in a responsible way by choosing the best possible environmentally friendly options. However, it is important to avoid 'green washing' and to be transparent about the products or how the service is designed. This is fundamental to maintain the confidence of buyers. Moreover, the company will benefit from informing customers about how to increase product life or the destination to select after its end of life.

12.3 Tools for eco-design

Different methods and tools have been developed to incorporate eco-design into the conception of products, such as the aforementioned LCAs, a MET Matrix (Brezet and van Hemel 1997), the ECODESIGN Checklist Method (Wimmer 1999), and MATto (Allione *et al* 2012). The implementation of an eco-design approach requires an environmental impact assessment in order to predict the potential impacts associated with different solutions and its improvement options (Kemna 2011). An LCA is a methodology that requires conducting a quantitative assessment of the environmental impacts associated with the full life cycle of a product or service, thus describing the production, distribution, use, and end-of-life phase (including recycling and reuse) of the product in terms of its impact over the environment and human health (International Organization for Standardization 2006a). An LCA is an excellent tool for eco-design because

(i) it provides information that can be verified and that supports the measures implemented to improve the environmental profile of products or services;

(ii) its life-cycle perspective, which aids in avoiding impact shifting (environmental burdens between life-cycle stages and/or between environmental impact categories), as the influence of some modification to the product or service can be determined; and
(iii) it helps in identification of the main contributors to the impact, which enables the environmental improvement of products or services by reducing or even avoiding that impact (Blanca-Alcubilla *et al* 2020, Ribeiro *et al* 2013). In fact, the potential enhancements to the environmental profile of a product or service, originated from new design measures, can only be properly achieved if the main contributing factors responsible for the impacts are well known (Blanca-Alcubilla *et al* 2020, Bassani *et al* 2022, Chun *et al* 2018, de Otazu *et al* 2022, Ribeiro *et al* 2013).

An LCA is organised in four main phases: goal and scope definition, life-cycle inventory, life-cycle impact assessment, and interpretation. In the goal and scope phase, the main objective of the analysis is defined as well as the boundaries of the system (what is included in the study). In this phase, the functional unit is also defined. This is a critical part of an LCA study, as this functional unit is a reference for the system that will allow the comparison of different products or systems with the same function. In the second phase, life-cycle inventory (LCI), a compilation of the emissions and resources or materials consumption per individual process is performed. This inventory is developed based on a chain of processes, where emissions and resources, partitioned as outputs of some processes, are then used as inputs in subsequent processes within one life-cycle phase. This inventory is also developed based on the functional unit defined in the previous LCA phase. The third phase, life-cycle impact assessment (LCIA), the emissions released and the resources consumed are converted into specific impact categories. The conversion factors are unique for each impact category and are developed based on the state-of-the-art insights into the effect of the emissions on ecosystems, humans, and/or the scarcity of the resources (Kemna 2011, International Organization for Standardization 2006a). However, it is important to mention that some uncertainty is associated with the impact assessment phase due to different reasons such as lack of data, lack of scientific consensus, and the time horizon employed (Kemna 2011). There are several impact categories, in which some examples are global warming, acidification, eutrophication, depletion or resources, water scarcity, and human health. The impact categories that will be used are selected by practitioners based on their experience or on the relevance for the study. The last phase of an LCA study is the interpretation, in which recommendations or suggestions are presented based on the results obtained. This last phase is iterative for all LCA phases, as the findings of the LCA can lead to the organisation of the goals, system boundaries, or inclusion of new impact categories. A detailed description of the LCA methodology applied to energetic materials can be found in Ferreira *et al* (2019a).

The MEErP also gives guidelines to implement a life-cycle cost analysis, allowing the assessment of the monetary consequences associated with the design options established, in order to identify their environmental potential monetary benefits and find solutions with the so-termed least life-cycle costs (Kemna 2011). The assessment of monetary life-cycle costs is important in order to select design solutions with a positive impact over the total product life (e.g. purchase and running costs), while also taking into account the research and development and investment costs (Kemna 2011).

12.4 Application of life-cycle assessment to energetic materials

The introduction of eco-design principles and strategies (e.g. less materials, use of different materials or energy sources, defining new end-of-life procedures) for the design of energetic materials needs to be done quantitively and with a holistic view in order to identify the potential environmental but also technical, economic, and social benefits from those strategies (Margallo *et al* 2019). Based on the identification of the potential environmental performance improvements, it is then possible to ascertain with confidence which strategies can be applied.

This section presents an example of an LCA applied to energetic materials, a generic 40 mm ammunition, in order to illustrate the application of the four phases of the methodology and the advantages arising from its use. The ascertainment of the best eco-design strategies will be done mainly by identifying the major contributors to the environmental impacts associated with the production of a generic 40 mm unit. Beyond the example mentioned before, other examples from the literature which have applied LCA to energetic materials are briefly discussed to present other advantages, as well as some of the limitations of the methodology.

12.4.1 Life-cycle assessment of a 40 mm generic ammunition

The four main phases of the LCA methodology referred in section 12.3 will be followed to perform the assessment of the environmental impacts associated with a 40 mm generic ammunition.

Phase 1. Goal and scope.

The main goal of this study is to assess and compare the environmental and toxicological impacts associated with the life-cycle of a 40 mm generic ammunition filled with two different kind of explosives. The two explosives used in this case were Composition B and a polymer-bonded explosive (PBX). All other components were the same for both munitions. This research also aims to identify opportunities to improve the environmental performance of ammunition by identifying the environmental hot spots. Figure 12.1 shows a representation of the generic ammunition considered for the analysis.

A life-cycle model is developed for the production of the 40 mm ammunition, with different explosive charges, based on literature data. Figure 12.2 shows the system boundaries of the life-cycle model for the 40 mm ammunition. Direct emissions associated with the production phase (e.g. liquid or gaseous effluents) and the energy required for the assembly of the ammunition are not included in the model due to lack of data. The life-cycle model, the LCI, and the calculation of the life-cycle impacts were conducted using SimaPro 9.1.1 software. A 40 mm

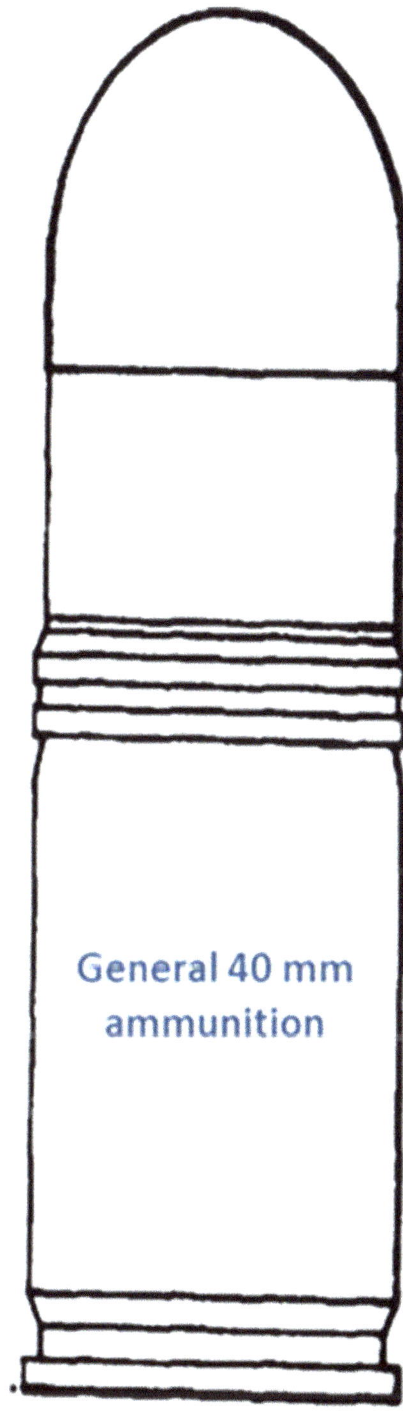

Figure 12.1. Representation of a general 40 mm ammunition (adapted from Department of the Army 1994).

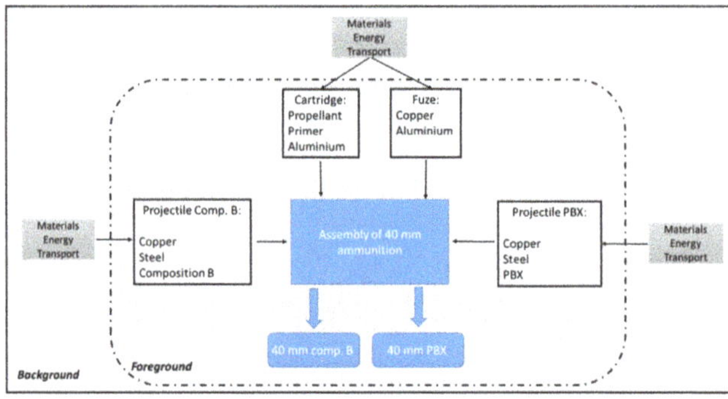

Figure 12.2. Life-cycle model developed for the production of the two types of 40 mm ammunition.

Table 12.1. Materials used for life-cycle inventory referrent to the production of a general 40 mm ammunition.

	Components	Amount	Unit
Cartridge	Aluminium alloy	0.067	kg
	Propellant charge	0.004	kg
	Primer	0.004	kg
Projectile	Steel	0.071	kg
	Copper liner	0.019	kg
	Copper rotating band	0.005	kg
	Explosive charge	0.038	kg
Fuze	Aluminium ogive	0.038	kg
	Copper spring and firing pin	0.007	kg

ammunition is defined as a functional unit, so all the inputs and outputs, as well as the impacts calculated with this model, are relative to this unit.

Phase 2. Life-cycle inventory.

A detailed LCI for the production phase of the ammunition is implemented based on ammunition data sheets (Department of the Army 1994) and ammunition producers' pamphlets. The remaining data is based on the Ecoinvent database v3.8. Ecoinvent is a consistent and transparent LCI database that supports environmental assessments of products and processes worldwide (http://www.ecoinvent.ch). This database facilitates life-cycle modelling as various products, energy sources, and processes have already been studied and can be used as a base to construct the model.

Table 12.1 shows the materials and components used for the production of the mentioned 40 mm ammunitions. The only difference between the two 40 mm ammunitions is the explosive charge: Composition B (60% TNT + 40% RDX + 0.001% wax), and PBX (90% RDX + 9.1% polystyrene + 0.5% dioctyl phthalate + 0.4% resin). For the primer, in both ammunitions a mixture of 60% lead styphnate, 25% barium nitrate, 10% antimony sulphide, and 5% tetrazene was considered.

Some assumptions were made due to method limitations or lack of data. As some substances used in the propellant and primer composition are not available in life-cycle databases (including Ecoinvent), its models for their production were developed following the recommended approach for LCA studies when this limitation is present. The data related to production of lead styphnate were provided by a Romanian ammunition producer (Ferreira *et al* 2016), whilst for tetrazene the data are taken from the literature (Urbanski 1968). For the other primer substances (barium nitrate and antimony sulphide) the models were developed assuming their production from materials available in the databases and stoichiometric conditions.

The lack of real data for the production of the energetic materials is an important constraint of an LCA. The difficulty to obtain data is significantly noticeable for the production of energetic materials, principally due to information confidentiality. Companies are not comfortable with providing data that can be covered by intellectual property rights, which leads to scarcity of information regarding raw materials, energy consumption, and emissions. Furthermore, the most common life-cycle databases do not cover these types of products, so data (including the ingredients) regarding the production and use of explosives are not available, hindering the environmental impact assessment of explosives. The demonstration of the advantages arising from the use of this tool in the improvement of the environmental profile of the ammunition, and the emphasis on the potential costs savings, can be seen as a way to obtain the data required to carried out LCA studies without the need of increased simplifications, and, at the same time, to maintain the confidence in the results obtained. The transport for raw materials is included in the database; however, the transport of the main components is not included due to uncertainty in the provenance of the materials and the locality of ammunition production.

Phase 3. Life-cycle impact assessment.

Two complementary LCIA methods are used: the CML for six environmental impact categories (abiotic depletion, acidification, eutrophication, global warming, and ozone layer depletion) and USEtox for three toxicity impact categories (human toxicity with cancerous and non-cancerous effects, and eco-toxicity). Table 12.2 shows a brief description of the impact categories selected as most significant for this study, and their corresponding units.

Firstly, the LCIA comparative results are presented for the two 40 mm ammunitions. The main impact contributors for each one of the impact categories are then detailed. Figure 12.3 shows the comparative environmental analysis of the two 40 mm munitions. The results are converted to values between 0 and 1 (1 being the highest impact value for each category), in order to simplify the representation and comparison of the impact categories with different units in the same graph; however, the total impact for each impact category is also represented in the graph. The total impacts are similar for the two 40 mm ammunitions for almost all the impact categories. However, for both the human toxicity impact categories the 40 mm ammunition filled with Composition B presents lower impacts (40% less for human toxicity with cancerous effects, and 20% less for human toxicity with non-cancerous effects) than the one filled with PBX.

Table 12.2. Description of the selected six environmental impact categories (CML method), and three toxicological impact categories (USEtox method) (adapted from PRé 2014, Ferreira *et al* 2019b).

Impact category	Description	Unit
Abiotic depletion	Determines the extraction of minerals and fossil fuels, based on concentration of reserves and rate of de-accumulation	kg Sb eq
Acidification	Describes the fate and deposition of acidifying substances	kg SO_2 eq
Eutrophication	Describes the impacts due to excessive levels of macronutrients in the aquatic environment	kg PO_4^{3-}eq
Global warming	Describes the impacts associated with emissions of greenhouse gases	kg CO_2 eq
Ozone layer depletion	Defines the impacts from the emission of gases that have an effect in the destruction of the stratospheric ozone layer	kg CFC-11 eq
Human toxicity, cancerous	Describes the cancerous effects of chemical emissions on human health	Cases
Human toxicity, non-cancerous	Describes the non-cancerous effects of chemical emissions on human health	Cases
Eco-toxicity	Describes the effects of chemicals on ecosystems	PAF m^3 day

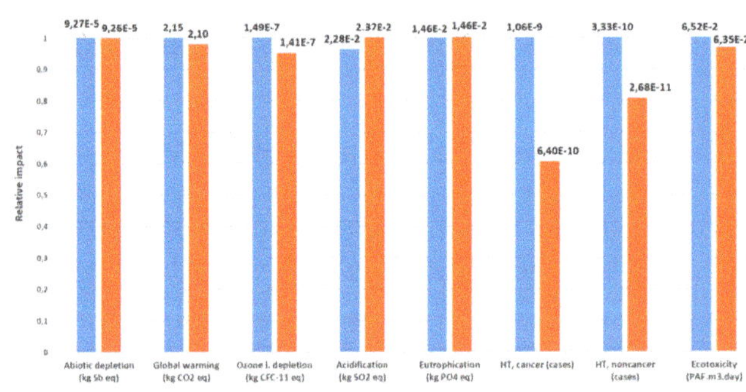

Figure 12.3. Comparison of the total environmental and toxicological impacts associated with the production of two 40 mm ammunitions with different explosive fills: Composition B and PBX.

As the analysis carried out with LCA is not a black box analysis, the reason for these differences can be analysed by identifying the contributors to those impacts. Table 12.3 shows the constituents of the ammunition that contribute most for the impact categories, whilst figures 12.4 and 12.5 show the impact contribution for each impact category associated with the components of the production of 40 mm ammunition with the PBX and with Composition B, respectively. The major impacts for both munitions are associated with the main components, with slight variations,

Table 12.3. Contribution, in percentage, of the principal components of two types of 40 mm ammunition to the total environmental and toxicological impact.

	40 mm PBX (%)			40 mm Composition B (%)		
Impact Category	Cartridge	Fuze	Projectile	Cartridge	Fuze	Projectile
Abiotic depletion	35.698	13.772	50.528	35.734	13.786	50.479
Global warming	45.279	22.389	31.964	46.243	22.866	30.516
Ozone layer depletion	46.782	24.217	28.665	49.231	25.485	24.931
Acidification	19.155	26.652	53.921	18.446	25.665	55.627
Eutrophication	19.738	20.463	59.697	19.747	20.473	59.677
Human toxicity, cancerous	15.410	8.544	76.009	25.427	14.098	60.415
Human toxicity, non-cancerous	93.895	1.077	5.023	95.313	1.094	3.589
Eco-toxicity	97.886	0.196	1.903	98.686	0.198	1.101

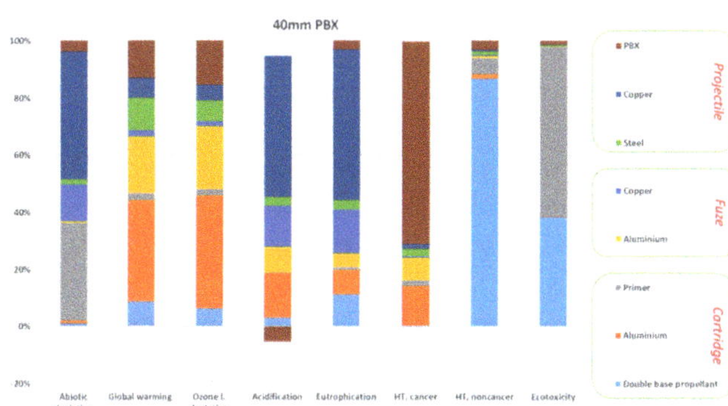

Figure 12.4. Analysis of the contributors, in percentage, to the environmental and toxicological impact categories associated with the production of a 40 mm ammunition with PBX.

which explains the similar total impacts. An exception is observed for the human toxicity categories, where in comparison with the projectile with PBX the one with Composition B presents a lower contribution to the impact value. As result, for these impact categories, and for the ammunition filled with Composition B, higher contributions of the other components for the impacts values are observed.

This analysis also shows that the constituent having a major influence on the total impact depends of the impact category. The projectile has a higher influence in the impact categories of abiotic depletion, acidification, eutrophication, and human toxicity (cancerous); it is responsible for more than 50% of the impact value in all of these categories. The cartridge has a higher importance for the impact categories of global warming, ozone layer depletion, human toxicity (non-cancerous) and eco-toxicity; for the two latter impact categories mentioned, the impact is completely

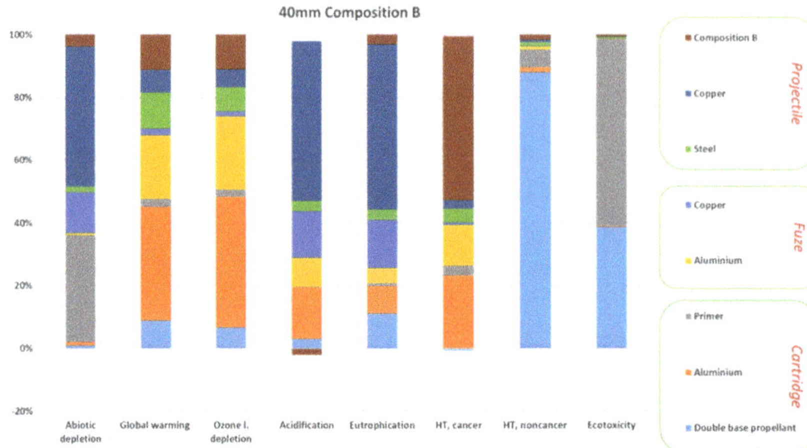

Figure 12.5. Analysis of the contributions, in percentage, to the environmental and toxicological impacts categories associated with the production of a 40 mm munition with Composition B.

dominated by the cartridge components. The Fuze also produces a noteworthy contribution for all the majority of the impact categories.

The major constituents that contribute to the impact categories are also similar for both of the munitions. For the abiotic depletion category, copper in the projectile (44%) and the primer in the cartridge (33%) are the main contributors to the impact. For the impact categories of global warming and ozone layer depletion, the constituents that are responsible for the majority of the impacts are the aluminium used in the fuze (37%) and the cartridge (20%). Copper, used in the projectile and fuze, presents the higher impact for the impact categories of acidification and eutrophication. Regarding eco-toxicity, the primer dominates the impact, with more than 60% of the impact contribution (the main contributor is tetrazene, which dominates the impact for primer production). For this impact category, the propellant also has a great influence, with approximately 40% of the total impact for both munitions. The propellant also dominates the impact relevant to the human toxicity (non-cancerous) impact category.

For the particular case of the human toxicity (cancerous) impact category, the explosive fill of the projectile has a major role. In the case where a PBX is used as the fill 70% of the impact in this category is associated with its production. This value, however, decreases to 53% when, instead of PBX, Composition B is used as the explosive charge. The same trend is observed for the human toxicity (non-cancerous) impact category. For this category, even if the difference is not so prominent as the one observed for the human toxicity with cancerous effects, the impact associated with the production of Composition B is minor when compared with those arising from the production of the PBX, as observed in figure 12.2.

Phase 4. Interpretation.

In this phase, results and data are analysed according to the goal and scope defined in the first phase, in order to draw conclusions, recommendations, and decisions that are relevant for all those involved. These include stakeholders, decision makers, range managers, and manufacturers who are interested in the environmental issues related to ammunition. The conclusions drawn should always consider the limitations of the life-cycle model developed. Therefore, in this phase some modifications of the model, or the search for new data to improve the analysis or the confidence in the results obtained, can be proposed. Where no option is left to improve the life-cycle model, such as lack of data referring to the production of energetic materials, the LCA practitioners need to share their conclusions with great care.

With the analysis carried out the hot spots (i.e. the major contributors) to the environmental and toxicological impacts are possible to determine. This information allows the designers to understand the most important processes or materials that should be changed to enhance the environmental performance of the energetic materials, and to compare different alternative strategies to ascertain the best option from an environmental perspective. However, these hot spots vary with the environmental impact categories, so the practitioners or stakeholders need to clarify which impact category is more relevant for the product or activity being studied.

12.4.2 Eco-design of small-calibre ammunition

The previous section showed an example of the assessment of the life-cycle impacts and identified hot spots for the most relevant ammunition components. An LCA can also be used to identify the best eco-design solutions for products with the same function and avoid the shifting of impacts between different life-cycle stages or impact categories. Ferreira *et al* (2016) presented an eco-design study for 9 mm calibre ammunition whose objective was to identify the potential benefits from lead removal. In this section, the major results from that study are presented. Four different compositions with and without lead in the primer and projectile were considered in order to ascertain, with the application of an LCA, the solutions with lower environmental and toxicological impacts:

1. Steel-lead projectile + lead primer.
2. Steel-lead projectile + non-lead primer.
3. Nylon-copper projectile + lead primer.
4. Nylon-copper projectile + non-lead primer.

In this case, the life-cycle model was developed for the whole life cycle of the ammunition (raw material extraction and procurement, production of ammunition, use of ammunition). The functional unit was defined as one 9 mm small-calibre ammunition.

The results showed that the environmental impact is mostly associated with the production of the ammunition components and with the energy required for their assembly. Despite the differences in the components and in the materials of the different munitions, the impacts obtained for each of the four different munitions are similar. This means that the modifications to the projectile and primer are not reflected in the overall values of the impacts for the environmental impact categories. The most significant results are obtained for the toxicological impact categories, in

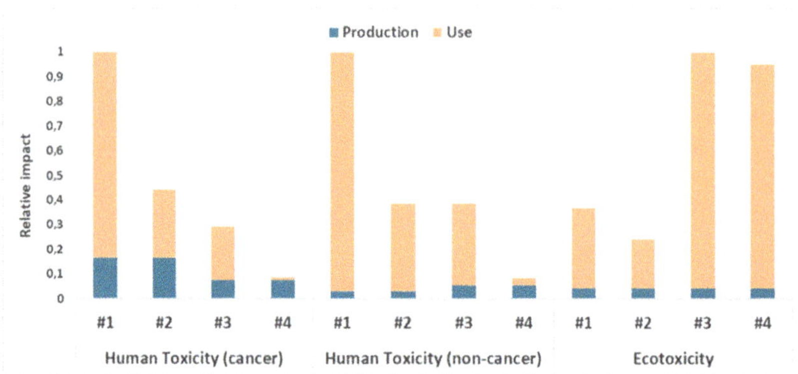

Figure 12.6. Life-cycle impact comparison between four 9 mm small-calibre ammunitions: #1 Steel-lead projectile + lead primer, #2 Steel-lead projectile + non-lead primer, #3 Nylon-copper projectile + lead primer, #4 Nylon-copper projectile + non-lead primer (based on Ferreira *et al* 2016).

which not only are the major benefits from removing lead shown, but also the unexpected results that can be assessed and identified with the use of an LCA. Figure 12.6 shows a comparison of the toxicological impacts for the four 9 mm ammunition types previously described.

For the human toxicity categories, the substitution of lead produces an important decrease in the total impact. In fact, the 9 mm ammunition without lead in the projectile and in the primer almost does not have any impact for the use phase, and the impacts are dominated by the production phase instead. Even the substitution of lead in the primer is sufficient to significantly decrease the total impact for these categories. However, impact shifting is observed for the eco-toxicity category, which results in higher impacts for the ammunition with a composite projectile than those from lead projectiles. This negative consequence associated with an eco-design solution and when observed for the ecosystems is essentially associated with copper emissions from the projectile erosion during firing. This assessment shows that, for this particular case, further improvements are required to reduce the environmental and toxicological impacts and to avoid the negative consequences for ecosystems. Ferreira *et al* (2016) proposed the substitution of a traditional cartridge (in which brass is used) by plastic, cardboard, or, eventually, the use of caseless ammunitions, as well as the replacement of copper by another metal in the composite projectile. Moreover, the discovery of this trade-off is relevant for range managers who need to be aware of the potential consequences of using these munitions on the ecosystems of the ranges for which they are responsible; and these conclusions are also significant for ammunition producers in order to persuade them to search for alternatives that mitigate the adverse consequences.

12.5 Concluding remarks

This chapter is designed to demonstrate that eco-design strategies or solutions allows improvement of the environmental profile of products or services, when considered from a life-cycle perspective. These strategies need to be considered from the design phase onwards in order to capture the consequences for the whole life cycle. An LCA is an appropriate tool to quantitatively assess the environmental impacts of conventional products or services, and in addition the consequence of proposing new strategies for their design. This type of assessment shows that tailored and robust decisions can be provided with the aim of mitigating (or decreasing) the impacts from a life-cycle perspective. Therefore, an LCA is a reliable tool (i) to quantitatively assess the impacts associated with energetic materials; (ii) to identify which life-cycle phase produces a higher impact in the total life cycle; (iii) to identify the main contributors to the impact and allow for designers to act accordingly to decrease or mitigate that impact; (iv) to avoid trade-offs or impact shifting for other life-cycle phases or impact categories; (v) to compare products with the same function; and (vi) to assist in decision making in order to select an appropriate product or technology from an environmental perspective.

References

Allione C, De Giorgi C, Lerma B and Petruccelli L 2012 From ecodesign products guidelines to materials guidelines for a sustainable product. Qualitative and quantitative multicriteria environmental profile of a material *Energy* **39** 90–9

Bassani F, Rodrigues C, Marques P and Freire F 2022 Ecodesign approach for pharmaceutical packaging based on life cycle assessment *Sci. Total Environ* **816** 151565

Blanca-Alcubilla G, Bala A, de Castro N, Colomé R and Fullana-i-Palmer P 2020 Is the reusable tableware the best option? Analysis of the aviation catering sector with a life cycle approach *Sci. Total Environ.* **708** 135121

Brezet H and van Hemel C 1997 *Ecodesign, a Promising Approach to Sustainable Production and Consumption.* (Paris: United Nations Environmental Program)

Chun Y, Lee K, Lee J S, Lee J Y, Lee M H, Mishima N and Tahara K 2018 Identifying key components of products based on consumer- and producer-oriented ecodesign indices considering environmental impacts, costs, and utility value *J. Clean. Prod.* **198** 1031–43

Department of the Army 1994 *Army Ammunition Data Sheets for Artillery Ammunition Guns, Howitzers, Mortars, Recoilless Rifles, Grenade Launchers and Artillery Fuzes* TM 43-0001-28 Headquarters Department of the Army no. 11

European Commission 2006 No 1907/2006 of the European Parliament and of the Council of 18 December 2006 concerning the Registration, Evaluation, Authorisation and Restriction of Chemicals (REACH), establishing a European Chemicals Agency, amending Directive 1999/45/EC and repealing Council Regulation (EEC) No 793/93 and Commission Regulation (EC) No 1488/94 as well as Council Directive 76/769/EEC and Commission Directives 91/155/EEC, 93/67/EEC, 93/105/EC and 2000/21/EC

European Commission 2009 Directive 2009/125/EC of the European Parliament and of the Council of 21 October 2009 establishing a framework for the setting of ecodesign requirements for energy-related products

European Commission 2017 Regulation (EU) 2017/1369 of the European Parliament and of the Council of 4 July 2017 setting a framework for energy labelling and repealing Directive 2010/30/EU

European Commission 2021 Regulation (EU) 2021/1119 of the European Parliament and of the Council of 30 June 2021 establishing the framework for achieving climate neutrality and amending Regulations (EC) No 401/2009 and (EU) 2018/1999 ('European Climate Law')

Ferreira C, Ribeiro J, Almada S, Traian Rotariu T and Freire F 2016 Reducing impacts from ammunitions: a comparative life-cycle assessment of four types of 9 mm ammunitions *Sci. Total Environ.* **566-567** 34–40

Ferreira C, Ribeiro J and Freire F 2019a Environmental assessment of military systems with the life-cycle assessment methodology *Energetic Materials and Munitions: Life Cycle Management, Environmental Impact and Demilitarization* ed A Cumming and M Johnson (New York: Wiley-VCH) pp 169–97

Ferreira C, Ribeiro J, Clift R and Freire F 2019b A circular economy approach to military munitions: valorization of energetic material from ammunition disposal through incorporation in civil explosives *Sustainability* **11** 255

International Organization for Standardization 2006a ISO 14044: *Environmental Management—Life Cycle Assessment—Principles and Framework* (Geneva, Switzerland)

International Organization for Standardization 2006b ISO 14044: *Environmental Management—Life Cycle Assessment—Requirements and Guidelines* (Geneva, Switzerland)

de Otazu R L L D, Akizu-Gardoki O, de Ulibarri B, Iturrondobeitia M, Minguez R and Erlantz Lizundia E 2022 Ecodesign coupled with life cycle assessment to reduce the environmental impacts of an industrial enzymatic cleaner *Sustainable Production and Consumption* **29** 718–29

Kemna R 2011 *Methodology for Ecodesign of Energy-related Products* MEErP Methodology Report, Part 1: Methods

Margallo M, Dominguez-Ramos R and Aldaco A. 2019 A Incorporating life cycle assessment and ecodesign tools for green chemical engineering: A case study of competences and learning outcomes assessment *Educ. Chem. Eng.* **26** 89–961

PRé 2014 SimaPro Database Manual Methods Library, Report Version 2.7; Available online: https://support.simapro.com/articles/Manual/SimaPro-Methodsmanual (Accessed: 19 February 2018)

Polverini D 2021 Regulating the circular economy within the ecodesign directive: progress so far, methodological challenges and outlook *Sustainable Production and Consumption* **27** 1113–23

Ribeiro I, Peças P and Henriques E 2013 A life cycle framework to support materials selection for ecodesign: a case study on biodegradable polymers *Mater. Des.* **51** 300–8

Wimmer W 1999 The ecodesign checklist method: a redesign tool for environmental product improvements *Environmentally Conscious Design and Inverse Manufacturing, 1999. Proc. EcoDesign'99: First Int. Symp. on IEEE* pp 685–8

Urbanski T 1968 *Chemistry and Technology of Explosives* vols 1–4 (Oxford: Pergamon)

IOP Publishing

Energetics Science and Technology
An integrated approach
Adam S Cumming

Chapter 13

Sustainable disposal solutions for weapons, ordnance, munitions, and explosives (WOME)

Rebecca Yates, Adam Cumming, Tracey Temple and Melissa Ladyman

13.1 Introduction

Worldwide more than 130 countries have committed to Net Zero Carbon Emissions by 2050 (NZ50). So far, a handful of these countries (including the UK) have made NZ50 targets legally binding. The current disposal methods used in the UK for weapons, ordnance, munitions, and explosives (WOME) are open burning/open detonation (OBOD), cage burning, incineration, or transporting overseas for dismantling or disposal. All of these methods are associated with high carbon emissions via direct air emissions or require high energy consumption. Alternative technologies to the current status quo are available; the main constraint is the investment needed to adapt and upscale them. Social changes (defence direction, attitudes, and training) are also necessary to effectively implement sustainable design and disposal options where possible.

13.2 Design and manufacture

13.2.1 Intelligent design for demilitarisation

13.2.1.1 Historic context

Separating explosives from shells/munitions is not a new concept: it has been around for over 100 years (a US patent was obtained in 1924 for the method and equipment design to extract high-explosive content from a shell [1]).

The 'polluter pays' principle was first introduced in 1972. The UK Ministry of Defence (MOD) has accepted this principle in its Joint Service Publication 418 leaflet 02 [2] on pollution prevention. Despite this, it remains an issue that, in practice, little consideration is given to the end-of-life phase during concept and design. This is beginning to change, with design for disposal becoming a general requirement to help transition to a more sustainable, circular economy in order to meet the UK's NZ50 target.

doi:10.1088/978-0-7503-3943-8ch13

13.2.1.2 Approaches

According to NATO Allied Ordnance Publication (AOP) 4518 [3] and the book on *Life Cycle Management Environmental Impact and Demilitarisation* [4], the basic principles to enable design for demilitarisation and disposal to work are as follows:

- Components can be disassembled easily.
- Energetic materials can be removed.
- Demilitarisation processes can be used efficiently.
- Munitions are safe to handle by operators throughout demilitarisation processing.
- Reusable or recyclable components/materials can be economically recovered.
- That there is minimal environmental impact.
- That, ultimately, cost savings to future development programs may possibly be realised through the use of recovered components and materials in new production.

Early design for demilitarisation can lead to low-cost alterations that do not affect the item's performance [5].

Another important area to consider is a reduction in manufacturing by only producing what is needed. This is recognisably challenging in the defence sector with national security being of paramount importance and the potential need to access WOME of varying types, for a wide range of potential future conflict scenarios, when weapon production may have long lead times and they may be required at short notice.

Advanced decision modelling software and tools are now widely available commercially to model the necessary requirements based on a number of variables and to predict the quantities required. There is therefore a need to establish whether these techniques can improve current planning practices, leading to fewer weapons systems being procured and therefore less waste produced by disposing of unexpended items.

13.2.1.3 Technologies for dismantling

To enable increased reuse, recovery, and recycling (R3) of WOME, dismantling into component parts is absolutely necessary. This is understandably a hazardous activity with significant safety considerations.

WOME could be designed so that they can be easily dismantled once they reach end of life. There are a number of challenges associated with this, including the following:

- To function effectively in potentially challenging environments (excessive heat, cold, dust, wet, etc) WOME must be robust.
- Easily dismantling presents increased security/terrorist risks.
- Designers are keen to protect intellectual property and easily dismantling would make reverse engineering easier.

Therefore, there is a need for WOME to be designed to facilitate demilitarisation and disposal but not be dismantled easily.

Remote mechanical disassembly (which may include the use of robotics) is a favourable option, as it minimises safety concerns considerably by removing the human element. Over a decade ago it was considered (as part of the Canadian Green Technologies Concept RIGHTTRAC demonstration) that many technologies were available at high technology readiness level for dismantling munitions and separating and reusing/recycling the explosive fill [5]. Technology has since moved on considerably, and the US Army currently uses mechanical disassembly at a variety of locations.

Despite the proven technologies mechanical disassembly has not yet been implemented in some countries (including the UK). This is due to the high upfront costs required for building mechanical disassembly equipment and its relatively small and diverse inventory. The UK does, however, send ammunition overseas for breakdown (e.g. to the Nammo site in Vingäker, Sweden). NATO [6] recommends investment for design improvements in mechanical disassembly for the following reasons:

- Easy dismantling and easy access to pyrotechnic fillings are necessary.
- Removal of explosives and propellants should be easy, i.e. melted explosives may easily be taken out or have enough access for easy washout.
- Benefits of having good documentation for the munition such as a demilitarisation and disposal plan, if available, technical drawings, description of parts, etc.

NATO [6] states that mechanical disassembly to separate explosive fills was identified as being commercially available for most calibre munitions.

13.2.2 Chemical design solutions for future energetics

13.2.2.1 Green chemistry

Green chemistry is defined as

'... efficiently using (preferably renewable) raw materials, eliminating waste and avoiding the use of toxic and/or hazardous reagents and solvents in the manufacture and application of chemical products' [7].

13.2.2.2 Green energetics

Green energetic materials are those that are manufactured using green chemistry principles, which have limited environmental impact during their use and disposal. A journal article, 'Environmentally compatible next generation green energetic materials (GEMs)' [7], describes novel advances in lead-free primary explosives and cross-linked thermoplastic elastomers. In particular, cross-linked thermoplastic elastomers show potential for recycling of the energetic component and the polymer component. Nitration chemistry also presents an area for further research in this area.

13.2.2.3 Flow chemistry

Flow chemistry is a chemical reaction that is run via a continuous flowing stream rather than using standard batch production. Flow chemistry is showing promise for energetics as a synthesis option by changing the composition of energetic material at the design stage [8]. This method is simpler for producing the quantities, required therefore supporting just-in-time production, compared to oversupply resulting in surplus stockpiles. This area of research is still developing and would require further development with the potential for associated high costs.

13.2.2.4 Energetics predictive toxicological modelling

Advances in the development of computational methods to predict properties relating to performance, sensitivity, and environmental impact of a new energetic material have been utilised by the US Army Research Laboratory since 2007 [9]. This presents a potentially interesting approach to energetic material design that includes predictive toxicological modelling, progressive standards for conducting environmental health assessments, high-level efforts to foster a fundamental green chemistry ethic, and projects to seek alternative energetic materials. There is limited further information available on this.

13.2.2.5 Lead alternatives

Traditionally, small arms ammunition (SAA) has contained lead or lead alloys. For example, SAA usually contains a primer cap containing, amongst other compounds, lead styphnate, and a standard projectile usually consists of a lead/antimony bullet sheathed in a protective copper alloy jacket. The ammunition has a high environmental impact during manufacture, producing toxic byproducts that must be disposed of as hazardous materials. Great lengths are taken to recover lead fired on training ranges, however this is not always possible for all practices or during operations. In these scenarios lead bullets will be discharged to the environment (ultimately land or water). Because of this introducing lead alternatives is a growing area of research, especially in relation to ammunition design. Alternative designs must provide better or similar performance to current ammunition without being detrimental to the life of the weapon (e.g. not causing excessive barrel wear), exposing the firer to harmful gasses or smoke, and be benign when released to the environment. In our research we found reference to a number of lead alternatives. These were as follows:

- A method for manufacturing lead-free ammunition has been developed [10] in Europe. It comprises a plurality of alloying elements in the chemical composition selected from the group formed by zinc, tin, and aluminium. The invention is primarily focused on manufacturing nontoxic, lead-free ammunition for firing. It is assumed that once fired the ammunition residue deposited on land or in water will be much less toxic than lead ammunition. There is limited information available in relation to disposal, therefore this is an area for further research.
- A patent has been obtained for synthesis of basic bismuth salts [11], in the form of pentahydrate of disodium salt as an alternative to lead styphnate due to its improved environmental profile during manufacture. Experimental work completed suggests scale-up to industry may be possible. The material is non-soluble so may be extractable for reuse. Extensive studies would have to

be completed on the long-term stability prior to use. The fact it has been patented suggests it is a viable option. The patent presents problems in terms of considering this method, as it would need to be licensed and this would have high associated costs.

13.3 Processing and recycling materials

13.3.1 Non-WOME processing and recycling techniques

Designing and processing materials effectively to enable increased recycling is integral to moving towards a more sustainable circular economy. This is demonstrated effectively in the automotive and electronics industries with high rates of R3 in Western countries.

13.3.1.1 Vehicle processing and recycling

When a vehicle reaches the end of its life it is normally dismantled, crushed, and shredded. High-value items are reused and almost all of the vehicle is recycled. Auto shredder residue is often the most difficult item to be recycled but it can be treated via specialised washing. This high percentage of reuse and recycling has been heavily driven by legislation, implemented due to poor and unregulated practices historically in automotive disposal. As per the UK End-of-Life Vehicle (ELV) Regulations 2005 [12] and EU Directive 2000 [13] producers must meet annual targets for their brands. The target is currently 95% recovery and 85% recycling by average weight of each ELV. This presents an example of where the failure of the industry to voluntarily adapt resulted in statutory requirements.

13.3.1.2 Electrical and electronic equipment processing and recycling

Waste electrical and electronic equipment (WEEE) treatment varies depending on the category and technology. Broadly, it includes large-scale shredding technologies and disassembly processes (manual, automated, or a combination thereof). It is typically treated as follows:

- Broken down, often using a Querstromzerspaner (a type of mechanical cross-flow shredder) or electro-mechanical separation technology.
- Cleansed/washed.
- Current or gravity-separation techniques to segregate, then pyrolysis/thermal treatment if necessary for extra separation.
- Output product cleaned.
- Finally, further treatment prior to reuse in manufacturing.

This extensive process, to enable reuse and recycling, again has been heavily driven by legislation due to historically bad practices. The UK Waste from WEEE Regulations 2013 [14] and EU Directive 2012 [15] cover the recycling of a wide range of WEEE products that are typically made up of a complex mix of product types and materials.

The WEEE industry uses mass spectrometry, near-infrared, thermal analysis, inductive coupled plasma mass spectrometry, and inductive coupled plasma optical emission spectrometry for reliable identification of materials [16]. Laser-induced breakdown spectroscopy (LIBS) technology is also a promising analytical technique still in its infancy [16]. It is being considered for use in the WEEE recycling industry to use lasers to chemically inspect electronic waste using the LIBS technique. These techniques may become relevant to demilitarisation in the future should it become necessary to identify unknown material.

13.3.1.3 Lessons to be learnt from non-WOME industries
Worldwide there are vast quantities of vehicles and WEEE that are often made up of high-value substances. This has made it economically viable for investors to finance setting up ELV or WEEE disposal/recycling plants. The MOD has a responsibility to dispose of end-of-life WOME in an ethical way regardless of economic viability, although it is recognised that resale and recouping costs where possible is optimal. Lessons can be learnt from the ELV and WEEE industries; in particular, changes to legislation, regulation, and policy generate significant positive shifts in a relatively short space of time.

13.3.2 WOME disposal process

The order of the different demilitarisation and disposal process steps, as well as the demilitarisation techniques applied to munitions depend mainly on the following factors [3]:

- Available means.
- Physical state.
- Munition quantities to facilitate economies of scale, i.e. the quantity that allows for the most economical method of destruction.
- The demilitarisation techniques available and their capacities.
- Safety, security, and environmental regulations.

13.3.3 Disassembly, pre-treatment, and extraction

13.3.3.1 Manual and mechanical disassembly
Where possible the disassembly process generally follows reverse-order assembly procedures. Often munitions are transported to a demilitarisation and disassembly location. In addition to this, organisations such as European Metals Recycling (EMR) provide mobile plant facilities for disassembly. EMR uses a mobile shear and baler for larger items, typically over 12 inches in length, such as missiles and torpedoes which have already been declared free from explosives (FFE). Needing to be FFE at present is a limitation; an increase in the type of mobile plant facilities, in particular if they were developed further for demilitarisation, would minimise risk and documentation in relation to transporting WOME as a Dangerous Goods Class 1 hazardous material. From an environmental perspective this would mean broken down components can enter the appropriate R3 or disposal streams directly and help to reduce unnecessary transport emissions.

Manual disassembly presents safety issues and automated mechanical disassembly is a more favourable option. Mechanical disassembly is currently in use within the US Army and its contractors across various locations [17]. Due to the high volumes of WOME in the US it has historically been economically viable to build bespoke facilities for disassembly. Despite this, manual disassembly will still be something that is needed for complex munitions such as sea mines, at least for the foreseeable future.

13.3.3.2 Water-jetting

Water-jetting is a process of forcing high-pressure water through a nozzle to form a thin jet capable of cutting materials including metals. If an abrasive is entrained into the water-jet, the method is called abrasive water-jet cutting. Metallic or hard materials such as metal plates, ceramics, or glass can then easily be cut [3]. Water-jetting has been employed to wash out energetics from munition bodies and water-jet cutting used for size reduction of WOME and accessing explosive fillings. This aids separation and material recovery for reprocessing. Water-jet cutting or abrasive water-jet cutting for size reduction is currently in use in the US at the Crane Army Ammunition Activity and at the Hawthorne Army Depot [17].

13.3.3.3 Cryo-fracturing

Cryo-fracturing is the process of cooling a munition, usually in a bath of liquid nitrogen. At such temperatures a steel projectile case becomes brittle and can be shattered by a hydraulic press to expose its contents [3]. The extreme cold desensitises the explosive, making this a safe process.

Cryo-fracturing is used in France by AF Demilitarisation for opening ICM M42 and M46 grenades, and by ISL in Germany. AF Demilitarisation was the first working cryo-fracture plant in Europe, which opened in 2000 at Précigné, near Le Mans. This technology has been used for a long time in the US and it continues to be developed by General Atomics in the US and at McAlester AAP, where a plant has been built for the automated destruction of 9.2 million area denial anti-personnel mines [5, 17]. There was no evidence found in the literature we reviewed to suggest any current use in the UK.

13.3.3.4 Mechanical cutting and mechanical punching

Mechanical cutting and/or punching can be used to break down WOME. Band saws are used in the US [17] during demilitarisation and disposal. Band saws use a continuously looped blade to cut material such as metal. They are widely available commercially and do not have significant upfront costs. Commercially off the shelf equipment would need to be adapted so that the inherent risk of unintentional initiation of the explosive fills can be managed.

13.3.3.5 Laser cutting

Laser grooving/cutting for WOME uses a laser to score followed by mechanical fracturing a munition casing to expose the filler. This method is currently in development and could be applied to a large range of munition types [3].

13.3.3.6 Chemical and biological degradation

13.3.3.6.1 Green chemistry

Green chemistry has the potential to drive a transition to a more circular economy. In particular, green chemistry has the ability to reduce and modify materials so they are no longer toxic. This means they can either follow normal waste streams (via water or air) or be transformed into new or secondary raw materials. Chemical and biological degradation has historically been difficult in relation to maintaining the consistency and purity of the secondary product. Whilst this does not preclude its use for demilitarisation processes, it can make it harder to generate a saleable product.

13.3.3.6.2 Alkaline hydrolosis

Alkaline hydrolysis has been identified as useful and commercially viable for degradation of surplus munitions [18, 19]. Aqueous alkaline hydrolysis is a chemical reaction that breaks down organic molecules by reaction with a hydroxide ion, often used to convert solid matter into liquid. This method is preferable in comparison to OBOD. Though it does result in the complete destruction of the explosive material, it can require disposal of the resulting byproducts, which reduces the sustainability of this option and can also have associated costs. The byproducts are dependent on the alkaline hydrolysis method used. For example, alkaline hydrolysis can be used for the disposal of high explosives such as Research Department eXplosive (RDX) and High Melting eXplosive (HMX), which are both chemically classified as nitroamine. However, to result in complete disposal hydrogen peroxide must be added to the process to initiate the mineralisation of byproducts by hydrothermal oxidation.

13.3.3.6.3 Biodegredation

Biodegradation (sometimes called composting) in the WOME industry is not a new concept: it has been around since at least the 1990s (a 1996 US Army Environmental Centre report discussed the feasibility of composting [20]). It details an example method of biodegradation by breaking down nitrocellulose fines (NC fines) using composting. The production of nitrocellulose (NC) for munition purposes (propellants and rocket motors) also results in the formation of NC fines. In their solid form NC fines are not considered toxic. Under certain conditions (burning) they become reactive and produce toxic oxides of nitrogen. The report claims studies have shown that NC fines can be treated by composting to yield a nonreactive soil amendment suitable for beneficial use. It also states testing has shown that a compost matrix containing approx. 10%–35% NC by weight, with the remaining percent made up of organic amendment with a moisture content of 30%, may be handled safely [20]. Biodegradation is a government-preferred route but there are doubts in relation to the possible toxicity of the resultant compost. Another drawback is that this method is appropriate for single-based propellants but not for double- or triple-based propellants, which form the bulk of propellants in gun ammunition and within complex weapons.

13.3.3.6.4 White phosphorus

Red and white phosphorus are used in ammunition to provide rapid smoke or obscurants. Disposal of phosphorus, in particular white phosphorus, is an issue due to its toxicity. Event Horizon, who are predominantly involved in pyrotechnics for the film industry, have built a UK-based underwater facility to process/break down military WOME, including items containing white phosphorus. The site is able to recover waste phosphorous effectively and has previously sold the white phosphorus to an Italian match company for reuse. As white phosphorus is relatively pure it could also be well placed for repurposing in the chemical industry.

13.3.4 Reuse, recovery, and recycling (R3)

13.3.4.1 Overview of recycling, reuse, and resale

> *'R3 methods break down the munitions into their basic, recyclable component parts and compounds, which can then be sold to offset processing costs and thus reduce the overall cost of demilitarization'* [3].

Although recovered explosives may be reused in munitions or recycled into commercial products, in current UK practice they are more frequently used as donor explosives in OD operations [9]. This is partly because the risk appetite (in relation to safety and quality) for military reuse of energetic material is much lower in the UK compared to other countries due to more stringent quality standards[1]. The recovery of value of OME fits into two broad areas, comprising military reuse and commercial reuse.

Considering end-of-life disposal at the design stage of WOME presents significant environmental benefits. It also brings financial benefits, as resaleable options for munitions include the following:

- Scrap metal recovery.
- Energetics recovery.
- Reuse as commercial explosive.
- Chemical breakdown for saleable product.

A particularly favourable option is the reuse of high explosives into civil explosives (e.g. quarrying and mining). This is because the purity for reuse does not need to be as high as it does for recycling back into munitions. A circular economy (valorisation of energetics) approach to military munitions [21] demonstrated that reuse reduced impact across all areas studied by approximately 80%, including reductions in the following:

- Nonrenewable primary energy use.
- Use of raw materials (in production of civil explosive).
- Acidification.

[1] Energetic materials within WOME may be required to function precisely as originally designed many years after they have been manufactured. In the intervening period they may have been subjected to excessive environmental stimuli such as changes in temperature or exposure to prolonged or excessive vibration. These properties are vital if user safety and confidence are to be maintained.

- Eutrophication.
- Global warming.
- Ozone depletion.
- Photochemical oxidation.
- Human toxicity.
- Eco-toxicity.

This study was based on trinitrotoluene (TNT) equivalents, therefore it is cross-applicable to a variety of WOME. A drawback is that the benefits are greater where processing can be carried out in bulk to minimise transport requirements and maximise the mass of the civil explosive produced. This means it could be impractical for nations with smaller disposal requirements. There could also be objection from civil manufacturers unless they were part of the solution. Overall, it provides evidence that disassembly, followed by recovery and reuse, has the potential to deliver significant environmental improvements (especially when compared to OBOD and incineration), as well as producing a saleable product.

13.3.4.2 Reuse of TNT, RDX, and HMX

The reuse of TNT, RDX, and HMX has been reviewed extensively in the literature. The key findings from our review are summarised below:

- Recovering HMX/TNT by dissolution and filtration has been successfully employed in the US, where HMX is recovered for reuse.
- Melt filtration of compositions based on TNT may offer a way forward for TNT-based compositions, specifically RDX/TNT formulations as long as RDX could be requalified or a civil use/market identified.
- Recovery and reuse of HMX/RDX has been explored by a US programme in 2001–2002 [22, 23]. In comparison to OBOD, recovery of HMX showed cost savings (in the first year) in excess of $300,000. RDX was less ($90,000).
- HMX recovery was only employed at one plant in the US. The technology was not routinely employed due to high labour costs. If automation was developed it could solve this issue and make this a viable option.

13.3.4.3 Energy recovery

It is widely known that all kinds of waste can be used as a heat and/or electricity source. At present this energy is not widely captured in the WOME industry. By using closed incineration, electricity can be harnessed from the burning of waste energetics. EST Steinbach, in Germany, have as part of their facilities a demilitarisation incinerator that produces electricity [6]. This incinerator is capable of burning ordinary waste as well as energetic material. Although a feasible sustainable disposal method, electricity and heat from energetics has a surprisingly low value. This is because the energy density of explosives is considerably lower than other readily available fuels. For example, TNT has an approximate energy density of 4.18 MJ kg^{-1}, diesel fuel oil is 32 MJ kg^{-1}, coal is 23 MJ kg^{-1}, wheat/straw is 18.9 MJ kg^{-1} and wood ranges from 10 MJ kg (if green) and up to 20 MJ kg^{-1} if oven dried [6].

The Nammo Norwegian Ammunition Disposal (NAD) Company has recently implemented a sustainability programme. Part of this includes recovering hot air from a detonation chamber and redistributing the recovered heat for energy use on site. At present the key performance indicator for heat energy recovery is approximately 3500 kW per year. This demonstrates a good use case for energy recovery from detonation, but it must be recognised that the Nammo NAD site in Norway is unique and this may present challenges in being replicated elsewhere.

13.3.5 Last-resort destruction

13.3.5.1 Destruction overview

Not all munitions are suitable for R3. In many cases insufficient quantities exist to develop economically sustainable demilitarisation solutions and operations [3] or it is simply not safe enough, in which case closed detonation or incineration should be considered; OBOD should only be considered as a last resort if all other options are not fit for purpose. Incineration is not considered as a sustainable option due to the high energy demands required; however, at least air emissions from incineration and closed detonation can be controlled, which is why they are more favourable options in comparison to OBOD.

13.3.5.2 Incineration and detonation

Typical detonation/closed incineration methods include the following [3]:

- Closed detonation: involves placing munitions in a closed chamber and initiating them via explosive charge or external heat source.
- Incineration (static kiln): involves a sealed chamber where items can burn or detonate.
- Incineration (rotary kiln): involves items slowly being moved through a rotating kiln.
- Incineration (car bottom furnace): uses a moveable 'car' that inserts the waste into a furnace.
- Incineration (plasma arc): molten slag is heated by a plasma arc and destroys munitions.

Plasma incineration is probably the most recent development in munition processing. It can be used on explosive waste slurry or granular solids to deal with pyrotechnics, and slag encapsulates the hazardous waste. Items may require pre-treatment, such as sectioning before incineration, and it is deemed relatively expensive for bulk energetics [3]. The NATO AOP 4518 states it is 'in use' across NATO nations, although it is not widespread. This method is not currently in use in the UK.

Most of these methods are fit for purpose for disposal of the majority of munitions, with the exception of pyrotechnics as they block incinerator filters. Therefore, at present the only option in the UK for the disposal of pyrotechnics is plasma arc incineration (but this is not currently used in the UK) or OB/cage burning, which is widely used but is a limitation. Pyrotechnics can also be disposed of via closed detonation at the Nammo NAD site in Norway due to the unique nature of the natural detonation chamber, a considerable amount of donor charge and the bespoke 8 hr cleaning/filtration system. A downside is there are transport emissions associated with shipping to Nammo sites from overseas. On the plus side, these closed incineration and detonation methods do allow the recovery of metals following destruction, and air emissions can be controlled.

13.3.5.3 Use of fuels

The high energy demands of incineration or emissions associated with transporting to disposal locations (in particular shipping overseas) are a drawback in terms of sustainability, unless they are run on or transported using green alternatives (i.e. biofuels, renewables, or green hydrogen) or even nuclear power.

Many shipping companies worldwide are currently trialling low-carbon solutions for both shipping (e.g. flettner rotors, hydrogen, ammonia, biofuels, and sails) and port infrastructure (e.g. low-carbon refuelling and shoreside power). In April 2021 the vessel Stena Immortal (a passenger ship) completed a successful voyage fuelled 100% by biofuel [24]. Also, in March 2021 the UK government launched a £20 million competition designed to lower carbon emissions from the global shipping industry [25]. Further to this, in May 2021 a partnership between the Scottish Port of Cromarty, Firth and Norwegian Gen2 Energy AS was signed stating this 'will guarantee green hydrogen to those who want access to clean energy by mid-2023'. The Firth will be the UK transhipment hub for hydrogen, which will then be distributed to customers across the UK by road, rail, and sea [26].

13.3.5.4 Air quality regulation

There is a range of air quality regulations aimed at reducing emissions which facilities must abide by, including the UK Air Quality Standards Regulations [27], the Environmental Permitting Regulations (England and Wales) [28], the EU Industrial Emissions Directive [29], the international Montreal Protocol [30], and the international Kyoto Protocol [31]. This helps to minimise the air emissions associated with closed detonation and/or incineration.

13.3.5.5 Variables affecting operation

Variables that affect incinerator operation can be controlled by the operators, so combustion conditions controlling emission rates can be substantially affected by operator decisions [32]. This can be improved by training, monitoring and data collection, optimising methods and technologies to control process emissions specifically adapted to the specific type of waste.

13.3.5.6 Control measures

CO_2, SO_2, NO_x, VOCs, dioxins, and heavy metals are some of the emissions that are released during incineration of WOME. These emissions can be controlled by a variety of technologies (filters, secondary abatement, scrubbers, activated carbon injection, and modified oxygen concentrations are some examples), alongside the ability to vary the incinerator temperature. Since the 1990s incinerators have been capable of complete destruction with minimised emissions to the atmosphere, as all emissins are trapped by the furnace, which 'has atmospheric control of fumes from incinerated baskets that can be cooling while fresh baskets are fired' [33].

Methods employed by industry for limiting particulate matter, acid gases, mercury, lead, dioxins and furans, and products of incomplete combustion are discussed in detail in chapter 5 of the *Waste Incineration & Public Health* book [32]. The Waste Incineration EU Best Available Techniques reference documents (BREF) [34] and associated Best Available Techniques (BAT) [35] documents describe the optimal techniques for waste incineration that can be followed to help minimise emissions and demonstrate compliance with BAT. These kinds of best practice methods could be drawn into the defence/WOME industry.

Unfortunately, control measures do not resolve the issue of the high energy demands of incinerators. Rotary kiln incineration (e.g. AVOCET at Shoeburyness) necessitates extremely high temperatures and is powered by large hydrocarbon burners producing an ignition source for the munition or explosive. This also limits options for alternative power sources (e.g. nuclear, energy from waste, and renewables).

13.3.5.7 OBOD

In some cases, OBOD is the only safe viable option (e.g. pyrotechnics), though these methods should only be considered as a last resort if all other options are not fit for purpose. In the US OBOD has been significantly reduced since 1998, when the Department of Defense (DoD) published the 'OBOD Optimization Study'. By 2019 the DoD had achieved the following [36]:

- Actively reduced the number of permitted OB/OD units by 44% (from 61 to 34).
- Reduced the total quantity of waste military munitions treated at DoD facilities by OB or OD from 20,326 tons net explosive weight in 1998, to a current average of 8630 tons net explosive weight per year (a 58% decrease).
- Decreased the number of permitted OB/OD units and the quantity of waste military munitions treated by OB or OD across each of the four DoD military services.

The US DoD has for decades increased use of R3 to reduce the quantity of waste military munitions stockpiles treated by OB or OD. The US DoD has also implemented the use of viable contained disposal technologies to reduce further reliance on the use of OB or OD.

These actions have allowed the US DoD to attain large reductions in the quantity of waste military munitions treated by OBOD.

The US DoD works closely with the US Environmental Protection Agency and state regulators to set operational constraints and limits in relation to OBOD to protect human health and the environment [36].

The US DoD has the capability to destroy their own munitions due to government funding. In the UK the MOD generally contracts out the logistical disposal of stockpiles to QinetiQ via the Long Term Partnering Agreement, which enables the MOD to utilise facilities previously owned or developed by it at reduced cost, or Nammo facilities. A contracting structure that favoured contractors who would use non-OBOD techniques would significantly help to minimise the quantity of material disposed of via OBOD.

13.4 Sustinable options assesment

It is clear that a universal design and/or disposal technique is not fit for purpose given the wide variety of WOME. To address this an options assessment was undertaken (results summary provided in table 13.1). The options assessment listed all reviewed design and disposal examples with an accompanying summary of the usage, relative maturity, pros/cons, and a red, amber, green (RAG) assessment.

To review the sustainability of the identified options we have considered the three pillars of sustainability (environmental protection, economic viability, and social equity) against each option. This has been based upon the following definitions:

- **Environmental sustainability** means consuming natural resources at a sustainable rate, minimising waste, considering if resources can be kept within a circular economy, and aiming for NZ50 (i.e. reducing carbon emissions).
- **Economic sustainability** requires a country or business to consume resources efficiently and responsibly whilst still producing an operational profit to sustain activities long term. Managerial, organisational, and economic challenges are the most critical to overcome when it comes to sustainable supply chain management and promoting a more circular economy [37].

To consider economic sustainability in the demilitarisation process NATO AOP 4518 recommends following the 'best available techniques not entailing excessive costs' (BATNEEC[2]) principle [3]:

(a) 'Best' means most effective in achieving a high level of physical and environmental protection.
(b) 'Technique' includes both the technology used and the way in which the installation is designed, built, maintained, operated, and decommissioned.
(c) 'Available technique' means those developed on a scale which allows economically and technically viable implementation.
(d) 'Not entailing excessive costs' implies where the benefits gained are worth more than the costs of obtaining those benefits.

- **Social sustainability** is the ability of people/society to achieve a good level of well-being and maintain this in the long term. Security, from having a robust defence force, provides one level of maintaining well-being, but this needs to be balanced against public attitudes, perception, and acceptability. The majority of the public in the UK are behind the fight against climate change and NZ50 targets; this has been catalysed in recent years by David Attenborough, Greta

[2] It should be noted that BATNEEC has been widely replaced by BAT as the preferred approach for permitting and pollution control across the UK.

Thunberg, and (the controversial) Extinction Rebellion. Due to shifting public sentiment it is likely that OBOD will become increasingly unacceptable.

Table 13.1 provides a summary of the sustainable options assessment results, and table 13.2 details the RAG assessment definitions.

Public perception is backed by UK [31, 32], EU [33], and international policy [34, 35]. Historically, MOD exemptions, derogations and dis-applications have meant defence is not fully accountable to these standards. The Secretary of State's policy was updated in April 2020 [38] and still requires that where exemptions, derogations and dis-applications exist, that legislation is followed to an equal standard, and that specific exemptions are only sought when it is essential for operational capability. The levels within this structure include international law, conventions, protocols, and treaties (to which the UK is a signatory); EU and UK legislation; and MOD/corporate policy and requirements and industry standards.

Table 13.1. Sustainable options assessment summary.

Technique	Sustainability RAG
Intelligent Design for Demilitarisation and Disposal	
Decision Modelling	Green
Designing for Mechanical Disassembly	Amber
Chemical Design Solutions	Green
Disassembly and Pre-treatment for Disposal	
Manual Disassembly	Green
Mechanical Disassembly	Amber
Cutting, Abrasive Water or Slurry Jet	Amber
Laser Grooving/Cutting	Amber
Cryofracture	Amber
Removal for Disposal	
Mechanical Machining	Amber
Washout with High Pressure Water	Amber
Melt out Steam	Amber
Melt out Autoclave	Amber

Technique	Sustainability RAG
End of Life Destruction	
Open Burning	Red
Open Detonation	Red
Closed Detonation	Amber
Incineration Static Kiln	Amber
Incineration Rotary Kiln	Amber
Incineration Car Bottom Furnace	Red
Incineration Plasma Arc	Amber
Biodegradation	Green
Aqueous/Slurry	Green
Disposal Alternatives: Reuse, Recovery and Recycling	
Resale	Amber
Energy Recovery	Green
Scrap Metal Recovery	Green
Chemical Conversion	Green
Energetics Recovery	Green
Reuse as Commercial explosive	Green

Table 13.2. RAG assessment definitions.

Green	Considered sustainable taking generic account of environmental, economic, and social sustainability as defined above.
Amber	Contributes to environmental, economic, and social sustainability as defined above.
Red	Not sustainable. Alternative methods should be considered.

13.5 Conclusions

13.5.1 Overarching conclusions

When it comes to sustainable disposal of any item it is important to consider the whole product life cycle. The future of designing for disposal is fundamental in moving towards a more sustainable economy and meeting NZ50 targets. Regulations are placing a greater emphasis on this, and a variety of options for intelligent design are already available.

Based on our review of the literature and our options assessment the most sustainable options for WOME disposal are the implementation of eco-design principles and the development of mechanical disassembly. This is summarised in table 13.1. These options all provide significant environmental benefits but also bring the following additional benefits:

- Decision modelling: a potentially promising technique for managing the quantities for future design and manufacturing with widespread uses across all WOME.
- Mechanical disassembly: reduces personnel exposure to munitions. Investing smartly in adaptable machinery to deal with a wide variety of WOME would help to address economic concerns.
- Chemical design solutions: a wide variety of possibilities across all WOME.

We recognise that there is a large inventory of WOME that have not been subject to eco-design. The most sustainable options for disposal of legacy WOME are summarised in table 13.1. This highlights that eight of the techniques identified are considered sustainable. Twelve contribute towards sustainability and eight of these twelve could be adapted to be measured as sustainable.

Technologies and a variety of options are evidently available; the main constraint is the investment needed to adapt and upscale these technologies to enable them to be widely used in the UK. Technology and investment alone will not generate meaningful change to achieve NZ50: it will also require changes in defence direction, attitudes, and training to effectively implement sustainable disposal options. The effectiveness of a top-down approach by changing policy and regulations is demonstrated in both the ELV and WEEE industries.

Acknowledgements

Dstl originally commissioned this work as part of a wider project via the Weapons, Sector, Research, Framework (WSRF). The project undertook research into economically viable and sustainable solutions for WOME disposal. The literature review from this project has been adapted to form this chapter. We would like to thank both Dstl and the WSRF for granting permission to use the research for the basis of this chapter.

References

[1] Knight T F 1924 *Method and Apparatus for Unloading High-explosive Shells* US Patent 1,492,924
[2] MOD Joint Service Publication – JSP 418 2010 Management of Environmental Protection in Defence. Leaflet 02 Pollution Prevention (revised September 2010)
[3] NATO 2018 *Safe Disposal of Munitions, Design Principles and Requirements, and Safety Assessment: Edition A Version 1 (May 2018)* AOP-4518 NATO
[4] Cumming A S and Johnson M S 2019 *Energetic Materials and Munitions – Life Cycle Management, Environmental Impact and Demilitarization* (New York: Wiley VCH)
[5] Poulin I 2010 *Literature Review on Demilitarization of Munitions (Document Prepared for the RIGHTTRAC Technology Demonstration Project)* DRDC-VALCARTIER-TM-2010-213 Defence Research and Development, Canada
[6] NATO 2014 *Design for Disposal of Present and Future Munitions and Application of Greener Munition Technology* AVT-179 NATO
[7] Talawar M B, Sivabalan R, Mukundan T, Muthurajan H, Sikder A K, Gandhe B R and Rao A S 2009 Environmentally compatible next generation green energetic materials (GEMs) *J. Hazard. Mater.* **161** 589–607
[8] Chavez D E 2014 The development of environmentally sustainable manufacturing technologies for energetic materials *Green Energetic Materials* ed T Brinck (New York: Wiley) pp 235–58
[9] Hurley M 2007 Intelligent design of energetic materials *SERDP/ESTCP Partners in Environmental Technology Technical Symp. and Workshop (Washington, DC, 4–6 December)*
[10] Andres G B, Rodriguez S and Dario E 2015Chemical Composition for Manufacturing Ammunition European Patent EP 2878917 A1
[11] Nesveda J 2019 *Method for Producing Ecological Explosive for Primer Compositions of Ammunition* International Patent WO 2020030203 A1
[12] UK Gov 2005 End-of-Life Vehicles (Producer Responsibility) Regulations 2005 (UK Statutory Instruments No.263) (as amended) https://legislation.gov.uk/uksi/2005/263/data.pdf
[13] EU 2000 European Directive 2000/53/EC On end-of-life vehicles (as amended) https://eur lex.europa.eu/legal-content/EN/TXT/?uri=LEGISSUM%3Al21225
[14] UK Gov 2013 The Waste Electrical and Electronic Equipment Regulations 2013 (UK Statutory Instruments No.3113) (as amended) https://legislation.gov.uk/uksi/2013/3113/contents
[15] EU 2012 European Directive 2012/19/EU on Waste Electrical and Electronic Equipment (as amended) https://eur-lex.europa.eu/legal-content/EN/TXT/?uri=CELEX:02012L0019–20180704
[16] Costa V, Castro J, Andrade D, Victor Babos D, Garcia J, Sperança M, Catelani T and Pereira-Filho E 2018 Laser-induced breakdown spectroscopy (LIBS) applications in the chemical analysis of waste electrical and electronic equipment (WEEE) *TrAC, Trends Anal. Chem.* **108** 65–73
[17] National Academies of Sciences, Engineering, and Medicine 2019 *Alternatives for the Demilitarization of Conventional Munitions* (Washington, DC: The National Academies Press)
[18] Heilmann A, Wiesmann U and Stenstrom M 1996 Kinetics of the alkaline hydrolysis of high explosives RDX and HMX in aqueous solution and adsorbed to activated carbon *Science & Technology.* **30** 1485–92

[19] Balakrishnan V K, Halasz A and Hawari J 2003 Alkaline hydrolysis of the cyclic nitramine explosives, RDX, HMX and CL-20: new insights into degradation pathways obtained by the observation of novel intermediates *Environmental Science & Technology.* **37** 1838–43
[20] US Army Environmental Centre 1996 Composting Nitrocellulose Fines: Regulatory and Logistical Feasibility RAAP Installation Position Paper (May 1996)
[21] Ferreira C, Ribeiro J, Clift R and Freire F 2019 A circular economy approach to military munitions: valorization of energetic material from ammunition disposal through incorporation in civil explosives *Sustainability* **11** 255
[22] Poehlein S, Wilharm C, Sims K and Burch D 2002 Recovery and reuse of HMX/RDX from propellants and explosives. Final report from sub-scale (150 lb/day) plant for ESTCP
[23] US DoD 2001 *Recovery and Reuse of HMX/RDX from Propellants and Explosives (150 lbs/day Pilot Plant)* ESTCP Cost and Performance Report
[24] The Maritime Executive 2021 Low carbon shipping options after successful biofuel trial (April 2021) https://maritime-executive.com/article/low-carbon-shipping-options-after-successful-biofuel-trial
[25] E&T 2021 UK launches £20 m competition to develop low-carbon shipping https://eandt.theiet.org/content/articles/2021/03/uk-launches-20m-competition-to-develop-low-carbon-shipping/
[26] Port of Cromarty, Firth 2021 Port of Cromarthy Firth signs historic green hydrogen MoU with Norwegian clean energy firm Gen2 Energy AS (24th May 2021) https://pocf.co.uk/2021/05/24/post-entry-a/
[27] UK Gov 2010 The Air Quality Standards Regulations 2010 (UK Statutory Instruments No.1001) (as amended) https://legislation.gov.uk/uksi/2010/1001/contents/made
[28] UK Gov 2016 The Environmental Permitting (England and Wales) Regulations 2016 (UK Statutory Instruments No.1154) (as amended) https://legislation.gov.uk/uksi/2016/1154/contents/made
[29] EU 2010 European Directive 2010/75/EU Industrial Emissions Directive (as amended) https://eur-lex.europa.eu/legal-content/EN/TXT/?uri=CELEX:32010L0075
[30] UN 2020 *Handbook for the Montreal Protocol on Substances that Deplete the Ozone Layer* 14th edn (Nairobi: UNEP)
[31] UN 1998 *Kyoto Protocol to the United Nations Framework Convention on Climate Change adopted at COP3 in Kyoto, Japan, on 11 December 1997* UNFCCC https://unfccc.int/resource/docs/convkp/kpeng.pdf
[32] National Research Council (US) Committee on Health Effects of Waste Incineration 2000 *Waste Incineration & Public Health* (Washington, DC: National Academies Press)
[33] Bishop J L, Higley L J and Dell'Orco P 1996 Combustive demilitarization of explosives, toxic coloured smokes, chemical agents and related munitions by a military incinerator; destruction of fillers by hydrothermal oxidation *27th DoD Explosives Safety Seminar (Las Vegas, NV, 20–22 August)*
[34] EU 2019 *BREF Best Available Techniques (BAT) Reference Document for Waste Incineration* EUR 29971 EN European IPPC
[35] EU 2019 *BREF Best Available Techniques (BAT) Conclusions Document for Waste Incineration* EUR 29971 EN European IPPC
[36] Munitions Demilitarization/Disposal Subgroup and Environmental Subgroup for the Joint Ordnance Commanders Group Office of the Secretary of Defence. Optimization of Department of Defense Open Burning/Open Detonation Units. March 2019

[37] Yadav G, Luthra S, Jakhar S, Mangla S and Rai D 2020 A framework to overcome sustainable supply chain challenges through solution measures of industry 4.0 and circular economy: an automotive case *J. Clean. Prod.* **254** 120112
[38] Secretary of State for Defence policy statement on health, safety, and environmental protection (April 2020)

Chapter 14

Small-calibre gun propellants

Alain Dejeaifve and Rowan Dobson

14.1 Introduction

Small-calibre gun propellants produce gas to propel projectiles from small-calibre firearms, such as pistols, rifles, machine guns, and shotguns. The concept originates from the 14th century and has remained valid over the centuries, with numerous improvements with respect to performance, handling, and safety. Propellants fill the interior of an ammunition cartridge or the chamber of a gun or cannon, leading to the expulsion of a projectile. Solid propellants generate gas to accelerate a projectile so that it achieves the desired muzzle velocity. Small-calibre ammunition is .50 calibre and below, for example NATO 5.56 and 7.62, and 9 mm.

Propellants or low explosives are combustible materials having a very rapid burning rate, able to deflagrate rather than detonate and containing the oxygen necessary for their combustion. A gun is a device of precision engineering used to convert the chemical energy contained in a propellant into kinetic energy through a projectile. For safety and efficiency reasons, the powder must carefully match the performance requirements of the weapon (e.g. produce a peak pressure below the mechanical strength limit, avoid excessive barrel erosion, etc). There are slow-burning powders, for example, for a military rifle, and fast-burning powders, such as for a shotgun [1].

The first propellant employed for firearms was black powder (or gunpowder), which is typically a mixture of 15% charcoal (carbon), 10% sulphur, and 75% potassium nitrate. Smokeless propellants were developed toward the end of the 19th century and are based on nitrocellulose (NC), which is produced by the nitration of cotton linters or wood pulp using a mixture of nitric and sulphuric acids. Modern propellants are not basic powders anymore—they have very specific shapes and chemical compositions, allowing precise burning properties.

The burning rate, vivacity, temperature, and force are important properties of a propellant, which strongly influence gun performance and projectile range. While the burning rate and vivacity are usually measured experimentally (i.e. using a closed

doi:10.1088/978-0-7503-3943-8ch14

Figure 14.1. Extruded and spherical nitrate-ester-based propellant powders.

vessel test), the adiabatic flame temperature and force of a propellant can be calculated thermodynamically and used to design new propellants. The burning surface at each point of a solid propellant grain recedes in the direction normal to the surface at that point. The rate of propellant consumption depends on the burning surface, the chemical composition, the pressure, and other internal ballistics variables [2].

In this chapter, nitrate-ester-based propellants (NEBPs) will mainly be discussed. However, in the section on future axes for gun propellants, other energetic bases will be discussed.

Figure 14.1 shows extruded propellants, which are semi-industrial 'green' propellants, containing alpha-tocopherol (vitamin E) [3, 4]. Spherical propellant powders for 7.62 mm are shown on the right of figure 14.1. The black appearance of the propellants is due to graphite on the surface, mainly to improve flowability.

14.2 Current formulations and processes

Propellants can be classified according to the number of energetic bases they contain. Single-base propellants have NC as the sole explosive ingredient. Usually on the fast-burning side, the energy content is between 3100 and 3700 J g^{-1} and they are typically used in pistols, shotguns, or artillery weapons. Single-base propellants are basically stabilised NC that has been made colloidal by the action of a solvent [5].

Double-base propellants contain two energetic ingredients, NC and generally nitroglycerine (NGL). They have an energy content between 3300 and 5200 J g^{-1}, depending mainly on the amount of NGL present, and are used in ammunition for (sub-)machine guns. NGL is used to plasticise the high-molecular-weight NC.

Triple-base propellants contain up to 55% by weight nitroguanidine, or another energetic complement, as well as NC and a limited amount of NGL. The triple-base propellants have energy values ranging from 3200 to 3700 J g^{-1} [6] and are used for larger-calibre applications. They are also usually more difficult to ignite due to the presence of nitroguanidine, a white crystalline solid that remains as a suspension in the NC/NGL but yields a lower flame temperature of the propellant and reduced erosion of gun barrels, which is vital, especially for large-calibre applications.

Other nitrate esters, such as diethylene glycol dinitrate (DEGDN), can also be used in place of NGL as explosive plasticisers for NC.

Composite propellants are mainly used for rocket propulsion and generally consist of a mixture of a fuel such as metallic aluminium, a (synthetic) polymer binder such as hydroxyl-terminated polybutadiene (HTPB), polybutadiene-acrylic acid or polybutadiene-acrylonitrileacrylic acid, and an inorganic oxidizing agent, commonly a perchlorate or nitrate. These are heterogeneous systems, whereas single- and double-base propellants are homogeneous physical structures.

In addition to the compounds and chemicals mentioned in the four types of powders, all gun propellant compositions include smaller amounts of additives in order to either adjust some inherent properties (e.g. flowability or chemical stability) or to impart other desirable properties (e.g. burning rate or de-coppering ability). In small-calibre powders, the additives are either added before the graining operation, as is the case of the stabiliser which is added during the preparation of the lacquer or the dough, or applied to the surface of finished propellant grains during the glazing operation.

It is interesting to note that despite the large number of compositions which have been manufactured and tested over the years, NC remained the ingredient of choice for the preparation of small-calibre propellants. This nitrated natural polymer gives the mechanical strength and resilience necessary to prevent the breaking up of the grains during handling and firing. The chemical energy of the NC is adjusted by carrying the degree of nitration achieved during its manufacture and influence the solubility of this fibrous material in organic solvents such as acetone, ethanol, ethyl acetate, and NGL. The main ingredients in fielded propellants have remained the same for over 30 years. A summary is given in table 14.1.

In addition to the chemical composition, the grain size and shape of the propellant are important as they can influence the rate of burning and have significant impact on the internal ballistics of the gun. As a rule of thumb, one can consider the need for constant and longer thrust increasing with the calibre—the

Table 14.1. Composition of gun propellants.

Component	Function	Example
Energetic matrix	Propulsion of projectiles from firearms.	NC
Energetic complement	Provides more energy to the powder.	RDX
Energetic complement and plasticiser	Provides more energy to the powder while acting as a plasticiser.	NGL, DEGDN
Deterrent/surface moderant	Slows the burning rate to compensate for the decreasing burning surface area..	DBP*, CI
Stabiliser	Slowing down of self-decomposition, by reaction with NOx.	DPA, AK-II
De-coppering additive	Hinders the buildup of copper residues from the gun barrel rifling.	Bi_2O_3, SnO_2
Flash reducer	Reduces the brightness of the muzzle flash.	KNO_3, K_2SO_4
Lubricant	Improves flow and packing density.	Graphite

*Note that the the use of DBP has been restricted in the European Union by the REACH legislation (REACH is a regulation of the European Union, adopted to improve the protection of human health and the environment from the risks that can be posed by chemicals, while enhancing the competitiveness of the EU chemicals industry; http://echa.europa.eu/web/guest/regulations/reach/).

ideal cycle would be where the gas production increases in such a way that the force behind the projectile remains constant for the complete duration of the process, i.e. the bullet exiting the weapon's tube. A fast-burning spherical propellant would be appropriate for a pistol application, whilst a more complex shape, such as a multi-tubular geometry, would be required for a 30 mm gun. Single- and double-base propellant grains can therefore have various morphologies, including spheres, single- or multi-perforated cylinders, rosettes, etc, as illustrated in figure 14.2.

Each one of these geometries has a specific surface area that evolves during the combustion process. Solid propellant grains in the form of spheres, for example, are called 'degressive propellants' due to the fact that the surface available for the combustion decreases with time and causes a decrease in the rate of the production of gas at constant pressure. On the other hand, the surface of a multi-tubular grain (i.e. rosette) increases during the combustion process, therefore causing an increase of the production rate of gases at constant pressure. They are called 'progressive propellants'. In between, mono-tubular propellants having a 'neutral' burning profile are found. This behaviour can be calculated and rationalised as a form function coefficient Φ, which is the ratio between the propellant surface area at a certain time and its initial value (S/S_0), as depicted in figure 14.3.

More recently, some more exotic form functions, such as four-perforated cubic propellants [7], have been proposed. This type of propellant has a good progressivity according to the authors, and much work is being undertaken in that field in order to propose the ultimate form function, as detailed further in this chapter.

If the combustion properties of the propellant mainly depend on the composition and the form function, deterrents are often used to improve the gas production rate of neutral or degressive propellants and maximise the thrust time. They are found to a certain depth, on the schematised propellant powder in figure 14.4, in a gradient-wise fashion (with the concentration decreasing from the exterior surface towards the interior), especially for spherical powder. This allows a balance between the surface area available for combustion and the rate of combustion that eventually

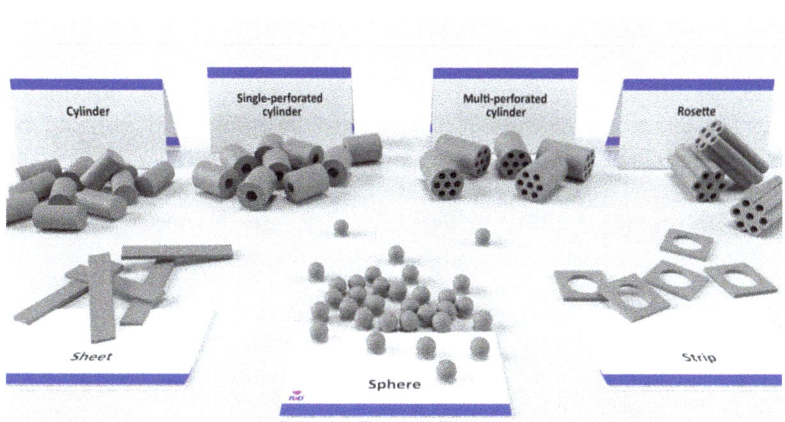

Figure 14.2. Shapes of single- and double-base propellants.

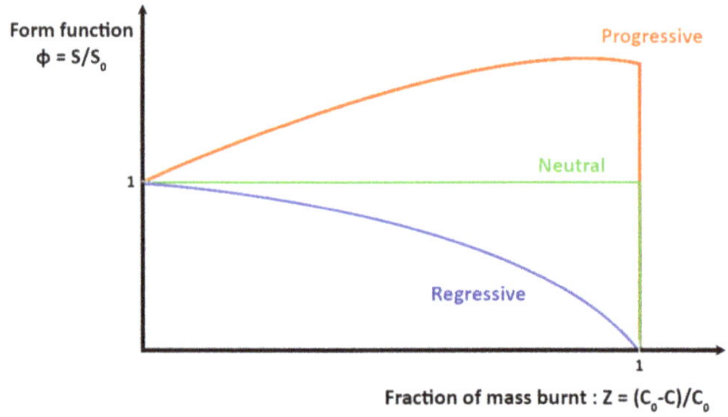

Figure 14.3. Form function as a function of fraction of mass burnt.

Figure 14.4. Propellant powder, showing the deterrent on the exterior part of the grain.

provides a less regressive profile. On the right of figure 14.4, a real propellant grain was stained, showing the position of the deterrent.

Improving the burning progressivity of propellants can increase the energy efficiency and the initial velocity of projectiles, and reduce the maximum bore pressure [8, 9]. Improving the muzzle kinetic energy, whilst not increasing or decreasing the chamber pressure too much, is the main goal pursued by charge designers. Too high a chamber pressure can affect the mobility and safety of the gun, causing safety issues [10] (figure 14.5).

No single formulation fulfils all user requirements. An ideal propellant should have the following properties (although the list is not exhaustive):

- Low flame temperature.
- Predictable burning rate over a range of pressures.
- Long shelf life.
- Not have a muzzle flash.
- Not produce toxic combustion gases.
- Be easy ignitable, with a consistent and reliable performance.

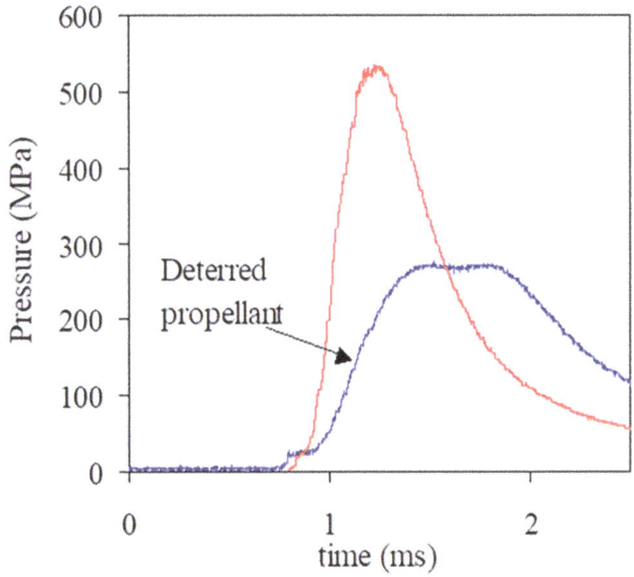

Figure 14.5. Pressure–time graph for a spherical propellant, with and without a deterrent [11].

- Have a high density (1.6–1.75 g cc^{-1}).
- Uninterrupted regular and continuous burning for the duration that the projectile is in the barrel.
- Less corrosive combustion products to reduce gun barrel wear.
- Low vulnerability to external stimuli, such as heat, impact, etc.
- The propellants, components/ingredients and their daughter products, and their products of combustion should not be toxic [12].

14.3 Testing of propellants

The testing of energetic materials and propellants can be divided into two parts: the sensitivity assessment and the performance evaluation. The closed vessel test is routinely used to evaluate the performance and ballistic properties of a propellant. The pressure over time signal is measured when a propellant powder is burned in a reinforced vessel able to sustain pressures of the range of 500 MPa. The analysis of the signal and its first derivative provides important ballistics parameters such as the force, co-volume, vivacity, and burning rate of the propellant, which are all critical for the design of a propellant charge. The technique, based on the form function of the propellant and Vieille's law, is well described in the literature [13–15] and in STANAG 4115, and is out of the scope of this book. A recent publication proposed a simplified approach using a model based on relative pressure impulse, allowing the prediction of the performances of complex charges [16]. Boulkadid *et al* confirmed the method for the estimation of the burning rate of a solid propellant [58].

The sensitivity tests are fundamental when working with energetic materials and a thorough characterisation of energetic materials is essential to ensure human safety.

A range of NATO documents, such as Standard NATO Agreements (STANAGs) and Allied Ordnance Protocols (AOPs), cover the technical procedures, methods, equipments, and processes for military personnel and member countries of the alliance. Amongst the numerous standardisation agreements are specific documents for small arms ammunition and NATO documents such as AOP-7 (Data Requirements and Tests for The Qualification of Explosive Materials) [17] and AOP-4170 (Principles and Methodology for The Qualification of Explosive Materials for Military Use) [18] which serve as guidance to ensure that the necessary tests are completed [19]. Test standards specify a common approach to conduct the qualification tests [20]. The transport of these so-called dangerous goods is strictly regulated and guidelines are provided by the United Nations to harmonise this transport (e.g. 'Recommendations on the Transport of Dangerous Goods') [21]; however, this is beyond the scope of this chapter.

Many studies have been performed with the required stability tests, described in AOP-7 [17]. Accelerated ageing allows an increase in the decomposition rate in a shorter time, however it is not always representative of the natural ageing processes at lower temperatures. It is necessary to routinely monitor, using various testing procedures, and to carefully control the stored energetic formulations to ensure safety. [22] These stability tests include conventional techniques such as the weight loss test, Bergmann–Junk (BJ) method, the Abel test, and the Methyl Violet test. However, more recent techniques, such as heat flow calorimetry (HFC), are also described.

This topic has always been important and will always be so. Recent research is regularly published on this subject, and as analytical techniques evolve safety will improve even more. Parate *et al* describe the design qualification testing, evaluation, and test methods of gas generators using double-base propellants for improvised explosive devices. Various qualification tests were performed, including closed vessel firings at hot and cold temperatures [20].

Davenas *et al* found three degradation stages for double-base rocket propellants (DBRPs) using the vacuum stability test (VST). Kinetic modelling indicated a decrease in the activation energy value with ageing progress and a change in the most probable reaction model after 10 months of artificial ageing [23]. Chelouche *et al* also found that there were three degradation stages during the ageing of the DBRPs. Kinetic modeling indicated a decrease in the activation energy value with ageing progress and a change in the most probable reaction model after 10 months of artificial ageing [24].

14.3.1 Sensitiveness tests

Energetic materials are sensitive compounds by nature to allow on-demand initiation and release of their energy to perform whatever action the material was designed to achieve. They are also able to react to accidental external stimuli such as indirect fire or impact from a fragment. Their ability to react to these stimuli and trigger combustion is defined as sensitivity and is a key property for the use of propellants.

Energetic materials and propellants can be initiated by applying an external stimulus, often of a mechanical nature, creating a so-called 'hot spot' under the action of local forces of strain and stress in the materials, ultimately causing an increase of temperature. Thermal energy (i.e. a flame or a spark, etc) can obviously be used to ignite propellants. The autoignition of conventional gun propellants for small-calibre applications usually occurs above 160 °C.

Sensitivity tests can be categorised by the nature of the stimulus, such as impact, friction, shock, electrostatic discharge, and thermal impulse, and are well described in the previous literature [25].

14.3.2 Stability of nitrate-ester-based propellants

Nitrate ester propellants undergo chemical and physical degradation, affecting their chemical and ballistic stability, therefore affecting the service life during storage; and this can change the burning rate and/or vivacity, increasing the safety risk to the operator during transportation and handling.

The stability of a propellant is generally evaluated according to an artificial ageing at high temperatures (i.e. 70 °C to 90 °C), whereas a property of the propellant such as nitric oxide release (determined by a BJ or Holland test), stabiliser consumption (measured by high-performance liquid chromatography, HPLC), energy release (measured by HFC), or peak pressure increase (determined using instrumented firearms) is measured. Procedures and methods can be found in the corresponding STANAGs, AOPs, or Manual of Proof and Inspection Procedures (MOPIs) [26]. While kinetics at such temperatures may differ from the actual ageing of a propellant in storage or in the field, these methods offer a good estimation of the chemical stability of propellants. Trache *et al* provide an excellent review of the various analytical techniques (chromatography, spectroscopy, microscopy, thermal, and mechanical analysis) to study the ageing and stability of NEBPs [22, 27–31].

Chemical stability is concerned with the chemical reactions taking place in the NEBPs, mainly by avoiding autocatalytic runaway reactions, by stabilisers capturing the reactive species, which prevent further reactions of these reactive species.

Ballistic stability is concerned with obtaining optimised and safe pressure–time graphs, dependant on the gas production rate from the propellant for the projectile and weapon system. The ballistic stability is mainly affected by the migration of components, such as the deterrent and age-related physical changes.

14.3.2.1 Chemical stability

The inherent instability of nitrate esters was illustrated very rapidly after the introduction of NC-based propellants and caused numerous accidents, even recently. Pioneers such as Lenk or Vieille reported that NC powders deteriorate during storage, even at room temperature. This tendency increases with the temperature and can be catalysed by the degradation products and the exothermic nature of the phenomena. This is well documented and the stabilising mechanisms have been widely discussed for many years [30, 32–41].

Nitrate esters, like other esters, undergo hydrolysis to provide an alcohol and nitric acid. However, hydrolysis is not the only reaction influencing the stability of the nitrate esters. Bond scission can also occur, providing alkoxy and NO_x radicals, which are able to react further by hydrogen (H) abstraction or oxidation pathways, leading eventually to decarboxylation and chain scission reactions, in a mostly exothermic fashion, as shown in figure 14.6. Hydrolysis and homolysis are both responsible for the decomposition of propellants and the possibility of unwanted autoignition. Stabilisers consisting of secondary amines, such as diphenylamine (DPA), 2-nitrophenylamine or urea derivatives (e.g. ethyl or methyl centralite), or more recently 'green' stabilisers, absorb the NOx gases released during the decomposition of the nitrate esters, therefore delaying autocatalytic degradation [36, 39, 42, 43].

The slow degradation of propellants and the release of nitrogen oxides and nitric acid cause degradation of the physical and ballistic properties of propellants [44]. The degradation of nitrate esters is affected by several factors, including water content, oxygen (O_2), chemical composition, etc [22, 38, 45–49]. An exothermic reaction of NC is very unlikely to take place without O_2, emphasising the greater importance of O_2 to nitrogen dioxide (NO_2) in the heat-release mechanism and spontaneous ignition [50].

Figueirôa de Souza Defanti *et al* found that the combustion properties of single-base NC propellants are affected by degradation by ageing. A single-base gunpowder sample containing methyl centralite and ethyl centralite as stabilisers was artificially aged for 10, 20, and 25 years [51]. Namir *et al* analysed a 42-year-old single-base propellant. The heat flow limit of $\leqslant 114\ \mu W\ g^{-1}$ for HFC was not exceeded and the sample met the requirements laid down in STANAG 4582. They found that for samples containing very low effective stabiliser contents (2 mL g^{-1}) the remaining daughter products continued to be effective [44].

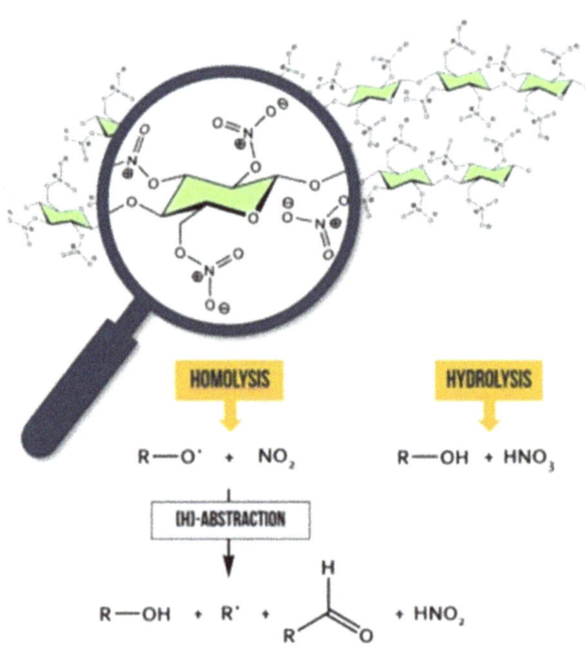

Figure 14.6. Degradation pathways of nitrocellulose: homolytic and hydrolytic decomposition [4].

Due to the importance of these phenomena for safety and the widespread use of NC-based propellants, work is still being published in this domain.

14.3.2.2 Ballistic stability

As mentioned earlier, the combustion properties of propellants can be adjusted by the diffusion of a deterrent into the propellant's grain down to a certain depth. The concentration profile of the burning-rate modifier has a great effect on the burning rate and vivacity of the powder. One of the main aims of interior ballistics is to attain a higher muzzle velocity with a smaller maximum chamber pressure. The burning rate of the propellant directly affects the formation rate of burning gases, which affects the ballistic parameters.

Ballistic stability is therefore concerned with ageing-related physical and chemical changes. The surface area of the propellants increases, and pores and cracks are created due to the migration of the stabiliser and NGL and to the escape of NO_x gases from the propellant grains. The dynamic vivacity and pressure $p(t)$ therefore increase in the high-pressure environment that exists during gun firing [43, 52, 53]. Figueirôa de Souza Defanti *et al* report two phases (intensification and passivation) for the burning rate (r_{max}) during ageing. The first occurs while the propellant is chemically safe and the second when the propellant becomes unsafe for handling and storage [54].

During ageing, and particularly at higher temperatures, the deterrent diffuses further in the grain causing significant changes to its initial concentration profile. This phenomenon causes an increased burning rate at the surface of the propellant with a direct consequence on the peak pressure in firearms that can cause safety issues for the shooter.

NGL forms weak bonds with NC, making it more mobile [55]. Suceska *et al* found that an increase of the modulus of elasticity and Young's modulus, an increase of the glass transition, and a decrease of the toughness were observed for DBRPs during ageing [56]. Musanic *et al* found that using kinetic parameters, such as the activation energy and pre-exponential factor, that they could predict the time necessary to reach a certain change of properties, such as the loss modulus [57].

Under certain pressure conditions, the burning rate and the slope of the burning-rate pressure curve of the modified single-base propellant decrease as the nitrogen content of the NC decreases. The burning-rate pressure index is less affected by the nitrogen content of NC. As the nitrogen content of NC decreases, the force capacity and co-volume of the modified single-base propellant decrease [58]. It was previously shown that, depending on the propellant formulation, ageing can increase or decrease the maximum pressure in the combustion chamber [51]. Tirak *et al* found that the dynamic vivacity of the 5.56 mm × 45 mm and 7.62 mm × 51 mm calibre propellants tested increased proportionally to the ageing time [43]. Molecular dynamics simulations can be used to study plasticiser migration [59, 60]. A calculation method was provided by Zhang *et al* for the abnormal chamber pressure caused by fractured propellant charges [61]. An automatically-generated detailed kinetic model of NGL decomposition was developed using the open source software package Reaction Mechanism Generator [62]. Spherical double-base propellants that were artificially aged at 71 °C and 80 °C were investigated. Towards low pressures, the burning rate increased with ageing, and towards high pressures the burning rate decreased as a function of ageing [63].

14.3.2.3 Ammunition surveillance and service life

The service life of propellants comprises the chemical shelf life and the functional or ballistic shelf life [64]. The chemical shelf life deals with the interval of time during which the propellant can be stored without showing a hazard to its environment. The burning rate influences the generation rate of combustion gases [63, 65], which influences the ballistic parameters. As discussed in section 14.3.2.1 (chemical stability), nitrate esters, even if stored at ambient temperature and acceptable humidity level, self-decompose accompanied by self-heating, and in extreme cases can result in autocatalytic runaway reactions [66]. In section 14.3.2.2, ballistic stability was discussed, whereby the pressure safety margins for the firearms must be respected.

Extending the service life of propellants can reduce stockpile-management costs, however it is important to well understand and master the safety margins and shelf lives [43, 67]. The Small Arms Survey provides expertise on all aspects of small arms and resources about accidents like unplanned explosions at ammunition sites and subsequent casualties. The website can be accessed online and a recent report (2020) indicates that the total casualties of all unplanned explosions at ammunition sites from January 1979 to December 2019 is close to 30,000[1].

Unfortunately, many accidents have been observed due to self-ignition due to degraded NC-based propellants, such as in India (2001, 2016), Syria (2007), Tanzania (2009), and Cyprus (2011) [38, 54]. In Cyprus, a naval base had 120, 122, 125, and 160 mm high-explosive artillery shells, 7.62 mm shell casings, compressed gunpowder, etc, in open storage. A fire and explosions killed 12 people on the spot and injured a further 62 [68].

The NATO working group AC326 SGA discussed the unintended ignition of energetic materials. Defects in energetic materials, inspection capabilities, subscale testing, modelling and simulation, propulsion accelerations, live fire with induced defects, and statistical analysis were discussed [19].

Roduit *et al* investigated the determination of the shelf life, by including kinetic parameters into data loggers, allowing the continuous, on-line monitoring of the change of material properties as a function of temperature [69]. The aim of this work was mainly to prevent accidents, by understanding the extent of degradation, therefore intervening before autocatalytic runaway reactions occur.

14.4 Future axes for gun propellants

The domain of small-calibre gun propellants has been somewhat conservative for many years. However, in the past few years, research has been moving more rapidly and many new axes for research have been advanced. Military systems have become

[1] http://www.smallarmssurvey.org/home.html

more complex, and most of the military domains have visibly made improvements at an outstanding pace, with new materials and technologies such as rail guns, ultra-high-speed missiles, etc.

This section has been divided up into formulations and future charges with topics that we considered relevant for gun applications. The former deals mainly with the ingredients at the heart of formulations, based on recent developments made in the field of advanced materials, the latter covering the processability aspects of these formulations and new materials.

Bird *et al* describe future axes such as formulation development, three-dimensional (3D) printing, real-time process monitoring, and formulation techniques in manufacturing [70]. New formulations, including high-nitrogen compounds (HNCs), for example, and modelling were discussed by the National Research Council in 2004 [71].

Development programmes are active in promising insensitive high-energy materials with higher performances. New materials and modelling techniques are available [72].

14.4.1 New formulations

There is continuous research in the field of energetic materials and propellants that aims to improve different aspects of current formulations. The development of future propellant formulations should be based on the following key concepts: environmentally friendly formulations, insensitive munitions (IMs) properties with less sensitive ingredients and with higher energy, providing military personnel with safe and more efficient products.

14.4.1.1 Nitrocellulose and new binders

NC has been an outstanding raw material (binder/matrix) for many years in gun propellants. Having been utilised for more than 150 years, it continues to dominate the field, though it is still difficult to characterise completely and master, partly due to the fact that cellulose occurs in nature, usually acquired from wood or cotton. NC is obtained by nitrating high-quality cellulose with a mixture of nitric and sulphuric acids. The control of the concentration and ratio of these acids allows the production of various degrees of nitration of the NC. NC presenting a nitrogen content of 12.6%, up to 13.45%, is used in the manufacture of propellants. Cellulose is an extremely useful natural polymer and an essential raw material for gun propellants, however it also comes with uncertainties and a lack of control concerning the chemical structure inherent to such a class of materials. New research is therefore required in order to find a potential replacement. It is not surprising therefore that much effort has been made in the past to improve NC or to substitute it by a synthetic energetic polymer.

Two new types of celluloses and potential raw materials for military applications have been discussed recently: bacterial cellulose and nanocellulose.

1. Bacterial cellulose

Bacterial cellulose (BC) is a biomaterial with high purity compared to plant-based cellulose and does not contain cellulose, lignin, hemicelluloses, or pectin [73–75].

BC possesses excellent mechanical and chemical properties, including high crystallinity, high tensile strength, high water retention capacity, high elasticity, and good biocompatibility. BC is produced by certain types of bacteria to form protective envelopes around the cells; research is ongoing to enhance cellulose growth over other extracellular polysaccharides. A review of BC production methods and applications with different culture methods is provided by Wang *et al* [76]. Low-cost substrates and large waste biomass byproducts from the agricultural, brewery, sugar, and other industries are also being evaluated to develop potential industrialisation and commercialisation of BC, which remains a challenge due to its high fermentation cost, low productivity, and expensive culture media [77].

Wang *et al* opened the door to the applicability of nitrated BC in propellant formulations by producing a triple-base gun propellant containing 5% of nitrated BC, conventional NC, and nitramine by a semi-solvent method. A distinct 3D special network structure was observed, providing an equally energetic material with enhanced impact sensitivity and moderate burning-rate impact [78].

2. Nanocellulose

Nanocellulose is a form of cellulose which comprises nano-sized cellulose fibrils. The lateral dimensions of nanocellulose range from 5 to 20 nm and the longitudinal dimensions range up to hundreds of nanometres. Nanocellulose can be extracted from traditional wood pulp or from a variety of plant biomass (banana rachis, sisal, pineapple leaf and coir) [79] or fibre wastes from forestry or agricultural industries, offering a great recycling opportunity into high product values. Purified cellulose is treated mechanically, chemically, or biologically to produce the nanocellulose, which is classified into microfibrillated cellulose, nanocrystalline cellulose, and bacterial nanocellulose [80]. Nanocellulose, with its low thermal expansion coefficient, light weight, high strength, high recyclability, biocompatibility, low toxicity, high crystallinity, transparent and birefringence properties, is a promising material, considered a new 'wonder material', and is being investigated in numerous applications and composite materials that could affect numerous fields and industries, from water and water treatment [81], wood and paper manufacture [82], drug delivery agents [83] and 3D bioprinting [84, 85] to the automotive industry [86].

Nanocellulose also emerged in the defence industry as a new material for military textiles (bulletproof vests), packaging, energy devices (supercapacitors, batteries, solar-energy-based units) and fire-retardant material [87].

Nitrated nanocellulose is currently also being investigated in propellant formulations as a binder or an energetic carrier. Its high strength could be an inconvenience here, as the mechanical properties might be impacted; however, functionalisation and/or additional plasticisers could help overcome this weakness. The use of nitrated nanocellulose was additionally investigated to reduce the sensitivity of propellants, and will be discussed in section 14.4.1.4.

The replacement of NC is no easy task as the ideal synthetic polymer must fulfil many essential conditions, such as having a sufficient amount of energy to generate gases able to propel the projectile, good thermal, mechanical and environmentally friendly properties, etc, so that the produced propellant can offer optimal burning characteristics, and chemical and ballistic stability. In the field of larger-calibre propellants, a mixture of (energetic) synthetic polymers and fuels/oxidants, providing a class of composite propellants or gas generators, is often preferred. Energetic polymers, just like NC, contain energetic groups (azide, nitrate ester, etc) in their backbone or their side chains and can be seen as a strategy to improve propellant performance. Moreover, energetic binders contributing to the energy content of the formulation can be used to reduce the amount of crystalline high explosive (HE) and is also a method of choice to limit the sensitivity of propellants.

Nitrated hydroxyl-terminated polybutadiene (NHTPB) is a liquid polymer, the viscosity and glass phase transition of which depends on the degree of polymerisation and level of nitration of its initial double bonds. It can be prepared in a one-pot procedure [88] and is miscible and compatible with most energetic plasticisers. It can be incorporated into propellant formulations [89], providing more energetic propellants with little influence on the burning rate. Its mechanical properties are degraded due to the nitration of the double bonds and loss of mobility of the molecular chain, however this can be mitigated by the use of an additional energetic plasticiser, such as butyl nitroxyethylnitramine (NENA). Other energetic polymers are currently being investigated in propellant formulations, and a detailed review is given by Gayathri *et al* [90] on the use of the ester-nitrate-based polymers nitrato HTPB (NHTPB), poly(3-nitratomethyl-3-methyloxetane) (PVN), PolyNIMMO, PolyGLYN, and polyglycidyl nitrates in propellant formulations (figure 14.7).

14.4.1.2 Nitroglycerine and plasticisers

NGL is a blasting oil obtained by direct nitration of glycerine under controlled temperature and acid concentration conditions. It is important to use a very pure form of NGL (i.e. free of residual acids) to provide long-term stability of the propellant. NGL has two roles in gun propellants: the energetic complement and the plasticiser. Gelatinising helps to reduce the overall sensitivity of propellant mixtures. NGL has a few disadvantages, however, such as sensitivity and toxicity, mainly related to its vasodilatory potential [91]. NGL is, though, relatively cheap and easy to produce with well-mastered and controlled processes, and its replacement is no easy task, particularly in the field of small-calibre propellants where the cost plays an important role due to the large amounts of product used and deployed worldwide. On the other hand, better propellant properties, such as the temperature coefficient,

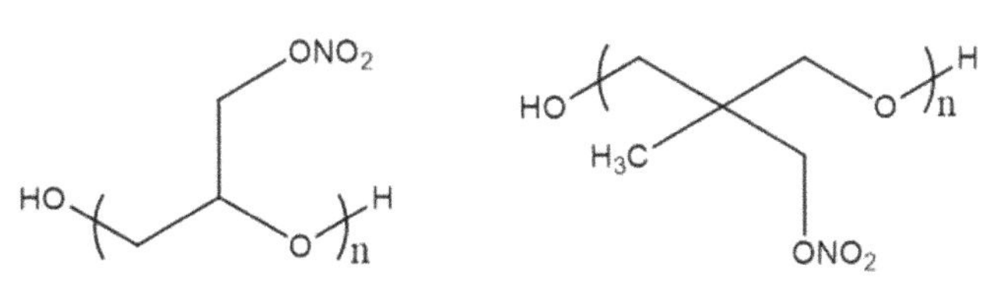

Figure 14.7. Structures of polyGLYN and polyNIMMO

resistance to mechanical stresses, or decreased sensitivity, could be considered with the use of new energetic plasticisers.

Replacements for NGL usually belong to the nitrate ester family, such as ethylene glycol dinitrate (EGDN), DEGDN, triethylene glycol dinitrate (TEGDN), triethylene glycol hexanitrate, tetraethylene glycol octanitrate (TetraEGON), pentaethylene glycol decanitrate, and hexaethylene glycol dodecanitrate. These molecules have often lower performances and oxygen balances, resulting in the necessity of the use of another energetic source, such as 1,3,5-Trinitro-1,3,5-triazinane (RDX) or hexanitrohexaazaisowurtzitane dodecane (CL-20), increasing the cost of the solution by several orders of magnitude in the case of CL-20. As an example, Nichele *et al* proposed an alternative to the conventional M-14 based on DEGDN, TEGDN, and RDX [92].

A theoretical density functional theory study at the B3LYP/6-31G* level of various polyethylene glycol polynitrates as potential energetic plasticisers for propellants was done by Wang *et al* [93]. They found that increasing the chain length (i.e. $-OCH_2CH_2-$, $-OCH(ONO_2)-$) will decrease the oxygen balance, energy, and standard enthalpy of formation while increasing the sensitivity. That behaviour was unexpected compared to conventional energetic materials. Moreover, the $O-NO_2$ bond is confirmed to be the 'trigger' bond and its energy or strength directly influences the thermal stability of the parent molecule. An increase in the number of $-O-CH_2-CH_2-$ groups will cause a lower density and specific impulse, while an increase in the number of $-O-CH(ONO_2)-CH(ONO_2)-$ groups will cause the exact opposite effect.

Liao *et al* achieved better temperature coefficients (behaviour of a propellant in terms of velocity and pressure when fired in cold and hot environments) by partial replacement of NGL by DEGDN and RDX to produced a mixed charge offering opposite behaviour in terms of combustion at extreme temperatures [94]. Improved temperature coefficients were also achieved by the use of linear dinitro-diaza-alkanes (DNDA) plasticisers. A mixture of 2,4-dinitro-2,4-diazapentane (DNDA-5), 2,4-dinitro-2,4-diazahexane (DNDA-6), and 3,5-dinitro-3,5-diazaheptane (DNDA-7) used in a gun propellant formulation reduced the dependence of the propellant on the operational temperature [95]. Another class of alkyl nitramines, NENA derivatives are often investigated as a replacement of or a supplement to NGL in propellants. They naturally impact their performance and stability, however. The IM nature of the charge and will be discussed later in section 14.4.1.4.

Recently, Wang *et al* reported dinitro ester energetic placticisers (DNEPs) as novel energetic substrates with potentially higher energies and low glass transition temperatures. 2,2-dinitropropyl acetate (DNPAc), 2,2-dinitropropyl propionate (DNPPc), and 2,2-dinitropropyl propionate (DNPBc) were synthesised and proved to possess better plasticising properties than NENAs, while offering better energetic contents. DNPBc, in particular, could partially replace NGL or NENAs in propellant formulations [96].

14.4.1.3 'Green' considerations

Taking care of the environment and human health will clearly continue to be an important path for the future, whether it be for army personnel dismantling ammunition, soldiers firing, or manufacturing personnel in contact with large quantities of raw materials. There is a need to produce more environmentally friendly propellants. It is necessary to consider the toxicity aspect of the combustion gases, which is another facet of the toxicity of propellants. Training sessions or special operations forces fire routinely large numbers of hand-held weapons, often in enclosed and restricted spaces regularly confronting the end user with substantial amounts of fumes.

Safety must be ensured at all stages of the life cycle of propellants, whether it be during manufacture, disposal, or firing [97]. Moreover, with Registration, Evaluation, Authorisation, and Restriction of Chemicals (REACH) legislation (1907/2006), industry in Europe has made progress in using more environmentally friendly compounds. Phthalates such as bis (2-ethylhexyl) phthalate, dibutyl phthalate (DBP), benzyl butyl phthalate, and diisobutyl phthalate, which are still used in many countries throughout the world, were examined by the European Chemicals Agency in 2006 and their use was confirmed to be restricted at a European Union level in 2014 by the Commission [98]. In the propellant industry, DBP was rapidly replaced by accepted solutions such as ethyl centralite even if the 'green factor' could not be considered satisfactory. Additional work was and is still ongoing to provide better solutions, i.e. in the form of citrate plasticisers, providing improved ballistic stability compared to phthalates.

Mendonça-Filho *et al* proposed 10 new formulations by replacing the original ingredients DPA, DBP, and dinitrotoluene of M-14 by RDX, DEGDN, TEGDN, acetyl tributyl citrate, ethyl centralite, and Akardite II (AK-II). The replacements focused on the reduction of the toxicity, the maintenance of the ballistic performance, and the reduction of gun barrel erosion [92].

Liang *et al* designed and prepared energetic composite-deterring agents (ECDAs), such as TEGDN and poly(neopentyl glycol adipate) for spherical propellants. The aim was to achieve ideal burning progressivity and reduce harmful phenomena such as muzzle flame and smoke. The maximum pressure in the closed vessel bore of the ECDA-deterred propellants was 2.29 MPa higher than that of the DBP-deterred propellants. The burning properties and harmful phenomena of the deterred propellants were improved by replacing traditional DBP with ECDAs [9].

In addition to the replacement of the deterrent by a safer alternative, and while reducing the toxicity of the combustion gases seems relatively complex without changing the nature of the energetic matrix, other ingredients such as the stabiliser should be replaced to ensure a better green impact. Traditional stabilisers for NC-based propellants are known to have carcinogenic and mutagenic properties and toxicity for reproduction [99]. The action of the stabiliser being to slow down the degradation of the propellant, it reacts with radicals arising from its degradation to form stabilised nitrate ester derivatives that cannot react further, increasing the shelf life of the propellant. The stabiliser counterpart, however, continues to react with other species present in the matrix and forming N-nitroso derivatives, which are probable carcinogenic agents. DPA, akardite, and ethyl centralite are well known and documented [100–105]. DPA has been used in NC-based propellants since 1910 [38]. DPA residues have been found in soil and groundwater, representing a hazard to aquatic life [106], bacteria, and animals. It is used widely, for example, to preserve post-harvest qualities of apple and pear crops [107]. Based on the work of Lussier, the daughter products of AK-II are presented in figure 14.8.

Greener alternatives to conventional stabilisers have been proposed: natural molecules, such as alpha-ionone, partially responsible for the scent of violets, alpha-tocopherol (vitamin E), and curcumin [3]. These molecules do not produce any toxic daughter products, as well [108], and compare well to conventional stabilisers, such as DPA or AK-II, as shown by BJ test/Holland test (HT)/HFC studies.

Other green alternatives, such as lignins, have been shown to inhibit the autocatalytic decomposition of NC and been compared to the widely used DPA in VSTs and BJ stability tests [109, 110].

These stabilisers seem to have efficiencies similar to or greater than those of current stabilisers [4]. Performance-wide, these stabilisers can stabilise NEBPs for hundreds of years, shown by an iso-conventional approach using HFC studies and accelerated ageing (figure 14.9).

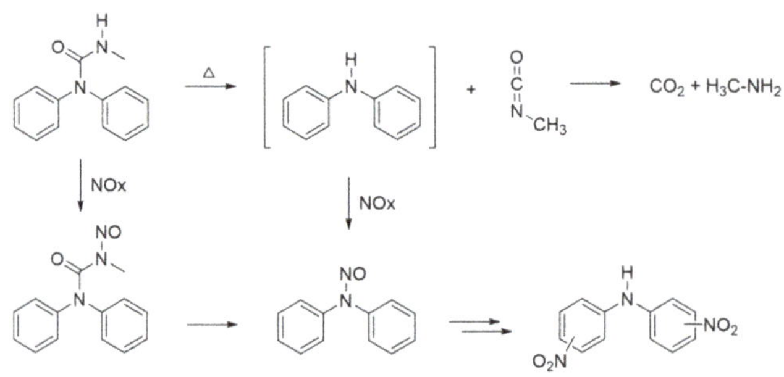

Figure 14.8. Carcinogenic N-nitrosamines produced by AK-II [3].

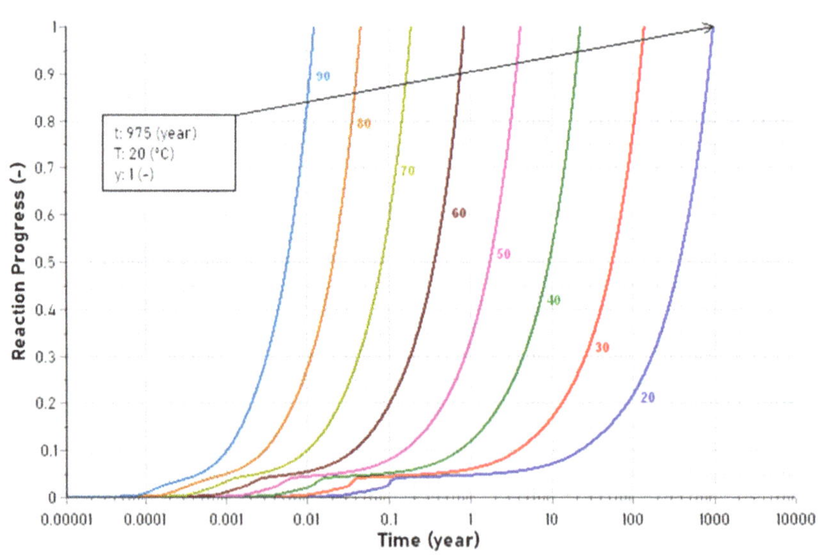

Figure 14.9. Reaction progress vs. time (years) for four 'green' stabilisers: #2: curcumin; #49: 2,3,5-trimethylphenol; #67: alpha-ionone; and #83: alpha-tocopherol [4].

While alpha-tocopherol mixtures seem to provide the best performance, curcumin has been investigated further by other groups [111] as a potential replacement for conventional stabiliser, and possible derivatives are currently being investigated [112, 113]. It was found that curcumin can work as well as DPA as a stabiliser for single-base propellants, and an effective and accurate method for the analysis of curcumin-based propellants in regards to the AOP-48 standard was proposed. A complementary *in silico* analysis was also performed using toxicity prediction software (Lazar, Toxtree, VEGA, and TEST), based on the similarities with substances contained in their databases. The results concluded that the stabiliser curcumin presents no toxicity.

More work is still required to be able to introduce the new environmentally friendlier stabilisers into propellants, and a collaborative work at the ministries of defence (MoD) level is necessary to assess the safety, which is directly impacted by the substitution of the stabiliser.

De Araújo *et al* discussed the synthesis and characterisation of green polyesters based on readily available and environmentally friendly diols for use as polymeric binders in propellants composites. Better thermal stability and good burning performances were obtained, without the need for a metal catalyst as an additional burning-rate modifier, when compared to conventional propellants. The use of the proposed green polyesters as polymeric binders could therefore be advantageous for formulations of new green propellants [114].

The major combustion products are carbon monoxide (CO), carbon dioxide (CO_2), hydrogen (H_2), water (H_2O), and nitrogen (N_2) [115]. Other gases are produced in smaller quantities, such as ammonia (NH_3) and hydrogen cyanide (HCN). The following gases are the most dangerous: CO, NH_3, and HCN. Computer codes, such as the ICT code, are able to predict the composition and quantities of each gas [116]. Moxnes *et al* studied the compositions of some of the main decomposition products, experimentally as well as theoretically. They provided good descriptions of the explanations for the toxicities of the different combustion gases [117].

Meuken *et al* investigated the effects of combustion products of small-calibre ammunition (5.56, 7.62, and 9 mm). The NOx emissions were higher for the 9 mm calibre than for the 5.56 and 7.62 mm, and the 7.62 mm showed the lowest CO_2 emissions. HCN was mainly produced with the 7.62 mm [118].

After the HK416 (Heckler & Koch, Germany) was introduced for the Norwegian army in 2006, soldiers became ill with flu-like symptoms after training. Repeated exposure to high concentrations of emitted fumes, consisting of particles and gases, gave acute respiratory symptoms. It was found that the newly introduced steel core ammunition, NM229, had generated more copper and zinc than the lead core ammunition, SS109, that it had replaced [119, 120].

Mariussen *et al* investigated the geno- and cyto-toxic potential of gunshot fumes from three types of ammunition. The particles of copper, zinc, and lead and gases (CO_2, CO, NH_3, HCN and NOx) were quantified [121].

Lead is, of course, a hot topic due to its toxicity to people and wildlife. Whilst slightly out of the scope of this chapter on small-calibre propellants, due to the importance of the topic a few words shall be mentioned. In primarily medium- and large-calibre applications, lead has been used as a de-coppering agent [97, 122–125].

Lead has also been much used in small arms ammunition projectiles, which have typically consisted of a copper outer shell with a lead/antimony core material. Around 100,000 tonnes of lead is dispersed into the EU environment, of which 44,000 tonnes originates from sports shooting, hunting, and fishing activities each year[2].

People, especially hunters and sports shooters, are exposed to lead mainly through inhalation and ingestion. Exposure to lead has resulted in various illnesses, such as cardiovascular disease (CVD) and peripheral arterial disease [126]. Ninety-nine percent (99%) of fired lead shot ends up in the environment [127]. Lead is dissolved by groundwater and absorbed into flora as well as contaminating surface water [128]. Dire consequences have been observed for wildlife, such as earthworms, [129] and scavenging or predatory animals, such as bald eagles [130–133].

In humans, exposure to lead is associated with a wide range of negative health effects, including reduced fertility, CVDs, damage to organs, and cancer. The European Chemicals Agency (ECHA) estimates that in any given year about one million children are vulnerable to lead due to game meat consumption. Exposure to lead is especially harmful to children's neurological development [134]. The ECHA would like to reduce lead emissions by approximately 1.5 million tonnes over 20 years. Copper-based bullets have been developed to replace lead [127]; however, in one study it was found that lead-free bullets produced poorer animal welfare outcomes, and were more expensive per killed rabbit than lead-based bullets [135].

14.4.1.4 Insensitive behaviour

IMs are chemically stable and can withstand unplanned external stimuli (thermal, mechanical, or electrical) during storage and transportation, and satisfy performance and operational requirements only on demand, but which minimise the probability of inadvertent initiation and the severity of subsequent collateral damage to weapon platforms, logistic systems, and personnel when subjected to unplanned stimuli [72].

The incorporation of insensitive energetic materials, such as thermoplastic elastomers or insensitive HEs as replacements to existing energetic bases (NC, NGL, etc) is probably necessary to obtain top-of-the-line IMs. The study of insensitive propellants has been the centre of interest of many groups and researchers, with many achievements and much recent progress. Much investment was required for the development of insensitive yet sufficiently energetic compounds. One critical aspect of IMs and where the biggest benefits were obtained at low expenses is the packaging and logistics of ammunitions and explosives.

While most of the work focused on larger-calibre applications, several approaches are worth mentioning in the scope of a modified double-base propellant that could be used in small-calibre guns.

The energetic plasticiser N-butyl-N-(2-nitroxy-ethyl) nitramine (Bu-NENA), shown in figure 14.10, has been used in the preparation of low-sensitivity and high-energy propellants [136, 137]. The replacement of NGL by Bu-NENA usually results in a decrease in friction sensitivity while increasing the impact sensitivity.

[2] https://echa.europa.eu/hot-topics/lead-in-shot-bullets-and-fishing-weights.

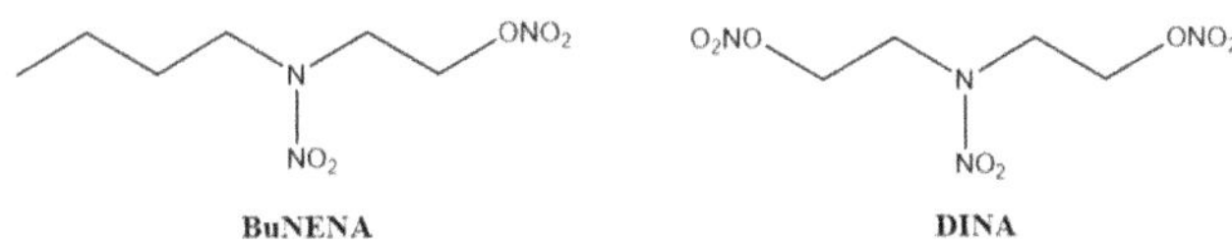

Figure 14.10. Structures of Bu-NENA and DINA.

One of the major drawbacks of NENAs or similar plasticisers, in addition to a reduced energetic contribution, is the potential diffusion of the energetic substance in the charge. Different approaches were developed to circumvent this issue and the use of energetic polymers, cross-linkage, or functionalisation of the NC were investigated. Wu *et al* produced and characterised a series of spherical propellants containing glycidyl azide polymer (GAP) and NC [138]. GAP showed good processability and plasticising properties in addition to reduced sensitivity and a slightly better onset decomposition temperature. Xiao *et al* cross-linked NC molecular chains using isocyanate chemistry in combination with polyethylene glycol to produce a step-ladder-like NC morphology that was proven to possess a quasi-spherical shape and regular crystal structure. The resulting NC showed a reduced sensitivity to thermal stability and should be evaluated in propellant formulations in order to assess its potential for insensitive propellants [139]. Comtois *et al* evaluated propellant formulations using NC plasticised with NGL and containing azidodeoxycellulose as a solid filler. They concluded that the maximum pressure of a propellant is increased and does not generate a linear burn rate behaviour (LBR) outside of that, which can be used for a propellant application [140]; however, it is currently impractical for large-scale production [140]. Meng *et al* [141] discussed the preparation and sensitivity characterisation of nano-sized spherical NC composite crystal from microcrystalline cellulose, as a way to reduce the sensitivity of propellants. A somewhat expected higher crystallinity was confirmed for the nitrated nanocellulose, however a reduced impact sensitivity of 45% was also observed, while the nitrogen content was increased to 14.89%, leading to an earlier thermal decomposition. Further studies should be undertaken to investigate the effects of the nitrated nanocellulose in propellant formulations in terms of stability, performance, sensitivity, and mechanical properties.

The replacement of the energetic base by a less-sensitive HE was studied, and Badgujar *et al* describe a series of new explosives which have been developed with promising blends of insensitivity and performance, such as 1,1′-diamino-2,2′-dinitroethylene (FOX-7). The application of these materials is due to their special properties, such as insensitivity to accidental impact and ability to withstand high temperature [72]. Anniyappan *et al* discuss potential insensitive energetic materials and plasticisers for the development of futuristic IMs, insensitive energetic gun propellant formulations based on FOX-7 and FOX-12. They reported a lower flame temperature and reliable performance, that they are eco-friendly in nature, combined with improved mechanical properties [142].

14.4.1.5 High-energy bases

Much research for new energetic materials for propellants concerns obtaining a higher performance or enhanced insensitivity to thermal or shock stimuli.

Klaptoke's group has been particularly active in this domain. The use of highly nitrated moieties, such as the di-nitro or tri-nitro gem [143–147], or the use of compounds with high standard enthalpies of formation, such as high-nitrogen-content heterocycles [148–151] or a combination of the two [147, 152, 153], are interesting potential strategies to improve the performance of energetic materials and therefore propellants.

Tetrazole-functionalised microcrystalline cellulose has recently been prepared and investigated [154] as a potentially new insensitive, nitrogen-rich, and dense energetic material for high-performance propellants. Compounds with high nitrogen contents are challenging, since a high thermal stability and low sensitivity are needed. Stierstorfer *et al* found that several nitrogen-rich compounds, such as 5,5′-bis(tetrazolyl)amine (H2bta), ammonium 1,5-bistetrazolate (NH_4–1,5BT), ammonium bis(1-methyltetrazolyl)triazenate (NH_4–1MeBTT), and hydrazinium 5-aminotetrazolate (N_2H_5At) may prove to be good candidates for additives in solid propellants [155]. Theoretical calculations made on examining more than 20 interesting ingredients reveal that a force constant up to 1400 J g^{-1} could be accessible [156]. Szala *et al* discuss the synthesis of high-nitrogen energetic compounds, such as azotetrazole salts, which have a long combustion-to-detonation transition. The decomposition of HNCs is characterised by a lower adiabatic combustion temperature [157].

Concerning small calibres, there is always the question of the availability and the price of the new solutions and formulations, and this is often a limiting factor for new products as the cost of small arms ammunition should remain as low as possible.

14.4.1.6 Low erosivity

Barrel wear caused by thermal and chemical erosion is well known in the field of firearms and is a complex process resulting from the friction of the projectile and from hot gases and particles, arising from the combustion of the primer and propellant at high pressures. Barrel wear is a critical aspect of weaponry as it has a direct impact on the operational effectiveness and accuracy of the weapon, and also on the overall cost of maintenance over the life cycle of the system.

Chemical erosion is caused by chemical reactions between the metal of the barrel and the combustion gases (and hot particles) produced during the combustion of the energetic material. The most abundant gases produced during firing are CO and CO_2, H_2, N_2, and H_2O. Several theories have been proposed to try and rank the contribution of each of these gases in the erosion phenomena [158–160].

While these theories do not absolutely converge, there is a global agreement that carbon monoxide and hydrogen will lead to an increased erosion potential while nitrogen could lead to a reduced erosivity. Carbon oxides will react with iron present at the gas–bore interface yielding solid carbon. The high temperature observed during the firing contributes to the dissolving of the carbon in the gun steel, which in turn cools down and precipitates out as iron carbide, causing a reduced local melting point, therefore a softer steel, potentially inducing more erosion [161, 162].

Decarburisation can also occur when hydrogen reacts with steel leading to the formation of carbides. Nitrogen, on the other hand, reacts with steel to form nitride that can harden the steel and increase the resistance to erosion.

Thermal erosion takes place when the steel of the barrel has been sufficiently warmed to the point of partially melting. Repeated or automatic firings obviously escalate this increase in temperature and play a crucial role in the erosion of barrels, as temperatures up to 1800 K are often reached during the ballistic cycle. Double-base propellants containing NGL as an additional energetic component usually possess higher erosivity potentials than single-base propellants due to their higher potentials, energy contents, and flame temperatures. Nitroguanidine is often used in composite propellants in order to reduce the barrel wear by reducing the flame temperature. The downside of the use of nitroguanidine is the drastic decrease of the force of the propellant, having a direct impact on the performance of the weapon.

Erosivity is an issue for any weapon system, based on the propulsion of a projectile through a barrel, and several groups have researched how to reduce gun barrel erosion [163–165]. Additives in the form of high-nitrogen-content molecules having a relatively high standard enthalpy of formation are often used in that endeavour. They have the advantage of producing larger amounts of nitrogen in their combustion gases. Nitrogen has the advantage of a neutral contribution to the adiabatic flame temperature, the high standard enthalpy of formation still contributing to the energy content of the formulation and limiting the loss of energy and performance.

While making use of high-nitrogen-content molecules in formulations seems appealing, it often has other consequences that should be taken into account. The compatibility of such materials, firstly, is not guaranteed as the presence of nitrogen and amine derivatives might influence the chemical stability of the formulation. While tetrazine and ionic tetrazoles often show poor chemical stability, the use of neutral tetrazoles seems compatible with NC- and NGL-based propellants. Secondly, the modification of the combustion properties of the propellant is the key parameter assessing the performance and usability of the propellant. Nitrogen-rich materials have been investigated in that perspective as burning-rate enhancers [166]. Damse *et al* [137] reported a suitable formulation containing guanidinium-5-aminotetrazolate as a partial replacement for RDX, providing a decrease of the propellant flame temperature by about 120 K, without adversely affecting the burning-rate characteristics. Lavoie *et al* [167] could achieve a flame temperature reduction of 461 K with a formulation containing 25% of a bistetrazole, unfortunately at the cost of important consequences for the combustion behaviour as demonstrated in the study of the influence of two bistetrazoles, 5,5′-bis(1H-tetrazolyl)-amine (BTA) and 5,5′-hydrazinebistetrazole (HBT), on the burning rates and thermal behaviour of a typical triple-base gun propellant. Excellent propellant forces but high burning rates, up to 93% higher than the reference propellant, were found as well as a higher dependency on pressure, with a pressure coefficient exponent as high as 1.4 where values close to 1 are expected for a reference propellant. The addition of 25% 3,6-dihydrazino-s-tetrazine (DHT) or 5-aminotetrazolium nitrate (HAT-NO_3) to NC-based propellants were also investigated by

Lavoie *et al.* Again, an increase of 80% to the burning rate was observed compared to a reference propellant not containing any additive. Moreover, they had very poor compatibility with the nitrated esters used in the formulation and caused rapid autocatalytic decomposition [167]. Recently, Comtois *et al* demonstrated that a mixture of NGL and GAP can be added to NC in order to influence both its erosivity and LBR properties [168]. A cord propellant was tested using the closed vessel technique and confirmed the trend of an increase of the burning-rate characteristics, while relative erosion data obtained from a vented vessel test are in agreement with the Arisawa–Kimura models and tends to indicate that the gas concentration ratio CO/CO_2 is a better erosivity indicator than the N_2/CO ratio.

14.4.2 New charges for gun propellants

With so many possibilities for formulations, the door is now open to new technologies or processes to fully exploit the potential of the research being done on the ingredients. There are many future options for future gun ammunition, from ambitious systems using liquid gun propellants or 3D-printed charges to more conventional approaches and ingenious process modifications such as the production of gradiently denitrated gun propellant to eliminate the need of a burning-rate modifier in gun applications [169]. One of the potential candidates for liquid gun propellants is a mixture of hydroxyl ammonium nitrate, triethanol ammonium nitrate, and water [156]. Liquid propellant guns could offer some improvements over solid propellant systems, such as the avoidance of cartridge disposal, and high-energetic liquid propellants could provide higher performance [170].

This section will be limited to high-output gun propellants and systems able to achieve significant velocity increases, additive manufacturing of energetic charges, and foamed propellants.

14.4.2.1 High-energy gun propellants and systems

Research is taking place worldwide on a new generation of projectiles with the ability to reach speeds of Mach 5 and above. Such breakthroughs might be regarded as a 'game changers' for defending armies and could be deployed over multiple weapon systems—from the small arms perspective to the artillery and large-calibre guns. The use of hypervelocity projectiles, however, requires the development of a new propulsion system together with an adequately formulated charge.

While more energetic fillers, such as RDX or new powerful highly nitrated substrates, can be used to increase the performance, this strategy often leads to problems associated with high temperatures and rapid gun erosion. Novel energetic ingredients are required to provide more energy while keeping the disadvantages to a minimum, but improved processes are mandatory to shape the formulation into a practical charge with maximum progressivity.

Fu *et al* found that phase-change-material-based nanocomposites improve the thermal stability and decrease the erosion performance. They also adjust the burning process and improve the progressive burning of high-energy propellants containing RDX [171].

The US Army patented a method of manufacturing propellant grains including generating an optimal surface area to mass fraction burned ratio profile for a predetermined solid structure [172].

14.4.2.2 Additive manufacturing

Additive manufacturing (AM), commonly referred to as 3D printing, involves building objects using digital computer-aided design data [173]. AM can be used to deposit energetic materials into complex geometries, allowing for performance improvements. Other advantages include minimising wastage, miniaturisation, weight reduction, and integration of propellant and components becoming possible [8, 174]. Small series of specialised ordnance 'on demand' could be produced [175]. Some of the disadvantages include the presence of voids [176] and the time necessary to produce large amounts of propellants, for example. AM is mainly suited for medium- and large-calibre applications, however there may be some niche applications for small calibres.

The techniques for the 3D printing of energetic materials are described elsewhere, and is out of the scope of this chapter. However, briefly, some of the following techniques for AM can be used for energetic materials: extrusion 3D printing [177, 178], selective laser sintering, and vat photopolymerisation [179]. The choice of method primarily depends on the type of material, resolution, structural integrity, costs, etc [8].

A direct ink-writing method was developed to fabricate complex 3D gradient structures by printing explosives using two different compositions of inks [180]. A direct extrusion technique was adopted to additively manufacture composite propellant grains. The propellant slurry included ammonium perchlorate as an oxidiser, ferric oxide as a catalyst, HTPB as fuel/binder, dioctyl adipate as a plasticiser, and isophorone di-isocyanate as a curing agent [173].

Concerning 3D-printing technologies using photo-curable matrices, some of the main components are the energetic filler, such as RDX, an acrylate binder, plasticiser, photo-initiator, dispersant, colorant, and photopolymer. A photo-curable binder matrix can consist of a combination of (multi-)functional monomers, oligomers, and a photo-initiator [181]. Straathof *et al* developed a photo-curable energetic resin for photopolymerisation AM, consisting of 50 wt % RDX, 25 wt% acrylate binder, and 25 wt% energetic plasticiser. The 3D-printed propellant was fired in a 30 mm Gau-8 gun [179]. Similar work was carried out by Yang *et al.* The formulation was composed of 50 wt% RDX, 25 wt% epoxy acrylate, 12.5 wt% Bu-NENA, 12.5 wt% reactive diluent, and additives. A 30 mm gun test was also performed on the propellant, unfortunately resulting in low muzzle velocity [8]. Li *et al* developed a composite propellant composed of poly-3-nitratomethyl-3-methyloxetane (APNIMMO) and CL-20 was 3D printed by stereolithography (SLA), a type of vat photopolymerisation, offering an improved thermodynamic energy and burn rate with respect to the inert binder [181].

McClain *et al* identified a photopolymer that is compatible with ammonium perchlorate composite propellant and Vibration Assisted Printing (VAP) [182].

McClain *et al* determined the effect of curing parameters on solid propellant properties. The photo-initiator, phenylbis(2,4,6-trimethylbenzoyl) phosphine oxide, is soluble in hexanediol diacrylate (HDDA) polymer, which promotes uniform mixing. The polybutadiene urethane acrylate served as the oligomer [178]. A novel energetic acrylate-terminated oligomer was designed and prepared. A new 3D-printed gun propellant was fabricated by SLA [181].

However, so far with these examples of 3D printing of propellants, the energy is below the traditional level required. There is a need for special photo-curable energetic resins without degrading the performance of the resulting compositions.

14.4.2.3 Foamed propellants

Several groups have described the production of foamed propellants [183–187].

The presence of the cell structure enhanced the specific surface area, promoting the combustion properties of the gun propellants. Supercritical carbon dioxide (SC-CO_2) foaming technology is available for the preparation of microcellular gun propellants.

The idea of foamed propellants is to be able to adjust the combustion properties through the micropore structure. Some experimental variables include the desorption time, foaming temperature, poly(neopentanediol adipate) (NA) content, desensitization time, and desensitization temperature

SC-CO_2 is the physical foaming agent and it escapes from the propellant at the end of the foaming process. However, it was not simple to control the cell nucleation and growth.

Gu *et al* and Yajun *et al* [188] made microporous oblate spherical powder foamed within the cell structure by SC-CO_2, and then NA was employed as the deterrent to modify the energy-releasing rates in the surface. Both groups found that the desired progressive combustion and the combustion properties could be adjusted by controlling the desorption time and foaming temperature [189]. The burning behavior of foamed propellants deviates from Vieille's law, however [190].

Chai *et al* [191] analysed the macro- and microstructures of NC samples with different nitrogen contents. In the main stage of pyrolysis, NC with higher nitrogen content will produce more kinds of gas, and more cyclization and recombination forms will occur in the post-reaction stage of pyrolysis [58].

14.4.2.4 Caseless ammunition

With caseless ammunition the metal cartridge case, typically made from brass or steel, is eliminated. This could reduce the weight and cost of ammunition as well as eliminate the need to extract and eject the empty case after firing, but it often comes at the cost of several changes in the weapon's system to ensure sufficient air tightness of the chamber and avoid unwanted cook off of the charge when the gun is hot after having already fired several rounds. Broadly speaking, caseless ammunition is not a new concept, and the fragility of the caseless ammunition is also an issue; probably the origin of semi-caseless concepts using either a light polymeric case or combustible cartridges was using a cartridge case as a propellant, which left no significant residue in the action after the gun had been fired. One of the military weapons, an early rifle used in the 19th century, which loads the cartridge in the back of the barrel, is a gun called the 'Dreyse'. The cartridge contains powder, a percussion cap, and the bullet wrapped in a paper cartridge [192]. From the early '60s, many manufacturers have taken a particular interest in the concept, and caseless or semi-caseless guns have been proposed, e.g. the Russian pistol VAG-73 in 1973, the German machine gun 4.92 mm in 1970, or the Italian Benelli CB-M2 submachine gun. More recently, in the 2000s, the US Lightweight Small Arms Technologies project focused on the reduction of the volume and weight of ammunition by 50%, targeting 5.56, 7.62, and 25 mm applications using polymer Cased Telescoped Ammunition and caseless ammunition variants [193–195].

Rivera *et al* describe 7.62 mm caseless-coreless ammunition with similar ballistic parameters to those of the ones produced by the 9 mm × 19 mm full metal jacket ammunition [196].

Driscoll *et al* describe a caseless ammunition design for a medium-calibre (40 mm) air-to-air cannon deployed in January 1939. The propellant charge and initiator were incorporated into the rear section of the projectile [197].

Polymeric case ammunition are commercialised today for civilian and military markets and seems to be an attractive alternative to conventional ammunitions.

14.5 Conclusion and perspectives

The domain of small-calibre gun propellants has been conservative for many decades. In the past few years, research has been moving more rapidly and many new axes for research have advanced. We have reviewed new formulations, such as green propellants and new charges for gun propellants, with new and improved techniques and processes. From the selected developments presented, the computation of optimised charges combined with the capabilities of AM might certainly be one of the avenues with the most appeal for the future, having the potential to establish new foundations for different concepts (i.e. hypervelocity) offering disruptive progress and advantage for our armies in terms of performance. Moreover, this applies not only to small-calibre applications but the whole range of projectiles and guns for which the impact might be even more important.

The testing of propellants is crucial, especially to ensure chemical and ballistic stability as these can affect the safety of military personnel, for example. Autocatalytic runaway reactions or unwanted ignition of stockpiles by external stimuli must be prevented. The safety of use of future formulations or charges will obviously have to be assessed and will require the development of new techniques and thorough characterisation of the kinetics involved. With an increase of consideration for the environment and green chemistry, not only the end product and its evolution during the lifetime of the product should present friendly profiles but, rather, the whole life cycle should be considered and intermediate preparations, synthesis, or sourcing of raw materials taken into account when assessing said 'green' factor. The implementation of European legislation, namely REACH, is a positive step and helps with the protection of human health and the environment, and should be supported as to improve its efficiency.

In conclusion, the domain of small-calibre propellant powders should evolve in the coming years, with greener and more performant propellant powders. New processes, analytical techniques for testing, and charges are being developed, and the safety of military personnel, for example, should be greatly improved, with a reduction in the number and severity of accidents.

References

[1] Burtsell A T 1941 The chemistry of powder and explosives *J. Chem. Educ.* **18** 500

[2] Brooks W, Keller R, Douglass H and Collins J March 1972 Solid propellant grain design and internal ballistics *NASA Space Vehicle Design Criteria (Chemical Propulsion)* NASA SP-8076 National Aeronautics and Space Administration

[3] Dejeaifve A, Fantin A, Monseur L and Dobson R 2018 Making progress towards «green» propellants *Propellants Explos. Pyrotech.* **43** 831–7

[4] Dejeaifve A, Sarbach A, Roduit B, Folly P and Dobson R 2020 Making progress towards "green" propellants—part II *Propellants Explos. Pyrotech.* **45** 1185–93

[5] Fordham S 1966 *High Explosives and Propellants* (Oxford: Pergamon)

[6] Bailey A and Murray S 1989 *Explosives, Propellants and Pyrotechnics(Land Warfare)* vol 2 (London: Brassey's)

[7] Vogelsanger B, Schädeli U, Von Mühlenen U and Antenen D 2013 *New Propellant Technologies for Small Calibre Ammunition* (https://ndiastorageblobcoreusgovcloudapinet/ndia/2013/IMEM/W15931_Vogelsangerpdf)

[8] Yang W, Hu R, Zheng L, Yan G and Yan W 2020 Fabrication and investigation of 3D-printed gun propellants *Mater. Des.* **192** 108761

[9] Liang H, Ding Y, Li S and Xiao Z 2021 Combustion performance of spherical propellants deterred by energetic composite deterring agents *ACS Omega* **6** 13024–32

[10] Chen S, Feng Q, Nan F, He W, Liu Z and Ma Y 2020 Optimization design of propellant charge for a medium and large caliber high chamber pressure gun *IOP Conf Series: Earth and Environmental Science.* **571** 012143

[11] Jeunieau L, Lefebvre M H, Papy A, Pirlot M C, Guillaume P and Reynaud C 2002 Closed vessel test: influence of the ignition method on the combustion rate *33rd Int. Annual Conf. of ICT*

[12] Parate B A 2020 Propellant actuated device for parachute deployment during seat ejection for an aircraft application *HighTech Innov.* J **1** 112–20

[13] Boulkadid K M, Trache D, Krai S, Lefebvre M H, Jeunieau L and Dejeaifve A 2020 Estimation of the ballistic parameters of double base gun propellants *Propellants Explos. Pyrotech.* **45** 751–8

[14] Leciejewski Z and Surma Z 2019 On a certain method of determining the burning rate of gun propellant *Cent. Eur. J. Energ. Mater.* **1.6** 433–48

[15] NATO S 1997 *Definition and Determination of Ballistic Properties of Gun Propellants: edition 2* STANAG 4115 NATO (MAS/257-LAND/4115)

[16] Xiao Z, Ying S and He W 2015 Interior ballistic prediction of gun propellants based on experimental pressure-apparent burning rate model in closed vessel *Sci. Technol. Energetic Mater.* **76** 1–7

[17] NATO 2003 *Data Requirements and Tests for the Qualification of Explosive Materials: edition 2* AOP-7 NATO

[18] NATO 2008 *Principles and Methodology for the Qualification of Explosive Materials for Military Use: edition 3* STANAG 4170 NATO

[19] Swaszek S and Baker E 2019 Energetics and munitions suitability for gun launch (unclassified IMEMG report) https://imemg.org/wp-content/uploads/2019/11/22229-Paper-Energetics-and-Munitions-Suitability-for-Gun-Launch.pdf

[20] Parate B A, Deodhar K D and Dixit V K 2021 Qualification testing, evaluation and test methods of gas generator for IEDs application *Def. Sci. J.* **71** 462–9

[21] United Nations 2009 UN Recommendations on the Transport of Dangerous Goods—Model Regulations 21st revised version

[22] Trache D and Tarchoun A F 2019 Analytical methods for stability assessment of nitrate esters-based propellants *Crit. Rev. Anal. Chem.* **49** 415–38

[23] Davenas A 2012 *Solid Rocket Propulsion Technology* (Oxford: Pergamon)

[24] Chelouche S, Trache D, Tarchoun A F, Khimeche K and Mezroua A 2021 Stability assessment for double base rocket propellant during long natural/artificial aging using various methods and kinetic modeling ed D Trache, F Benaliouche and A Mekki *Materials Research and Applications: Select Papers from JCH8–2019* (Singapore: Springer Singapore) pp 93–122

[25] Sućeska M 1995 Combustion of explosives *Test Methods for Explosives. High-Pressure Shock Compression of Condensed Matter* (New York: Springer) https://doi.org/10.1007/978-1-4612-0797-9_3

[26] NATO 2020 *Multi-Calibre Manual of Proof and Inspection (M-CMOPI) for NATO Small Arms Ammunition* AEP-97 NATO

[27] Tunnell R 2017 Overview and appraisal of analytical techniques for aging of solid rocket propellants.in chemical rocket propulsion *Chemical Rocket Propulsion.* (Cham: Springer) pp 743–69

[28] Zeng J, Qi J, Bai F, Yu J and Shih W 2014 Analysis of ethyl and methyl centralite vibrational spectra for mapping organic gunshot residues *Analyst* **2014** 4270–8

[29] Lopez-Lopez M and Garcıa-Ruiz C 2014 Infrared and Raman spectroscopy techniques applied to identification of explosives *Trends Anal. Chem.* **54** 36–44

[30] Heil M, Wimmer K and Bohn M A 2017 Characterization of gun propellants by long-term mass loss measurements *Prop. Explos. Pyrotech.* **42** 706–11

[31] Bladek J, Cudzilo S, Pietrzyk S and Wilker S 2010 A novel method for testing propellant stabilizers *Cent. Eur. J. Energetic Mater.* **7** 281–87

[32] Bohn M A and Eisenreich N 1997 kinetic modelling of the stabilizer consumption and of the consecutive products of the stabilizer in a gun propellant *Propellants Explos. Pyrotech.* **22** 125–36

[33] Bohn M A 2002 Kinetic description of mass loss data for the assessment of stability, compatibility and aging of energetic components and formulations exemplified with epsilon-CL20 *Propellants Explos. Pyrotech.* **27** 125–35

[34] Bohn M A and Volk F 1992 Aging behavior of propellants investigated by heat generation, stabilizer consumption and molar mass degradation *Propellants Explos. Pyrotech.* **17** 171–8

[35] Bohn M A 2017 *Principles of ageing of double base propellants and its assessment by several methods following propellant properties* STO-MP-AVT-268 NATO

[36] Trache D and Tarchoun A F 2018 Stabilizers for nitrate ester-based energetic materials and their mechanism of action: a state-of-the-art review *J. Mater. Sci.* **53** 100–23

[37] Coen E 2012 *Investigation of Alternative Stabilisers for Nitrocellulose Gun Propellants* (Shrivenham: The Defence Academy Year Book) pp 199–259

[38] De Klerk W 2015 Assessment of stability of propellants and safe lifetimes *Propellants, Explos Pyrotech* **40** 388–93

[39] Krumlinde P, Ek S, Tunestal E and Hafstrand A 2016 Synthesis and characterization of novel stabilizers for nitrocellulose-based propellants *Propellants Explos. Pyrotech.* **42** 78–83

[40] Wilker S, Heeb G, Vogelsanger B, Petržílek J and Skládal J 2007 Triphenylamine—a 'New' stabilizer for nitrocellulose based propellants—Part I: chemical stability studies *Propellants Explos. Pyrotech.* **32** 135–48

[41] Trache D, Benaliouche F and Mekki A 2021 *Materials Research and Applications: Select Papers from JCH8-2019* (New York: Springer)

[42] Kadiresh P N and Sridhar B T N 2008 Experimental evaluation and simulation on aging characteristics of aluminised AP-HTPB composite solid propellant *Mater. Sci. Technol.* **24** 406–12

[43] Tırak E, Moniruzzaman M, Degirmenci E and Hameed A 2019 Closed vessel burning behavior and ballistic properties of artificially-degraded spherical double-base propellants stabilized with diphenylamine *Thermochim. Acta* **680** 178347

[44] Namir H, Mojca K, Zahida A, Azra B H and Rasim O 2021 Analysis of stability of naturally aged single base propellants *Afr. J. Pure Appl. Chem.* **15** 1–7

[45] Druet L and Asselin M J 1988 A review of stability test methods for gun and mortar propellants, II: stability testing and surveillance *Energy Mater.* **6** 215–54

[46] Guo S, Wang Q, Sun J, Liao X and Wang Z 2009 Study on the influence of moisture content on thermal stability of propellant *J. Hazard. Mater.* **168** 536–41

[47] Wilker S, Guillaume P, Lefèbvre M, Chevalier S, Jeunieau L and Pantel G *et al* 2003 Stability analyses of rolled ball propellants. Part 1: microcalorimetric studies and stabilizer depletion *Int. Annu. Conf. ICT* 34

[48] Sovizi M, Hajimirsadeghi S and Naderizadeh B J 2009 Effect of particle size on thermal decomposition of nitrocellulose *Hazard Mater* **168** 1134–9

[49] Tomaszewski W, Cieślak K and Zygmunt A 2015 Influence of processing solvents on decomposition of nitrocellulose in smokeless powders studied by heat flow calorimetry *Polym. Degrad. Stab.* **111** 169–75

[50] Katoh K, Le A, Itoh M, Arai M and Tamura M 2003 Study on the spontaneous ignition of cellulose nitrate *Sci. Tech. Energ. Mater.* **64** 236

[51] Vogelsanger B, Ossola B, Schädeli U, Antenen D and Ryf K 2001 Ballistic shelf life of propellants for medium and small calibre ammunition—influence of deterrent diffusion and nitrocellulose degradation *19th Int. Symp. of Ballistics (Interlaken, Switzerland, 7–11 May)*

[52] Quinchon J, Tranchant J and Nicolas M 1987 *Les poudres pour armes Les poudres, propergols et explosifs* vol 3 (Paris: Tec & Doc)

[53] McClain M S, Afriat A, Montano B J, Ray S, Rhoads J and Gunduz I E *et al* 2020 Investigation of additively manufactured layered composite solid propellant *AIAA Scitech 2020 Forum*

[54] Figueirôa de Souza Defanti B, Mendonça-Filho L G d and Nichele J 2020 Effect of ageing on the combustion of single base propellants *Combust. Flame* **221** 212–18

[55] Bobic N, Simonovic R, Drmanic S, Milic S, Nikolic J and Stoiljkovic S *et al* 2017 The influence of migration processes in gunpowder charge on the quality of mortar ammunition *Hem. Ind.* **71** 231–40

[56] Suceska M, Mušanic S, Fiamengo I, Bakija S, Bakic A and Kodvanj J 2010 Study of mechanical properties of naturally aged double base rocket propellants *Central Eur. J. Energ. Mater.* **7** 47–60
[57] Musanic S and Suceska M 2013 Dynamic mechanical properties of artificially aged double base rocket propellant and the possibilities for the prediction of their service lifetime *Central Eur. J. Energ. Mater.* **10** 225–44
[58] Zhang H, Xu B, Liu Z and Liao X 2020 Effect of nitrate nitrogen content on properties of modified single-base propellant *J. Phys. Conf. Ser.* **1676** 012011
[59] Richards L A, Nash A, Phipps M J S and de Leeuw N H 2018 A molecular dynamics study of plasticiser migration in nitrocellulose binders *New J. Chem.* **42** 17420–8
[60] Yu Z, Wang W, Yao W, Zhang W, Xie W and Liu Y *et al* 2021 Simulation for the migration of nitrate ester plasticizers in different liners contacting with propellant by molecular dynamics *J. Energetic Mater.* **39** 74–84
[61] Zhang R, Rui X, Li C, Wang Y, Zhao X and Dong X 2020 A calculation method of interior ballistic two-phase flow considering the compression and fracture process of propellant bed *Int. Commun. Heat Mass Transfer* **115** 104601
[62] Glorian J and Baschung B 1721 Detailed kinetic modeling of nitroglycerin low-temperature decomposition *J. Phys. Conf. Ser.* **2021** 012002
[63] Boulkadid M K, Lefebvre M H, Jeunieau L and Dejeaifve A 2020 Burning rate of artificially aged solid double-base gun propellants *J. Energetic Mater.* **38** 1–19
[64] Jeunieau L, Lefebvre M H and Guillaume P 2007 Ballistic stability of a spherical propellant: comparison with a flattened spherical propellant *Cent. Eur. J. Energetic Mater.* **4** 31–44
[65] Wang Y-w, Zhu W-f, Di J-w and Hu X-h 2018 Study on the analysis method on ballistic performance of deterred propellant with large web size in large caliber artillery *Defence Technol.* **14** 522–6
[66] Chelouche S, Trache D, Pinho S P and Khimeche K 2019 Experimental and modeling studies of binary organic eutectic systems to be used as stabilizers for nitrate esters-based energetic materials *Fluid Phase Equilib.* **498** 104–15
[67] Boers M N and De Klerk W P C 2005 Lifetime prediction of EC, DPA, Akardite II and MNA stabilized triple base propellants, comparison of heat generation rate and stabilizer consumption *Propellants Explos. Pyrotech.* **30** 356–62
[68] BBC News 2011 Cyprus: Zygi naval base munitions blast kills 12 https://wwwbbccom/news/world-europe-14102253
[69] Roduit B, Luyet C A, Hartmann M, Folly P, Sarbach A and Dejeaifve A *et al* 2019 Continuous monitoring of shelf lives of materials by application of data loggers with implemented kinetic parameters *Molecules* **24** 2217
[70] Bird D T and Ravindra N M 2021 A review: advances and modernization in U.S Army gun propellants *JOM* **73** 1144–64
[71] NAP National Research Council 2004 *Advanced Energetic Materials* (Washington, DC: The National Academies Press)
[72] Badgujar D M, Talawar M B and Mahulikar P P 2017 Review of promising insensitive energetic materials *Cent. Eur. J. Energ. Mater.* **14** 821–43
[73] Buldum G, Bismarck A and Mantalaris A 2018 Recombinant biosynthesis of bacterial cellulose in genetically modified *Escherichia coli Bioprocess. Biosyst. Eng.* **41** 265–79
[74] Torres F G, Arroyo J J and Troncoso O P 2019 Bacterial cellulose nanocomposites: An all-nano type of material *Mater. Sci. Eng.* C **98** 1277–93

[75] Garland N T, McLamore E S, Gomes C, Marrow E A, Daniele M A and Walper S *et al* 2017 10—Synthesis and applications of cellulose nanohybrid materials ed V K Thakur, M K Thakur and A Pappu *Hybrid Polymer Composite Materials* (Sawston: Woodhead Publishing) pp 289–320

[76] Wang J, Tavakoli J and Tang Y 2019 Bacterial cellulose production, properties and applications with different culture methods—a review *Carbohydrate Polym.* **219** 63–76

[77] Tsouko E, Kourmentza C, Ladakis D, Kopsahelis N, Mandala I and Papanikolaou S *et al* 2015 Bacterial cellulose production from industrial waste and by-product streams *Int. J. Mol. Sci.* **16** 14832–49

[78] Wang Y, Jiang L, Dong J, Li B, Shen J and Chen L *et al* 2020 Three-dimensional network structure nitramine gun propellant with nitrated bacterial cellulose *J. Mater. Res. Technol.* **9** 15094–101

[79] Deepa B, Abraham E, Cordeiro N, Mozetic M, Mathew A P and Oksman K *et al* 2015 Utilization of various lignocellulosic biomass for the production of nanocellulose: a comparative study *Cellulose* **22** 1075–90

[80] Islam M, Patrucco A, Montarsolo A and Alam M 2014 Preparation of nanocellulose *AATCC J. Res.* **1** 17–23

[81] Liang H and Hu X 2016 A quick review of the applications of nano crystalline cellulose in wastewater treatment *J. Bioresour. Bioprod.* **1** 199–204

[82] Vilarinho F, Sanches Silva A, Vaz M F and Farinha J P 2018 Nanocellulose in green food packaging *Crit. Rev. Food Sci. Nutr.* **58** 1526–37

[83] Salimi S, Sotudeh-Gharebagh R, Zarghami R, Chan S Y and Yuen K H 2019 Production of nanocellulose and its applications in drug delivery: a critical review *ACS Sustain. Chem. Eng.* **7** 15800–27

[84] Yadav C, Saini A, Zhang W, You X, Chauhan I and Mohanty P *et al* 2021 Plant-based nanocellulose: a review of routine and recent preparation methods with current progress in its applications as rheology modifier and 3D bioprinting *Int. J. Biol. Macromol.* **166** 1586–616

[85] Athukoralalage S S, Balu R, Dutta N K and Roy Choudhury N 2019 3D bioprinted nanocellulose-based hydrogels for tissue engineering applications: a brief review *Polymers* **11** 898

[86] Ferreira F V, Pinheiro I F, de Souza S F, Mei L H I and Lona L M F 2019 Polymer composites reinforced with natural fibers and nanocellulose in the automotive industry: a short review *J. Compos. Sci.* **3** 51

[87] Norrrahim M, Kasim N, Knight V F, Ujang F, Janudin N and Razak M *et al* 2021 Nanocellulose: the next super versatile material for the military *Mater. Adv.* **2** 1485–506

[88] Pant S, Santosh M, Banerjee S and Khanna P 2013 Single step synthesis of nitro-functionalized hydroxyl-terminated polybutadiene *Propellants Explos. Pyrotech.* **38** 748–53

[89] Abdullah M, Gholamian F and Zarei A 2014 Investigation of composite solid propellants based on nitrated hydroxyl-terminated polybutadiene binder *J. Propul. Power* **30** 862–4

[90] Gayathri S and Reshmi S 2017 Nitrato functionalized polymers for high energy propellants and explosives: recent advances *Polym. Adv. Technol.* **28** 1539–50

[91] Kim K, Kerndt C, Adnan G and Schaller D 2021 Nitroglycerin *StatPearls* (Treasure Island, FL: StatPearls Publishing)

[92] Mendonça-Filho L G, Rodrigues R L B, Rosato R, Galante E B F and Nichele J 2019 Combined evaluation of nitrocellulose-based propellants: toxicity, performance, and erosivity *J. Energetic Mater.* **37** 293–308

[93] Wang G, Xu Y, Zhang W and Gong X 2019 A theoretical study of polyethylene glycol polynitrates as potential highly energetic plasticizers for propellants *Cent. Eur. J. Energet. Mat.* **16** 194–215

[94] Fu Y, Zhu Y, Liu Z, Xu B, Shen J and Wang Y *et al* 2021 Temperature sensitivity coefficients of RDX-based propellants and their mixed charges *Propellants Explos. Pyrotech.* **46** 1589–97

[95] Bohn M A 2006 Insensitivity aspects of NC bonded and DNDA plasticizer containing GPs *Int. Annual Conf.*

[96] Wang Y, Liu Y, Zhao B, Gao F, Liu W and Chen B 2021 Novel geminal dinitro esters as energetic plasticizers for GAP binder *Ind. Eng. Chem. Res.* **60** 3992–8

[97] Cumming A S and Johnson M S 2019 *Energetic Materials and Munitions: Life Cycle Management, Environmental Impact, and Demilitarization* (New York: Wiley)

[98] EU 2014 *On the finalisation of the restriction process on the four phthalates (DEHP, DBP, BBP and DIBP)* C 260/1 European Commission Communication

[99] Agency for Toxic Substances and Disease Registry (https://wwwatsdrcdcgov/2017)

[100] Rodrigues R, Nichele J, França T and Mendonça Filho L 2018 Predicao de toxicidade dos estabilizantes usuais em propelentes à base de nitrocellulose e de seus principais produtos de degradacao *Quim. Nova* **41** 867–73

[101] Lussier L-S, Bergeron E and Gagnon H 2006 Study of the daughter products of akardite-II *Propellants Explos. Pyrotech.* **31** 253–62

[102] Lussier L S and Gagnon H 1996 Development of modern methods for determination of stabilizers in propellants ADA637697 Defence Research Establishment Report

[103] Apatoff J and Norwitz G 1973 *Role of diphenylamine as a stabilizer in propellants; analytical chemistry of diphenylamine in propellants (a survey report)* AD0783499 Frankford Arsenal Test Report

[104] Curtis N J and Berry P 1989 Derivatives of ethyl centralite in australian gun propellants *Propellants Explos. Pyrotech.* **14** 260–5

[105] Low H 1974 Nitroso compounds *Arch. Environ. Health* **29** 256–60

[106] Drzyzga O, Jannsen S and Blotevogel K H 1995 Toxicity of diphenylamine and some of its nitrated and aminated derivatives to the luminescent bacterium *Vibrio fischeri Ecotoxicol. Environ. Safety* **31** 149–52

[107] Drzyzga O 2003 Diphenylamine and derivatives in the environment: a review *Chemosphere* **53** 809–18

[108] Damseaux C, Scholl G, Damblon C, Dejeaifve A, Dobson R and Ma X *et al* 2021 Identification of the degradation products from α-ionone used as stabiliser in 'green' propellants through its lifetime *Propellants Explos. Pyrotech* **46** 1653–61

[109] Fodil Cherif M, Trache D, Benaliouche F, Chelouche S, Tarchoun A F and Mezroua A 2020 Effect of Kraft lignins on the stability and thermal decomposition kinetics of nitrocellulose *Thermochim. Acta* **692** 178732

[110] Fodil Cherif M, Trache D, Benaliouche F, Tarchoun A F, Chelouche S and Mezroua A 2020 Organosolv lignins as new stabilizers for cellulose nitrate: thermal behavior and stability assessment *Int. J. Biol. Macromol.* **164** 794–807

[111] Graham D 2020 Biological production of thermal stabilizers *SERDP and ESTCP Symp. (Oak Ridge, TN)*

[112] Rodrigues R L B, Gomes Buitrago P A, Nakano N L, Peixoto F C, Lemos M F and França T C C *et al* 2021 Can green nitrocellulose-based propellants be made through the replacement of diphenylamine by the natural product curcumin? *J. Energetic Mater.* **40** 218–41

[113] Rodrigues R, Lemos M, França T and Mendonça Filho L 2019 Development of nitrocellulose-based propellants with natural stabilizers *J. Aerosp. Technol. Manag.* **11** 3–6

[114] de Araújo N A, Pinto J C, Caldeira A B, dos Santos Cople Lima K and dos Santos Lima A L 2019 Synthesis and characterization of binders for propellants *Macromol. Symp.* **383** 1800062

[115] Kirchner D B, Gaydos J and Battigelli M 2006 Combustion products of propellants and ammunition *Textbook of Military Medicine: III. Disease and the Environment* ed R Zajtchuk (Washington, DC: Borden Institute) ch 10 pp 359–96

[116] Bathelt H and Volk F 1996 The ICT thermodynamic code (ICT-Code) *Energetic Materials —Technology, manufacturing and processing: 27th Int. Annual Conf. of ICT (June 25–28)*

[117] Moxnes J F, Jensen T L, Smestad E, Unneberg E and Dullum O 2012 Lead free ammunition without toxic propellant gases *Propellants Explos. Pyrotech.* **38** 255–60

[118] Meuken D 2013 Effects of combustion products of small caliber munitions *HETMC (Human and Environmental Toxicity of Munition-related Compounds) Symp. (TNO, Rijksweg)*

[119] Stromseng A E, Ljønes M, Bakka L and Mariussen E 2009 Episodic discharge of lead, copper and antimony from a Norwegian small arm shooting range *J. Environ. Monit.* **11** 1259–67

[120] Mariussen E, Fjellsbø L, Frømyr T, Vaa Johnsen I, Karsrud T and Voie O 2019 *Toxic effects of gunshot fumes from different ammunitions for small arms* STO-MP-AVT-322 NATO

[121] Mariussen E, Fjellsbø L, Frømyr T, Johnsen I, Karsrud T and Voie Ø 2021 Toxic effects of gunshot fumes from different ammunitions for small arms on lung cells exposed at the air liquid interface *Toxicol. in Vitro* **72** 105095

[122] Ferreira C, Ribeiro J, Almada S and Freire F 2017 Environmental assessment of ammunition: the importance of a life-cycle approach *Propellants Explos. Pyrotech.* **42** 44–53

[123] Dahiwale S M, Bhongale C, Roy S, Navle P B and Asthana S N 2019 Studies on ballistic parameters of di-butyl phthalate-coated triple base propellant used in large caliber artillery gun ammunition *J. Energetic Mater.* **37** 98–109

[124] Thiboutot S, Ampleman G, Brochu S, Diaz E, Poulin I and Martel R *et al* 2012 Environmental characterization of military training ranges for munitions-related contaminants: understanding and minimizing the environmental impacts of live-fire training **11** 17–57

[125] DiGiulian T 2009 Naval propellants: a brief overview (NavWeaps: Naval Weapons, Naval Technology and Naval Reunions) http://www.navweaps.com/index_tech/tech-100.php

[126] Rahman H, Niemann D and Bugajski A 2021 Association of environmental toxic metals with high sensitivity C-reactive protein: a cross-sectional study *Occup. Dis. Environ. Med.* **9** 173–84

[127] Treu G, Drost W and Stock F 2020 An evaluation of the proposal to regulate lead in hunting ammunition through the European Union's REACH regulation *Environ. Sci. Europe* **32** 68

[128] Karta Technologies I 1999 Lead-free small arms ammunition study *United States Air Force IERA*

[129] Sujetovienė G and Česynaitė J 2019 Assessment of toxicity to earthworm *Eisenia fetida* of lead contaminated shooting range soils with different properties *Bull. Environ. Contam. Toxicol.* **103** 559–64

[130] Manning L K, Wünschmann A, Armién A G, Willette M, MacAulay K and Bender J B *et al* 2018 Lead intoxication in free-ranging bald eagles (*Haliaeetus leucocephalus*) *Vet. Pathol.* **56** 289–99

[131] Eid R, Guzman D, Keller K, Wiggans K, Murphy C and LaDouceur E *et al* 2016 Choroidal vasculopathy and retinal detachment in a bald eagle (*Haliaeetus leucocephalus*) with lead toxicosis *J. Avian. Med. Surg.* **30** 357–63

[132] Golden N H, Warner S E and Coffey M J 2016 A review and assessment of spent lead ammunition and its exposure and effects to scavenging birds in the United States *Rev. Environ. Contam. Toxicol.* **237** 123–91

[133] Jacobson E, Carpenter J W and Novilla M 1977 Suspected lead toxicosis in a bald eagle *J. Am. Vet. Med. Assoc.* **171** 952–4

[134] ECHA 2021 Towards sustainable outdoor shooting and fishing—ECHA proposes restrictions on lead use (ECHA/NR/21/07) https://echa.europa.eu/-/towards-sustainable-outdoor-shooting-and-fishing-echa-proposes-restrictions-on-lead-use

[135] Hampton J O, DeNicola A J and Forsyth D M 2020 Assessment of lead-free .22 LR bullets for shooting European rabbits *Wildlife Soc. Bull.* **44** 760–5

[136] Chakraborthy T K, Raha K C, Omprakash B and Singh A 2004 A study on gun propellants based on Butyl-NENA *J. Energetic Mater.* **22** 41–53

[137] Damse R S and Sikder A K 2009 Suitability of nitrogen rich compounds for gun propellant formulations *J. Hazard. Mater.* **166** 967–71

[138] Wu Y, Yi Z, Luo Y, Ge Z, Du F and Chen S *et al* 2017 Fabrication and properties of glycidyl azide polymer-modified nitrocellulose spherical powders *J. Therm. Anal. Calorim.* **129** 1555–62

[139] Meng X and Xiao Z 2018 Synthesis, thermal properties and sensitivity of ladder-like nitrocellulose grafted by polyethylene glycol *Propellants Explos. Pyrotech.* **43** 300–7

[140] Comtois E, Dubois C and Favis B D 2020 Linear burning rate of double base propellant containing azidodeoxycellulose *J. Energetic Mater.* **40** 136–52

[141] Meng X, Pu C, Cui P and Xiao Z 2020 Preparation, thermal and sensitivity properties of nano-sized spherical nitrocellulose composite crystal *Propellants Explos. Pyrotech.* **45** 1194–203

[142] Anniyappan M, Talawar M B, Sinha R K and Murthy K P S 2020 Review on advanced energetic materials for insensitive munition formulations *Combust. Explos. Shock Waves* **56** 495–519

[143] Unger C C, Holler M, Krumm B and Klapötke T M 2020 Oxygen-rich bis(trinitroethyl esters): suitable oxidizers as potential ammonium perchlorate replacements *Energy Fuels* **34** 16469–75

[144] Abd-Elghany M, Klapötke T M, Krumm B and Elbeih A 2018 Higher performance and safer handling: formulation based on 2,2,2-trinitroethyl formate and nitrocellulose *ChemPlusChem* **83** 128–31

[145] Yin P, Zhang Q, Zhang J, Parrish D A and Shreeve J M 2013 N-trinitroethylamino functionalization of nitroimidazoles: a new strategy for high performance energetic materials *J. Mater. Chem.* A **2013** 7500–10

[146] Baxter A F, Martin I, Christe K O and Haiges R 2018 Formamidinium nitroformate: an insensitive RDX alternative *J. Am. Chem. Soc.* **140** 15089–98

[147] Kumar D, Imler G H, Parrish D A and Shreeve J n M 2017 3,4,5-Trinitro-1-(nitromethyl)-1H-pyrazole (TNNMP): a perchlorate free high energy density oxidizer with high thermal stability *J. Mater. Chem.* A **5** 10437–41

[148] Tang J, Yang P, Yang H, Xiong H, Hu W and Cheng G 2020 A simple and efficient method to synthesize high-nitrogen compounds: incorporation of tetrazole derivatives with N5 chains *Chem. Eng. J.* **386** 124027

[149] Gao H, Zhang Q and Shreeve J 2020 Fused heterocycle-based energetic materials (2012–2019) *J. Mater. Chem.* A **8** 4193–216

[150] Yongjin C and Shuhong B 2019 High Energy Density Material (HEDM)—progress in research azine energetic compounds *Johnson Matthey Technol. Rev.* **63** 51–72

[151] Singh R, Gao H, Meshri D and Shreeve J 2007 Nitrogen-rich heterocycles *Struct. Bond.* **125** 35–83

[152] Chavez D E, Parrish D A and Mitchell L 2016 Energetic trinitro- and fluorodinitroethyl ethers of 1,2,4,5-tetrazines *Angew. Chem. Int. Ed.* **55** 8666–9

[153] Yu Q, Wang Z, Wu B, Yang H, Ju X and Lu C *et al* 2015 A study of N-trinitroethyl-substituted aminofurazans: high detonation performance energetic compounds with good oxygen balance *J. Mater. Chem.* A **3** 8156–64

[154] Tarchoun A F, Trache D, Klapötke T M and Khimeche K 2020 Tetrazole-functionalized microcrystalline cellulose: a promising biopolymer for advanced energetic materials *Chem. Eng. J.* **400** 125960

[155] Stierstorfer J, Klapotke T and Karaghiosoff K 2009 Advanced Energetic Materials based on 5-aminotetrazole—Synthesis, characterization, testing and scale-up *Dissertation zur Erlangung des Doktorgrades*

[156] Darnse R and Singh A 2003 Advanced concepts of the propulsion system for the futuristic gun ammunition *Def. Sci. J.* **53** 341–50

[157] Szala M, Szymańczyk L, Surma Z and Hara M 2020 Synthesis of nitrogen-rich compounds and their use in novel composite propellants *AIP Conf. Proc.* **2272** 120028

[158] Lawton B 2001 Thermo-chemical erosion in gun barrels *Wear* **251** 827–38

[159] Jaramaz S, Micković D and Elek P 2010 Determination of gun propellants erosivity: experimental and theoretical studies *Exp. Therm. Fluid Sci.* **34** 760–5

[160] Kimura J 1996 Hydrogen gas erosion of high-energy LOVA propellants *16th Int. Symp. on Ballistics (San Francisco, CA)*

[161] Cote P and Wear R 2000 Gas–metal reaction products in the erosion of chromium-plated gun bores *Wear* **241** 17–25

[162] Sopok S, O'Hara P, Pflegl G, Dunn S, Coats D and Nickerson G 1999 Unified computer model for predicting thermochemical erosion in gun barrels *J. Propul. Power* **15** 601–12

[163] Walsh C M, Knott C D and Leveritt C S 2006 *Improved gun propellant with lower flame temperature for reduced erosion of gun barrels* US Patent 6984275 B1

[164] Desgardin N 2014 Poudres propulsives pour armes à tube à force élevée et effet érosif réduit *Europe*

[165] Desgardin N and Perut C 2002 *Gun propellant powders of high power and reduced erosive action* EP1364931 A3 Eurenco France, SNPE

[166] Lavoie J, Petre C-F, Paradis P-Y and Dubois C 2017 Burning rates and thermal behavior of bistetrazole containing gun propellants *Propellants Explos. Pyrotech.* **42** 149–57

[167] Lavoie J, Petre C-F, Durand S and Dubois C 2019 Stability and performance of gun propellants incorporating 3,6-dihydrazino-s-tetrazine and 5-aminotetrazolium nitrate *J. Hazard. Mater.* **363** 457–63
[168] Comtois E, Favis B D and Dubois C 2021 Linear burning rate and erosivity properties of nitrocellulose propellant formulations plasticized by glycidyl azide polymer and nitroglycerine *Propellants Explos. Pyrotech.* **46** 494–504
[169] Li S, Tao Z, Ding Y, Liang H, Zhao X and Xiao Z *et al* 2020 Gradient denitration strategy eliminates phthalates associated potential hazards during gun propellant production and application *Propellants Explos. Pyrotech.* **45** 1156–67
[170] Maag G and Klingenberg H-J 1996 Gun propulsion concepts. Part 11: solid and liquid propellants *Propellants Explos. Pyrotech.* **21** 1–7
[171] Fu E-f, Sun N and Xiao Z-g 2021 Effects of phase change material (PCM)-based nanocomposite additives on thermal decomposition and burning characteristic of high energy propellants containing RDX *Defence Technol.* **18** 557–566
[172] Conroy P, Wildman R, Robinette E, Schmidt J and Gaynor A 2017 *Propellant grain for optimizing the interior ballistic performance of a weapon* US 2018/0273438 A1 US Army Research Laboratory
[173] Chandru R A, Balasubramanian N, Oommen C and Raghunandan B N 2018 Additive manufacturing of solid rocket propellant grains *J. Propul. Power* **34** 1090–3
[174] Groven L and Mezger M 2017 Printed energetics: the path toward additive manufacturing of materials *Energetic Materials: Advanced Processing Technologies for Next-Generation Materials* (Boca Raton, FL: Taylor & Francis, CRC Press) ch 7
[175] Muravyev N V, Monogarov K A, Schaller U, Fomenkov I V and Pivkina A N 2019 Progress in additive manufacturing of energetic materials: creating the reactive microstructures with high potential of applications *Propellants Explos. Pyrotech.* **44** 941–69
[176] Uosukainen M 2021 Design for additive manufacturing : utilization study for a novel application area *MSc Thesis* LUT University, LUT School of Energy Systems
[177] McClain M S, Gunduz I E and Son S F 2019 Additive manufacturing of ammonium perchlorate composite propellant with high solids loadings *Proc. Combustion Institute* **37** 3135–42
[178] McClain M S, Afriat A, Rhoads J, Gunduz I and Son S 2020 Development and characterization of a photopolymeric binder for additively manufactured composite solid propellant using vibration assisted printing *Propellants Explos. Pyrotech.* **45** 853–63
[179] Straathof M H, van Driel C A, van Lingen J N J, Ingenhut B L J, ten Cate A T and Maalderink H H 2020 Development of propellant compositions for vat photopolymerization additive manufacturing *Propellants Explos. Pyrotech.* **45** 36–52
[180] Zhou X, Mao Y, Zheng D, Zhong L, Wang R and Gao B *et al* 2021 3D printing of RDX-based aluminized high explosives with gradient structure, significantly altering the critical dimensions *J. Mater. Sci.* **56** 9171–82
[181] Li M, Yang W, Xu M, Hu R and Zheng L 2021 Study of photocurable energetic resin based propellants fabricated by 3D printing *Mater. Des.* **207** 109891
[182] McClain M S, Afriat A, Rhoads J F, Gunduz I E and Son S F 2020 Development and characterization of a photopolymeric binder for additively manufactured composite solid propellant using vibration assisted printing *Propellants Explos. Pyrotech.* **45** 853–63
[183] Bohnlein-Mauss J, Eberhardt A and Fischer T S 2002 Foamed propellants *Propellants Explos. Pyrotech.* **27** 156–60

[184] Bohnlein-Mauss J and Krçber H 2009 Technology of foamed propellants *Propellants Explos. Pyrotech.* **34** 239–44

[185] Ping W, Lei Z, Xiangyang L and Wenfang Z 2005 Preparation of nitrocellulose-based micro-pores spherical powder by solvent leaching method *Chin. J. Explos. Pyrot.* **23** 1107–10

[186] Yang W, Li Y and Ying S 2014 An investigation of the preparation and performance of microcellular combustible material *Cent. Eur. J. Energ. Mater.* **11** 257–69

[187] Li Y, Yang W, Ying S and Peng J 2015 Combustion of gas-permeable gun propellants *J. Energetic Mater.* **33** 167–79

[188] Yajun D, Sanjiu X Z Y, Wenlong W, Chunzhi L and Yun H 2020 Microcellular oblate propellant with skin-core structure deterred by poly(neopentanediol adipate *Cent. Eur. J. Energetic Mater.* **17** 49–65

[189] Gu H, Ding Y J, Xiao Z L, Ying S J and Wu W L 2020 A novel method for oblate spherical powder with adjustable combustion property *J. Phys. Conf. Ser.* **1507** 022007

[190] Yang W T, Hu R, Zhao Y H, Xiao X and Yang J X 2020 Constant-volume combustion performance of mixed charge of foamed propellant with nitramine propellant *J. Phys. Conf. Ser.* **1507** 022002

[191] Chai H, Duan Q, Cao H, Li M and Sun J 2020 Effects of nitrogen content on pyrolysis behavior of nitrocellulose *Fuel* **264** 116853

[192] Iacob A, Pielmu C and Toader C 2013 Relevant matter on caseless ammunition *Forensic Sci.* **87** 1353–6

[193] Schatz J 2012 Caseless ammunition small arms. The good, the bad, and the ugly 2012 NDIA Joint Armaments Conf. *(Seattle, WA)*

[194] Spiegel K and Shipley P 2007 Lightweight small arms technologies *National Defense Industrial Association Joint Services Small Arms Systems Annual Symp.*

[195] Spiegel K and Shipley P 2005 Lightweight small arms technologies *NDIA Int. Infantry and Joint Service Small Arms Systems Annual Symp.*

[196] Rivera U, Pineda J, Susarrey-Huerta O, Herrera L and Meza P 2015 Numerical modelling of caseless ammunition with coreless bullet in internal ballistics *Def. Sci. J.* **65** 203–7

[197] Driscoll K 2021 Japanese Ho-301 40 mm cannon and ammunition *Sabretache* **62** 15–23

Chapter 15

Recent advances in solid and hybrid rocket propulsion

Rajiv Kumar and Mohan Varma

Chemical rocket propulsion systems are utilized for propelling launch vehicles to send payloads into space or in tactical missile systems. The space achievements accomplished so far have largely been attained by employing chemical rockets of different types and calibers. Such systems have proved to be highly efficient, reliable, and capable in attaining the objectives of a variety of space missions, and provide a mature technology for next-generation tactical missile development programs. Solid and hybrid rocket propulsion systems have advanced appreciably in the last few decades. Homogeneous and heterogeneous solid propellants have extensively been used in gun and rocket applications. In this chapter, the next-generation high-energy propellants, eco-friendly solid propellants and their combustion phenomena are discussed along with recent advancements in the field. The major constraint with solid rocket motors (SRMs) remains the inability to terminate their thrust and operation at will and to restart combustion in order to attain precise mission objectives. A simple solution to this problem is offered by hybrid rocket engines, where fuel and oxidizer are stored separately and the combustion process may be terminated just by cutting the supply of the oxidizer. The advantages and disadvantages of this technology are here discussed. The research related to improvement of the mechanical properties of the fuel grain and enhancement of the regression rate of fuels using different approaches is further presented and analyzed. The possibility of employing cryogenic fuels in advanced propulsion systems is also addressed. The scalability issues and recent advances in hybrid engines are presented to illustrate new approaches and solutions for its effective utilization in the near future.

15.1 Solid rocket propulsion

15.1.1 Introduction

The recorded history of solid rocket propulsion may be traced back nearly 2000 years. The Chinese used gunpowder in projectiles in 220 BC, and subsequently the

doi:10.1088/978-0-7503-3943-8ch15

practice of employing such devices in contemporary aggression and wars moved to Korea, India, the Middle East and then to Europe through traders using land and sea routes. Later developments in the field gave birth to various kinds of solid propellants with improved energetics that were utilized in solid rocket propulsion.

An improvised version of rocketry in warfare was revived with the creation of iron-cased Indian Mysorean rockets that were used against the British during the middle of the 18th century. After the First World War, many countries recognized the importance of developing the capability to launch lethal projectiles at greater distances in order to have an edge over adversaries. Premier educational and research institutions such as Caltech, CA, in the USA, and the Institute of Chemical Physics, Moscow, in Russia, as well as military-led establishments in Germany and in the UK carried out research and development (R&D) work in the 1930s and 1940s [1–3].

The development in the field of solid rocket propulsion were intimately linked with the discovery and production of high-energy materials that could be used as propellants in SRMs. In 1863–1864, nitrocellulose-based propellants were made available in Europe. In 1888, Nobel invented the smokeless double-base (DB) ballistite and, later with the production of cyclotrimethylene trinitramine (RDX) and cyclotetramethylene tetranitramine (HMX) at industrial scales in the 1940s, composite modified double-base propellants were utilized. Subsequently, the first castable potassium-perchlorate-based composite solid propellant became available in 1942 and a novel operational composite solid propellant (CSP), comprising ammonium perchlorate (AP), polyurethane, and aluminum, was tried in 1960 in USA. Thereafter, numerous advances in CSP technology including metallized CSPs been made to grant larger solid propellant booster rockets producing huge thrust levels. This approach has been taken forward by several countries such as the USA, Russia, China, the European nations (via the European Space Agency), and India. Several books [4–9] cover the area of solid rocket propulsion, and readers are encouraged to refer to these for more details. The content presented here is focused on solid propellants which are currently in use, present trends in research work, and future direction in their advancement.

15.1.2 Propellant technology

15.1.2.1 Homogeneous solid propellants

This class of solid propellants is the oldest and simplest self-combustion sustaining system as it contains the oxidizing and reducing (fuel) entities in a single chemical compound or molecule. The typical example is nitrocellulose (NC), which is used as a single-base gun propellant. The propellant powder is generally processed by a solvent extrusion process to prepare cartridges for different applications in pistols, guns, and rifles. To improve the energetics, ease of processing, storability, and mechanical properties of the propellant, nitroglycerine (NG) is used to plasticize and to get a castable/moldable thermoplastic material. Such propellants are termed DB propellants and can be used in guns and rocket motors.

DB propellants show controlled deflagration characteristics and good mechanical properties, and energetics and large-sized grains can be prepared using different processing techniques like solvent extrusion, solventless extrusion, casting (slurry/ interstitial), etc. The mechanical and ballistic properties of DB propellants may further be improved by adding a stabilizer (ethyl centralite/2 nitrodiphenylamine), a coolant (oxamide), an opacifier (carbon black), a flame suppressant (potassium nitrate/potassium sulfate), a metal fuel (aluminum), a burning-rate catalyst (lead salicylate/lead stearate/copper salicylate/copper stearate), etc. DB propellants generally have a pressure exponent *n* value in the range of 0.5–0.8 (burn rate $r = a\ P_c^n$). At high operating pressures, this value may even shift from 0.7 to ≈1. The progressive increase in pressure with increasing burning rate poses problems in the design of the rocket motor, and so the value of *n* is to be kept at a reasonable level. The pressure index may be suitably tailored by incorporating certain additives in propellant formulation, known as platonizing agents, that can bring *n* closer to 0 (plateau burning); this approach is referred as platonization. Lead salts are commonly used as platonizing agents to control the value of *n* to a reasonable level in the range of the operating chamber pressure of the rocket motor. Besides their use in gun and rocket motors, DB propellants are also used in airbags, aircraft seat ejection, propellant-assisted ship engine starters, and aircraft takeoff, deceleration and braking systems. The details of their preparation methodology, mechanical and thermal properties, aging behavior, combustion and degradation characteristics with and without burn-rate catalysts, performance and application are available in textbooks [10–12]. Typical composition used for the homogeneous propellants is given in table 15.1.

Table 15.1. Composition of homogeneous propellants.

S. No.	Propellant type	Typical final composition* (wt.%)
1	Single base	Nitrocellulose 90–95 Diphenyl amine (stabilizer) 2–4 Alcohol (solvent) 1–2 KNO_3 (leeching agent) 0–5
2.	Double base	Nitrocellulose 65–70 Nitroglycerine 15–30 Stabilizer (ethyl centralite) 2–5 Plasticizer (mineral gelly) 7–10 Platonizing agent (lead stearate) 0–3 Solvent (acetone/water mix) 0.5–2
3	Triple base	Nitrocellulose 40–50 Nitroglycerine 15–25 Nitroguanidine 25–40 Stabilizer (ethyl centralite) 1–3 Solvent (acetone) 0.1–0.2
4	Composite modified double base	Nitrocellulose (globular) 40–45 Nitroglycerine 30–35 Stabilizer (diethyl phthalate) 5–10 Ammonium perchlorate 15–20 Aluminum powder 15–20 Burn-rate modifier (centralite) 1–3

*Initial composition of mix has large quantity of solvent for homogenization.

The mechanical properties of DB propellants are highly sensitive to temperature and the age of the propellant. NG tends to migrate with time and temperature, which may adversely affect the mechanical and ballistic properties of the propellant. The mechanical properties of DB propellants are also dependent on the process adopted for their manufacture. Table 15.2 presents a glimpse of the changes in mechanical properties with the different processes of manufacturing adopted for a DB propellant. In solid rocketry, the mechanical properties of the propellant are affected by static and dynamic combustion within the propellant grain. The composition of the propellant influences the burn rate, and the variation of steady-state pressure across the grain affects the stress and strain distribution within the propellant grain. The grain design and selection of a suitable configuration remains the key feature to realize the desired pressure/thrust-time profile during operation of the rocket motor. A DB propellant is preferred in applications where low smoke signature and no hydrogen chloride (HCl) gas in the exhaust plume is required, and where a medium performance ($I_s \leqslant 235$ s) is acceptable.

As aforementioned, DB propellants have ingredients that contain both fuel and oxidizer species in a single molecule and are premixed. The generated flame is, thus, controlled by one-dimensional chemical processes rather than three-dimensional complex diffusion processes, as observed in the case of CSPs. The combustion process remains highly pressure dependent, and ignition behavior, visibility, and the thickness of various zones may change with operating conditions. The primary flame (fizz zone) is separated from the secondary flame (luminous zone) at low pressures (figure 15.1), and as the pressure increases, secondary flame perturbation starts before merger of the two flames. Beginning from the subsurface, different combustion zones produce products after initial thermal degradation of NC and NG, and subsequent chemical reactions occur. The products of the preceding zone become the reactants in the following zone till the completion of the combustion process.

The overall burning characteristics of DB propellants have been well understood and are summarized in reference [13]. Research work on the degradation characteristics of ingredients individually and in propellant compositions, details of degradation products and their mass fraction in combustion zones, with temperature profiles of combustion waves, provides the basic understanding to formulate combustion models. The influence of additives on the thermal decomposition of a DB propellant in condensed and gas phases and their role in changing the mechanism of combustion with or without additives has been studied by many researchers. Comprehensive models to predict the burn rates of DB propellants are reported in the literature, and detailing of this aspect is beyond the scope of this chapter.

Table 15.2. Mechanical properties of DB propellant at 300 K.

S. No.	Propellant preparation process	Tensile strength (N m^{-2})	Elongation (%)
1	Solventless extrusion	3.04×10^6	11
2	Interstitially cast (modified)	1.76×10^6	16
3	Slurry cast	1.57×10^6	25

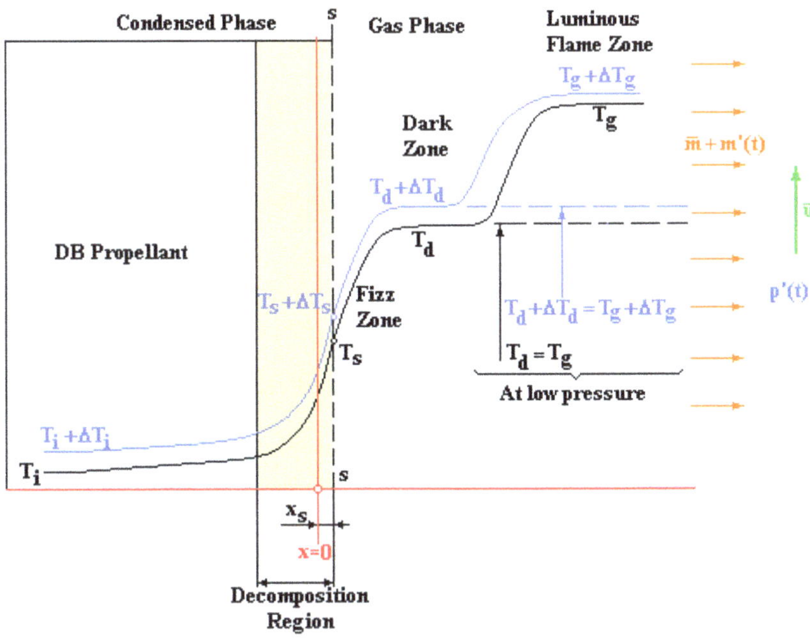

Figure 15.1. Various zones in combustion of a DB propellant and wave structure at different initial temperatures and pressures [14].

As mentioned earlier, additives play a very important role in controlling the storage and ballistic characteristics of DB propellants. Plasticizers, stabilizers, energy enhancers, coolants, opacifiers, flame suppressants, and combustion instability suppressants are commonly used in different concentrations, as per requirements, along with burn-rate catalysts. The DB propellants exhibit a large variation in their burning rate at different pressures depending on the burn-rate modifier used. One may get a linear increase in burn rate in a particular pressure segment with a typical pressure index (n) value that may be reduced with change in composition. On the other hand, some additives may cause super burn rate (higher than expected), plateau (pressure independent), and mesa (lower burn rate at increased pressure) behavior (figure 15.2).

This feature of DB propellants is normally obtained by using lead salts. It has been investigated by various researchers and explanation for it has been given [16, 17]. Alternatives are being sought given the problem of lead toxicity. The catalytic influence of nano-catalysts and other new materials on thermal decomposition and combustion of DB propellants has also been investigated [18, 19]. The understanding of manufacturing processes, mechanical properties at different temperatures, thermal decomposition, flame structure, and combustion of DB propellants has matured over many decades of their use in industry. However, the prediction of ballistic properties with aging of the propellant, the migration rate of NG with aging of the propellant under thermal cycling, and the mechanism of second super-rate effect at high pressure ranges, as well as the total life-cycle management, including impact sensitivity details,

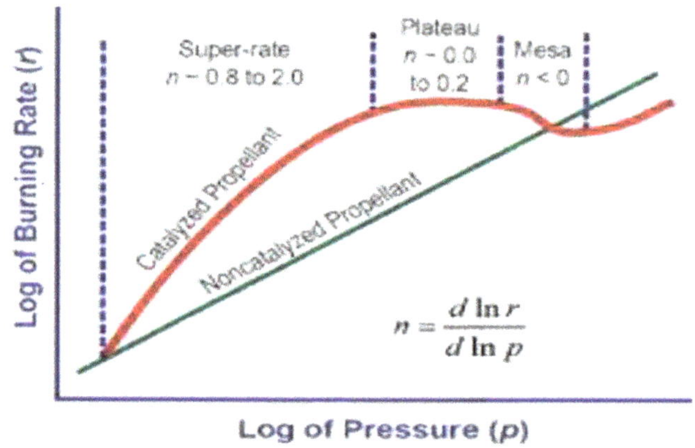

Figure 15.2. Pressure dependence of burn rate of non-catalyzed and catalyzed DB propellant [15].

remain some important areas for future investigations. Some conjectures about the flame attachment, conductivity, and heat flux in relation to the observed burn rate (particularly at high pressures) with and without additives and metals, presently used in combustion models, also need to be validated. Synergistic (combined) catalytic effects of potential nano-catalysts and conventional catalysts for DB propellants that may provide controlled characteristics and a wide range of burn rates with low combustion index also need to be explored.

15.1.2.2 Composite solid propellants

The CSPs presently used in SRMs generally comprise hydroxyl-terminated polybutadiene-ammonium perchlorate (HTPB-AP) binder-oxidizer system of different compositions with and without aluminum metal powders of varying particle sizes (from nanometer to micron sized). Such propellants may contain several other ingredients like curing/cross-linking agents, a bonding agent, plasticizer, burn-rate catalysts, metal fuel, combustion instability suppressant, etc, in different quantities as per requirements. Common examples from each category of ingredient are listed in reference [20]. The new requirements dictated by various missions call for CSPs with high-energy and eco-friendly ingredients, reduced levels of pollutants in rocket exhaust with minimum signature profiles to alleviate the risk of detection of the launch site, improved mechanical and ballistic properties, long and safe storage life, and high-density specific impulse. The burning rate of a CSP can be modified and tailored by incorporation of transition metal oxides or nano-catalysts at different concentration levels (0.5–3 wt.%) in the formulation.

Recent R&D has largely been focused on the replacement of AP, which produces hazardous HCl gas upon combustion. This can cause acid rain in the vicinity of the launch site testing or firing of large solid rocket boosters. Green oxidizers such as phase-stabilized ammonium nitrate (PSAN) [21], ammonium dinitramide (ADN) [22], hydrazinium nitroformate (HNF) [23], and ammonium ozonide [24] have been explored for application in next-generation CSPs. HNF appears to be most promising considering its high specific impulse, burn rate, and benign combustion products. However, CSPs individual properties contain a mixed bag when examined from the viewpoint of overall desirable benefits. All the abovementioned oxidizers are chlorine-free but, when checked on ballistic performance, physico-chemical properties including hygroscopicity, oxygen content, thermal stability, etc, the mechanical properties of the propellant, preparation and costing, etc, there are individual disadvantages and limitations. HNF, for example, appeared to promise a high I_s ($\geqslant$316 s), burn rate, and low-signature chlorine-free exhaust, but it has higher sensitivity in comparison to AP, RDX, and HMX, higher oral toxicity, and its propellants exhibit a high n value (0.81). It has harmful effects on skin and must be handled with protection and care. The lethal dose 50 (LD_{50}) value for said dermal exposure shows toxic characteristics. Further, its compatibility with presently used binders, like HTPB and glycidyl azide polymer (GAP), is poor and a high solid loading in the binder cannot be achieved due to the needle-like crystals of its high aspect ratio. HNF attacks the backbone of the binder and breaks down the binder chain. This results in formation of gaseous pockets that may lead to volume change and softer propellant. The C-C bond of the binder gets oxidized by HNF, and it may also react with common curing agents leading to deterioration in the mechanical properties of the propellant. It also has poor thermal stability and a low decomposition temperature ($\approx$393 K). The oxidizer is difficult to use in slurry cast composites and has lower oxygen balance (+13% as against 34% of AP). The lack of availability of oxidizer at industrial scale and its high cost pose limitations in its wider use. The above constraints need to be addressed for its larger acceptance to replace AP in the space industry. Some efforts are being made to solve these problems.

The combustion characteristics and non-steady burning at lower chamber pressures of eco-friendly oxidizers may be altered by incorporating metal powders in the propellant composition [25]. Additives have been explored to reduce the high-pressure exponent issue. Addition of nitroguanidine has been recommended in the formulation to reduce reactivity of oxidizers like HNF. Methods to control crystal structure, size and shape of new oxidizer have also been reported to achieve high solid filler loading in the CSPs. Research on new oxidizers with positive oxygen balance and molecules containing trinitromethyl functionality and analog fluorodinitromethyl derivatives has been recently reported [26]. A review on new oxidizers and energetic binders is presented in reference 27.

Recent efforts at incorporating energetic binders in CSPs have mainly considered GAP [28, 29], 3,3-bis-azidomethyl oxetane (BAMO) [30], 3-azidomethyl-3-methyl oxetane (AMMO) [31], tetrahydrofuran (BAMO-THF) copolymer [32], and nitrate ester plasticized polyether (NEPE): all have been seen to enhance the propulsive gains significantly.

The salient characteristics of these binders are presented in table 15.3. These binders, however, have technical issues relating to their efficient processing, optimum solid filler loading, and the mechanical properties of the prepared propellants. The use of nitrated hydroxy-terminated polybutadiene (NHTPB) [33] has also shown energy gains in comparison to the conventional resin. An addition of 2.7 wt.% in hydroxyl-terminated polybutadiene (HTPB) resin has shown an increase in calorific value (5.5%) of the propellant with reduced friction sensitivity. Higher concentrations of

Table 15.3. Characteristics of energetic binders.

Sl. No.	Binder	Burn rate (mm s^{-1}) at 6.89 MPa	Activation energy* kJ mol^{-1}	Oxygen balance	Density (g cm^{-3})	ΔH_f (kJ mol^{-1})
1.	HTPB	6.9	127	−65	0.92	−52
2.	PAMMO	7.9	152	−35	1.06	+179
3.	GAP	11.6	173.6	−45	1.30	+117
4.	P BAMO	–	150 (1st stage decomposition)	−45	1.30	+413
5.	BAMO-THF (28+AP52+ RDX 15+Al5 IN + %)	10.22	192.1 (BINDER gel)	−56	1.18	+189
6.	NEPE (32) NG: BTTN (50:50) +AP (20) +HMX (48%)	7.5 (at 2.25 MPa)	160.4	−16.6	1.52	–

*May vary from experimental method used.

(Nomenclature: PAMMO, 3-mesy loxymethyl-3-methyl oxetane; RDX, hexahydro-1,3,5-trinitro-1,3,5-triazine); BTTN, 1,2,4-butanetriol trinitrate; HMX, octahydro-1,3,5,7-tetranitro-1,3,5,7-tetrazocine.

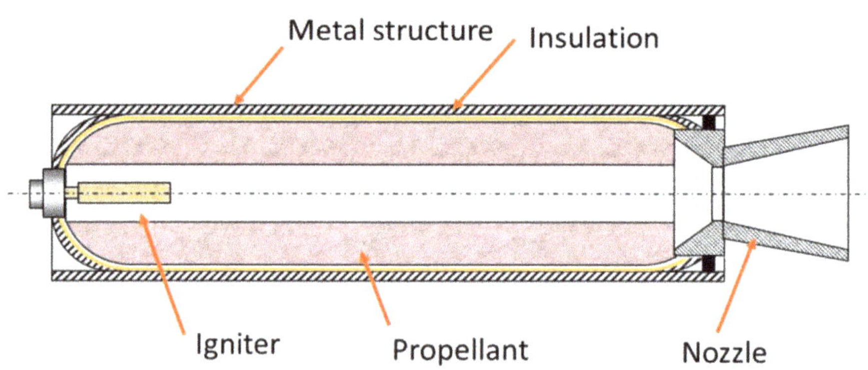

Figure 15.3. A typical solid rocket motor.

NHTPB in a composition may prove to be beneficial from an energetics point of view but adversely affect the viscosity build-up of the mix and its pot life.

Metallized CSPs have been utilized in SRMs (figure 15.3) for many years to achieve high-density specific impulse and better overall energetics [34]. Metals with low molecular weights like aluminum, magnesium, boron, and beryllium are preferred to keep the mean molecular weight of the combustion products (M_c) under check and yet gain the significant increase in combustion temperature (T_c), leading to a higher specific impulse ($I_s \approx \sqrt{T_c/M_c}$). Metals up to 16–18 wt.% in varying ranges of particle sizes (micrometer to nanometer) [35] are incorporated in propellant formulations depending on the mission requirements. The selection of metal also depends on the mechanism and efficiency of combustion desired; generally, metals undergoing vapor-phase combustion like aluminum and magnesium are preferred. Toxicity and associated hazards are also important criteria in its selection. Metallized propellants need to be tailored using suitable fuel binder-oxidizer combinations that can produce the required temperatures under motor operating conditions to ensure higher combustion efficiencies of the metal (95% and above).

Metallized propellants also pose several challenges to a rocket motor designer. Factors like the mechanical erosion of the propellant surface due to cross-flow conditions of combustion gases as they move towards the nozzle, varying combustion characteristics of single particles of metals and their agglomeration on the propellant surface and in the combustor, and altered burn rates due to blowing effects need to be addressed [36]. The puddling effect on the propellant surface due to spread of molten metal and its resultant impact on heat transfer from the established flame to the burning surface may lead at times to combustion instability and even quenching of the propellant grain, as shown in table 15.4. Current R&D works are aimed towards finding solutions to the above problems, selection of suitable oxidizer loading and particles sizes (bimodal/trimodal), and use of nano aluminum particles in the formulations to achieve improved combustion characteristics. The proper choice of high-energy binders and combustion catalysts may enhance combustion efficiency. Efficient motor design may be targeted at enabling the condensation of produced metal oxides just at the nozzle throat to harness maximum propulsive gains and yet minimize the dynamic nonequilibrium two-phase flow losses [37] due to the heterogeneous nature of combustion products.

The attainment of higher specific impulse with advanced energetic materials remains the focus of current R&D work. Amongst the advanced fuels, metallic hydrogen [38] promises I_s in the range of 500–1450 s but must yet pass through many stages of technological development relating to its synthesis, characterization, and allied technology demonstrations for material compatibility and pilot engine tests. Similar issues need to be resolved with potentially attractive high-nitrogen fuels (N_5^+ and N_5^- compounds). Some work [39] has been reported on this subject, however it is still at the synthesis level and would require detailed study on its characterization, storage, processing, and other ballistic and combustion features before use in practical propellant formulations.

The search for high-performance and safe solid propellants has generally aimed at replacing AP with a suitable oxidizer that can deliver higher specific impulse and

Table 15.4. Advantages and disadvantages of metallized solid propellants.

Advantages	Disadvantages
• Higher density to propellant. • Higher heat of combustion and • Higher I_s due to increased combustion chamber temperature. • Low-molecular-weight metals keep the mean molecular weight of combustion product lower to improve performance. • Increment of burn rate and suppression of combustion instability.	• Possibility of mechanical erosion of propellant surface. • Two-phase flow in the nozzle. • Agglomeration tendency of molten metal over the propellant surface. • Low combustion efficiency and combustion instability or even quenching of flame.

chlorine-free exhaust products. However, a change to the energetic binder in the propellant composition also offers a potential approach to prepare propellants that can possess the benefits of both DB propellants and CSPs. NEPE propellants seem to offer good energetic properties with negligible detectable exhaust plume. These propellants are high-energetic CSPs and incorporate a polyether, such as polyethylene glycol, or an ethylene oxide-tetrahydrofuran co-polyether as a polymeric binder and NG as plasticizer (22–27 wt.%) in their formulation. The solid phase may contain aluminum powder, octagon or HMX, and AP in the range of 73–78 wt.%. Nitramines like hexogen or RDX and triaminoguanidine nitrate may also be employed to replace AP to increase energy and reduce the signature of the exhaust plume [40, 41]. Other additives and catalysts can be added as per requirements of the ballistic and mechanical properties for a given mission. This type of propellant is particularly attractive in missiles where high performance, the reduced possibility of tracking, but smaller weight is needed. For launch vehicles that use large booster stages, a less-sensitive propellant without RDX/HMX is preferred.

The physical and chemical aging characteristics of NEPE propellants are complex as various chemical ingredients are used in their formulation; this may influence the shelf life of the rocket motors containing them. Information on the degree of moisture absorption and dehumidification at the ingredient interface, the nature of debonding between propellant and liner, and the appearance of cracks and voids due to local stress and strain, which affects the physical aging, is limited. Similarly, details on chemical changes during storage such as the decomposition of the oxidizer, hydrolysis, solid-state interactions due to thermal cycling, and the interaction between various ingredients of the propellant, need to be thoroughly investigated. Studies on combustion characteristics, thermal behavior, and decomposition kinetics [42, 43]; the effect of solid particle size on mechanical properties [44]; accelerated aging and storage life assessment [45]; the structural integrity of propellant grains [46–48]; and the influence of additives on the sensitivity and stability of NEPE propellants [49, 50] have been conducted recently to assess their suitability of use in different missions.

15.1.3 Combustion of composite solid propellants

The combustion behavior of CSPs has been investigated by various researchers since the 1960s to obtain better understanding of the subsurface [51–53], surface [54–56], and gas-phase [57–59] combustion details of the propellants. The temperature profiles [60, 61] in these zones, chemical kinetics and associated flames [62], reaction chemistry [63] of major ingredients and their interaction with other chemical entities, including metal additives [64] present in the formulation, have also been examined in detail. The combustion properties of AP used in CSPs has been determined under different conditions of pressure [65, 66] and heating rates [67]. This has been extended to investigate the effect of particle size on deflagration rate [68]. The burn rate of propellants containing various fuel binders and plasticizers [69, 70] has been thoroughly investigated with or without burn-rate catalysts and reviewed [71, 72]. Catalysts of various types added at different concentration levels influence the burning rate with pressure, as presented in table 15.5.

Table 15.5. Burning-rate data of CSPs showing influence of nano-scale metal catalysts [72].

			Burn rate ($\dot{r}$, mm s^{-1}), pressure regimes (MPa)								
Propellant	Metal nano-catalysts	Weight % loading	Ambient	0.098	0.98	1.37	3.92	5.88	6.86	7.84	9.80
AP/HTPB	–	–	–	1.63	–	–	–	–	–	–	–
Do	NTO	0.5	–	1.92	–	–	–	–	–	–	–
Do	$Fe(NTO)_2$	0.5	–	2.04	–	–	–	–	–	–	–
Do	CuO	0.5	–	2.02	–	–	–	–	–	–	–
Do	$Cu(NTO)_2$	0.5	–	2.51	–	–	–	–	–	–	–
Do	Fe_2O_3	0.5	–	1.94	–	–	–	–	–	–	–
AP/Al/HTPB	–	0.5	–	1.26	2.79	3.58	4.95	–	6.03	–	–
Do	NTO	0.5	–	1.43	3.55	4.44	5.20	–	–	–	–
Do	$Cu(NTO)_2$	0.1	–	1.67	5.09	6.90	6.31	–	7.90	–	–
Do	$Cu(NTO)_2$	0.5	–	–	–	–	6.91	–	8.70	–	–
Do	$Cu(NTO)_2$	2.0	–	–	–	–	11.3	–	14.0	–	–
Do	$Fe(NTO)_3$	0.5	–	1.56	4.81	5.71	6.80	–	8.20	–	–
Do	$CuCr_2O_4$	0.5	–	–	–	–	8.00	–	10.5	–	–
Do	$CuCr_2O_4$	2.0	–	–	–	–	13.8	–	17.1	–	–
AP/Al/HTPB	#$CuCr_2O_4$	2.0	–	–	–	–	–	–	9.4	–	–
Do	#$CuCr_2O_4$	2.0	–	–	–	–	–	–	10.4	–	–
Do	#$CuCr_2O_4$	2.0	–	–	–	–	–	–	15.5	–	–
Do	n-Fe_2O_3	2.0	–	–	–	–	–	–	12.7	–	–
AP/Al/HTPB	n-Fe_2O_3	–	–	9.27	–	–	20.8	–	29.1	–	32.8
Do	n-Fe_2O_3	1.0	–	9.80	–	–	21.8	–	30.1	–	35.4
Do	n-Fe_2O_3	2.0	–	10.1	–	–	22.9	–	31.7	–	38.2
AP/HTPB	–	–	2.04	–	–	–	–	–	–	–	–
Do	ZnO_2	1.0	3.18	–	–	–	–	–	–	–	–
Do	Cr_2O_3	1.0	2.70	–	–	–	–	–	–	–	–
Do	TiO_2	1.0	3.03	–	–	–	–	–	–	–	–

(*Continued*)

Table 15.5. (*Continued*)

			Burn rate ($\dot{r}$, mm s^{-1}), pressure regimes (MPa)								
Propellant	Metal nano-catalysts	Weight % loading	Ambient	0.098	0.98	1.37	3.92	5.88	6.86	7.84	9.80
AP/HTPB	MnC_2O_4	1.0	2.50	–	–	–	–	–	–	–	–
AP/HTPB	–	–	–	–	–	–	–	6.00	6.11	6.55	–
Do	n-Cr_2O_3	0.25	–	–	–	–	–	7.16	7.55	8.00	–
Do	n-Cr_2O_3	0.50	–	–	–	–	–	7.19	7.59	8.11	–
Do	n-Cr_2O_3	0.75	–	–	–	–	–	7.20	7.72	8.30	–
Do	n-Cr_2O_3	1.00	–	–	–	–	–	7.22	7.88	8.40	–
AP/HTPB	–	–	1.22	–	–	–	–	–	–	–	–
Do	Ni–Co	1.0	2.08	–	–	–	–	–	–	–	–
Do	Ni–Cu	1.0	1.77	–	–	–	–	–	–	–	–
Do	Ni–Zn	1.0	1.59	–	–	–	–	–	–	–	–
AP/Al/HTPB	n-MnO_2	0.25	–	–	–	–	–	6.82	7.18	7.64	–
Do	n-MnO_2	0.50	–	–	–	–	–	6.89	7.36	7.76	–
Do	n-MnO_2	0.75	–	–	–	–	–	6.96	7.43	7.95	–
Do	n-MnO_2	1.00	–	–	–	–	–	7.10	7.54	8.02	–
AP/HTPB	#$CuCr_2O_4$	2.00	–	–	–	–	6.40	–	–	–	–
Do	#$CuCr_2O_4$	2.00	–	–	–	–	7.91	–	–	–	–
AP/Al/HTPB	–	–	–	–	–	–	–	–	–	7.51	8.73
Do	Bu*/HTPB	1/9	–	–	–	–	–	–	–	11.44	12.15
Do	Bu*/HTPB	3/7	–	–	–	–	–	–	–	13.54	14.65
Do	Bu*/HTPB	5/5	–	–	–	–	–	–	–	17.41	19.03
Do	Bu*/HTPB	7/3	–	–	–	–	–	–	–	19.69	21.20
AP/Al/HTPB	Graphene	1.00	–	–	–	–	–	4.20	–	5.50	6.40
Do	n-TiO_2	1.00	–	–	–	–	–	5.25	–	6.80	7.90
Do	**GTNC	1.00	–	–	–	–	–	6.80	–	8.40	9.10

Note:
prepared by different methods,
* Bu-Butacene,
** GTNC - graphene decorated titanium dioxide.

Several combustion models have been proposed to predict the effects of the various parameters that influence the burning rates of CSPs and subsequently reviewed [73–75]. However, due to the heterogeneity and intricacies of the complex interactions of the chemical species involved, and the complex mechanism of heat transfer, intricate flame types and structure involved in the variety of compositional variations, it is not yet possible to evolve a satisfactory combustion model that addresses all the details of the combustion processes involved and predicts the burn rate of propellants under given experimental or motor operating conditions accurately.

Recent advances in solid rocket propulsion encompass the development of high-performance solid rocket engines using highly energetic new fuels and oxidizers, metal additives, and burn-rate enhancers to improve I_s and overall density specific impulse. Earlier major challenges, such as scaling-up of SRM production to produce efficient boosters and enhanced range of tactical missiles, have largely been addressed by improved grain designs, use of lightweight alloys or composites that can withstand very high temperatures and pressures, fabrication of rocket motors employing efficient liners, insulators, and inhibitors in case-bonded motors, and efficient nozzle designs with next-generation lightweight throat inserts. Better understanding and control of the erosive burning in radial burning grains and improved combustion efficiency in well-designed SRMs, in addition to effective control of combustion instability, has enabled the realization of advanced missions today with greater reliability and safety. The experiences and technology development in allied domains have led to significant advancements in hybrid rocket, ramjet, and scramjet propulsion.

SRMs of various calibers and boosters capable of generating sufficient thrust to lift-off modern launch vehicles and tactical missiles have existed for quite some time. More recently, micro-thrusters and micro-propulsion devices using reliable and safe solid propellants have also been developed for small satellites and spacecrafts. The limitation of the chamber volume available poses new challenges in ignition-related parameters like ignition delay, ignition probability, and requirement of ignition energy that are much different than typical SRMs. As combustion process and flame spreading along the exposed surface of the propellant charge involves a combination of different disciplines such fluid dynamics, heat and mass transfer, chemical kinetics and multiple chemical species interaction, complex transport phenomena, and thermodynamics with their intricate coupling, a clear theoretical or experimental approach is difficult to select. However, advances in computation fluid dynamics (CFD) have given an impetus to theoretical modeling and helped to develop numerical techniques that are used extensively in chemical propulsion. Further, advanced diagnostic techniques and nonintrusive methods are efficiently used to understand the combustion processes and mechanisms involved, and they can be gainfully used to optimize design methodology for a specific application.

The current focus of investigation is on combustion problems related to two-phase flow losses during the combustion of virgin and metallized solid propellants [76, 77], the effect on combustion and extinction of nano-sized metal particles in the propellant formulation [78–81], and oscillatory and turbulent combustion in large SRMs [82–84].

New challenges for SRMs operating at high chamber pressures (≈15 MPa) and temperatures (≈3600 K) include the complexities of radiative heat transfer in nozzles [85] leading to crucial performance-related issues. Modeling and experimental research on laboratory-scale SRMs is regularly carried out to understand the important features of the solid rocket exhaust plume [86–88]. This involves investigation of the supersonic after-burning flame of the exhaust combustion products emanating from the rocket nozzle and their interaction with atmospheric gases. This aspect has gained significance in efforts to reduce the infra-red and radar signature of a projectile to avoid its detection. Other aspects of the study of exhaust plumes relate to the need to control pollution diffusion in the atmosphere after vehicle liftoff.

Recent advances in high-speed imaging, laser diagnostics, and numerical data processing have led to significant advances in the field of computational combustion. Our understanding of many key features associated with flame structures, the interaction of different flame types, turbulent combustion, combustion dynamics, radiation processes and ionization in flames, and soot formation, etc, has advanced to a much higher level than in the past. This enables us to find new solutions to existing problems. Advances in these subjects are reported in the open literature.

As mentioned earlier, efforts are being directed to realize high-energy, eco-friendly composite solid rocket propellants to minimize the environmental damage caused by emittance of tons of HCl in the near-Earth environment, when tons of AP-containing solid propellants are burnt in the large booster motors of present-day launch vehicles. This may cause acid rain and lead to hazardous effects on flora and fauna in the vicinity of the launch site. Global awareness and concern about the environment have led to stringent environmental laws all over the world. Keeping this in mind, AP is being replaced with safer oxidizers in CSPs. Several approaches have been made to develop eco-friendly propellants, and their combustion characteristics have been studied [89].

Ammonium nitrate (AN) has often been used as a cheap oxidizer in CSPs to produce chlorine-free combustion products, and an exhaustive review on the various aspects of its physico-chemical properties, decomposition, burn rate, and safety aspects is available [90]. The practical problems in its use are it hygroscopicity, crystal phase transitions in the operational temperature range leading to net volume change, low burning rate, and poor overall performance. The net decomposition of AN at lower pressures remains endothermic; however, at higher pressures (>2.068 MPa) it becomes exothermic in nature [91]. Many additives have been found to have a positive influence on the decomposition and thermal degradation of AN and to be effective in masking its phase transitions. [92, 93]. Efforts have also been made to acquire PSAN for use in preparing CSPs in the past as well as more recently [94, 95]. Catalyzed AN/PSAN-based CSPs with different fuel binders have been investigated [96, 97] for many applications where reduced smoke and safer exhaust is desired and high performance and high burn rate are not the critical criteria. The combustion of virgin and metallized AN-based propellants with compatible binders, including HTPB, is well understood [98, 99] but changes in the condensed-phase, surface, and gas-phase combustion with catalysts under the operating conditions of SRMs have yet to be investigated.

Another eco-friendly oxidizer that has attracted the attention of many research workers is ADN. After its discovery in 1971 in the USSR, it remained relatively unknown and was secretly used in solid propellants for intercontinental ballistic missiles. It was later synthesized at the Stanford Research Institute, USA, in 1988, and then in Sweden. A good description of green propellants based on ADN is available in the literature [100, 101]. Details of the decomposition pathways [102], thermal characteristics, burning behavior, and combustion mechanisms [103–108] of ADN and ADN-based propellants have also been reported. These propellants can easily be metallized using aluminum or magnesium and deliver an improved burning rate and performance. Some propellant compositions have been found to exhibit poor mechanical properties in comparison to AP-based systems. The major difficulty in its use in HTPB-based CSPs is its incompatibility with isocyanate-type cross-linking agents.

Isocyanates, like toluene-2,4-diisocyanate (TDI) and isophorone diisocyanate, are used as curing agents to achieve better mechanical properties for the grains. These produce urethane linkage by reacting with the terminal –OH groups of HTPB that leads to conversion of highly viscous pre-polymer into an elastic solid binder to give better mechanical properties. These agents are highly toxic in nature and possess a highly reactive isocyanate group that reacts with biochemical structures in the human body through the reactive hydrogen atoms present in its free amino and hydroxyl groups. Its vapors on contact may cause irritation to the eyes, skin dermatitis, and inflammation in the respiratory system (nose, throat, and lungs) upon inhalation, for short or longer durations. The irritation threshold for short-term exposure of TDI is between 0.05 and 0.1 ml m^{-3}. Longer exposure to the vapors even at threshold levels (>0.02 ppm) may cause irritation of the nose and throat, breathlessness, chest discomfort, choking, coughing, nausea, vomiting, and even abdominal pain. Extreme cases of exposure may lead to hardening of tissue. The carcinogenic nature of di-isocyanates is also suspected. The possibility of gene mutation and chromosomal damage cannot be ruled out. Necessary precautions in the use and handling and storage of isocyanates are mandatory.

Propellant compositions with dual oxidizers have also been investigated [109]. CSPs with 15% ADN give a higher burn rate in comparison to AP-based propellants at all pressures, though the pressure exponent also increases from 0.41 to 0.68. AP/ADN dual-oxidizer formulations lead to increased impact and friction sensitivities. Presently, several high-energy fuel binders such as GAP, poly BAMO, AMMO, and NIMMO, are being used to prepare CSPs, and many of the initial problems in processing, such as adequate solids loading and metallization, and so on, have been sorted out. ADN in liquid and gel states has also been investigated for monopropellant thrusters [110] and gel propellant [111].

Another oxidizer, HNF, has also been considered as a replacement to AP (fully/partly) in CSPs, but issues relating to cost, safety, handling (i.e. its carcinogenic nature), thermal stability, and friction sensitivity have caused impediments in its use. An earlier work [112] reported use of oxidizer with several energetic binders such as GAP, BAMO, and polyglycidyl nitrate, and polynitromethyloxetane. Further investigations on HNF-based CSPs were later carried out [113] to present an overview of the efficacy of this class of propellants. A review on HNF and its propellants, covering its synthesis, properties, thermal behavior, as well as the kinetics of decomposition, burn rate of propellants, and ballistic properties, etc, is presented in reference [114]. It appears that the method of preparation, purity, and crystal size, etc, have a marked influence on the ballistic and mechanical properties of the prepared propellants. N-alkyl derivatives of HNF can also be synthesized and may be used to tailor the energetics of the CSPs [115, 116]. Efforts at combustion modeling have also been undertaken [117] and a suitable approach to elucidate the combustion phenomena of HNF in solid and gas phases has been identified, which is in conformity with the observed experimental results. Recently, the effect of HNF on the energetic characteristics of prepared CSPs has been studied [118]. It was observed that the presence of HNF reduces the infra-red emissions of the rocket exhaust plume in comparison to AP-containing compositions. The major application of this oxidizer in solid rocket establishments would require new high-energy binders that are insensitive to reaction with HNF and compatible curing agents. There is little work now being done on HNF, mainly due to safety issues. Other improvisations required are passive crystal coatings to reduce impact and friction sensitivity, and economically viable manufacturing processes to produce spherical and fine crystal sizes. Additional research into mechanical properties, storage, and combustion characteristics would be required in the near future.

15.1.4 Recent advances in solid rocket propulsion

SRMs have been used extensively due to their low cost, reliability, safety, long shelf life, ease of design and fabrication, as well as multiple propellant options with variable energetics and relatively shorter times required for firing the motor. As a result, they are generally the first choice for field-operated tactical military applications. Over the last seven decades, enormous technological advancements have been made in each major domain of the design, development, and scaling-up of SRMs, which range from small-caliber sounding rockets to huge boosters delivering enormous thrust. The major achievements of the last two decades could be attained thanks to phenomenal advances in many allied areas that led to the availability of new composite materials for rocket motor casings, high-temperature resistant insulators, hybrid optimization algorithms to find global optimum design details including internal ballistics, grain geometry, nozzle design and throat inserts, and efficient and compact igniter systems.

Research and development work to realize reusable and reignitable SRMs is also in progress in various organizations. Our understanding of the burning behavior, transient burning, and dynamic extinction [119] of solid propellants has enabled us to use experimental observations and theoretical results to construct reliable models for behavioral predictions. These new models have moved towards realistic assumptions at the initial stages of formulation of the problem and subsequently led to their satisfactory validation. Advances in the field of CFD and availability of several commercial code, which include details of subsonic and supersonic flows, modes of heat transfer, chemical kinetics and multiple species interactions, and spatial temperature profiles across the combustor has led to the adoption of new approaches of problem solving that promise robust and reliable SRMs.

Recent innovations have sought to develop a SRM with the capability to start-stop-restart precisely. The perfection of which, when attained, will replace many presently used storable liquid propulsion systems that employ toxic and hazardous propellants, and have obvious safety issues besides the requirement of long pre-launch preparation time. The major difficulty in restarting an extinguished SRM remains its reignition, as the propellant grain surface becomes charred and the burning surface area changes after first operation. Conventional pyrotechnic igniters cannot be used for reignition to restart the motor. An innovative approach to realize a SRM that can restart as per requirement has been attempted at the Los Alamos National University and uses a reusable ignition system using water that is electrolyzed to its constituents and ignited with a spark. The approach to extinguish the burning surface involves a nozzle with a changeable choke area and then resetting it as required.

The scope of this section is not to provide details of the vast recent research work being done in the field. However, for the benefit of readers, some reviews and important developments are noted. The design of a SRM involves considerations of its subsystems such as the combustion chamber, the propellant grain characteristics, insulation, igniter, and nozzle. All involve complex solutions for optimization through intricate iteration processes. The overall approach encompasses structural issues imposed due to varying pressure and thermal loads in addition to the exterior ballistics.

A review of composite materials' application to SRM cases [120] and a new approach to multidisciplinary design optimization [121] of SRMs provide good insight into the work done in the past and current trends in motor case design. Further, work done on the performance assessment of ablative insulators under practical operating conditions [122, 123] in SRMs provides advanced inputs to realize reliable insulation systems for high-temperature operations. Recent advances relating to the design optimization of SRMs have been reviewed [124] and the most used algorithms are presented. It may be inferred that an advanced design approach should select a suitable optimization method that addresses the key aspects of the design problem, keeping in view the overall operation issues, cost, reliability, and safety features.

The work done so far on the development and use of pyrotechnic and pyrogen igniters in SRMs of various sizes has contributed immensely to the successful operation of numerous missiles and launch vehicles. Laser-induced ignition of solid propellants has been an area of interest in the recent past. The approach promises precise and tunable ignition and reignition, as well as a reduction in the hardware weight of the SRM. A review [125] is available on laser ignition for aerospace propulsion. Further, efforts are also being made toward miniaturization of the high-energy laser systems designed for ignition of rocket engines [126]. The size, weight, and cost of the laser system used for ignition are the critical aspects of their acceptance in modern SRMs. Once perfected, these systems will be much easier to use for reignition of extinguished solid propellants.

The history of R&D work on the design and development of nozzles is as old as the SRMs themselves. The nozzle is a critical part of a SRM that plays an important role in the expansion and acceleration of hot combustion gases produced by burning of the propellant in the combustion chamber. This part is subjected to a harsh high-temperature environment throughout the operation of the SRM that leads to high heat transfer due to the high-speed flow of combustion gases, which may cause nozzle throat erosion. Advanced materials and lightweight alloys are presently being used for construction of nozzles and throat inserts to avoid erosion and to reduce the effective weight of the nozzle. Convergent-divergent nozzles are generally used in SRMs and may have different details based on the mission requirement. The main considerations in the selection remain maximum performance, light weight, and ease of manufacture. Most large SRMs have multiple movable nozzles for thrust vector control. An exhaustive review on the subject has recently been published [127]. It may be noted that nozzles are not pre-tested under operational conditions like many other parts of SRMs as this may cause erosion of its throat and deposition of condensed particles, changing the effective throat diameter.

Several SRMs presently use metallized solid propellants to improve their density specific impulse. Metal oxides produce enormous heat on condensation during combustion that should be effectively utilized before the combustion products are exhausted through the nozzle, and yet minimize the two-phase flow losses that may cause unacceptable velocity lag and reduce the net possible gains of metallization of the propellant. Advanced testing of nozzles is done using CFD simulations and suitable software for selected nozzle convergence and divergence angles, throat and exit diameters, and throat lengths, etc, to optimize the design for required performance. Previous experience is also used to fine-tune the design. Small nozzles, as used in micro-thrusters and reaction control systems, require a high level of precision for efficient functioning and control.

15.2 Hybrid rocket propulsion

15.2.1 Introduction

A hybrid rocket is a type of chemical propulsion system where fuel and oxidizer are utilized in two different phases. Generally, fuels are used in a solid phase and oxidizers in a liquid or gaseous phase. The reverse of this is generally not feasible, as most solid oxidizers are in a crystalline state and preparation of a suitable fuel grain is difficult. A schematic of a hybrid rocket is shown in figure 15.4. It is in effect a combination of solid and liquid rocket engines. Like a solid rocket, the fuel is stored within the combustion chamber; the oxidizer is contained in a separate tank, as in a

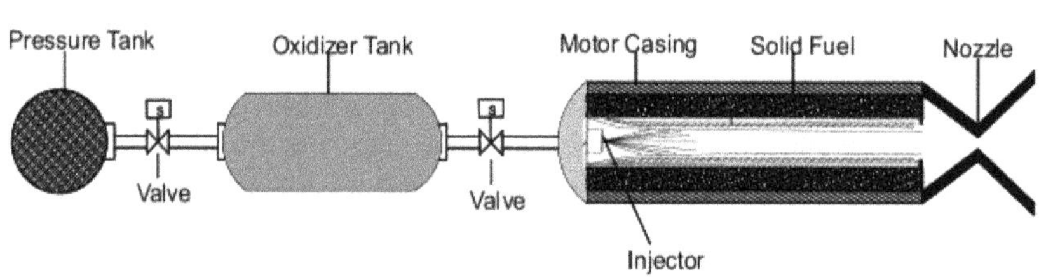

Figure 15.4. Schematic view of a hybrid rocket propulsion system.

liquid rocket engine, and is injected into the combustion chamber prior to the rocket's operation [128].

15.2.2 Combustion process in hybrid rocket motor

Although a hybrid rocket is a combination of solid and liquid rockets, the combustion behavior of the propellants in the combustion chamber is completely different. In a SRM, combustion takes place at the exposed surfaces of the propellant. The flame may be of a premixed or a diffusion type depending on the kind propellant being used [129]. Whereas, in a liquid rocket engine, the fuel and oxidizer are mixed near the injector face to generate a combustible mixture. In a hybrid rocket motor (HRM), the combustion process is controlled by the rate at which the heat from the diffusion flame that exists within the boundary layer is transferred back to the fuel surface [130–133]. Various research efforts have been made to understand the fundamentals of hybrid rocket combustion processes. A detailed combustion model for hybrid rockets was given by Marxman and Gilbert [134] and Marxman [135] in the early 1960s. According to Marxman [134–136], the physical processes involved in hybrid rocket combustion are illustrated in figure 15.5.

It can be seen from figure 15.5 that when oxidizer is injected into the combustion chamber, a boundary layer is formed over the fuel surface. This boundary layer is assumed to be turbulent in nature due to the high injection velocities. After ignition, a diffusion flame sheet is developed within the boundary layer, where the fuel and oxidizer are present within the combustible limit. The heat from the flame zone is transferred to the fuel surface, causing decomposition of the solid fuel through pyrolysis processes. The zone above the flame is oxidizer-rich and the zone below the flame is fuel-rich. The vaporized fuel from the fuel-rich zone and oxidizer from the oxidizer-rich zone are diffused into the flame by diffusion and convection process.

The rate at which the solid fuel recedes due to pyrolysis is known as the regression rate (r, mm s^{-1}) [137]. It is one of the key parameters in a HRM. The regression rate in a hybrid rocket is a strong function of the total mass flux (G). The total mass flux

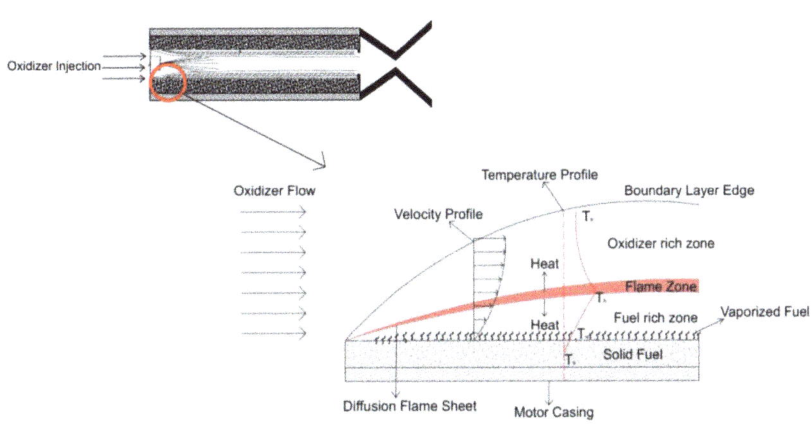

Figure 15.5. Hybrid rocket combustion process.

is a function of the mass flow of oxidizer and the mass flow of fuel. The mass flow of the oxidizer would be constant within the combustion chamber, but the mass flow rate of the fuel changes with the burn time. The mass flow rate of fuel would be relatively smaller compared to the burnt fuel flow rate at any cross-sectional area. Hence, for simplicity, the regression rate in a hybrid rocket is generally taken as a function of the oxidizer mass flux (G_{ox}). The empirical correlation between the regression rate and the oxidizer mass flux is given in equation (15.1). In equation (15.1), 'a' and 'n' are the empirical constants for a given hybrid rocket system. As the combustion process in the hybrid rocket is dominated by the diffusion process and chemical kinetics, the chamber pressure has negligible influence on the regression rate, unlike in SRMs, unless the fuel is highly metallized [137].

$$\dot{r} = a \quad G_{ox}{}^{n} \tag{15.1}$$

15.2.3 Advantages and disadvantages of hybrid rocket motors

15.2.3.1 Advantages

The typical feature of a hybrid rocket is having the fuel and oxidizers in different phases and storing them separately. This feature gives a hybrid rocket advantages over solid and liquid rocket engines, as fuel and oxidizers may be selected from a wide range of potential ingredients. Some of the key benefits of a HRM are discussed below.

Safety. In a solid rocket, the propellant grain is a blend of the fuel and oxidizer; any spark within the motor can lead to unintentional combustion or explosion. However, this is not the case with a hybrid rocket as the propellant components are stored separately and are in two different physical states. Even in the case of spillage of oxidizers, the oxidizer could not easily mix with the solid fuel and thus it would be difficult to cause combustion [128, 129, 138]. Hence, a higher level of safety may be maintained in the manufacturing, transportation, and handling of the inert fuel grains. Moreover, the diffusive nature of the combustion makes the hybrid rocket insensitive to cracks and voids in the fuel grain. Unlike in solid rockets, the regression rate in hybrid rocket is independent of the chamber pressure. This provides additional safety to the hybrid rocket.

Performance. The specific impulse of some of the propellant combinations in a hybrid rocket is potentially better than in solid rockets and equivalent to liquid semi-cryogenic rocket engines (kerosene-liquid oxygen) [129, 139]. The specific impulse data for frequently used common propellant combinations for solid, liquid, and hybrid rockets are given in table 15.6. As can be seen, the density specific impulse of hybrid rockets is lower than the solid rocket and higher than liquid rocket engines.

Propellant selection. Options to select the fuel and oxidizer combinations are abundant for hybrid rockets when compared with solid and liquid rockets. This is mainly due to the possibility of using two different phases of fuel and oxidizer combinations. Energetic additives may be added easily to the fuel grain to enhance the performance [128].

Simplicity in design. In a hybrid rocket, only one propellant constituent, i.e. the oxidizer, is in liquid or gaseous phase that needs to be stored in a tank. Therefore,

Table 15.6. Propellant combinations frequently used in solid, liquid, and hybrid rockets [129, 139].

	Propellants	I_s (s)	$(O/F)_{opt}$	A_e/A_t
Solid rocket	–	–	–	–
–	HTPB(15)/AP(70)/Al(15)	270	2.33	8
Liquid rocket	–	–	–	–
–	UDMH/N_2O_4	304	3.3	31
–	Kerosene/LOX	286	2.1	16
–	LH_2/LOX	453	5	51
Hybrid rocket	–	–	–	–
–	HTPB/LOX	280	1.9	–
–	HTPB/N_2O_4	258	3.5	–
–	PMMA/LOX	259	1.5	–
–	Paraffin Wax/LOX	281	2.5	–
–	Paraffin Wax/N_2O	248	4	–

(Nomenclature: HTPB: hydroxyl-terminated polybutadiene; AP: ammonium perchlorate; Al: aluminum; UDMH: unsymmetrical di-methyl hydrazine; N_2O_4: dinitrogen tetroxide; LOX: liquid oxygen; LH2: liquid hydrogen; PMMA: polymethyl methacrylate; N_2O: nitrous oxide).

the complexity in designing the propellant feed system and hardware weight is reduced in half in comparison to a liquid rocket engine. The solid fuel grain itself acts as the thermal protection system for the combustion chamber casing and, thus, need of cooling is eliminated as is required in a liquid rocket engine.

Flexibility in engine operation. Like a liquid rocket engine, throttling, start-stop, and restart operations are possible in a hybrid rocket and are much simpler when compared to a liquid rocket engine. In a hybrid rocket, these operations can be achieved by just controlling the oxidizer mass flow rate into the combustion chamber [128, 137, 139]. The thrust termination and restart of the combustion process in a SRM is difficult; it is normally essential to wait till the whole propellant has been consumed [129].

Environmental friendliness. The combustion products of hybrid propellant combinations are generally chlorine-free and nontoxic [128, 137].

Low cost. As the possibility of explosion hazards is lower in hybrid rockets, the cost involved in manufacturing, transporting, storing, and handling of the propellant constituents is less compared to solid and liquid rockets [129, 140]. Apart from these, even the overall cost of the propellant combination is much lower compared to solid and liquid rocket engines. For example, the cost of paraffin wax fuel per kilogram is around $ 0.88, whereas the solid propellant (HTPB/AP/aluminum) and liquid fuel (kerosene) are $3.64 and $5.51, respectively [140].

15.2.3.2 Disadvantages

Following the previous discussion on advantages, hybrid rockets seem an attractive alternative to solid and liquid rocket engines. However, certain drawbacks restrict the usability of hybrid rockets in practical applications. More details on the drawbacks associated with hybrid rockes are well addressed in reference [141]. Some of these major drawbacks are given below.

Low regression rate. The low regression rate is a result of the diffusion-controlled combustion process, which is sustained by pyrolysis of the fuel and heat feedback from the flame to the fuel surface through the convection and radiation processes [141, 142]. The regression rate in a hybrid rocket is lower, typically of the order of 1 mm s^{-1} (for HTPB-based fuel) [142].

Low combustion efficiency. In a solid rocket the mixing of fuel and oxidizer are much easier during combustion as both are stored together within the combustion chamber, while in a liquid rocket engine combustion occurs on a droplet-sized scale in the combustion chamber [128, 141]. In a hybrid rocket, the mixing between solid fuel vapors and liquid or gaseous oxidizer occurs in a combustion chamber dominated by the diffusion process within the boundary layer. Due to this the mixing rate is lower, and this results in lower combustion efficiency compared to solid and liquid rocket engines. Normally, the combustion efficiencies in hybrid rockets vary between 70% and 95% depending on the size of the rocket motor and the type of the propellant combinations used [143].

Shift in oxidizer-to-fuel ratio. The oxidizer-to-fuel (O/F) ratio in hybrid rockets is not constant throughout the combustion duration, which contrasts with solid and liquid rocket engines [129, 141]. The O/F ratio has great dependency on the port diameter as it changes during the combustion process. This change in the port diameter of the fuel grain during the combustion process leads to a change in the mass flow rate of the fuel burning, while the mass flow rate of the oxidizer remains constant. The shift in O/F ratio causes a change in the specific impulse during the flight.

Low volumetric efficiency. To achieve sufficient fuel flow rates for higher thrust levels, multi-port grains are used in hybrid rockets [143]. Further, to stabilize the flow entering the combustion chamber and improve the mixing of gaseous products, normally pre- and post-combustion chambers are used in a hybrid rocket. These techniques for attaining higher efficiency cause a lower propellant mass fraction in a typical HRM. The structural integrity of the multi-port grain is another issue that is encountered while designing the solid fuel grain.

Sliver loss. The regression rate in a hybrid rocket is a function of the total mass flux. This mass flux changes from the head end to the nozzle end of the HRM due to the continuous mass addition of the fuel. The change in mass flux leads to an increased regression rate towards the nozzle end and results in the uneven burning of the grain from the head end to the nozzle end. Further, this uneven burning leads to a higher sliver loss compared to a solid rocket engine. Apart from these, the boundary layer thickness also increases from head end towards nozzle end. This increase in boundary layer also changes the flame height from the fuel surface and, in turn, the heat transfer also changes along the length of the grain. This effect also adds to the uneven fuel surface burning from the head end to the nozzle end.

Combustion instabilities. Low-frequency instabilities are dominant in HRMs. Apart from these, instabilities due to longitudinal acoustics and vortex shedding in the pre- and post-combustion chambers are also frequently observed [144]. The combustion instabilities in hybrid rocket are not catastrophic as compared to solid and liquid rocket engines. However, the instabilities in a hybrid rocket can produce thrust oscillations, and may damage the payload and other control systems of the vehicle.

15.2.4 High-regression-rate fuels for hybrid rockets

As discussed earlier, one of the drawbacks of hybrid rockets is their low regression rate. A high regression rate is very important in achieving the desired mass flow rate such that the required thrust can be obtained. To achieve this, various types of fuels have been discovered by researchers. The most common classifications of fuel are liquefying and non-liquefying fuels. A liquefying fuel is that which forms a melt layer during combustion, while a non-liquefying fuel is that which does not form a melt layer at its surface during combustion.

15.2.4.1 Liquefying fuel

A liquefying fuel, such as paraffin wax, forms a melt layer during combustion. This melt layer formation helps in achieving a higher regression rate. The melt layer becomes unstable due to the oxidizer flow in the port, which gets sheared into fine droplets, as shown in figure 15.6. These fine droplets vaporize and mix with the oxidizer and burn in addition to the boundary layer combustion as seen with typical polymeric fuels. This is the reason for the high regression rate observed with liquefying fuel. However, wax or liquefying fuels generally have poor mechanical properties. Hence, in order to make them suitable for practical purposes various attempts have been made by researchers, as will be discussed in the next section.

A. Methods to improve mechanical properties

Due to the various merits of paraffin wax, such as its higher regression rate, ready availability, cheaper cost, safe manufacturing process, and eco-friendly combustion, it is preferred as a hybrid fuel. These merits of such a liquefying fuel have attracted researchers to explore its effectiveness. It was initially studied by Karabeyoglu [145, 146], and research work is still continuing on this fuel. The combustion theory for these fuels was also proposed by the same author using Marxman's theory [134–136]. The main stated drawback, of its poor mechanical properties, may lead the fuel grain to become damaged under various kinds of loads during combustion. High shrinkage during the manufacturing of the wax fuel grains causes a complexity in the grain processing, especially for a large-sized fuel grain. Internal voids and cracks are

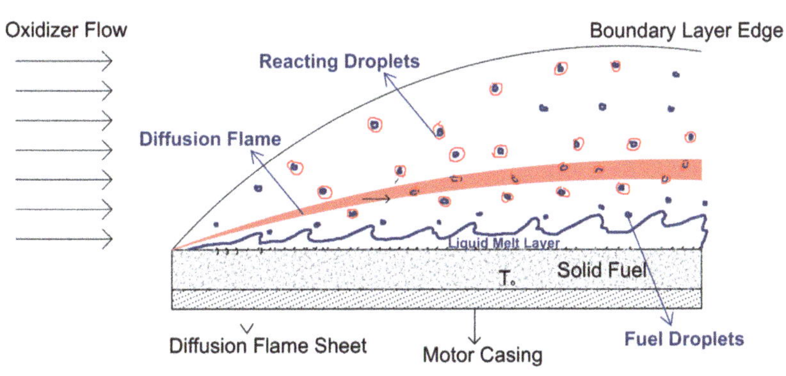

Figure 15.6. Combustion process of liquefying fuel.

common in wax fuel grain [147]. These drawbacks of wax-based fuel restrict its uses for practical applications, even though it gives a higher regression rate.

Mukunda [148] previously reported that the mechanical properties of propellant grains to be used in the form of a cartridge grain should have an ultimate tensile strength of 1 to 2 MPa, percentage elongation at break between 6% and 25%, and a Young's modulus (*E*) of around 30–40 MPa. Unfortunately, wax-based fuels have mechanical properties below these critical values. Their mechanical properties are generally enhanced by adding polymeric additives. The most frequently used polymeric additives are poly propylene, polyethylene (PE), HTPB, ethylene vinyl acetate (EVA), and styrene–ethylene–butylene–styrene (SEBS) copolymer grafted with 2% maleic anhydride. DeSain *et al* [149] obtained a tensile strength from 0.9 MPa to 2.5 MPa and a percentage elongation from 0.4% to 1% with the addition of 4% low-density polyethylene (LDPE) to wax. Kim *et al* [150] also achieved an around 42% increment in maximum tensile strength by adding 10% LDPE to wax. Murayama *et al* [151, 152] utilized a wax-EVA combination for an improvement in tensile strength and percentage elongation. They observed a tensile strength of around 1.5 MPa and a 3.8% increment in elongation, respectively, with use of wax in 20% EVA having 28% vinyl acetate. Kumar and Ramakrishna [153] used 20% EVA with 18% VA and employed a proper solidification for the improvement of the mechanical properties of wax. They achieved a maximum tensile strength, percentage elongation, and Young's modulus of around 3.5 MPa, 17%, and 63.2 MPa, respectively.

When compared with the work of DeSain *et al* [149] and Murayama [151, 152], the mechanical properties obtained by Kumar and Ramakrishna [153] are significantly improved. However, the percentage elongation required for practical applications still needs to be improved further to meet the target properties values stipulated by Mukunda [148]. Recently, Paravan *et al* [154] utilized a SEBS polymer to improve the mechanical characteristics. By adding 20% SEBS to wax they obtained a 48% increase in percentage elongation and a value of 3.5 MPa for tensile strength at 8 °C under a 50 mm min^{-1} strain rate. It was also observed from these studies that use of these additives reduces the regression rate compared to the base fuel. Hence, attempts were made by Dinesh and Kumar [155] to utilize a combination of EVA and SEBS as additives in the wax so that, apart from the improvement in mechanical properties, the net reduction in regression rate remains lower. It was reported that among the various fuel combinations developed, wax mixed with a combination of 10% SEBS and 5% EVA showed better improvement in mechanical properties (maximum tensile strength 4.26 MPa and percentage elongation 24.6%) with less reduction in regression rate. This combination has the potential to be used in practical applications, as it has the required mechanical properties and an improved regression rate, as is evident from the data given in table 15.7.

B. Additives used to improve the regression rate of liquefying fuel

It can be seen from the previous discussion that the regression rate of liquefying fuel is reduced with attempts to improve its mechanical properties. Hence, attempts were

Table 15.7. Various methods attempted by researchers to improve mechanical properties.

Author	Fuel	Tensile strength (MPa)	Percentage elongation (%)	Reduction in regression rate (%)
Murayama *et al* [151, 152]	80% Wax, 20% EVA	1.2	3.7%	35%
Kim *et al* [150]	90 Wax, 10% LDPE	2.2	–	53%
Kumar and Ramakrishna [153]	80% Wax, 20% EVA	4.6	16%	73%
Dinesh and Kumar [155]	85% Wax, 10% SEBS, 5% EVA	4.2	24%	36%
Ishigaki and Nakagawa [156]	20% EVA	3.4	5%	75%
	15% Stearic acid	2.8	3.5%	20%

made to improve the regression rate by adding certain energetic additives such as aluminum, boron, magnesium diboride (MgB_2), sodium amide ($NaNH_2$), potassium bis(trimethylsilyl)amide (PBTSA), sodium borohydride ($NaBH_4$), etc.

Bertoldi *et al* [157] added MgB_2 to paraffin wax in order to enhance the mechanical properties and the combustion characteristics of the hybrid fuel. Experiments were conducted with varying percentages of MgB_2 in the paraffin wax, and an improvement was obtained in the performance of their hybrid motor. $NaNH_2$, PBTSA, and $NaBH_4$ were used by Carayon *et al* [158] as hypergolic solid fuel with MON-25 (25 wt.% nitric oxide, 75 wt.% nitrogen tetroxide) as the oxidizer. The main objective of using these fuels was to utilize them as a hypergolic combination so that start and restart feasibility becomes easier. It was reported that only a few experiments were successful in terms of self-ignition, and that further studies are required to achieve the goal of reliable ignition. Experiments were also conducted at DLR, German Aerospace Center and at the Space Propulsion Laboratory by Kobald *et al* [159] to understand the regression rate of the various kinds of wax available. Paraffin wax, microcrystalline wax, and two other types with different melting points were used. Apart from studies focused on regression rate, varioius additives such as stearic acid up to 10% stearic acid, up to 10% polymer and 2% nano clay along with 5% polymer were incorporated for the improvement of mechanical properties. The addition of these additives showed an improved regression rate as well as the improvements in mechanical properties compared to the paraffin wax used by Karbeyoglu *et al* [145, 146]. The use of metal additives has also been tried to enhance the regression rate in a hybrid rocket. Apart from improving the regression rate, addition of metal also increases the fuel density. Common metal additives that were identified by Ishihara and Brewster [160] include aluminum, boron, lithium, magnesium, and beryllium. The gravimetric and volumetric heat release of boron, aluminum, and beryllium are the highest among all additives [161]. Aluminum and boron have been studied as fuel additives in hybrid fuel by several researchers to improve the energy density, specific impulse, regression rate, performance, and combustion temperature [162–164]. It may be seen from the literature that the improvement in regression rate with the use of metal additives is minimum. Shark *et al* [165] reported that the use of metal hydride leads to a reduction of the aging characteristics of fuel. Hence, researchers studied a new fuel combination comprising dicyclopentadiene with a HTPB-based polymer. The metal hydrides used were $NaBH_4$ and aluminum hydride (AlH_3). The improvement in regression rate recorded was up to 85%, as shown in table 15.8.

Research has also been conducted on hypergolic fuels for hybrid rocket applications. The potential of getting a high regression rate and start and restart capability has attracted researchers to explore such fuels; details are available in reference [169]. Table 15.9 shows the various combinations of hybrid propellant that could be used as hypergolic system. The short ignition delay of a solid-liquid hybrid system is important for their effective utilization in hybrid rockets. Thus, currently efforts are being made by researchers to identify systems with minimum ignition delay for steady combustion in hybrid rockets.

C. Cryogenic fuel as a hybrid fuel

Cryogenic fuels have also been utilized by various researchers. These are generally in a gaseous phase at ambient conditions. A cryogenic fuel is cast as a grain at cryogenic temperature to get the required fuel grain. As it is in gaseous phase at ambient temperature, it gives around an 8–10 times higher regression rate when compared to any polymeric fuel. Several researchers [181–186] have conducted studies on the regression rate using lab-scale rocket motors. Although these fuels showed very high regression rates, they could not be used for any practical applications due to the difficulty in fabricating a large fuel grain.

15.2.4.2 Non-liquefying fuel

The combustion behavior of non-liquefying fuel is different from that of liquefying fuel (figure 15.5). In this case, the combustion takes place within the boundary layer, and the regression rate is highly dependent on the heat transfer from the flame to the fuel surface. As discussed earlier, various additives have also been added to polymeric fuel to achieve a higher regression rate, as is presented in table 15.10.

Connell *et al* [187] investigated experimentally the combustion behavior of solid fuels containing nano aluminum and titanium-aluminum-boron reactive mixed-metal particles (RMNPs) using a static-fired cylindrical motor. It was reported that the addition of 10 wt% RMNPs to HTPB solid fuel increased the measured regression rate by approximately 20% for all oxidizer mass flow rates. Dossi *et al* [188] attempted to utilize activated additives such as activated aluminum and PTFE with HTPB fuels. Activated Al-PTFE ingredients exhibited a very good capability in increasing the regression rate of the hybrid fuel with respect to pure HTPB formulations. Increments were observed ranging from 12.3% at 400 kg m^{-2} s^{-1} to 29.4% at 150 kg m^{-2} s^{-1}. Further, to get a higher regression rate, oxidizer particles were added in a limited quantity with the fuel to enhance the rate. AP is a widely used oxidizer used to

Table 15.8. Various additives utilized in liquefying fuel to get enhanced regression rate.

Authors	Fuel and oxidizers	Additives used	Regression rate ($mm\ s^{-1}$) at given Gox ($kg\ m^{-2}\ s^{-1}$)	Percentage improvement (%)
Bertoldi *et al* [157]	Paraffin wax and nitrous oxide	Magnesium diboride (MgB_2)	2.99	21
Carayon *et al* [158]	Paraffin wax and mixed oxides of nitrogen (MON-25)	Sodium amide ($NaNH_2$), potassium bis(trimethylsilyl)amide (PBTSA), and sodium borohydride ($NaBH_4$)	2.00 (116)	11
Kobald *et al* [159]	Paraffin wax and gaseous oxygen	Stearic acid (SA)	7.20 (100)	137
		Nano clay	6.71 (100)	121
		Polymer	4.96 (100)	63.16
Shark *et al* [165, 166]	Dicyclopentadiene and hydrogen peroxide (H_2O_2)	Sodium	1.37 (295)	47
		borohydride ($NaBH_4$)	2.06 (478)	85
		Aluminum hydride (AlH_3)		
Scaramuzzino *et al* [167]	HTPB, paraffin wax, and gaseous oxygen, N_2O	Carbon black	1.11 (96)	8.82
		Nano aluminum	1.23 (102)	20.61
		Micro aluminum	1.09 (101)	6.86
Karakas *et al* [168]	Paraffin wax and N_2O	Aluminum (Al), magnesium (Mg), lithium aluminum hydride ($LiAlH_4$), ammonia borane (BH_3NH_3)	1.80 (100)	40

Table 15.9. Hypergolic hybrid rocket fuel and oxidizer combinations [169].

Authors	Fuels	Oxidizers
Schmucker [170], Magnusson [171]	Tagaform	White fuming nitric acid (WFNA)
Ankarsward [172]	Sagaform A	Red fuming nitric acid (RFNA)
Schmucker [170], Chiaverini and Kuo [128]	p-toluidine/p-aminophenol	White fuming nitric acid (WFNA)
Pugibet and Moutet [173], Nadaud and Baisini [174]	Metatoluenediamine/nylon	RFNA/hydrogen peroxide
Osmon [175]	Lithium aluminum hydride/ polyethylene	Hydrogen peroxide
Austin *et al* [176], Wernimont *et al* [177]	Manganese dioxide/sodium borohydride/polyethylene	Hydrogen peroxide
Paul *et al* [178]	Difurfurylidenecyclohexanone/ polyisoprene	RFNA
Munjal and Parvatiyar [179]	Aniline formaldehyde/magnesium	99% RFNA/1% ammonium vanadate
Bhosale *et al* [180]	Ammonia borane (AB), 1-ethyl-3-methyl imidazolium borohydride ([EMIM][BH_4]), methyl imidazolium borane (MIMB), and 40% AB-paraffin wax	Hydrogen peroxide

improve regression rate of the fuel [189–191]. The maximum improvement in regression rate obtained was around 200%–400%. The drawback of this approach is that the in-built safety of the hybrid rocket is compromised.

Various methods of improving regression rate

Apart from the development of energetic fuel to achieve the enhanced regression rate in hybrid rockets, as discussed in the previous section, various other techniques have also been attempted by researchers. These include controlling the oxidizer flow pattern into the combustion chamber, using multi-port grain, vortex flow pancakes, etc.

As discussed earlier, the regression rate in any hybrid rocket is lower due to the growth of the boundary layer thickness along the length of the fuel. This causes the flame to move away from the fuel surface and, in turn, reduces the heat transfer to the fuel surface. Thus, various attempts have been made to enhance the regression rate by enhancing the heat transfer to the fuel surface. Various techniques utilized for this purpose use the selection of different injection patterns such as swirl injection, radial injection, convergent-divergent injection, etc. The various injection techniques used to achieve enhanced regression rate are summarized in table 15.11. It may be noted that the improvement in regression rate is remarkable, i.e. up to

Table 15.10. Various additives utilized in non-liquefying fuel to get enhanced regression rate.

Authors	Fuel and oxidizers	Additives used	Regression rate (mm s^{-1}) at given Gox (kg m^{-2} s^{-1})	Percentage improvement (%)
Connell *et al* [187]	HTPB and gaseous oxygen (GOX)	Titanium-aluminum-boron (Ti-Al-B) nanopowder	1.20 (100)	20
Dossi *et al* [188]	HTPB and GOX	Activated aluminium-polytetrafluoroethylene (Al-PTFE)	0.9 (150)	29.40
Venugopal *et al*[189]	HTPB and RFNA	Ammonium perchlorate (AP)	–	100
George *et al* [190]	HTPB and GOX	AP aluminum (Al) particles	–	–
Frederick *et al* [191]	HTPB and GOX	AP and ferric oxide	4.25 (218)	447

Table 15.11. Various injection techniques utilized to get enhanced regression rate.

Authors	Fuel and oxidizers	Techniques utilized	Regression rate (mm s^{-1}) at given Gox (kg m^{-2} s^{-1})	Percentage improvement (%)
Kumar and Kumar [195]	HTPB and gaseous oxygen (GOX)	Swirl injection with different swirl number	2.4 (132)	270
Quadros and Lacava [196]	Paraffin wax and LOX	Swirl injection of different swirl number	6.41 (31.40)	240
Kitagawa *et al* [197]	PMMA and LOX/GOX	Swirl injection of different swirl number	0.9 (180)	–
Hitt and Frederick [192]	PE and GOX	Axial injection and end burning	7.5 (12 bar pressure)	–
Gomes *et al* [198]	High-density polyethylene (HDPE) and GOX	Swirl injection	1.70 (79.9)	50
Kumar and Kumar [199]	HTPB and GOX	Swirl injection with diaphragm	2.8	300
Heydari and Massoom [193]	HTPB and N_2O	Axial, tangential, and swirl injection	0.85 (0.78)	33
Berwal and Biswas [194]	Paraffin wax, polyvinyl chloride-dibutyl phthalate (PVC-DBP) and GOX	Multi-angle diverging injector	3.12 (65.68)	36.14
Lestrade *et al* [200]	HDPE and hydrogen peroxide (H_2O_2)	End-burning swirling flow	0.243	–
Li *et al* [201]	PE and H_2O_2	Multi-location swirl injection	4.83	830
Ohyama *et al* [202]	HDPE and GOX	Multi-location swirl injection	0.69 (50.9)	–
Vignesh and Kumar [203]	PVC-DBP and GOX	Multi-location swirl injection	2.5 (160)	200

around 830%. The use of an axial injector with the end-burning techniques has also been tried by Hitt and Frederick [192]. To generate end burning with an axial injector, porous types of fuel hwre used. Other injectors that have been utilized are axial-tangential injectors [193], multi-angle diverging injectors, [194] etc. These injectors had shown lower regression rates in comparison to swirl injectors.

It has been observed that a single swirl injection is not sufficient for the enhancement of regression rate, and thus attempts are being made to use multi-location swirl injection. In the multi-location swirl injection method, the swirling injection is created at various locations in the grain, as shown in figure 15.7.

Apart from experimenting with injection techniques, attempts have also been made by researchers to disturb the flow internally within the combustion chamber. For this purpose, protrusions or diaphragms are placed in the fuel grain or a bluff body is introduced at the head end of the hybrid motor. As a single protrusion or diaphragm is not sufficient to get an improved regression rate in a long motor, efforts have been made to utilize multi-protrusions or diaphragms at various locations within the fuel grain. Among these, the use of a bluff body has been found to give the greatest improvement in regression rate. A bluff body is generally placed near the injector end to divert the oxidizer flow towards the fuel surface, as shown in figure 15.8.

The recent R&D work on the enhancement of fuel regression rate by using different flow-controlling techniques is summarized in table 15.12. The techniques employed include the use ofa bluff body at head end of the motor, single- and multi-hole diaphragms, and protrusions at multiple locations in the cylindrical port of the grain. The objective of using a bluff body is that it enhances the regression rate at the head end significantly due to an increase in the local oxidizer mass flux. The increased regression rate at the head end contributes to the increased mass flux downstream of the fuel grain. Hence, the overall regression rate is improved, perhaps by as much as 400%.

The use of a multi-port grain also presents an alternative method by which to increase the mass of fuel burning by having an increased number of fuel ports. It has

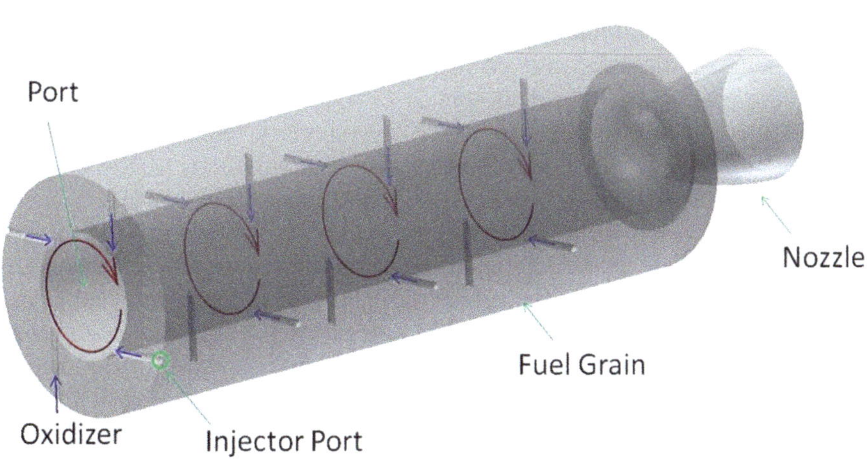

Figure 15.7. Swirl injection at multiple locations [202].

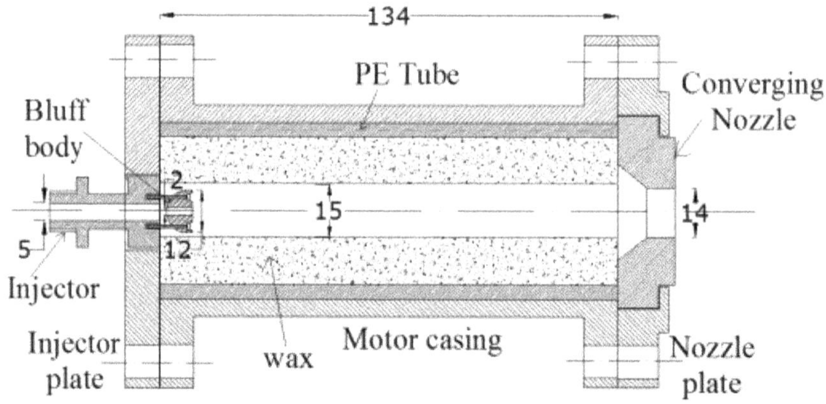

Figure 15.8. Schematic diagram of a hybrid rocket motor along with a bluff body assembly [204].

been utilized by Yakar and Gany [211] and Kearney *et al* [212]. The drawback of this method is that it creates a higher sliver loss, as well as the manufacturing of the grain being tedious. A vortex flow pancake technique has been suggested by Haag *et al* [213]. In this method, oxidizer is injected using a swirl injector into the fuel grain, which consists of two end-burning disks. This leads to a reduction in the volumetric loading of the rocket motor. The low regression rate issue may also be solved by using a fuel and oxidizer which gives optimum specific impulse at higher O/F ratio. Such fuel and oxidizer combinations are PE and HTPB with N_2O and H_2O_2. These studies were conducted by Kearney *et al* [212] and Casalino and Pastrone [214].

15.2.5 Scalability of hybrid rocket engine

One of the biggest challenges in the development of hybrid rockets is the lack of detailed information about their scalability. This is mainly because no generalized model has yet been developed for scaling. As reported in reference [215], hybrid propulsion systems are difficult to design due to the lack of availability of data for a specific configuration, scale, and propellant combination. A detailed description related to the scaling of hybrid motors is reviewed by Nguyen *et al* [216]. An assimilated method is proposed in which a systematic combination of various methodologies is utilized to understand the key fundamentals of the combustion process that can be applied to a variety of hybrid propulsion systems. Swami and Gany [217] conducted a theoretical similarity study by considering the three main similar processes, namely the geometrical similarity, similar fuel and oxidizer combinations, and the mass flow rate with a change in port diameter. For this analysis polymethyl methacrylate (PMMA) was selected as fuel and gaseous oxygen as the oxidizer at a chamber pressure of 25 bar. It was shown that the theoretical validation was in good agreement with test results data. It was also reported that smaller port diameters gave higher overall O/F ratios, the combustion efficiency remained nearly constant, and thrust had increased approximately linearly as the port diameter increased.

Table 15.12. Various flow-controlling techniques utilized to get enhanced regression rate.

Authors	Fuel and oxidizers	Techniques utilized	Regression rate (mm s^{-1}) at given Gox (kg m^{-2} s^{-1})	Percentage improvement (%)
Kumar and Ramakrishna [204]	Paraffin wax and gaseous oxygen (GOX)	Bluff body at head end of motor	8.12 (300)	400
Gany and Timnat [205]	Polyester and RFNA	Single diaphragm	0.8 (90)	50
Dinesh and Kumar [206]	Wax and GOX	Multi-location protrusion	4.40 (130)	29
Grosse [207]	Microcrystalline wax and N_2O	Single- as well as multi-hole diaphragm	3.8 (450)	80
Kumar and Kumar [208]	HTPB and GOX	Single and multi diaphragm	3.8 (500)	120
Kumar and Ramakrishna [209]	Wax and GOX	Single and multi diaphragm	6.2 (280)	45
Sun *et al* [210]	PE and H_2O_2	Single diaphragm	0.59	179

Story *et al* [218] conducted experiments with large-scale hybrid motors at the Hybrid Propulsion Demonstration Program. They conducted successive experiments of four tests to get 250,000 pound thrust in collaboration with NASA's Stennis Space Center. To control the combustion instability, two models were proposed. The first model describes the ramjet stability concept, wherein the American Rocket Company (AMROC) investigated the oxidizer feed line and found that it must be decoupled from any rocket motor oscillations just before the injector. This enabled the oxidizer to accelerate where the local fluid pressure was less than the vapor pressure. It also showed that the oxidizer flow rate is more controlled by vapor pressure than the downstream pressure. It was also observed that by adding a flat fuel deflector into the fuel grain a uniform regression rate was obtained along the length of the grain. The second model was based on AMROC's historical data. The main cause of the combustion instability was the flow behavior of unvaporized liquid oxidizer, which disturbs the boundary layer combustion. It was suggested that the liquid oxidizer disturbs the temperature of the reaction zone along the surface of the fuel grain, which causes an efficient combustion area towards the end of the rocket. It also causes an unwanted pressure gradient along the fuel grain that leads to unwanted low-frequency instability.

Story [219] reported large-scale motor testing from NASA's Marshall Space Flight Center. They considered various details on how fuel regression rates decrease with an increase in the port diameter by keeping the oxidizer flux constant. Combustion links with the boundary layer are the main method for keeping the boundary layer stable and appears to be the main requirement for a large-scale HRM.

Karabeyogly *et al* [220, 221] conducted scale-up studies using paraffin wax as the hybrid fuel. Grains of 19 mm diameter and 1.14 m length were fabricated in the laboratory at Stanford University. To demonstrate the scaling studies, a series of scale-up tests with gaseous oxygen were carried out using a new Hybrid Combustion Facility (HCF) at NASA's Ames Research Center. The data generated from these studies were in good agreement with the small-scale motor that was conducted at low pressure and low mass flux.

Paccagnella *et al* [222] conducted numerical studies with swirl injection at Università degli Studi di Padova, Padouva, Italy, along with the Space Propulsion Group Inc., Sunnyvale, California. The main objective of their studies was the examination of the various performance parameters relating to the various configurations. Three swirl injectors of different swirl numbers were used: an increased regression rate was obtained and the performance of the hybrid motor improved with the increase of the swirl number. Scaling studies with respect to the oxidizer mass flux have also been undetaken.

15.2.6 Developments in hybrid rocket motor technology and its status

Development of hybrid rockets started in the 1930s [223–225]. The first flight test of a hybrid rocket (GRID-09, gelled gasoline with liquid oxygen, LOX) was conducted by Korolev and Tikhonravov. Several flight tests were performed in the early 1950s. Among them, the most successful one was XDF-23, which used a rubber-based fuel and reached 9 km altitude. The major developments in our understanding of hybrid rocket propulsion took place during the period from the mid-1950s to the early 1970s. Over this period, various combustion models (including Marxman's combustion model [134–136]) for hybrid rockets were developed. Several fuel and oxidizer combinations were tested by the United Technologies Corporation, NASA, Office National d'Etudes et de Recherches Aérospatiales (ONERA), and the General Electric Company, among others. In the year 1967, the LEX sounding rocket (nitrous oxidize/meta toluene diamine nylon) that was developed by ONERA reached 100 km altitude. Research progress on hybrid rockets declined during the period 1970–80 due to the lack of state-of-the-art technologies to deal with hybrid rocket problems. In the 1980s, the USA developed several target drones such as Sandpiper (PMMA/Mg/MON), the High Altitude Supersonic Target (PB/PMMA/IRFNA), and Firebolt. Later, in the 1990s, researchers revived interest in hybrid rocket research as the safety of propulsion systems became a high priority. The availability of cutting-edge technologies relating to computational tools, measurement techniques, and development of various propellant combinations helped to gain insight into the fundamentals of hybrid rocket systems. In the span of 15 years, from 1985 to 2000, large-scale HRMs were built and tested by Starstruck, AMROC, and consortia of companies such as Space Dev and Thiokol. In most of these tests, LOX/HTPB was used due to its good physical properties, performance, and low cost. One common problem that was encountered by these research groups was combustion instabilities when the motor was scaled-up to a large size. So far, SpaceShipOne, containing a hybrid propulsion system (N_2O/HTPB), is the only vehicle built and tested for suborbital flights. This was built by Scale Composites and won the famous Ansari X prize in 2004. One of the greatest advancements in the past two decades of hybrid rocket research was utilizing paraffin wax as a solid fuel. Paraffin wax produces 3–4 times higher regression rate than the conventional polymer-based fuels such as HTPB, PMMA, and PE [144, 145].

In the past two decades, significant developments have been achieved in hybrid rocket propulsion in terms of performance and cost. This has attracted investors to utilize hybrid rockets in commercial space-exploration activities. In the present scenario, several private companies are focusing on the development of HRMs for sounding rocket applications, small satellites, and low Earth orbit transfer applications. Information on the companies currently working on HRMs for different applications is given in table 15.13.

Most of the working suborbital rockets utilize SRMs, but the recent requirements for economical sounding rockets have drawn attention towards hybrid systems. In most cases, their feasibility for use in launch vehicle technologies has been demonstrated. Although most sounding rockets utilize SRMs [130, 131], several new developments programs are being considered for suborbital missions using hybrid rocket propulsion.

Table 15.14 presents the details of modern small rocket demonstrators using hybrid rocket propulsion and summarizes the cumulative data of the companies that are currently engaged in developing sounding rockets and launch vehicles that use hybrid rocket propulsion. Most of them are at a developmental stage.

Table 15.13. List of companies currently working on the hybrid rocket propulsion [226].

Company	Country	Propellant Combination	Application
HyImpulse	Germany	Paraffin-LOX	Sounding rocket, launch vehicle
Nammo	Norway	HTPB-H_2O_2	Sounding rocket, launching vehicle
The Spaceship Company	USA	Polymide-N_2O	Suborbital manned
Space Propulsion Group	USA	Paraffin-LOX	Orbital
TiSpace	Taiwan	HTPB-N_2O	Sounding rocket, launch vehicle
Gilmour Space	Australia	3D printed fuel-H_2O_2	Sounding rocket, launch vehicle
Rocket Crafters	USA	3D printed ABS (pyrolysis of acrylonitrile-butadiene-styrene)-N_2O	Launch vehicle
T4I	Italy	Paraffin-H_2O_2	Orbital propulsion
Space Forest	Poland	Paraffin-N_2O	Sounding rocket
Space Link	Slovenia	Paraffin-LOX	Sounding rocket
Equatorial Space Ind.	Singapore	Paraffin-LOX	Launch vehicle

In 2021, Virgin Galactic successfully completed its first manned suborbital missions, known as VSS Unity. In this, a hybrid rocket propulsion system was used to reach a suborbital location (i.e. below the Karman line). A similar propulsion system had been used in the past in their SpaceShipOne.

Figure 15.9 represents the increasing activity of and demand for suborbital flights for the period 2000 to 2025. It can be seen that in the past 5 years the number of flights has increased significantly. These flights were only operated by Blue Origin (New Shepard) and Virgin Galactic (SpaceShipTwo). These companies are presently utilizing either liquid or solid rockets in most of their flights. This involves a higher cost regarding launch, propellant management, transportation, and safety. These drawbacks can be mitigated using hybrid rocket propulsion. Developing a launch vehicle for suborbital flights with hybrid rockets would be advantageous in terms of safety, cost, and nontoxicity. Therefore, many academic institutions and advanced projects are working on the use of HRMs for suborbital vehicles. The increase in the intensity of hybrid rocket propulsion research and the growing number of attempts to take the technology beyond the laboratory environment promise great prospects for the future, and wider successful utilization of HRMs in suborbital vehicles is likely. The recent flight by VSS Unity has further boosted confidence in the hybrid rocket community. If the number of launch vehicles or sounding rockets were to

Table 15.14. Data showing the various modern small hybrid rocket motors used as demonstrators [227].

Parameter	ILR-33 'Amber' [228, 229]	Nucleus [230, 231]	Atea 1 [232]	Peregine [233, 234]
Maximum diameter (mm)	230	356	150	406
Length (m)	5.00	9.00	6.00	9.75
Launch mass (kg)	<200	820	60	850
Hybrid propellant type	98% Hydroxytryptophan (HTP)/PE	87.5% HTP/HTPB+C	N_2O/polymer-based	N_2O/paraffin
Fuel grain geometry	Multi-port	Multi-port	Unknown	Single-port
Propellant mass fraction	Not released	0.58	0.75	0.52
Motor burn duration (s)	40.0	25.0	14.5	18.0
Rocket staging	Hybrid main stage with SRM boosters	Hybrid single stage rocket	Hybrid booster stage with dart	Hybrid single stage rocket
Maximum apogee (103 m)	100	120	150	100
Status	Test-flown	Development	Test-flown	Development

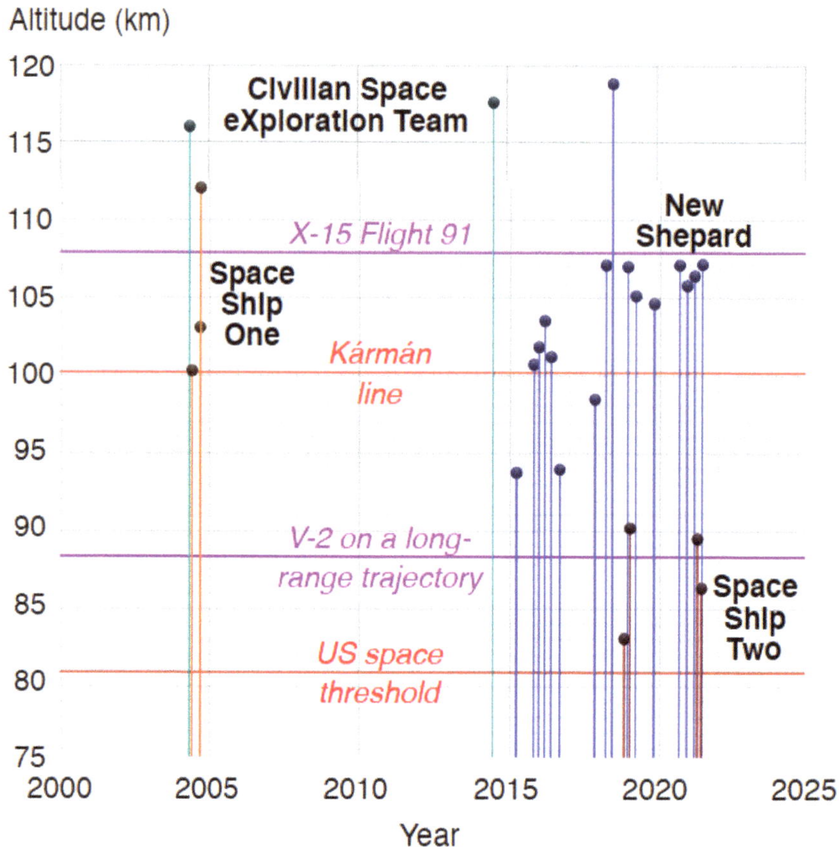

Figure 15.9. Suborbital flights over the period from 2000 to 2025.

increase, the cost per launch would also decrease. Therefore, space would become easy to access for several market players.

New Space Global [198] is a consulting company that provides financial analyses related to the various commercial exploitations of space. The company has identified eight pillars that could lead to success in the developing space market. The analysis starts from the already existing market segments, which include spacecraft manufacturers and launch vehicle providers offering the basic infrastructure requirements. The next two pillars deal with human spaceflight and microgravity research. Much of this is already in place but mainly under the direct management of governments. The remaining four verticals correspond to more ambitious developments, which include spaceport infrastructure, in-space services, space resources exploitation, and space-based energy. There is a lot of scope for human spaceflight and microgravity research that may generate revenue from space tourism. The development in the field of space tourism requires the development of improved technologies. It further requires a reduction in the cost-of-service price, along with rapid development of propulsion systems. The applications related to space propulsion are highly dependent on the cost reduction that again depends strongly on space vehicle design. Hence, any strategy for space propulsion cost reduction will rely mainly on two approaches. The first approach is based on the reduction of overall costs, and the second approach requires higher-performance rocket engines. Hybrid rocket engines hold viable promise towards meeting that demand.

As mentioned earlier, the importance of HRMs for suborbital systems is increasing with the progress being made in both commercial as well as public-funded efforts. Commercial vehicles such as Copenhagen Suborbitals' Tycho Brahe, Virgin Galactic's SpaceShipTwo and Whitting Hill Aerospace's mCLS are expected to be powered by HRMs [222]. Contemporary analysis suggests that the current launch vehicle capacities are not suitable to meet the required demands. The demand for the launch of micro/nanosatellites is also increasing. This will be only feasible by increasing the number of economical launches. In this situation, hybrid launch vehicles could be developed for small payloads based on the knowledge established by suborbital applications.

References

[1] Hunley J D 1999 The history of solid-propellant rocketry: what we do and do not know *35th Joint Propulsion Conf. and Exhibit, AIAA-99–2925 (LA, USA, June 20–24)*

[2] Umholtz P D 1999 The history of solid rocket propulsion and aerojet *35th Joint Propulsion Conf. and Exhibit, AIAA-99–2927 (LA, USA, June 20–24)*

[3] Price E W 1998 History of solid rocket motors *34th Joint Propulsion Conf. and Exhibit (OH, USA, July 13–15)*

[4] DeLuca L T 2017 Highlights of solid rocket propulsion history *Chemical Rocket Propulsion* ed L T DeLuca *et al* (Switzerland: Springer) pp 1015–32

[5] Geisler R L, Frederick R A Jr and Giarra M 2010 Historical overview and solid rocket motor fundamentals *Encyclopedia of Aerospace Engineering* (New York: Wiley)

[6] Davenas A (ed) 1992 *Solid Rocket Propulsion Technology* (Oxford: Pergamon)

[7] Sutton G P and Biblarz O 2016 *Rocket Propulsion Elements* 9th edn (New York: Wiley)

[8] Ramamurthi K 2016 *Rocket Propulsion* 2nd edn (Bangaluru: Trinity Press)

[9] Heister S D, Anderson W E, Pourpoint T L and Cassady R J 2019 *Rocket Propulsion* (Cambridge: Cambridge University Press)

[10] Singh H and Shekhar H 2017 *Solid Rocket Propellants: Science and Technology Challenges* (London: Royal Society of Chemistry)

[11] Hugget C, Bartley C E and Mills M M 2016 *Solid Propellant Rockets* (Princeton, NJ: Princeton University Press)

[12] Kubota N 2015 *Propellants and Explosives: Thermal Aspects of Combustion* (New York: Wiley)

[13] Lengelle G, Bizot A, Duterque J and Trubert J F 1984 Steady-state burning of homogeneous propellants *Fundamentals of Solid-Propellant Combustion Progress in Astronautics and Aeronautics* ed K K Kuo and M Summerfield vol 90 (New York: AIAA) pp 361–407

[14] Safta D, Vaisile T and Ion I 2012 Regarding the perturbed operating process of DB propellant rocket motor at extreme initial grain temperatures *INCAS Bull.* **4** 77–92

[15] Kubota N, Ohlemiller T J, Cavney L H and Summerfield M 1974 Site and mode of action of platonizers in double-base propellants *AIAA J.* **12** 1709–14

[16] Youfang C 1987 Combustion mechanism of double-base propellants with lead burning catalysts *Prop. Explos. Pyrotech.* **12** 209–14

[17] Warren L R, Wang Z, Pulham C R and Morrison C A 2020 A review of catalytic effects of lead-based ballistic modifiers on combustion of double-base propellants *Prop. Explos. Pyrotech.* **46** 13–25

[18] Wang Y, Yan Q L, An T, Chen B, Ji Y P, Wang W and Zhao F 2018 Unravelling the effect of anthraquinone metal salts as wide range plateau catalysts to enhance combustion properties of solid propellants *J. Energ. Mater.* **15** 376–90

[19] Wei T, Zang Y, Xu K, Ren Z, Gao H and Zhao F 2015 Catalytic action of nano Bi_2WO_6 on therma decomposition of AP, RDX, HMX and combustion of NG/NC propellant *RSC Adv.* **5** 70323–8

[20] Kubota N 1984 Survey of rocket propellants and their combustion characteristics *Fundamentals of Solid-Propellant Combustion Progress in Astronautics and Aeronautics* ed K K Kuo and M Summerfield vol 90 (New York: AIAA) pp 1–52

[21] |; Babuk V A, Glebov A, Arkhipov V A, Vorozhtsov A B, Klyakin G F, Severini F, Galfetti L and DeLuca L T 2005 Dual-oxidizer solid rocket propellants for low-cost access of space *10th Int. Workshop on Combustion and Propulsion* 15–20

[22] Kumar P 2018 An overview on properties, thermal decomposition, and combustion behavior of ADN and ADN based solid propellants *Defence Technol.* **14** 661–73

[23] Schoyer H F R, Welland-Veltmans W H M, Louwers J, Korting P A O G, van der Heijden A E D M, Keizers H L J and van den Berg R P 2002 Overview of the development of hydrazinium nitroformate-based propellants *J. Propul. Power* **18** 138–45

[24] Politzer P, Murray J S, Grice M E and Brinck T 2000 Two potential energetic compounds: ammonium superoxide and ammonium ozonide *J. Energ. Mat.* **18** 89–95

[25] Cristilli F, Weiser V, Maggi F, Imiolek A, Tagliabue C, Gettwert V and Dossi S 2017 Burning behavior of ADN-based propellants loaded with Al-Mg mechanically activated powders *7th European Conf. for Aeronautics and Space Sciences*

[26] Kettner M A and Klapotke T M 2017 Synthesis of new oxidizers for potential use in chemical rocket propulsion *Chemical Rocket Propulsion* ed L T DeLuca *et al* pp 63–88 (Switzerland: Springer International Publishing)

[27] Dey A, Sikder A K, Talawar M B and Chattopadhyay S 2015 Towards new directions in oxidizers/energetic fillers for composite propellants: an overview *Cent. Eur. J. Energetic Mater.* **12** 377–99

[28] Boshra I K, Elbeih A and Mostafa H E 2019 Composite solid rocket propellant based on GAP polyurethane matrix with different plasticizers *18th Int. Conf. on Aerospace Sciences and Aviation Technology IOP Conf. Series: Materials Science and Engineering*

[29] Komai I, Kobayasi K and Kato K 2000 Burning rate characteristics of glycidyl azide polymer (GAP) fuels and propellants *Solid Propellant Chemistry, Combustion, and Motor Interior Ballistics Progress in Astronautics and Aeronautics* ed V Yang, T B Brill and W Ren vol 185 (Reston, VA: AIAA) pp 455–64

[30] Kawamoto A M, Diniz M F, Lourenco V L, Takahashi M F K, Keicher T, Krause H, Menke K and Kempa P B 2010 Synthesis and characterization of GAP/BAMO copolymers applied at high energetic composite propellants *J. Aerospace Technol. Manag.* **2** 307–22

[31] Bazaki H 2000 Combustion mechanism of 3-azidomethyl-3-methyloxetane (AMMO) composite propellants *Solid Propellant Chemistry, Combustion, and Motor Interior BallisticsProgress in Astronautics and Aeronautics* vol 185 ed V Yang, T B Brill and W Ren (Reston, VA: AIAA) pp 439–54

[32] Xiao Z, He W, Ying S, Zhou W and Xu F 2014 Current trends in energetic thermoplastic elastomers as binders in high energy insensitive propellants in China *Sci. Tech. Energ. Mater.* **75** 37

[33] Bogdan F, Bogusz R, Skupinski W, Chmielarek M and Dzik A 2015 Study of the effect of nitrated hydroxyl-terminated polybutadiene (NHTPB) on the properties of heterogeneous rocket propellants *Cent. Eur. J. Energetic Mater.* **12** 841–54
[34] Price E W 1984 Combustion of metallized propellants *in Fundamentals of Solid-Propellant Combustion Progress in Astronautics and Aeronautics* ed K K Kuo and M Summerfield vol 90 (Reston, VA: AIAA) pp 479–514
[35] DeLuca L T and Galfetti L 2008 Burning of metallized composite solid propellants: from micrometric to nanometric aluminum size *Asian Joint Conf. on Propulsion and Power (Gyeongiu, Republic of Korea)*
[36] Alexey K, Vasiliy P, Ksenia M and Dmitry K 2020 Metalized solid propellant combustion under high-speed blowing flow *J. Mech. Sci. Technol.* **34** 2245–53
[37] Minkov L L, Shrager E R and Pikushchak E V 2014 Optimum disposition of metal particles in the propellant grain *Int. J. Aerospace Eng.* **2014** 958758
[38] Silvera I F and Cole J W 2010 Metallic hydrogen: the most powerful rocket fuel yet to exist *J. Phys.: Conf. Ser.* **215** 012194
[39] Zhang C, Sun C, Hu B, Yu C and Lu M 2017 Synthesis and characterization of pentazolate anion cyclo N_5^- in $(N_5)_6(H_3O)_3(NH_4)_4Cl$ *Science* **355** 374–6
[40] DeLuca L, Bohn M A, Gettwert V, Weiser V and Tagliabue C 2018 Innovative solid propellant formulations for space propulsion *Energetic Materials Research, Applications, and New Technologies* (Hershey, PA: IGI Global) pp 1–24
[41] Yan X, Xia Z, Huang L, Feng Y and Na X 2017 Experimental study on combustion process of NEPE propellant *7th EUCASS (Milan, Italy)*
[42] Pang W, Xu H, Li Y and Shi X 2012 Characteristics of NEPE propellant with ammonium dinitramide (ADN) *Adv. Mater. Res.* **399–401** 279–83
[43] Sun Y, Ren H and Jiao Q 2017 Comparison of thermal behaviors and decomposition kinetics of NEPE propellant before and after storage *J. Therm. Anal. Calorim.* **131** 1–11
[44] Masoodi M 2013 Preparation of NEPE polymeric missiles propellants and effect of particle size in mechanical properties *Polym. Res. J.* **7** 337–43
[45] Xiong Y, Sui X and Wang N 2015 Research on NEPE propellant life prediction model *51st AIAA/ASME/SAE/ASEE Joint Propulsion Conf. and Exhibit (FL, USA)* AIAA-2015–3872
[46] Lia H, Xua J, Liua J, Wanga T, Chena X and Hong W 2021 Research on the influences of confining pressure and strain rate on nepe propellant: experimental assessment and constitutive model *Defence Technol.* **17** 1764–74
[47] Han L, Chen X, Xu J, Zhou C and Yu J 2015 Research on the time-temperature-damage superposition principle of NEPE propellant *Mech. Time-Depend. Mater.* **19** 581–99
[48] Milekhin Y M, Koptelov A A, Shishov N I, Koptelov I A and Rogozina A A 2018 Evaporation of plasticizer from NEPE type propellant *Russ. J. Appl. Chem.* **91** 802–12
[49] Xie W, Zhao Y, Zhang W, Liu Y, Fan X, Wang B, He W and Yan Q 2018 Sensitivity and stability improvements of NEPE propellants by inclusion of FOX-7 *Propellants, Explos. Pyrotech.* **43** 1–8
[50] Wang Y, Zhang W, Xiao K, Yao W, Xie W, Liu Y, Yang J and Chen Y 2021 Experimental and simulation study of the phase separation of neutral polymeric bonding agent in nitrate ester plasticized polyether propellant and its application *Colloid Surf.* **610** 125665
[51] Kumar R N 1973 Condensed phase details in the time-dependent combustion of AP composite propellants *Comb. Sci. Tech.* **8** 133–48
[52] Wise H, Inami S H and McCulley L 1967 Role of condensed phase reactions in ignition and deflagration of AP propellant *Combust. Flame* **11** 483–8

[53] Kishore K, Pai Vernekar V R, Sunitha M R and Prasad G 1977 Contribution of condensed phase reaction in the combustion of solid rocket fuels *Fuel* **56** 347–52
[54] Beckstead M W, Derr R L and Price cf 1971 The combustion of solid monopropellants and composite propellants *Symp. (Int.) Combust.* **13** 1047–56
[55] Blair D W, Bastress E K, Hermance C E, Hall K P and Summerfield M 1960 Some research problems in steady-state burning of composite solid propellants *Solid Propellant Rocket Research* (New York: Academic) pp 183–206
[56] Cohen N S 1980 Review of composite propellant burn rate modeling *AIAA J.* **18** 277–93
[57] Kumar R N and New A 1973 Look at AP/composite propellant combustion *JPL Quart. Tech. Rev.* **3** 53–77
[58] Beckstead M W, Derr R L and Price cf 1970 A model of composite solid propellant combustion based on multiple flames *AIAA J.* **8** 2200–7
[59] Kishore K and Comprehensive A 1979 View on the combustion models of the composite solid propellants *AIAA J.* **17** 1216–27
[60] Kubota N 1984 Survey of rocket propellants and their combustion characteristics *in Fundamentals of Solid-Propellant Combustion Progress in Astronautics and Aeronautics* ed K K Kuo and M Summerfield vol 90 (Reston, VA: AIAA) pp 1–52
[61] Cai W, Thakre P and Yang V 2008 A model of AP-HTPB composite propellant combustion in rocket motor environments *Comb. Sci. Tech.* **180** 2143–69
[62] Massa L, Jackson T L and Buckmaster J 2005 New kinetics for a model of heterogeneous propellant combustion *J. Propul. Power* **21** 914–24
[63] Waesche R H W and Wenograd J 2000 Calculation of solid propellant burning rates from condensed-phase decomposition kinetics *Comb. Explos. Shock Waves* **36** 125–34
[64] Price E W 1984 Combustion of metallized propellants *Fundamentals of Solid-Propellant Combustion Progress in Astronautics and Aeronautics* ed K K Kuo and M Summerfield vol 90 (Reston, VA: AIAA) pp 479–512
[65] Boggs T L, Zurn D E and Netzer D W 1973 Ammonium perchlorate combustion: effect of sample preparation, ingredient type and pressure, temperature and acceleration environments *Comb. Sci. Tech.* **7** 177–183
[66] Peterson E D, Rodriguez F A, Thomas J C, Dillier C A and Peterson E L 2019 Combustion behavior of ammonium perchlorate at high pressures *AIAA Propulsion and Energy 2019 Forum* AIAA 2019-4366
[67] Kumari R and Prasad M R R 2020 Thermal decomposition of ammonium perchlorate: Part 1 effect of heating rate *J. Appl. Chem.* **9** 441–6
[68] Strunin V A, Shkadinskii K G and Nikolaeva L I 2010 Effect of ammonium perchlorate particle size on its burning rate and combustion limit in terms of pressure *Comb. Explos. Shock Waves* **46** 183–7
[69] Bazaki H and Kubota N 2000 Effect of binders on the burning rate of AP composite propellants *Propellant, Explos. Pyrotech.* **25** 312–16
[70] Enew A M, Abadir E, Elmarsafy S and Elsharkawy K K 2019 Effects of plasticizers, antioxidants and burning rate modifiers on aging performance of the HTPB/HMDI composite solid propellant *HannengCailiao/Chin. J. Energ. Mater.* **27** 21–7
[71] Kishore K and Gayathri V 1984 Ignition and combustion of AP based propellants *Fundamentals of Solid-Propellant Combustion Progress in Astronautics and Aeronautics* ed K K Kuo and M Summerfield vol 90 (Reston, VA: AIAA) pp 98–106

[72] Yadav N, Srivastava P K and Varma M 2021 Recent advances in catalytic combustion of AP based composite solid propellants *Defence Technol.* **17** 1013–31
[73] Hegab A M 2007 An overview of solid propellant combustion modeling *12th Int. Conf. on Aerospace Science and Aviation Technology*
[74] Maggi F, DeLuca L, Colombo G, Bandera A and Galfetti L 2009 Modelling of diffusion and pre-mixed flame of AP/HTPB laminate propellant *Prog. Propuls. Phys.* **1** 109–20
[75] Thomas J C, Peterson E and Walker J M 2019 Updated three-flame modelling of composite AP/HTPB propellants *27th ICDERS (Beijing, China, July 28–Aug. 2)*
[76] Vasenin I M, Narimanov R K, Glazunov A A, Kuvshinov N E and Ivanov V A 1995 Two-phase flows in nozzles of solid rocket motors *J. Propuls. Power* **11** 583–92
[77] Li Z, Wang N, Shi B, Li S and Yang R 2019 Effect of particle size on two-phase flow loss in aluminized solid rocket motor *Acta Astronaut.* **159** 33–40
[78] DeLuca L T, Galfetti L, Severini F, Meda L, Marra G, Vorozhtsov A B, Sedoi V S and Babuk V A 2005 Burning of nano-aluminized composite rocket propellants *Comb., Explos. Shockwaves* **41** 680–92
[79] Pang W, Li Y, DeLuca L T, Liang D, Quin Z, Liu X, Xu H and Fan X 2021 Effect of nanopowders on the performance of solid rocket propellants: a review *Nanomaterials* **11** 2749
[80] Krainov A Y, Poryazov V A and Krainov D A 2021 Mathematical modeling on extinction of metallized composite solid propellants under a sudden drop in pressure *Propellant, Explos., Pyrotech.* **47** e202100123
[81] Krainov A Y, Poryazov V A, Moiseeva K and Krainov D A 2020 Metalized solid propellant combustion under high-speed blowing flow *J. Mater. Sci. Technol* **34** 2245–53
[82] Emelyanov V N, Teterina I V, Volkov K N and Garkushev A U 2017 Pressure oscillations and instability of working processes in combustion chambers of solid rocket motors *Acta Astronaut.* **135** 161–71
[83] Albatai F A, Hegab A M, Redy M A, Abuhabaya A A and El-Behery S M 2021 Turbulent flow characteristics in a model of a solid rocket motor chamber with sidewall mass injection and end-wall disturbance *Math. Prob. Eng.* **2021** 9978102
[84] Xu G, Liu P, Ao W and Jin B 2020 Interactions between turbulent flow and diffusion flame of composite propellant in a solid rocket motor with sidewall mass injection *Aerospace Sci. Tech.* **102** 105872
[85] Cross P G 2019 Radiative heat transfer in solid rocket nozzles *J. Spacecraft Rockets* **57** 1–14
[86] Ma Y J, Bao F T, Xu H and Sun L 2015 Numerical simulation of solid propellant rocket exhaust plume considering two-phase flow effects *Appl. Mech. Mater.* **798** 185–9
[87] Zhou Z, Zhao C, Lu C and Le G 2021 Numerical studies on four-engine rocket exhaust plume impinging on flame deflectors with after-burning *Defence Technol.* **17** 1207–16
[88] Saile D *et al* 2021 Characterization of SRM plumes with alumina particulate in subscale testing *CEAS Space J.* **13** 247–68
[89] Varma M, Chatterjee A K and Pandey M 2002 Ecofriendly propellants and their combustion characteristics *Advances in Solid Propellant Technology* ed M Varma and A K Chatterjee (New Delhi: Tata McGraw-Hill) pp 144–79
[90] Oommen C and Jain S R 1999 Ammonium nitrate: a promising rocket propellant oxidizer *J. Hazard. Mater.* **67** 253–81
[91] Varma M, Pandey M and Gupta B L 2003 Pressure dependence of thermal decomposition of ammonium nitrate oxidizer *J. Ind. Chem. Soc.* **80** 569–74

[92] Varma M and Pandey M 2000 Thermal decomposition studies on catalyzed ammonium nitrate *Proc. 3rd Int. HEMCE (Thiruvananthapuram, India)* 182–8
[93] Varma M and Pandey M 2005 Thermal degradation of HTPB, AN, their virgin and catalyzed propellants under slow decomposition conditions *Proc. 5th Int. HEMCE*
[94] Engel W and Heinisch H 2003 *Process for Producing Phase-stabilized Ammonium Nitrate* US Patent 6,508,995 B1
[95] Kumar P 2019 Advances in phase stabilization techniques of AN using KDN and other chemical compounds for preparing green oxidizers *Defence Technol.* **15** 949–57
[96] Manelis G B and Lempert D B 2009 Ammonium nitrate as oxidizer in composite solid propellants *2nd EUCASS*
[97] Maggi F, Arosio L and Galfetti L 2017 Catalyst screening of ammonium nitrate oxidizer *7th EUCASS (Milan, Italy)*
[98] Kohga M, Naya T and Okamoto K 2012 Burning characteristics of ammonium nitrate based composite propellants with hydroxy-terminated polybutadiene/polytetrahydrofuran blend binder *Int. J. Aerosp. Engg.* **2012** 378483
[99] DeLuca L T, Galfetti L, Severini F, Galeotta M, De Amicis R, Babuk V A, Kondrikov B N and Vorozhtsov A B 2004 Low-cost and green solid propellants for space propulsion *Proc. 2nd Int. Conf. on Green Propellants for Space Propulsion (Sardinia, Italy)*
[100] Larsson A and Wingborg N 2011 Green propellants based on ammonium dinitramide (ADN) *Advances in Spacecraft Technologies* ed J Hall (London: Intech Open)
[101] Cican G and Mitrache A 2017 Rocket solid propellant alternative based on ammonium dinitramide *INCAS Bull.* **9** 17–24
[102] Lang J and Bohn M A 2020 Decomposition pathways of ammonium dinitramide (ADN) and its HNO_3 clusters eluciated by DFT calculations *Propellant. Explos. Pyrotech.* **46** 233–44
[103] Weiser V, Franzin A, Gettwert V and DeLuca L T 2013 Combustion of metallized ADN/GAP solid rocket propellants with focus on agglomeration effects *5th EUCASS (Munich, Germany)*
[104] Cristilli F, Weiser V, Maggi F, Imiolek A, Tagliabue C, Gettwert V and Dossi S 2017 Burning behavior of and based propellants loaded with Al–Mg mechanically activated powders *7th EUCASS (Milan, Italy)*
[105] Sinditskii V P, Egorshev V Y, Levshenkov A I and Serushkin V V 2006 Combustion of ammonium dinitramide, part 1: burning bevavior *J. Prop. Power* **22** 769–76
[106] Sinditskii V P, Egorshev V Y, Levshenkov A I and Serushkin V V 2006 Combustion of ammonium dinitramide, part 2: combustion mechanism *J. Prop. Power* **22** 777–85
[107] Kumar P 2018 An overview of properties, thermal decomposition and combustion behavior of ADN and ADN based solid propellants *Defence Technol.* **14** 661–73
[108] Weiser V, Franzin A, DeLuca L T, Fischer S and Gettwert V *et al* 2017 Combustion behavior of aluminum particles in ADN/GAP composite propellants *Chemical Rocket Propulsion* ed L T DeLuca *et al* (Berlin: Springer) pp 253–70
[109] Pang W, DeLuca L T, Xu H, Fan X, Zhao F and Xie W 2017 Effect of dual oxidizers on the properties of composite solid rocket propellants *Chemical Rocket Propulsion* ed L T DeLuca *et al* (Berlin: Springer) pp 403–23
[110] Zhang T, Li G, Yu Y, Chen J and Wang M 2016 Effect of catalytic bed thermal characteristics on liquid monopropellant decomposition and combustion characteristics with an ecofriendly thruster based on ammonium dinitramide *Comb. Sci. Technol.* **188** 910–23

[111] Guan H, Li G and Zhang N 2018 Experimental investigation of atomization characteristics of swirling spray by ADN gelled propellant *Acta Astronaut.* **144** 119–25
[112] Gadiot G M H J L, Mul J M, Meulenbrugge J J, Korting P A O G, Schnorkh A J and Schoyer H F R 1993 New solid propellants based on energetic binders and HNF *Acta Astronaut.* **29** 771–9
[113] Schoyer H F R, Welland-Veltmans W H M, Louwers J, Korting P A O G, van der Heijden A E D M, Keizers H L J and van den Berg R P 2002 Overview of the development of hydrazinium nitroformate based propellants *J. Prop. Power* **18** 138–45
[114] Dendage P S, Sarwade D B, Asthana S N and Singh H 2001 Hydrazinium nitroformate (HNF) and HNF based propellants: a review *J. Energ. Mat.* **19** 41–78
[115] Jadhav H S, Talwar M B, Dhavale D D, Asthana S N and Krishnamurthy V N 2005 Synthesis, characterization and thermal behavior of hydrazinium nitroformate (HNF) and its new N-alkyl Substituted derivatives *Ind. J. Chem. Technol.* **12** 187–92
[116] Dendage P S, Sarwade D B and Mandale A B 2003 Characterization and thermal analysis of hydrazinium nitroformate (HNF) *J. Energ. Mat.* **21** 167–83
[117] Louwers J, Gadiot G M H J L, Brewster M Q, Son S F, Parr T and Hanson-Parr D 1999 Steady state hydrazinium nitroformate (HNF) combustion modeling *J. Prop. Power* **15** 772–7
[118] Zhang L, Lin Q, Chang B, Wang P, Lu M and Lin Q 2021 Effect of hexanitroethane (HNE) and hydrazinium nitroformate (HNF) on energy characteristics of composite solid propellants *FirePhysChem.* **1** 116–22
[119] DeLuca L T 1984 Extinction theories and experiments *Fundamentals of Solid-Propellant Combustion Progress in Astronautics and Aeronautics* ed K K Kuo and M Summerfield vol 90 (Reston, VA: AIAA) pp 661–732
[120] Lee T 2012 Review of composite material application to the solid rocket motor cases *Compos. Res.* **25** 82–9
[121] Adami A, Mortazavi M and Nosratollahi M 2017 A new approach to multi-disciplinary design optimization of solid propulsion system including heat transfer and ablative cooling *J. Aerospace Technol. Manage.* **9** 71–82
[122] Martin H T, Cortopassi A C and Kuo K K 2017 Assessment of performance of ablative insulators under realistic solid motor operating conditions *Int. J. Energ. Mater. Chem. Propuls.* **16** 1–22
[123] Liu Y, Li X, Zhu P and Xi K 2021 Ablation characteristics of insulator under high-temperature gas dual-pulse erosion *Defence Technol.* **18** 1875–85
[124] Mahjub A, Mazlan N M, Abdullah M Z and Azam Q 2020 Design optimization of solid rocket propulsion: a survey of recent advancements *J. Spacecraft Rockets* **57** 3–11
[125] O'Briant S A, Gupta S B and Vasu S S 2016 Review: laser ignition for aerospace propulsion *Propuls. Power Res.* **5** 1–21
[126] Kroupa G and Borner M 2019 A miniaturized high-energy laser for ignition of rocket engines *Proc. Int. Conf. on Space Optics (Chania, Greece)*
[127] Khare S and Saha U K 2021 Rocket nozzles: 75 years of research and development *Sadhana* **46** 1–22
[128] Altman D and Holzman A 2006 Overview and history of hybrid rocket propulsion *Fundamentals of Hybrid Rocket Combustion and Propulsion Progress in Astronautics and Aeronautics* ed M J Chiaverini and K K Kuo vol 218 (Reston, VA: AIAA) pp 1–36
[129] Mukunda H S, Jain V K and Paul P J 1979 A review of hybrid rockets: present status and future potential *Proc. Indian Acad. Sci., C* **2** 215–42

[130] Mukunda H S 2004 *Introduction. Understanding Aerospace Chemical Propulsion* (Bangalore: Interline Publishing) pp 1–47

[131] Marquardt T and Majdalani J 2019 Review of classical diffusion-limited regression rate models in hybrid rockets *Aerospace* **6** 1–18

[132] Chiaverini M J 2010 Hybrid propulsion *Encyclopedia of Aerospace Engineering* (New York: Wiley)

[133] Krishnan S 2002 Hybrid rocket technology: an overview *Presented as a plenary paper at the 6th Asia pacific Int. Symp. on Combustion and Energy Utilization (Kualalumpur, Malaysia, May 20–22)*

[134] Marxman G A and Gilbert M 1963 Turbulent boundary layer combustion in the hybrid rocket *Proc. of the Ninth Int. Symp. on Combustion* (New York: Academic) pp 371–83

[135] Marxman G A 1967 Boundary-layer combustion in propulsion *Eleventh Symp. (Int.) Combust.* **11** 269–89

[136] Marxman G A, Wooldridge C E and Muzzy R J 1963 Fundamental of hybrid boundary layer combustion *AIAA Paper 1963–0505*

[137] Sutton G P and Biblarz O 2000 Hybrid propellant rockets *Rocket Propulsion Elements* (New York: Wiley) pp 417–607

[138] Pastrone D 2012 Approaches to low fuel regression rate in hybrid rocket engines *Int. J. Aerospace Eng.* **2012** 649753

[139] Estey N P and George R W 1992 Hybrid rocket motor propellant selectionAIAA Paper 92–3592

[140] Mazzetti A, Merotto L and Pinarello G 2014 Paraffin-based hybrid rocket engines applications: a review and a market perspective *Acta Astronaut.* **126** 286–97

[141] Kuo K K and Chiaverini M 2006 Challenges of hybrid rocket propulsion in the 21st century *Fundamentals of Hybrid Rocket Combustion and Propulsion Progress in Astronautics and Aeronautics* ed M J Chiaverini and K K Kuo vol 218 (Reston, VA: AIAA) pp 593–638

[142] Chiaverini M 2006 Review of solid-fuel regression rate behavior in classical and non-classical hybrid rocket motors *In Fundamentals of Hybrid Rocket Combustion and Propulsion* ed M J Chiaverini and K K Kuo vol 218 37–126 (Reston, VA: Progress in Astronautics and Aeronautics AIAA)

[143] Nguyen B *et al* 2017 Overview of current hybrid propulsion research and development *Proc. ASME 2017 Int. Mechanical Engineering Congress and Exposition (Tampa, Florida, USA)*

[144] Karabeyoglu A 2006 Combustion instability and transient behaviour in hybrid rocket motors *Fundamentals of Hybrid Rocket Combustion and Propulsion Progress in Astronautics and Aeronautics* ed M J Chiaverini and K K Kuo vol 218 (Reston, VA: AIAA) pp 351–410

[145] Karabeyoglu M A, Altman D and Cantwell B J 2002 Combustion of liquefying hybrid propellants: part 1, general theory *J. Propuls. Power* **18** 610–20

[146] Karabeyoglu M A and Cantwell B J 2002 Combustion of liquefying hybrid propellants: part 2, stability of liquid films *J. Propuls. Power* **18** 621–30

[147] https://iss.instructure/courses/6244/pages/density-slash-waterdisplacement

[148] Mukunda H S 2004 Fuels and propellants *Understanding Aerospace Chemical Propulsion* pp 243–88 (Bangalore: Interline Publishing)

[149] DeSain J D, Brady B B, Metzler K M, Curtiss T J and Albright T V 2009 Tensile tests of paraffin wax for hybrid rocket fuel grains *AIAA Paper 2009–5115*

[150] Kim S, Moon H and Kim J 2015 Evaluation of paraffin–polyethylene blends as novel solid fuel for hybrid rockets *J. Propuls. Power* **31** 215–26

[151] Maruyama S, Ishiguro T, Shinohara K and Nakagawa I 2011 Study on mechanical characteristic of paraffin-based fuel *AIAA Paper 2011–5678*

[152] Maruyama S, Ishiguro T, Shinohara K and Nakagawa I 2011 Study on the mechanical characteristics of paraffin-based fuel *8th Int. Conf. on Flow Dynamics (Sendai, Japan)*

[153] Kumar R and Ramakrishna P A 2016 Studies on EVA-based wax fuels for launch vehicle applications *Propellants Explos. Pyrotech.* **141** 295–303

[154] Paravan C, Galfetti L and Maggi F 2017 A critical analysis of paraffin-based fuel formulations *AIAA Paper 2017–4830*

[155] Dinesh M and Kumar R 2018 Development of EVA-SEBS based wax fuel for hybrid rocket applications *Acta Astronaut.* **152** 325–34

[156] Ishigaki T and Nakagawa I 2020 Improving physical properties of wax-based fuels and its effect on regression rate *J. Propuls. Power* **36** 123–8

[157] Bertoldi A E M, Bouziane M, Lee D, Hendrick P, Vandevelde C, Lefebvre M and Veras C A G 2018 Development and test of magnesium-based additive for hybrid rockets fuels *SpaceOps Conf. (Marseille, France, 28 May–1 June 2018)*

[158] Carayon A B, McCormick J, Yilmaz C, Gabl J R, Whitehead B E and Pourpoint T L 2020 Hypergolic ignition of a paraffin-based hybrid motor with MON-25 at 100,000 ft simulated altitude *AIAA Propulsion and Energy Forum*

[159] Kobald M, Schmierer C, Ciezki H K, Schlechtriem S, Toson E and De Luca L T 2017 Viscosity and regression rate of liquefying hybrid rocket fuels *J. Propul. Power* **33** 1245–51

[160] Ishihara A and Brewster M Q 1993 Combustion studies of boron, magnesium, and aluminium composite propellants *Combust. Sci. Technol.* **87** 275–90

[161] Risha G, Boyer E, Wehrman R, Evans B and Kuo K 2003 Nano-sized aluminum and boron-based solid fuel characterization in a hybrid rocket engine *39th AIAAASMESAEASEE Joint Propulsion Conf. and Exhibition, American Institute of Aeronautics and Astronautics, 2003 AIAA* 2003–4593

[162] Connell T L, Risha G A, Yetter R A and Young G 2013 Combustion of boron and fluorocarbon solid fuels in a hybrid rocket *AIAA Paper 2013-3970*

[163] Risha G, Boyer E, Wehrman R and Kuo K 2002 Performance comparison of HTPB-based solid fuels containing nano-sized energetic powder in a cylindrical hybrid rocket motor *38th AIAA/ASME/SAE/ASEE Joint Propulsion Conf. and Exhibition*

[164] Boiocchi M, Maggi F, Paravan C and Galfetti L 2015 Paraffin-based fuels and energetic additives for hybrid rocket propulsion *AIAA Paper 2015–4042*

[165] Shark S C, Pourpoint T L, Son S F and Heister S D 2013 Performance of dicyclopentadiene/ H_2O_2-based hybrid rocket motors with metal hydride additives *J. Propuls. Power* **29** 1122–9

[166] Shark S C, Zaseck C R, Pourpoint T L, Son S F and Heister S D 2016 Performance and flame visualization of dicyclopentadiene rocket propellants with metal hydride additives *J. Propul. Power* **32** 1–13

[167] Scaramuzzino F, Carmicino C, Festa G,, Sorge A R and Viviani A 2013 Fuel regression-rate characterization on a lab-scale hybrid rocket burning N_2O and paraffin-based propellants *49th AIAA/ASME/SAE/ASEE Joint Propulsion Conf. (San Jose, CA)*

[168] Karakas H, Kara O, Ozkol I and Karabeyoglu A 2019 Performance enhancing additives for hybrid rockets *AIAA Propulsion and Energy Forum (Indianapolis, IN)*

[169] Pfeil M A, Dennis J D, Son S F and Heister S D 2015 Characterization of Ethylenediaminebisborane as a hypergolic hybrid rocket fuel additive *J. Propuls. Power* **31** 365–72

[170] Schmucker R 1972 *Hybridraketenantriebe: eine Einführung in Theoretische und Technische Probleme* (Munich: Goldmann) pp 184–7

[171] Magnusson U 1996 A fuel for hybrid rockets *World Aerospace Systems* **2** 50–2

[172] Ankarsward B 1970 The SR-1 propulsion system *British Interplanetary Society Symp. on Sounding Rockets (Cardiff, UK)*

[173] Pugibet M and Moutet H 1969 Utilisation dans les systèmes hybrids de l'eau Oxygénée Comme Comburant *La Recherche Aerospatiale* **132** 15–31

[174] Nadaud L and Baisini J 1966 The ablation of solid fuels used in hybrid engines *La Recherche Aerospatiale* **115** 45–55

[175] Osmon R V 1966 An experimental investigation of a lithium aluminum hydride–hydrogen peroxide hybrid rocket *Aerospace Chem. Eng.* **62** 92–102

[176] Austin B L, Heister S D, Dambach E M, Meyer S E and Wernimont E J 2010 Variable thrust, multiple start hybrid motor solutions for missile and space applications *46thAIAA/ASME/SAE/ASEE Joint Propulsion Conf. and Exhibit* AIAA Paper 2010–7121, July

[177] Wernimont E J, Meyer S E and Ventura M C 1998 Hybrid Motor System with Consumable Catalytic Bed – A Composition of the Catalytic Bed and a Method of Using *US Patent No. 5,727,368*

[178] Paul P J, Mukunda H S, Narahari H K, Venkataraman R and Jain V K 1981 Regression rate studies in hypergolic system *Combust. Sci. Technol.* **26** 17–24

[179] Munjal N L and Parvatiyar M G 1976 Regression rate studies of metalized aniline formaldehyde hybrid fuel *J Spacecraft* **13** 572–3

[180] Bhosale V K, Jeong J and Kwon S 2019 Ignition of boron-based green hypergolic fuels with hydrogen peroxide *Fuel* **255** 115729

[181] Carrick P G and Larson C W 1995 Lab scale test and evaluation of cryogenic solid hybrid rocket fuels *AIAA Paper 1995–2948*

[182] Larson C W, Pfeil K L, DeRose M E and Garrick P G 1996 High pressure combustion of cryogenic solid fuels for hybrid rockets *AIAA Paper 1996–2594*

[183] Larson C W, DeRose M E, Pfeil K L and Carrick P G 1997 Tube burner studies of cryogenic solid combustion *AIAA Paper 1997–0307*

[184] Clair C P S, Rice E E, Knuth W H and Gramer D J 1998 Advanced cryogenic solid hybrid rocket engine developments: concept and testing *AIAA Paper 1998–3508*

[185] Gramer D J, Rice E E, Knuth W H and Clair C P S 1998 Experimental investigation of a metallized cryogenic hybrid rocket engine *AIAA Paper 1998–3509*

[186] Gramer D J and Taagen T J 2001 Low cost surface regression sensor for hybrid fuels, solid propellants and ablatives *AIAA Paper 2001–3529*

[187] Connell T L, Yetter R A, Risha G A, Huba Z J, Epshteyn A and Fisher B T 2019 Enhancement of solid fuel combustion in a hybrid rocket motor using amorphous Ti–Al–B nanopowder additives *J. Propul. Power* **35** 1–4

[188] Dossi S, Paravan C, Maggi F, Di Lorenzo M, Gibertoni S, Ardalic J and Galfetti L 2016 Novel activated metal powders for improved hybrid fuels and green solid propellants *52nd AIAA/SAE/ASEE Joint Propulsion Conf., Propulsion and Energy Forum (Salt Lake City, UT, July 25–27)*

[189] Venugopal S, Ramanujachari V and Rajesh K K 2011 Combustion experiments of HTPB/RFNA mixed hybrid propellants *Indian J. Sci Tech.* **4** 1267–72

[190] George P, Krishnan S, Varkey P M, Ravindran M and Ramachandran L 2001 Fuel regression rate in hydroxyl terminated polybutadiene/gaseous oxygen hybrid rocket motors *J. Propul. Power* **17** 35–42

[191] Frederick R A, Whitehead J J, Knox I R and Moser M D 2007 Regression rates study of mixed hybrid propellants *J. Propul. Power* **23** 175–80
[192] Hitt M and Frederick R A 2015 Testing and modeling of a porous polyethylene axial injection, end burning hybrid rocket motor *51st AIAA/SAE/ASEE Joint Propulsion Conf., Propulsion and Energy Forum (Orlando, FL)*
[193] Heydari M M and Massoo N G 2017 Experimental study of the swirling oxidizer flow in HTPB/N_2O hybrid rocket motor *Int. J. Aerospace Eng.* **2017** 3174140
[194] Berwal P and Biswas S 2020 Experimental investigation on combustion characteristics of hybrid rocket fuels with multi-angle diverging injector *Int. J. Turbomach. Propuls. Power* **5** 12
[195] Kumar C P and Kumar A 2013 Effect of swirl on the regression rate in hybrid rocket motors *Aerosp. Sci.Technol.* **29** 92–9
[196] Quadros F D A and Lacava P T 2019 Swirl injection of gaseous oxygen in a lab-scale paraffin hybrid rocket motor *J. Propuls. Power* **35** 1–10
[197] Kitagawa K, Mitsutani T, Ro T and Yuasa S 2004 Effects of swirling liquid oxygen flow on combustion of a hybrid rocket engine *40thAIAA/ASME/SAE/ASEE Joint Propulsion Conf. and Exhibit (Fort Lauderdale, FL)*
[198] Gomes S R, Rocco L and Rocco J A F F 2015 Swirl injection effects on hybrid rocket motors *J. Aerosp. Technol. Manag.* **7** 418–24
[199] Kumar P and Kumar A 2014 Combined effect of diaphragm and oxidizer swirl on regression rate in hybrid rocket motors *Def. Sci. J.* **64** 21–6
[200] Lestrade J Y, Anthoine J, Musker A J and Lecossais A 2019 Experimental demonstration of an end-burning swirling flow hybrid rocket engine *Aerospace Sci. Technol.* **92** 1–8
[201] Li C, Cai G and Tian H 2016 Numerical analysis of combustion characteristics of hybrid rocket motor with multi-section swirl injection *Acta Astronaut.* **123** 26–36
[202] Ohyama S, Hiratay Y, Araki K, Ohe K, Asoz S and Tani Y 2013 Effects of multi-section swirl injection method on fuel regression rate of high density polyethylene fueled hybrid rocket engine *49th AIAA/ASME/SAE/ASEE Joint Propulsion Conf., Joint Propulsion Conf. (San Jose, CA, July 14–17, 2013)*
[203] Vignesh B and Kumar R 2020 Effect of multi-location swirl injection on the performance of hybrid rocket motor *Acta Astronaut.* **176** 111–23
[204] Kumar R and Ramakrishna P A 2014 Enhancement of hybrid fuel regression rate using a bluff body *J. Propul. Power* **30** 909–16
[205] Gany A and Timnat Y M 1972 Parametric study of a hybrid rocket motor *Israel J. Technol.* **10** 85–96
[206] Dinesh M and Kumar R 2019 Utility of multiprotrusion as the performance enhancer in hybrid rocket motor *J. Propul. Power* **35** 1–13
[207] Grosse M 2009 Effect of a diaphragm on performance and fuel regression of a laboratory scale hybrid rocket motor using nitrous oxide and paraffin *45thAIAA/ASME/SAE/ASEE Joint Propulsion Conf. and Exhibit (Denver, CO, 2–5 August)*
[208] Kumar P and Kumar A 2013 Effect of diaphragms on regression rate in hybrid rocket motors *J. Propul. Power* **29** 559–72
[209] Kumar R and Ramakrishna P A 2014 Effect of protrusion on the enhancement of regression rate *Aerosp. Sci.Technol.* **39** 169–78
[210] Sun X, Tian H and Cai G 2016 Diameter and position effect determination of diaphragm on hybrid rocket motor *Acta Astronaut.* **126** 325–33

[211] Yakar A B and Gany A 1993 Hybrid engine design and analysis *29th AIAA, SAE, ASME, and ASEE, Joint Propulsion Conf. (Monterey, CA, June 28–30)*

[212] Kearney D A, Joiner K F, Gnau M P and Casemor M A 2007 Improvements to the marketability of hybrid propulsion technologies *AIAA Paper 2007–6144*

[213] Haag G, Sweeting M and Richardson G 2000 An alternative geometry hybrid rocket for spacecraft orbit transfer maneuvers *Proc. Int. Astronautical Federation*

[214] Casalino L and Pastrone D 2008 Optimal design of hybrid rocket motors for microgravity platform *J. Propul. Power* **24** 491–8

[215] Barato F, Bellomo N and Pavarin D 2016 Integrated approach for hybrid rocket technology development *Acta Astronaut.* **128** 257–61

[216] Nguyen B *et al* 2017 Overview of current hybrid propulsion research and development *Proc. ASME 2017 Int. Mechanical Engineering Congress and Exposition, IMECE2017 (Tampa, FL, November 3–9)*

[217] Swami R D and Gany A 2003 Analysis and testing of similarity and scale effects in hybrid rocket motors *Acta Astronaut.* **52** 619–28

[218] Story G, Zoladz T, Arves J, Kearney D, Abel T and Park O 2003 Hybrid propulsion demonstration program 250K hybrid motor *39thAIAA/ASME/ASEE Joint Propulsion Conf. and Exhibit* (Huntsville, AL: AIAA)

[219] Story G 2006 Large-scale hybrid motor testing *Marshall Space Flight Center Prepring* 20060047689 NASA

[220] Karabeyoglu M A, Zilliac G, Cantwell B J, De Zilwa S and Castelluci P 2003 Scale-up tests of high regression rate liquefying hybrid rocket fuels *41st Aerospace Sciences Meeting and Exhibit (Reno, NV, 6–9 January)*

[221] Karabeyoglu M A, Zilliac G, Cantwell B J, De Zilwa S and Castelluci P 2004 Scale-up tests of high regression rate paraffin-based hybrid rocket fuels *J. Propul. Power* **20** 1037–45

[222] Paccagnella E, Barato F, Pavarin D and Karabeyoğlu M A 2015 Scaling of hybrid rocket motors with swirling oxidizer injection *51st AIAA/SAE/ASEE Joint Propulsion Conf., Propulsion and Energy Forum (Orlando, FL, July 27–29)*

[223] Calabro M 2011 Overview on hybrid propulsion *Progress in Propuls. Phys.* **2** 353–74

[224] Altman D 2003 Highlights in hybrid rocket propulsion *In-space Propulsion. 10th Int. Workshop Combustion and Propulsion Proc., (La Spezia, Italy)*

[225] Calabro M 1991 History of the European hybrids *AIAA Propuls. Lecture Series (Reno, NV)*

[226] Schmierer C, Kobald M, Fischer U, Tomilin K, Petrarolo A and Hertel F 2019 Advancing Europe's hybrid rocket engine technology with paraffin and LOX *8th European Conf. for Aeronautics and Aerospace Sciences (EUCASS), EUCASS Association (July 2019)*

[227] Okninski A 2018 On use of hybrid rocket propulsion for suborbital vehicles *Acta Astronaut.* **145** 1–10

[228] Marciniak B *et al* 2018 Development of ILR-33 Amber sounding rocket for microgravity experimentation *Aerosp. Sci. Technol.* **73** 19–31

[229] Nowakowski P, Okninski A, Pakosz M, Cieslinski D, Bartkowiak B and Wolanski P 2017 Development of small solid rocket boosters for the ILR-33 sounding rocket *Acta Astronaut.* **138** 374–83

[230] Haemmerli B, Boiron A J and Verberne O 2016 The Norwegian initiative for a satellite nano-launcher *30th Annual AIAA/USU Conference on Small Satellites (Logan, UT, August 8–11)*

[231] Verberne O 2015 Development of the North Star sounding rocket: getting ready for the first demonstration launch *51st AIAA/SAE/ASEE Joint Propulsion Conf., AIAA Propulsion and Energy Forum* (AIAA 2015–4045)
[232] Wade M 2017 Atea-1 (*Encyclopedia Astronautica*) (http://astronautix.com/a/atea-1.html)
[233] Dyer J, Doran E, Dunn Z, Lohner K, Bayart C, Sadhwani A, Zilliac G, Cantwel B and Karabeyoglu M A 2007 Design and development of a 100km nitrous oxide/paraffin hybrid rocket vehicle *AIAA Paper 2007–5362*
[234] Dunn Z, Dyer J, Lohner K, Doran E, Bayart C, Sadhwani A, Zilliac G, Karabeyoglu M A and Cantwell B 2007 Test facility development for the 15,000lb thrust peregrine sounding rocket *43rd AIAA/ASME/SAE/ASEE Joint Propulsion Conf. and Exhibit, AIAA-2007–5358 (Cincinnati, OH)*

IOP Publishing

Energetics Science and Technology
An integrated approach
Adam S Cumming

Chapter 16

Current trends in liquid and gel rocket propulsion

Rajiv Kumar and Mohan Varma

Modern liquid rocket engines (LREs) deliver improved performance in several ways. These include specific impulse, engine start-stop-restart capability, the ability to control thrust on demand, and improved combustion efficiency to meet present-day mission requirements. These have gone through a series of technological advances in the last few decades. Liquid propellant engines may utilize hypergolic (self-igniting) or non-hypergolic fuel-oxidant combinations; semi-cryogenic engines use a high-density fuel and liquid oxygen (LOX), and cryogenic engines employ liquid hydrogen and LOX as propellants. These systems are complex in nature, as both the fuel and oxidizer are stored separately are required to be injected into the combustion chamber with precision. The combustion process of a LRE involves effective atomization, vaporization, and mixing of propellants before combustion. The advanced research and development work toward development of pumps and turbomachinery for handling large mass flow rates of propellants, evolution of efficient ignition systems for cryogenic propellants, and efficient cooling techniques for the combustion chamber is here briefly summarized. The injector design and precision in fabrication, in tune with the required engine performance, always remains a critical criterion, and an overview is presented. Recent trends toward the use of green liquid propellants that are environmentally benign and safe for liquid rocket applications are also highlighted. The major deficiency in LREs is the lower overall density specific impulse, and that new applications demand larger weights of propellant to be stored in volume-limited missiles. A proven approach to enhance the performance and bulk density of fuel is to incorporate metal particles. The metallized fuel may then be combusted using a conventional oxidizer. However, the homogeneity of metallized liquid over a long period of time remains a concern. To avoid this problem, gel propellants have been developed that use a small percentage of compatible chemical gellant to arrest the settling of metal powder on storage; the technology of gel propellants is here presented. Further, a suitably processed gel propellant exhibits thixotropic (shear-thinning) characteristics, which may ensure the liquification of gel to a free-flowing fluid when subjected to high shear rates through the propellant feed system and injector. Gel systems also have many application benefits such as negligible vaporization loss of generally volatile fuels, safe transportation and logistics, low losses through joints and flanges in the feed line, and reduced risk of accidental spillage; and, above all, metallized gels burn at a much-reduced oxidizer-fuel ratio that may lead to significant reduction in rocket size. The key requirements of gel propulsion systems like gel formulations, efficient ignition and combustion behavior, and the flow characteristics of propellants are also discussed.

16.1 Liquid rocket engines

16.1.1 Introduction

A liquid rocket is a chemical propulsion system, where fuel and oxidizer are in the liquid phase and they are stored outside the combustion chamber in separate tanks. These liquid propellants are injected into the combustion chamber using a suitable feed that may be either a pressure feed or a pump feed system. Liquid propellant rockets consist of a fuel tank and an oxidizer tank, feed arrangements, a combustion chamber, and a nozzle, as shown in figure 16.1. The propellant feed system supplies the propellant from the tank into the combustion chamber at the required flow rates and pressures. The fuel and oxidizer flow rate are metered, and atomized by passing through the injectors. The combustion takes place in the combustion chamber and hot gases are generated, which are then ejected through the nozzle. The injector, combustion chamber, and nozzle as a whole are termed the thrust chamber.

The thrust chamber is the key assembly of a LRE. Propellants are metered, injected, atomized, vaporized, mixed, and burned to generate hot combustion gas products, which in turn are expanded and ejected through the nozzle at high velocity. Figure 16.2(a) shows a thrust chamber assembly and its components. The basic components of a liquid rocket thrust chamber are as follows:

- Injector.
- Combustion chamber.

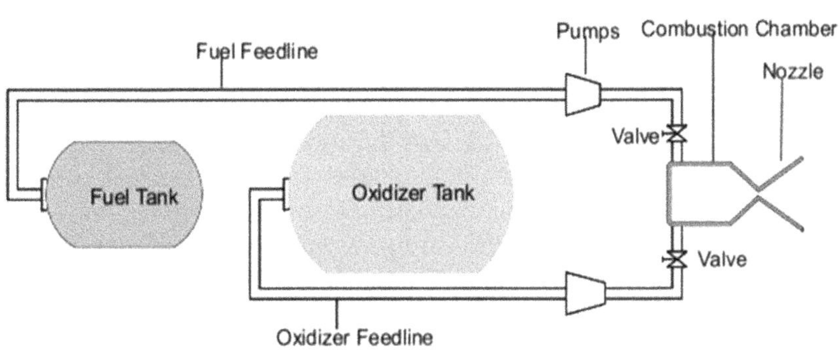

Figure 16.1. Schematic of a liquid rocket engine.

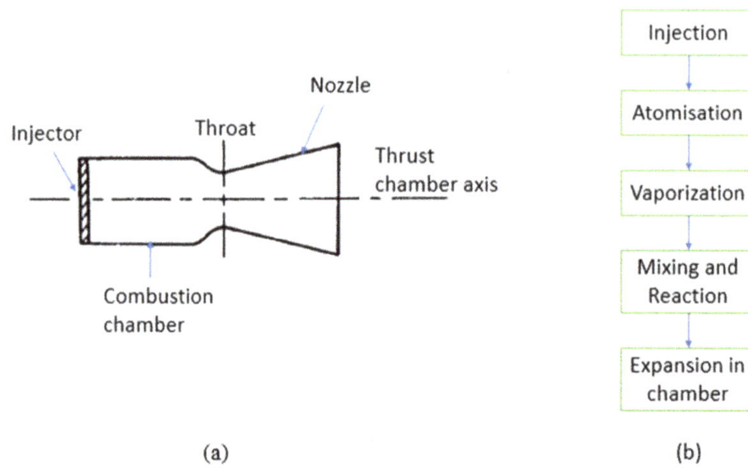

Figure 16.2. (a) Basic thrust chamber assembly [1], and (b) physical and chemical processes in the combustion of liquid propellants [2].

- Nozzle.
- Ignition device (for non-hypergolic propellants).
- Mounting provisions.

There are various physical and chemical processes associated with the combustion of liquid propellants. Figure 16.2(b) shows a flow chart of the processes involved in the combustion of liquid propellants. These are as follows.

- *Injection.* In this process, liquid propellants pressurized by either of the pressurization feed systems (gas pressurized or turbopump) are injected into the combustion chamber. The liquid streams are injected into the combustion chamber at relatively low velocities and are made to impinge. The liquids get partially vaporized in this process.
- *Atomization.* In this process, the impingement of jets or sheets leads to the formation of liquid fans. Liquid jets or fans become unstable due to oscillations during the breakup. Liquid fans break into ligaments, which further break down to form fine droplets. A secondary breakup of drops also takes place. Liquid mixing and vaporization begin in this process, and it leads to certain liquid–liquid and some vapor reactions to occur.
- *Vaporization.* Due to the certain liquid–liquid and vapor reactions during atomization, droplet gasification and diffusion also take place. This improves the mixing of fuel and oxidizer that leads to local chemical reactions and further heat release from the reactions. Gases gain low velocities, and some crossflow is also generated. Pressure or temperature oscillation and acoustic waves influence the rate of vaporization. Heat is absorbed from radiation and conduction from the blowback of turbulent gases from the hot reaction zone. This enhances the droplet gasification and diffusion process. The gases are further accelerated to higher velocities.

- *Mixing and reaction*. In this process, multiple chemical reactions take place due to turbulent mixing. This leads to high heat release. Turbulence causes interaction with the droplets and enhances chemical reactions due to which local mixture ratios, reaction rates, or velocities rapidly change with reference to the time and are not uniform across the combustion chamber. Due to the rapid chemical reactions, the temperature rises and density gets reduced.
- *Expansion in chamber*. Due to the chemical kinetics, the final combustion temperature and equilibrium reaction gas composition are attained in this process. Due to the temperature rise, gas expansion takes place and axial velocities increase. At the end of the combustion chamber, or near the nozzle end, streamlined high-velocity axial flow with very little crossflow is developed.

16.1.2 Feed system for liquid propellants

Liquid propellant is required to be injected into the combustion chamber at the required flow rates from the fuel and oxidizer storage tank. This can be via either a pressure feed or pump feed system. A pressure feed system is generally used for lower thrust and short-duration operations. They are further classified into gas-regulated and blow-down pressure feed systems, as shown in figures 16.3(a) and (b). In a blow-

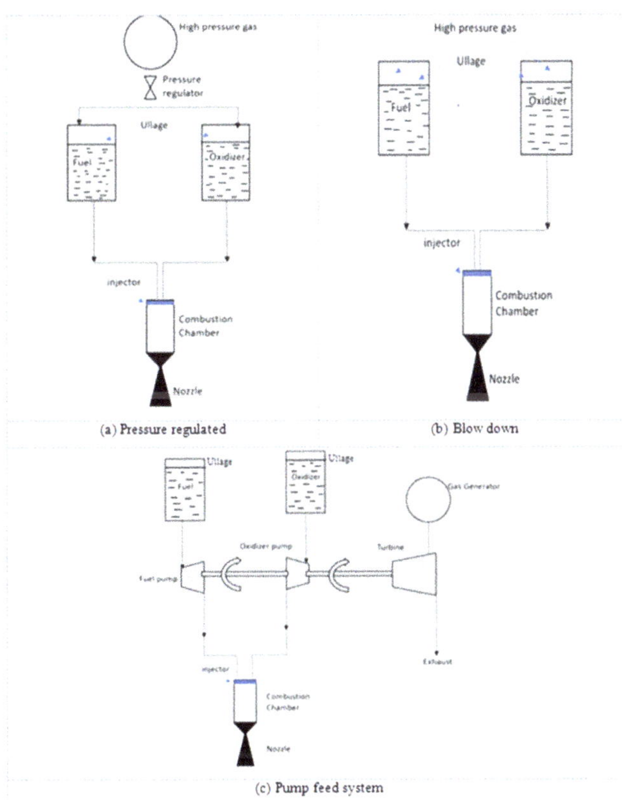

Figure 16.3. Schematic diagram of various types of propellant feed system.

down system, the ullage space of the propellant tank is directly pressurized. The tank pressure progressively decreases as the burning progresses; while in a gas-regulated pressure feed system, the gas is stored at sufficiently high pressure at ambient temperature in a separate tank, which is also known as the gas bottle. The liquid fuel or oxidizer is supplied to the combustion chamber using a high-pressure, gas-regulated feed system [3].

An alternate method of supplying the fuel and oxidizer into the combustion chamber uses a pump feed arrangement. The pump can be driven by mechanical or electrical energy, but the power required to drive the pump for higher mass flows of fuel and oxidizer to achieve higher pressure injection is significantly high. Thus, in a LRE a turbine is utilized to run the pump. A gas generator is used to provide high-temperature gas to run the turbine. It takes the fuel and oxidizer from the main fuel and oxidizer tank, which is then combusted in the gas generator. The combustion product from the gas generator is passed into the turbine to provide the shaft power to drive the pump. The temperature in the gas generator is maintained at around 900 K, which is feasible by combusting the fuel and oxidizer in either fuel- or oxidizer-rich conditions. The exhaust after expanding through the turbine could be utilized to pass into the combustion chamber, which enhances the performance of the propulsion system. In certain cases, the exhaust from the turbine may be released to the ambient atmosphere, as shown in figure 16.3(c).

16.1.3 Injectors utilized in liquid propulsion systems

Injectors are used in liquid rockets to inject the fuel and oxidizer into the combustion chamber in the correct proportions and at the required flow rate. They are also used to atomize the liquid propellant into fine droplets and mix them appropriately for proper combustion. The energy required for atomization is obtained by providing high pressure or high-velocity gases (i.e. gas-assist injector). In a pressure atomizing injector, the fuel and oxidizer are pressurized for injection into the combustion chamber; while in a gas-assist injector, high-velocity gas is made to impinge on the liquid sheet at very high velocities to atomize during the injection. Pressure atomizing injectors are classified into showerhead injectors, impinging-type injectors and coaxial-type injectors, as shown in figures 16.4(a), (b), and (c), respectively. These types of injectors are mainly used for fuel and oxidizer combinations that are in the liquid phase prior to the injector. Gas-assist injectors are used mainly for cryogenic or semi-cryogenic systems, where one of the

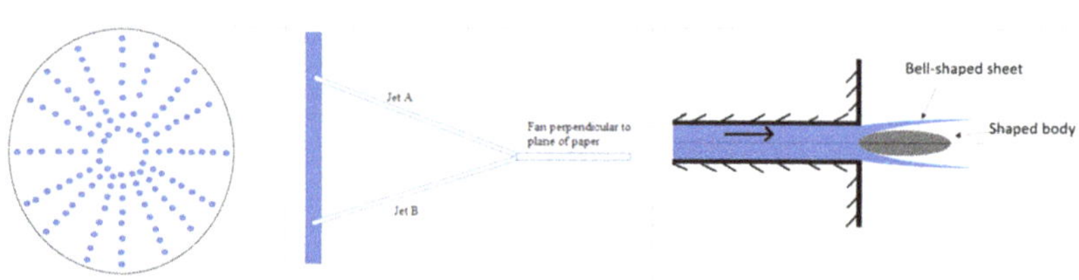

(a) Showerhead injector (b) Impinging Injector (c) Coaxial injector

Figure 16.4. Schematic diagram of various types of pressure atomizing injectors.

liquid propellants is in the gas phase prior to injection. The selection of the injector depends on the liquid propellant used, the application, and the manufacturing complexity associated with it [3].

16.1.4 Advances toward improved mixing and atomization

As discussed in the previous section, the function of an injector is to inject the fuel and oxidizer into the combustion chamber in the correct proportion of oxidizer-to-fuel (O/F) ratio and further atomize it for efficient mixing. Efficient mixing enhances the combustion reaction and leads to higher heat release. The high-temperature combustion gas then expands through the nozzle to generate the required thrust. To achieve this, various types of injectors have been developed such as showerhead, doublet, triplet, ring slot, coaxial, splash plate, etc. Selection of the appropriate injector is important for efficient mixing and atomization [4–9]. Plane injectors, which are circular in their orifice shape, have been widely used up until now. Some examples of plane atomizers, apart from the ones discussed earlier, are rotary atomizers [10–12], twin-fluid atomizers [13–15], and ultrasonic atomizers [16–19]. The main drawback of these plane injectors are their low spray cone angle. The breakup length, mean droplet diamete, and Sauter mean diameter are also generally higher. For better combustion of the propellant, fine sprays with smaller droplets are essential [20, 21]. Thus, the recent trend is toward selection of noncircular orifices such as elliptic, plus, etc, instead of a circular orifice [22, 23]. Research has been conducted toward developing noncircular orifices to improve the combustion efficiency by improving the atomization behavior, which also includes variation of the *L*/*D* (length-to-diameter) ratio [24–30]. A noncircular orifice possesses three-dimensionality of its flow behavior, which makes the jet far more unstable when compared to circular orifices [31]. These noncircular jets have been found to be an efficient passive flow control technique, which improves the performance of various practical systems only with a change in the geometry of the injector [24–26]. The major advantages of noncircular jets are improved mixing in low- and high-speed flows, increased combustion efficiency, noise suppression, thrust vector control, reduced combustion instabilities, and extended flammability limits [32].

16.1.5 Throttling in liquid rocket engines

The term ‘throttling’ is used to describe a varying thrust profile or thrust modulation. Regulation of propellant flow rates by control valves is the most common method of thrust control in a LRE. Throttling is used in space applications like planetary descent, boost-phase trajectory tailoring, orbital rendezvous, and reusable launch vehicles. Throttling has been demonstrated on many engines such as the Apollo Lunar Module Descent Engine (LMDE), the RL-10 and its derivatives, and the Space Shuttle Main Engine (SSME).

Liquid rocket engine throttling is defined as the varying of the thrust for a rocket engine in reference to the 100% rated power level (RPL). Throttling down to a ratio 4:1 thrust (25% RPL) is termed as shallow throttling. Throttling beyond 4:1 thrust is considered deep throttling. The throttling range of a LRE is dependent on the mission requirements. For more precise thrust control, the throttling ratio should be high.

The governing thrust equation is given below:

$$F_T = \dot{m} \cdot v_e + (p_e - p_e) \cdot A_e, \tag{16.1}$$

where

F_T is the thrust of the rocket engine,
$\dot{m}$ is the mass flow rate of the propellant,
v_e is the exit velocity of hot gasses,
p_e is the pressure at the nozzle exit,
p_a is the ambient pressure, and
A_e is the nozzle exit area.

The physical parameters that can be controlled to throttle a LRE are as follows:

- Propellant flow rates.
- Propellant type and composition.
- Nozzle exit area.
- Exit velocity.
- Exit pressure.

The thrust of a rocket engine can be decreased if ambient pressure is increased or if mass flow rate, exit velocity, exit pressure, or exit area is decreased. Due to practical restrictions, it is difficult to control or vary the propellants and nozzle exit area, while it is difficult to vary nozzle throat area if heat fluxes are high. Varying the propellant flow rates is a favorable way to vary the thrust. An adequate pressure drop across the injector is required for propellant atomization and mixing. Maintaining the pressure drop across the injector is the main difficulty in the throttling of LREs.

The critical issues with throttling are as follows:

- Combustion and system instabilities.
- Performance degradation.
- Excessive heat transfer to the chamber walls.

To achieve throttling, a suitable injector is required. Various type of injectors have been developed by researchers [10–19]. Among them, a pintle injector is a variable-area injector used in liquid propellant rockets [1, 33–38]. Variable-area injectors offer suitable choices to develop throttlable rocket engines because it is difficult to vary thrust efficiently when fixed-area injectors are used. They are constructed of a single-element injector mounted on the injector head that controls the mass flow rate of the propellants. Several research works have been conducted on the utilization of pintle injectors for throttling purposes [1, 33–38].

In a pintle injector, the interaction is between an annular sheet and a radial jet of liquid propellants, as shown in figure 16.5. One pintle injector can cover the entire combustion chamber because it has a wider spray angle. Hence, it can replace a large injector plate having several injector elements with a single injector unit. A pintle injector can control both the injection area and velocity to maintain good spray conditions at low rates of propellant flow. The first known flight usage of this type of

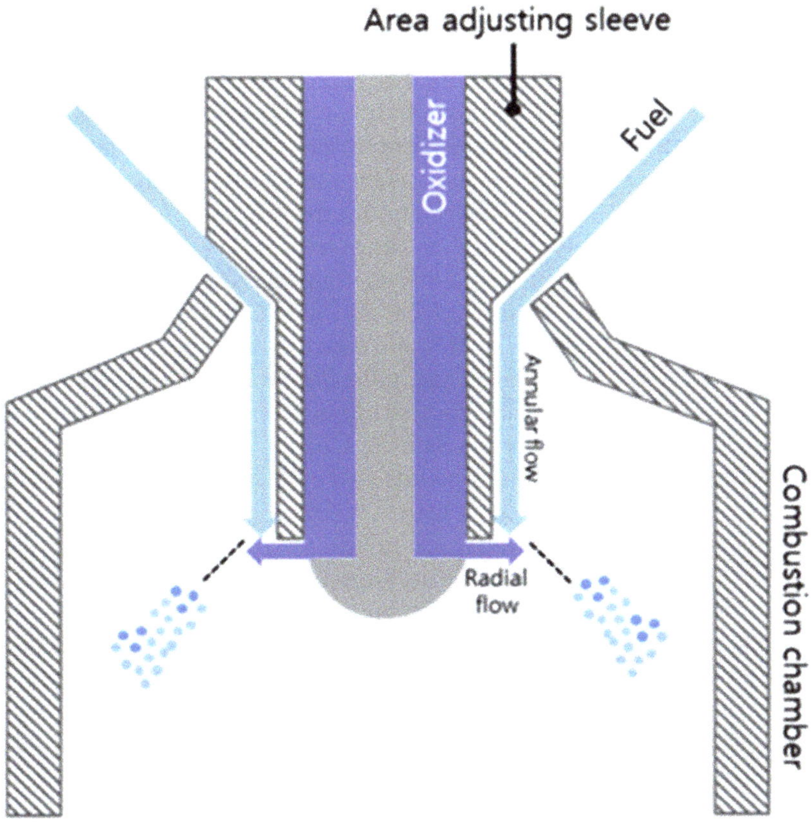

Figure 16.5. Pintle injector arrangement [37].

injectors is in the LMDE [1]. The LMDE has a 45 kN thrust engine with Aerozine-50/N_2O_4 propellant, delivering a specific impulse of 311 s. The engine has a throttling capability of 10%–60% of full thrust. Throttlability, scalability, and cost-effectiveness are the major advantages of a pintle injector. It has a simple design but complexity of its actuating and guiding elements. It can be used in LREs in missions such as planetary entry and descent, space rendezvous, orbital maneuverings including orientation and stabilization in space, hovering and hazard avoidance during planetary landing, stage recovery of reusable launch vehicles, and ballistic missile defense trajectory control.

16.1.6 Developmental trends of liquid propellant rockets

Liquid rocket systems are more versatile in space applications as they deliver higher specific impulse when compared to a solid rocket motor. They can be controlled easily by controlling the supply of fuel and oxidizer, and can therefore be used in the booster and upper stages of launch vehicles. In the booster stage, high-density fuel such as kerosene with LOX and, in the upper stages, unsymmetrical dimethyl hydrazine (UDMH) with nitrogen tetraoxide (N_2O_4 or NTO) are being utilized. This high-density fuel gives the high density impulse that is required for the booster stage. In the Apollo mission, which was designed for the landing of men on the Moon, a Saturn rocket was utilized, in which five F-1 rockets using Rocket Propellant-1 and LOX were used. It provided a thrust of 6.8 MN at a chamber pressure of 7 MPa. This engine used a gas generator arrangement for the pump feed system. The Atlas Centaur also used kerosene-LOX for the booster stage, and provided a thrust of 1.8 MN. It also utilized gas generator cycle arrangements for the propellant feed system. The RD-170 rocket is known to provide the highest thrust level among all the liquid propulsion system developed so far and it uses a staged combustion cycle. It was used as the booster stage for the Zenith launch vehicle that uses kerosene as fuel and LOX as oxidizer. It provided a thrust of 7.25 MN at a chamber pressure of 24.5 MPa. A LRE was used in the Ariane launch vehicles and in the Proton launch vehicle for the booster stage with UDMH-N_2O_4 propellant combinations. This engine again also used a gas generator cycle. The Geosynchronous Satellite Launch Vehicle also uses a LRE with similar propellant combination and the same feed cycle arrangement. Considering the higher cost and carcinogenic nature of hydrazine-based propellants, additional efforts are being made for its handing and safer utilization. Hence, this makes a kerosene-LOX, semi-cryogenic system advantageous for liquid rocket applications, although its use has most often led to combustion instability problems. This issue had prompted the development of high-performance cryogenic rocket engines employing liquid hydrogen and LOX in different stages of the launch vehicle. They have been used in various launch vehicles, such as the SSME and Ariane V launch vehicles in the booster stage. The SSME measured a chamber pressure of 19 MPa and a thrust of 1.8 MN. Apart from these examples, a cryogenic engine is also used in the launch vehicle known as Vulcan, operating at a chamber pressure of 11 MPa that produces a thrust of 1 MN. The RL-10 is another cryogenic rocket, used in the Titan and Delta launch vehicles, providing a thrust of 70 kN and a chamber pressure between 3 and 4 MPa.

Monomethyl hydrazine (MMH) and N_2O_4 are also used as Earth-storable and space-storable propellants in pressure fed arrangements in the upper stages of launch vehicles, propulsion, and control of satellites, and also for orbital transfer applications. The thrust level for these systems is very low, around 10–20 N for altitude and orbit control systems. The upper-stages rocket has a thrust level from a fraction of a kilonewton to a few tens of kilonewton. Satellites use a rocket generating thrust in a pulse for long durations of operation for the upper stages, including orbit transfer applications. LREs of different thrust levels continue to be utilized in various applications. New improvements in the injector designs are being made to control the atomization, distribution, and their mixing in the combustion chambers for future missions. Research and development work on advanced high-temperature-resistant materials, such as refractory materials, ceramics, and carbon-carbon materials, is being done to reduce the cooling requirement of the combustion chamber and the nozzle. The development of high-speed turbopumps and control systems are an integral part for large-thrust rocket engines, and continuous efforts are being made to maximize their performance and minimum weight. The aim of reducing the number of stages had led to developmental efforts toward adoptive nozzles, and interest in using multi-propellants such as tri-propellant systems is also being pursued [3].

16.1.7 Development of green liquid propellants

16.1.7.1 Hydroxylammonium nitrate propellants

Selection of the propellant in current scenarios should consider not only the overall mission performance but also environmental pollution concerns. In earlier missions, the propellant performance was assessed by achievable specific impulse and propellant density, but now several other factors such as hazards in handling and transportation, cost, and environmental pollution concerns need to be addressed. Hydrazine and its derivatives along with red fuming nitric acid (RFNA)/N_2O_4 as propellants, which were previously commonly used in LREs, are toxic and carcinogenic in nature and their combustion products are very hazardous. The search for environmentally safe fuels has attracted the use of monopropellant systems such as hydroxylammonium nitrate (HAN) and ammonium dinitramide (ADN)-based liquid propellants. These types of propellants provide benefits in terms of safety, low cost, and high performance. To decrease environmental pollution, research work is being undertaken to develop propellants that are environmentally benign and nonhazardous in nature [39]. Substitutes for hydrazine and N_2O_4 are being sought to attain cost reduction and lower environmental impact. High-energy green propellants such as ADN, HAN, and hydrazinium nitroformate (HNF) may offer attractive options, but the high molecular weight of their combustion products and high operational temperatures pose challenges for their utilization.

HAN as an ionic liquid is a substitute due to its improved thermal management, lower complexity, and minimized toxicity [40]. The requirements of appropriate ignition catalysts, high sensitivity, and compatible materials are major obstacles to its practical application. Catalytic ignition and combustion studies for a baseline thruster system for a HAN-based monopropellant with a platinum-based catalyst have been proposed by Jankovsky [41]. A very-high-temperature oxidizing environment is required for these types of thrusters. Alternate ignition systems such as laser ignition, electrical ignition, and chemical injection ignition can be considered for next-generation advanced monopropellant thruster systems. The effects of various types of parameters such as fuel type, catalyst bed material, and design for the combustion of HAN are presented in table 16.1.

Binary solutions of HAN and ADN have been synthesized and then thermally and catalytically decomposed by Amrousse *et al* [7]. Different concentrations of 95, 80, and 60 wt.% of binary HAN mixtures and 75, 60, and 50 wt.% ADN solution were studied. Alumina doped with lanthanum oxide was used as the active phase with iridium metals for HAN combustion and copper oxide for ADN combustion. The catalysts were characterized by using transmission electron microscopy, x-ray diffraction, and chemisorption. The combination of (10%) $Ir/Al_2O_3–La_2O_3$ and HAN 95% and (10%) $CuO/Al_2O_3–La_2O_3$ and ADN 75% gave lower decomposition temperatures and large reaction rates, leading to an efficient system.

The mechanism of high burning HAN-based ionic solutions and their combustion characteristics have been studied by Katsumi *et al* [52]. In the case of HAN-based aqueous solutions, three types of combustion wave structures were proposed; the gas temperature and the water content determine the combustion

Table 16.1. Effect of fuel, oxidizer composition, and catalyst type on the combustion of HAN propellants.

S. No.	Fuel and oxidizer used	Catalyst used for combustion	Important result	References
1.	79 wt.% HAN solutions	Platinum (Pt)/Al_2O_3–Si	Presence of theta alumina structure and the heterogeneous distribution of silica and platinum particles.	Courtheoux *et al* [42]
2.	HAN (95 wt.%)-water solution	High specific surface area (SSA)-activated carbons (1% catalyst)	A double-stage thermal decomposition of pure HAN (95 wt.-%)-water solution becomes single stage with a reduction of decomposition temperature by 40 °C–45 °C.	Atamanov *et al* [43]
3.	95, 80, and 60 wt.%. HAN solutions	Ir/Al_2O_3–La_2O_3	Lower decomposition temperatures, larger reaction rates.	Amrousse *et al* [44]
4.	HAN, H_2O, MeOH (60:30:10)	Pt/barium hexaaluminate	Tested the feasibility of the new catalyst at with catalyst bed temperature of 350 °C.	Kang and Kwon [45]
5.	HAN-water blends	Iridium on γ-alumina support	Significantly active in decomposing at low temperatures.	Oommen *et al* [46]
6.	Binary HAN-water mixtures	Alumina pellets impregnated with iridium (Shell 405)	Highest catalytic effect with an ignition temperature drop of about 100 °C.	Amariei *et al* [47]
7.	Different HAN concentrations of 20%, 60%, and 83% by weight	Pt deposited on alumina (Al_2O_3) and silica-doped alumina (Al_2O_3Si)	Reduction in the dissociation temperature of the 83% HAN solution to temperatures slightly above the boiling point of water, 110 °C–115 °C.	Courthéoux *et al* [48]

(Continued)

Table 16.1. *(Continued)*

S. No.	Fuel and oxidizer used	Catalyst used for combustion	Important result	References
8.	HAN, aminoethylamine-trinitrate (TRN_3), water mixture	Zirconium diboride (ZrB_2) and zirconium carbide (ZrC) as catalyst support with iridium	HAN thruster performance above 250 s measured Isp was achieved on a repeated basis without the handicap of frequent catalyst change.	Zube *et al* [49]
9.	HAN/H_2O/AN/MeOH with 95/8/5/21 as mass composition	Different supports (monolith and grains) with iridium as active phase	The reactivity of the grain catalyst (Shell 405) was more active than the monolith catalyst.	Amrousse *et al* [50]
10.	75% concentrated solution of HAN	Cobalt-doped cerium oxide-based catalyst	High temperature robustness, resistance to catalytic poisoning and thermal shock catalyst were obtained for HAN.	Agnihotri and Oommen [51]

wave structure and the linear burning rate. Hydrodynamic instability and the pressure dependence of the instability were estimated. Thermal and catalytic decompositions of HAN-based monopropellants in a dynamic reactor with on-line product analysis using mass spectroscopy were studied by Amariei *et al* [53]. The effect of catalyst shape (powder or spheres) was investigated. The results indicated that the powder catalyst led to the complete decomposition of the HAN solution, whereas the sphere-shaped catalyst led to partial decomposition of the HAN solution. The use of 10 wt% platinum on silicon-doped alumina (Pt/Si–Al_2O_3) for the combustion of HAN was carried out in a batch reactor [42]. The presence of low metal–silica interactions lead to migration of the metallic particles, which led to the sintering of the catalyst. Binary aqueous mixtures of HAN, AN, ADN, and HNF were prepared, and thermal decomposition studies were carried out by Amariei *et al* [47]. HAN-based solutions demonstrated the highest catalytic effect whereas ADN-based propellants displayed the highest decomposition rates. The challenges of this type of propulsion system are that they are difficult to ignite, each propellant formulation requires a different catalyst, and it is difficult to achieve stable combustion (causes detonation). These requirements need to be investigated for efficient rocket performance and stable combustion.

HAN-based monopropellant and a hydrazine monopropellant propulsion subsystem have been used for three NASA missions, the Total Ozone Mapping Spectrometer-Earth Probe, the Tropical Rainfall Measuring Mission, and the Microwave Anisotropy Probe [41]. Currently, the Air Force Research Laboratory, NASA Glenn Research Center, and the Swedish Space Corporation are working on energetic formulations that could replace the traditional storable propellants with higher performance/lower toxicity compounds that meet requirements for a next-generation green fuel. These efforts have focused on a new class of monopropellants consisting of fuel and an oxidizing salt dissolved in water to form a stable liquid monopropellant. The US has focused on using an energetic formulation of HAN with a fuel such as alcohol, ethanol, or glycine in a water-based solution, while Sweden's formula uses ADN with methanol, ammonia, and water to create an ionic liquid monopropellant. Sweden's PRISMA mission successfully demonstrated ADN's improved performance over hydrazine during its debut launch.

The JAXA group from Japan has also conducted experiments on HAN-based thrusters with a 20 N thruster and have studied the effect of catalyst type, and the shape and size of the catalyst on HAN combustion. The thermal decomposition of HAN 95% water solution, a 'green' alternative of a fuel/oxidizer formulation to hydrazine nitrate, as a rocket propellant in the presence of high specific surface area (SSA)-activated carbons was investigated by Atamanov *et al* from the same group [43]. The catalytic performance of an iridium-based catalyst for the decomposition of HAN was carried out in a monopropellant thruster [50]. Changes to the catalyst bed temperature, ignition delay, and chamber pressure confirmed the effectiveness of the catalyst. The influence of the catalyst support shapes such as monolith honeycombs and alumina grains, on the performance of liquid monopropellant thrusters has also been studied by the same research group [50].

16.1.7.2 H_2O_2 green propellants

Another low-toxicity liquid propellant as a possible substitute for hydrazine is hydrogen peroxide (H_2O_2), which can be an excellent alternative but has slightly inferior energy efficiency. Hydrogen peroxide was discovered by Thénard in 1818, but it was only made commercially available in the late 1930s [54]. In the 1930s, H_2O_2 as an oxidant or a monopropellant was used in the Soviet Union and Germany. In the 1940s–50s, 80%–85% H_2O_2 was used in many rockets and aviation engines such as the V-2, Redstone, Viking, Jupiter, Sea Slug, Natter, Blue Streak, Me-163, X-15, etc [55]. Later, it was used as an oxidizer in the Black Knight rocket (United Kingdom) in combination with kerosene [56]. The properties of H_2O_2 are presented in table 16.2.

The performance of propellants using hydrogen peroxide is dependent on their concentration as the oxygen content and decomposition temperature of H_2O_2 increases with concentration. The accessible oxygen content increases from 40.0% for 85% hydrogen peroxide to 46.1% for 98% hydrogen peroxide [57]. The decomposition temperature also increases from 907 to 1125 K. The versatility of hydrogen peroxide, its low toxicity, and its easy availability have been influential in developing an inexpensive and environmentally affable propellant system. Hydrogen peroxide decomposition can be initiated without using an igniter device. This allows the development of a multiple startup engine, but the higher temperatures generated require a high-heat-resistant and high-performance catalyst. The factors which affect the catalyst life are the catalyst material, chamber pressure, concentration of hydrogen peroxide, and mode of operation. The catalyst should have high activity, stability over a wide operating condition, should be shock and mechanical impact resistant, and possess high mechanical strength and optimum specific surface area [58].

To increase the specific surface area of the catalyst metals are used only in the active phase as a thin layer onto a highly porous support. The various types of catalyst used consist of gauze catalysts, granulated support catalysts, and monolithic supports. Despite the high cost of silver-based catalysts, they are widely used. Many present thrusters use packs of silver gauze pieces or silver deposited onto metal gauze, or ceramic supports as a catalyst for H_2O_2 decomposition [59]. The main advantages of silver catalysts are the high efficiency of their decomposition of

Table 16.2. Physical properties of different concentrations of H_2O_2.

	Property	70% by mass*	80% by mass*
1.	Density	1288 kg m^{-3}	1365 kg m^{-3}
2.	Molecular mass	27 g mol^{-1}	30 g mol^{-1}
3.	Melting point	233 K	255 K
4.	Boiling point	399 K	410 K
5.	Heat capacity (C_p)	82 J mol^{-1} kg^{-1}	85 J mol^{-1} kg^{-1}
6.	Enthalpy of formation (at 298 K)	−233 kJ mol^{-1}	−213 kJ mol^{-1}

* Taken from RPA Thermodynamic Database, January 2011 (http://software.lpre.de/thermo_data.htm).

hydrogen peroxide, their small size, easy fabrication, and availability in numerous shapes [59].

Catalysts using platinum on various supports have also been studied. The various supports which have been used are Al_2O_3 [60] and γ-alumina [61, 62]. A 1 N thrust in a two-component thruster with monolithic a platinum/cordierite catalyst showed satisfactory performance using kerosene or ethanol as fuel [63]. Catalysts based on manganese oxides were used for the decomposition of hydrogen peroxide in a hybrid rocket motor and were found to be efficient for combustion of rocket-grade H_2O_2 [64]. Table 16.3 presents the effect of various parameters on H_2O_2 propellant combustion, summarized from the work done by various researchers.

In the development of monopropellant thrusters, major attention has been paid to the influence of the catalyst and thruster design on the course of operation. Jo *et al* [69] studied the pulsating instability of monopropellant thrusters using granulated MnO_2–PbO and MnO_2 catalysts on an aluminum oxide support. The propellant feeding mode, catalyst activity, and catalyst bed size affected the instability in the combustion. Experimental studies revealed that the characteristics of hydrogen peroxide decomposition was a function of the pressure and time of the increase in temperature; however, the residual time depended on the catalyst activity [70]. Baek *et al* [71] investigated the decomposition of an ethanol-hydrogen peroxide

Table 16.3. Effect of fuel, catalyst, and conditions on the combustion of H_2O_2 in a liquid engine.

S. No.	Fuel and oxidizer used	Catalyst used for combustion	Important results	References
1.	Hydrogen peroxide (50%, 60%, and 70%) and ethanol	A refractory ceramic composition (Al_2O_3 (51%), SiO_2 (42%), Fe_2O_3 (1.6%), and CaO (5.4%)), impregnated with manganese dioxide (MnO_2)	Economically viable system obtained.	Miraglia *et al* [65]
2.	50% H_2O_2	Pt and Ag catalyst	Platinum catalyst thruster was found to be more stable.	Khaji *et al* [66]
3.	>90% H_2O_2	20-mesh pure-silver screens	Enhanced temperature and increased catalyst contact area required for successful firing.	Krishnan *et al* [67]
4.	30% hydrazine hydrate, 57% methyl alcohol, and 13% water with 80 to 83% H_2O_2	KCu(CN)6 ignition catalyst	A maximum thrust of 3770 lb at sea level was developed.	Bloom *et al* [68]
5.	Hydrogen peroxide (>70%)	Silver screen interspaced with disks of stainless steel (SUS 306) screen	High efficiency of decomposition of hydrogen peroxide.	Lee and Lee [59]

monopropellant with platinum on γ-alumina as catalyst in a 10 N thruster. A specific impulse of 214 s in vacuum and 95.3% efficiency was obtained. It may be concluded from the literature that silver catalysts are mainly used in the development of 0.5–1 N thrusters and platinum catalysts on aluminum oxide are used in the development of 5–50 N thrusters [72–75]. Catalysts based on the active phase of manganese oxides are widely used in the development of monopropellant thrusters [76, 77] and hybrid engines [64, 78].

Development tests of H_2O_2-oxidized bipropellant liquid engines and hybrid rocket engines have been conducted, and such engines are being developed for application in upper-stage propulsion. A H_2O_2-oxidized hybrid motor is being developed for upper-stage application in reusable launch vehicles and emerging defense applications [79]. ONERA, in France, have developed a polyethylene- or hydroxyl-terminated polybutadiene-fuel-based H_2O_2 hybrid propulsion system for micro-satellites and small tactical missiles [80]. Aerojet has developed a tri-fluid propellant injector for a H_2O_2-kerosene reusable bipropellant engine. Boeing Rocketdyne is working on the development of a H_2O_2 thruster for applications in orbital maneuvering systems, crew escape systems, and all upper-stage and on-orbit applications requiring lower life-cycle costs and improved safety [81].

16.2 Cryogenic rocket propulsion

Cryogenic engineering relates to the utilization of a low-temperature fluid that can be used in various engineering applications. The lower end of temperature is fixed at absolute zero, while its upper limit is not a fixed value. Typically, the normal boiling point at 1 atm of the fluid is considered as the upper limit. Scott [82] had given the upper temperature limit as 123 K; Bell [83] had reported 200 K as the upper limit; while Zabetakis [84] proposed that fluids with a critical temperature below room temperature may be considered cryogens, which are also known as cryogenic fluids. A common gas that condenses at −150 °C to form a liquid is a working value for cryogenic fluids. The upper limit generally depends on the application and the fluid requirements. Various unusual phenomena generally take place at low temperatures that make cryogenic engineering a special and unique field of engineering. It has various applications in the fields of science, energy, astronautics, space, etc [85].

16.2.1 Applications

16.2.1.1 Science

No less than 15 Nobel prizes have been awarded for research involving some cryogenic phenomenon or where cryogenic technology was required to induce certain process. A few among the Nobel winners are Van der Waals, Kammerlingh Onnes, Bardeen, Cooper and Schrieffer, Lee and Yang, Josephson, and Alvarez. Among them, Alvarez is known for the development of the liquid hydrogen bubble chamber [85].

16.2.1.2 Aeronautics

Cryogenics in the field of aeronautics includes research into the propulsion systems which are used in aircraft that fly at the border of space at hypersonic speeds. The earlier and most common use in astronautics was to provide breathing oxygen in high-altitude, high-performance military aircraft and the use of gaseous nitrogen for the inflation of aircraft tires. More recently, liquid nitrogen has been used in food preservation for aircraft. Presently, supersonic aircraft use hydrocarbon fuels, but the future hypersonic aircraft are expected to use liquid hydrogen. Liquid hydrogen has a higher specific impulse and, thus, it could be utilized for the cooling of the aerodynamic surfaces that are exposed to high frictional heating in high-speed flight.

16.2.1.3 Space

The first use of cryogenic systems took place on a large scale during the Second World War with the V-2 rocket, which used alcohol and LOX. It was followed by unmanned missions and then manned missions. The first manned space flight took off with the Air Force X-I airplane, which was similar to the V-2, and also propelled with alcohol and LOX. Later, the more advanced X-15 was launched with 254 kN thrust, achieving a speed of 2 km s^{-1} at over 108 km altitude. From this time onward, cryogenic systems were frequently utilized in manned and unmanned missions. At first, only LOX was used as the cryogenic fluid and the fuel used was either alcohol or kerosene; hence, this kind of propulsion system was also termed semi-cryogenic propulsion. The best-known launch vehicles of this type are Redstone, Atlas, and Saturn I. In the Saturn V, its first stage utilized kerosene-LOX while its upper stage utilized liquid hydrogen-LOX. It was the first attempt wherein fuel and oxidizer were used in a cryogenic condition to gain an increased specific impulse. More recently, the Space Shuttle used liquid hydrogen and LOX propellants and cryogenic oxygen for life support. The tank sizes for the LOX and liquid hydrogen were 530 and 1438 m^3, respectively. This represents the largest cryogenic tank that has been used in a launch vehicle.

16.2.2 Safety issues

To maximize the safety of a cryogenic system, it is important to manage all aspects of the system. This means that safety processes should be adopted in each phase of the system design, fabrication process, in the system checks, and throughout its operation. It is not only important to have an efficient system, but also important to operate the system in a safe manner. In the design process for any new system, concern for safety should be adopted from the conceptual design phase, starting from the basic design process, and maintained until the final design is completed.

16.2.3 Problems associated with cryogenic propulsion systems

Highly complex phenomena occur in cryogenic engines. The most common issues associated with their operation are subcritical or supercritical combustion, accompanying high-frequency oscillations, tribology, or cavitation for turbopumps. It should be noted that these engines are subjected to very challenging operating conditions, which include both thermal and mechanical operating conditions. Due to innovative and smart manufacturing processes, low-cost and robust systems have been developed for cryogenic engine injectors, tanks, feed lines, and even for the main combustors. The main design aspect is related to the physics behind the cryogenic phenomenon, and a deep and correct knowledge on this can be helpful in accurately predicting the performance of future cryogenic engines.

Understanding the combustion phenomenon is a crucial part in understanding a cryogenic engine. The combustion efficiency of the engine is directly related to the I_s of the engine. Thus, for the performance of the engine, the thermal flux ought to be precisely predicted, particularly for expander-type engines. Combustion instabilities, either high or low frequency, are also a critical problem in cryogenic engines as they can damage the engine. Research groups are working on such high-frequency combustion instabilities in France, Germany, and the European research group, REST (Rocket Engine Stability iniTiative). These research groups have worked on complex simulations to build simplified and industrial tools for the prediction of stable combustion in a given chamber geometry.

Turbopumps have been used in cryogenic engines for input of the fuel and oxidizer. Cavitation effects and mastering of rotor dynamics are the two most critical phenomena in cryogenic turbopumps. A lot of research has been carried out on the cavitation phenomenon as it can damage the turbopump's, inducer which will lead to flight failures. Again, too big cavitation margins are unfavorable to the performance of the launcher.

The rotor dynamics can predict the dynamic behavior of every component and its study has a decisive effect on the understanding of the whole machine dynamics. This is performed by conducting a water test. The dynamic characteristics of components such as seals, impellers, and bearings can only be evaluated using specific instrumentation. The valves for cryogenic engines are being converted from pneumatic to electric actuators, as they affect the heat energy consumption and thus will affect engine regulation. The essence of current research in this field is related to reduction in cost and size and enhancement of performance. Another important phenomenon for the cryogenic engines is the hydraulic behavior of the propellants in the tanks, which becomes critical during the coasting and reignition phases of launch missions.

16.2.4 Combustion instabilities in cryogenic rocket engines

Since the development of cryogenic engines, the problem of high-frequency combustion instability is a major problem faced by engineers [86]. In a LRE, fluids are filled within the rigid walls of the engine components. These engine components namely are the combustion chamber, the propellant tanks, and injectors. The geometry and fluid properties of the propellant lead to specific resonance frequencies. Due to random pressure oscillations formed due to the combustion and turbulence in the chamber, the related volumes oscillate with their resonance frequencies. This leads to a periodic modulation of the heat release rate of the combustion process. Rayleigh [87, 88] hypothesized that when the heat release rate oscillation is in phase with the pressure oscillation, an energy transfer from the heat release into the acoustic pressure oscillation occurs [89–91]. This will lead to a rapid increase of the oscillation amplitudes, which causes combustion instabilities [4, 92, 93].

Barrère and Williams [94] as well as Candel [95] categorized these coupling mechanisms for combustion devices into intrinsic instabilities, combustion chamber instabilities, and system instabilities. Intrinsic instabilities are related only to the reactants and do not have any effect on acoustic coupling. The interaction of combustion chamber acoustic resonance modes and the processes occurring after the addition of the propellant lead to the combustion chamber instability. The phenomena which take place after the entry of the propellant into the combustion chamber are propellant atomization, heating, vaporization, diffusion, turbulent mixing, and chemical reactions [96–99]. The feed system, the injectors, the combustion chamber, and the nozzle lead to system instability in LREs [100]. Hutt and Rocker [101] define injection coupling as a subtype of system instability. Oscillations inside the injectors cause heat release rate fluctuations in the engines.

The effect of low hydrogen injection temperature has a destabilizing effect on the combustion process, as was investigated by Pratt and Whitney in the RL-10 engine [4, 102, 103] that was meant for the Centaur upper stage [103]. The same effect was also obtained for Rocketdyne's J-2 engine for the Saturn IB and Saturn V launchers [102, 103] using coaxial injectors with a propellant combination of liquid hydrogen-LOX. Four instability mechanisms were proposed by the researchers for the combustion instabilities. The mechanisms can be described as follows [103]:

1. The LOX droplet size is influenced by the injection velocity ratio.
2. There is a decrease in pressure over the fuel inlet during the rate of change of temperature.
3. Coupling takes place of the resonance modes of the LOX post with that of the resonance modes of the combustion chamber.
4. There is an excitation of the LOX post inner acoustics due to the combustion in the recess volume of the injector.

Feiler and Heidman [104] conducted stability modeling work and observed using a response factor model that the hydrogen pressure drop was a crucial parameter in combustion chamber instabilities. The acoustic coupling between the combustion chamber and the hydrogen is attributed to manifold instabilities, as detected in the NASA LeRC test series [93, 105]. Experiments were conducted for ascertaining the instabilities detected at low methane temperature during test runs at Rocketdyne, and spontaneous instabilities of the 1T mode at ambient methane temperature were detected [106–108]. This type of instability was credited to the coupling of the 2L mode of the LOX post with that of the 1T mode of the combustion chamber. The same type of instability and oscillations were observed during the development of the J-2S program [101]

16.2.5 Effect of low temperature of hydrogen on cryogenic engines

The destabilizing effect of the low hydrogen injection temperature on rocket engines with a hydrogen-oxygen propellant combination has been investigated. It was observed that, for propellant combinations of hydrogen-oxygen, the hydrogen temperature of 95 K produces a first tangential mode leading to self-excited combustion instabilities. A research combustor was used to study the effect of hydrogen temperature ramping on the self-excited combustion instabilities. The temperature range studied was between 40 K and 135 K for hydrogen. It was observed that the combustor was found to be stable at low hydrogen temperatures, while at higher temperatures of around 100 K and above increased oscillation amplitudes of the first tangential mode were found. Further analysis showed that hydrogen temperature has a strong impact on the combustion instabilities.

There are two hypothesized mechanisms for this low-temperature destabilizing effect. The first one corresponds to the LOX post coupling and the other to the hydrogen-side injector coupling. Experiments were performed in a research rocket combustor, named BKD, at the Institute of Space Propulsion of the German Aerospace Center (DLR) [109]. It was observed from the test that BKD demonstrated self-excited high-frequency combustion instabilities of the 1T mode at a hydrogen injection temperature of 95 K. However, no instabilities were observed at a temperature of 50 K [110]. It was further postulated that the instabilities at 95 K was due to the alternations of the heat release rate, which was because of the resonance modes of the LOX posts [111]. The effect of hydrogen temperature on this mechanism was studied with the same BKD research combustor, and it was observed that the low hydrogen-side pressure drop could also affect the stability behavior.

Some frequencies which are not related to the combustion chamber resonance frequencies also contribute to the combustion instabilities. The 1T mode of frequencies corresponding to the OH* flame radiation intensity was observed as one of the dominant frequencies. When the mixture ratio was varied the 1T frequency mode varied in the intensity oscillation from a stable to an unstable state of operation. When an evaluation of the transitory variation of these dominant frequencies of intensity signal was carried out, it was found to be associated with the oxygen speed of sound signal, and it was inferred that these oscillations are associated with the oxygen system of the combustor. The effect of the variation of the LOX post length showed that these principal frequencies are due to LOX post resonance frequencies, which can be scaled relative to the LOX post length.

16.3 Gel propulsion systems

16.3.1 Introduction

Future space launch vehicles will be required to deliver massive payloads into distant Earth orbits with improved performance. The growing mission requirements also necessitate the augmentation of existing launch vehicles. The increase in payload mass inadvertently leads to larger stage volumes and overall rocket size. As mentioned earlier, the presently used storable LREs employ propellants like hydrazine-based fuels and nitrogen tetroxide that are hazardous and toxic in nature. The focus is now shifting toward green liquid propellants with benign combustion products; however, these fall short on performance criteria. There is also demand for propulsion systems that offer improved safety, high density impulse, and energy-management capability.

Advanced tactical missions require insensitive munitions, systems with safer storage and handling, easy disposal, and low overall life-cycle cost. A gel propulsion system utilizes a liquid fuel, oxidizer, or both in a semi-solid 'gel state' using a chemical or particulate gelling agent that can be safely stored in propellant tanks. These gels are injected through a suitably designed injector into the combustion chamber. This approach provides an attractive way to incorporate metal powder into the liquid fuel, which is otherwise impossible due to the high density of metals. The metallized gel propellant can facilitate combustion of low-molecular-weight metals in conjunction with the oxidant in the rocket engine, enhancing its performance. Some earlier reviews on the subject are available [112–116]. Recently, a very exhaustive review [117] covering all major areas related to gel propellants, such as the formulation of gel propellants and simulants, performance benefits, rheological characterization of heterogeneous fuels, spray characteristics using different types of injectors, ignition and combustion of the gel and metallized propellants, static tests of gel propulsion systems, and green gel propellants for future applications and their use in ramjet engines is presented. Efforts toward flight tests are also mentioned.

The main advantages of a gel propulsion system include the performance enhancement of a conventional liquid bipropellant system leveraging the high exothermicity of combustion of incorporated metal powders, and metallized gel propellants offer an excellent approach to cut down rocket sizes as these can burn at much-reduced O/F ratios. Gelled propellants would help rocket engine restart under zero-gravity conditions as gel will not disperse under minor accelerations. The normal vaporization losses and leakage issues of liquid propellants will be curtailed if a gel propellant is used. Similarly, logistics and handling problems are eased. Gel systems may prove advantageous in reduction of heat transfer from warm tank walls to low-temperature propellant in case of rocket flights of long duration. Such propulsion systems can be stored in proximity of personnel such as on a naval ship, as the risk of accidental hypergolic ignition will be greatly reduced. Cryogenic gel propellants are a good option and will be able to address the low-density issue of liquid hydrogen by converting it to a gel state by using a suitable gellant. The prevention of geysering and cavitation phenomea in storage tanks and the effective expulsion of propellant are added benefits.

However, gel propulsion systems also have several challenges in formulation, design, development, and in the operation of rocket engines. The major issue in formulation is to convert the liquid base fuel into a gel state by using a minor concentration of gellant that does not deteriorate the overall energetics of the system. The storage characteristics need to be acceptable as poorly constituted gels destabilize with time and temperature cycling. Another important feature desired from such systems is that they should exhibit thixotropic (shear-thinning) characteristics. This characteristic enables liquefication of gel by subjecting it to flow through the fine orifice of the injector plate to bring its viscosity as close as to the parent liquid fuel by choosing the right injection pressure and injector type. The rheological characterization of gels at low shear rates is essential to maintain their production quality, and flow characterization at high shear rates is required to utilize the results for feed system and injector design. Further, the ignition and combustion behavior of the metallized gel propellant becomes complex in nature and needs to be investigated on a case-to-case basis to ensure combustion efficiency, the envisaged higher performance, and optimized motor design. This can be done by understanding droplet size and distribution, and the temperature profile in the chamber as well as the required residence time so that the near-complete combustion of metal powder is ensured. Significant earlier investigations on the formulation of many fuels and oxidizer gels [118–121] and other simulants have been covered in earlier reviews [113, 117]. Details of performance benefits [122–124], rheological characterization [125–128], ignition and combustion [122–132], and static motor operation and evaluation [133–136], as well as flight tests of gel propulsion systems [137, 138] have been presented and provide information on the critical areas that have been intensively investigated. A brief description and recent important contributions relating to gel propellants and gel rocket motors, specifically employing practical propellant systems, are discussed below in each segment to provide a better explanation and understanding of the subject.

16.3.2 Formulation of gel propellants

Research done on the formulation of gel propellants has concentrated largely on both hydrazine-based as well as liquid-hydrocarbon-based fuels, and some oxidizers that may make a usable bipropellant hypergolic system have been reported in the open literature and summarized [112–114, 117]. Gel propellants can be prepared by using either chemical or particulate gelling agents by incorporating their suitable concentration in a base liquid. A limited number of gellants can form gels at an acceptable concentration level (<5 wt.%) by itself. Others may produce the gels by addition of different additives like solvents, surfactants, emulsifiers, thickening agents, dissolvers, stabilizers, polymeric thickeners, catalysts, etc. Most of these gellants and additives are detrimental to the net energetics of the gels prepared using them. The preferred gellant should be able to form stable thixotropic gel at very low concentrations of around 1 wt.% and ideally contribute to the exothermicity of the system. High concentrations of inert and noncombustible gellant not only adversely affect the fluidity of the gelled fuel but also reduce the specific impulse of the propellant system. Though numerous studies have been conducted on propellant gels, gellants with the desired characteristics that would be able to form energetic gels are yet to be identified or synthesized.

The production of gels in reasonable quantities and with consistent properties poses a challenge. There are many factors that influence the quality of the product. These include the concentration of gellant, particle size, chemical structure, viz. degree of substitution of effective functional group, mixing time, method used for mixing, dissolver speed, span and blade configuration, diameter of the container and temperature, etc. Most of these parameters influence the storage and aging characteristics besides the stability of the gel. Gels generally are very delicately balanced systems and are prone to changes in properties due to aging, thermal cycling, shear force exerted due to transportation and handling, etc. It is essential to investigate the impact of such factors on the gel behavior before their attempted application. The use of nonintrusive acoustic resonant mixers may prove beneficial in comparison to mechanical mixers to produce good-quality gels. Many gel systems show oozing of the base fuel on storage whereas others may exhibit shrinkage. This crucial aspect is rarely studied or reported in the literature. The change in gel volume due to shrinkage and presence of supernatant liquid in the storage tank will be detrimental to reliable engine performance. Another important parameter is the gelation time, i.e. the time required between mixing of the ingredients and setting of a stable gel for a particular formulation. The buildup of the gel network may be influenced by various conditions encountered during processing and the environment. Atmospheric carbon dioxide may also cause poisoning of the gel network. Nonintrusive methods for determining the gelation time in a reactor need to be established. It may be noted that the critical concentration of gellant required in a formulation may vary with ambient conditions, the cross-section area of the vessel used, production history of the gellant used, and other additives used in conjunction with the primary gellant.

Very little work has been done on the mechanism of propellant gel formation using different gellants with conclusive evidence. Particulate gelling agents like carbon black, colloidal silica, clays, etc, are believed to form gels at higher concentration levels by weak physical (Van der Waals) force. Chemical gellants like modified derivative of castor oil (Thixatrol), Carbopol, and cellulose derivatives (methyl, ethyl, hydroxypropyl, etc) are effective in producing aviation fuel gels and of hydrazine-based rocket fuels by swelling of the gellant. A good stable and reproducible gel with consistent properties should have a uniform three-dimensional gel network in the bulk. Another approach that may prove useful involves controlled *in situ* polymerization of a miscible monomer with the base fuel/oxidizer. Nano-gellants may also be effective due to their large surface area. The oxidizer RFNA gels prepared using sodium silicate lower the performance of the bipropellant system significantly [122]. An oxidizer gel should only be considered in cases where safety, logistics, and storage issues are weighed over the net performance.

The interest in gel propellants centers around the possibility of incorporating metal powders into the fuel to gain a high density specific impulse. These systems require a much-reduced oxidizer content for combustion as the metal burns in the flame generated by the bipropellants [123]. The dispersion of low-molecular-weight metals like aluminum, magnesium, and boron in the fuel becomes possible in such semi-solid gels. The addition of ultrafine or nano-sized metal powders itself shows the tendency to build-up the fuel viscosity and, as a result, may require much lesser concentrations of gellant to formulate usable heterogeneous fuel systems in a rocket or ramjet engine. To sum up, a desirable gel for propulsion application needs to satisfy many requirements and characteristics that may vary with operational needs. The ultimate objective remains the formulation of a stable and storable gel using the minimum gellant concentration of an energetic material. This gel should be able to liquify in order to flow through the feed system and injector to efficiently atomize and burn in the combustion chamber. Scaling-up issues of batch sizes during gel production may present new problems.

16.3.3 Performance of heterogeneous propellants

The performance predictions of virgin and metallized gel propellants with or without an oxidizer gel has been studied generally by using NASA programs and reported in the open literature. It is inferred [123] that the specific impulse (I_s) of a gel fuel (UDMH)-liquid oxidizer (RFNA) bipropellant system decreases marginally if a gellant with high negative heat of formation is used. Subsequent metallization can surpass the performance of virgin bipropellant system nearly at 5 wt.% level. The increase in metal concentration in the gel demonstrated an enhancement in specific impulse, bulk density, and density specific impulse up to a level of 40 wt.%, and beyond that the I_s registers a fall due to the nonavailability of the working fluid for expansion (table 16.4). Equation (16.2) gives information about the various parameters that influence the specific impulse (I_s). A maximum gain of over 3% can be achieved with aluminized propellant at a nominal chamber pressure of around 2 MPa. A further rise in chamber pressure to a level of 5 MPa may lead to an I_s rise of the order of over 14% under equilibrium flow conditions. This appears to be quite attractive, as even a 1% gain in I_s may increase the range of an intercontinental ballistic missile by 7%, if good combustion efficiency is maintained. Magnesium-containing gels deliver about 2% lesser I_s than their aluminized counterparts.

$$I_s = \left[\frac{2\gamma}{\gamma - 1} \frac{JRT_c}{g\overline{M}_c} \left\{ 1 - \left(\frac{P_e}{P_c} \right)^{\gamma - 1/\gamma} \right\} \right]^{1/2}, \tag{16.2}$$

where J is the mechanical equivalent of heat,
R is the universal gas constant,
γ is the specific heat ratio,
T_c is the combustion chamber temperature,
$\overline{M}_c$ is the mean molecular mass of burnt gases,
P_e is the nozzle exit pressure, and
P_c is the chamber pressure.

The individual performance of a gel fuel should be assessed by varying individual influencing formulation parameters such as gellant and metal concentration in the fuel and oxidizer. Further computations at varying equivalence ratios and chamber pressures for different compositions are necessary to evaluate the gain in I_s, characteristic velocity (C^*), reduction in O/F ratio for optimum performance, combustion temperature, and combustion product profile to further manipulate the rise in performance for a given mission requirement. Performance computation inputs using other metals such as magnesium, beryllium, and boron may be of interest for specific purposes.

The impact of metallized propellants on payload delivery capability in low Earth orbit has been assessed considering cryogenic, semi-cryogenic, and storable bipropellant systems [139, 140]. $O_2/H_2/Al$, O_2/RP-1/Al, and NTO/MMH/Al systems were considered for comparison of their respective potential. Metallized propellants, due to their high density, low oxidant requirement, and high I_s, offer a reduction in the

Table 16.4. Composition of metallized UDMH-MC fuel and equivalence ratio corresponding to maximum specific impulse with RFNA oxidizer.

S. No.	Fuel composition by weight fraction			Optimum ER		Maximum specific impulse (kgf s kg^{-1})	
	UDMH	MC	Metal	Equilibrium flow	Frozen flow	Equilibrium flow	Frozen flow
1.	1			1.20(2.726)	1.30(2.486)	238.9	232.9
2.	0.957 12	0.042 88	0.0	1.15(2.840)	1.30(2.456)	237.9	231.9
3.	0.861 41	0.038 59	0.1Al	1.25(2.427)	1.40(2..125)	239.7	231.9
4.	0.765 70	0.034 30	0.2Al	1.40(1.995)	1.60(1.714)	241.2	235.5
5.	0.669 98	0.3002	0.3Al	1.85(1.363)	1.90(1.323)	243.5	237.9
6.	0.574 27	0.025 73	0.4Al	2.45(0.931)	2.30(0.997)	246.3	240.5
7.	0.478 56	0.021 44	0.5Al	2.15(0.994)	2.10(1.020)	246.1	238.9
8.	0.861 41	0.036 59	0.1Mg	1.20(2.510)	1.40(2.100)	238.5	231.1
9.	0.765 70	0.034 30	0.2Mg	1.35(2.031)	1.65(1.616)	239.5	232.1
10.	0.669 98	0.030 02	0.3M	1.60(1.538)	2.00(1.199)	240.5	232.7
11.	0.574 27	0.025 73	0.4Mg	2.10(1.039)	2.25(0.963)	241.6	232.0
12.	0.478 56	0.021 44	0.5Mg	2.20(0.894)	2.30(0.851)	240.0	228.5

The figure in the parenthesis denotes the corresponding O/F ratio.

size of a stage for a particular mission, or the same volume-limited vehicle can achieve larger range or carry greater payload mass.

The use of storable propellants has added advantage as the boil-off losses are significantly reduced, leading to a simplified thermal design of the vehicle. For a mission to Mars, it has been evaluated that a nearly 20% additional payload can be delivered using metallized propellants. It can be inferred that metallized propellant technology is a cost effective and safe option for improving the performance of a chemical rocket engine. Replacement of the solid rocket booster and liquid sustainer stages in a space transportation system also has been evaluated, as these may offer enhanced controllability and reduced sensitivity and explosion hazard [141]. In that study, a NTO/MMH/Al system also showed potential to enhance the payload-carrying capability.

A promised increment of about 20% payload capability for the Indian Polar Satellite Launch Vehicle, using a UDMH/NTO system, has been shown if metallized gel is used [142]. It may be noted that theoretical estimates of reflected performance advantage are on the higher side as the assumptions made in the computation codes consider ideal conditions. Experimental realities should account for different kinds of losses due to nonoptimal mixture ratio, friction, and dispersion in a gas-film cooled rocket engine, turbulent mixing during combustion, inadequate residence time for reactants in the engine, two-phase losses during expansion in nozzle, and production of complex high-molecular-weight metal oxide species that may affect the mean molecular weight of the combustion products and attainment of maximum adiabatic flame temperature. There may also be a trade-off between the practically achievable gel composition and the one required for optimum engine performance.

16.3.4 Ignition and combustion of gels and metallized fuels

The ignition and combustion behavior of a gel and metallized fuel is much more complex than their base bipropellant system equivalents. The main reason is the conversion of the liquid fuel into a gel state, with its apparent viscosity becoming several orders of magnitude higher than the Newtonian base fluid. As a result, it is expected that gel propellants will take a longer time to ignite as the physical and chemical lags of the ignition process will significantly increase. Most gel propellants are seen to exhibit large ignition delays under stagnant and no-flow conditions due to a hindered mixing process of the fuel and oxidizer and utilization of some part of the oxidant for nitration as well as oxidative degradation of the gellant molecules. The initial diffusion process of the oxidant into the fuel gel remains slow and gets subsequently enhanced as the gel network gets broken and viscosity reduced. Some gel propellants, however, are found to exhibit a marginally reduced ignition delay in comparison to the parent bipropellant system around the critical O/F ratio that corresponds to the minimum ignition delay. This anomalous behavior may be due to the added heat of nitration of the gellant to the system when RFNA is used as an oxidant. A summary of the earlier work done on ignition of hypergolic gel propellants is presented in [117].

Metallized fuels with different extents of metal loading in the gel and metal particle size also influence their ignition and combustion characteristics. For a hypergolic bipropellant system, ignition delay is a very important property that affects the smooth start of the LRE. Heterogeneous gel and metallized propellant systems show an increased ignition delay due to very high viscosity that influences the diffusion, mixing, and reaction rates with the impinged oxidizer. The particle size of the metal powder used in the formulation also affects the ignition characteristics. Magnesium-containing gel propellants register a much-reduced ignition delay at corresponding O/F ratios than the aluminized ones under no-flow conditions in an ignition tester [129].

The pre-ignition and ignition process of virgin, gel (UDMH-HSMC), and metallized (aluminum and magnesium) fuel systems with RFNA oxidant, studied using a customized twin-jet combustor [130] (figure 16.6), showed that all systems spontaneously ignite in an injection pressure range of 1.37 to 2.06 MPa with a recorded ignition delay of less than 40 ms in a nozzle-less combustion chamber (figure 16.7). An injector orifice diameter of 1 mm and an impingement angle of 45° were used. It was, however, noticed that highly aluminized gels (40 wt.%) using a metal particle size of 20 μm do not readily flow through the injector below an injection pressure level of 4.13 MPa. This may indicate that a suitable metal particle size in a gel propellant formulation is an important criterion for their use in a gel propulsion system.

Efforts have been made in the recent past [143] to prepare gels of a green fuel, ethanolamine, that is inherently non-hypergolic in nature. These gelled fuels, which use a high percentage of inert gellants, could be made hypergolic with hydrogen peroxide as an oxidizer when certain concentrations of metal catalyst such as anhydrous copper chloride and manganese acetyl acetonate are incorporated in the

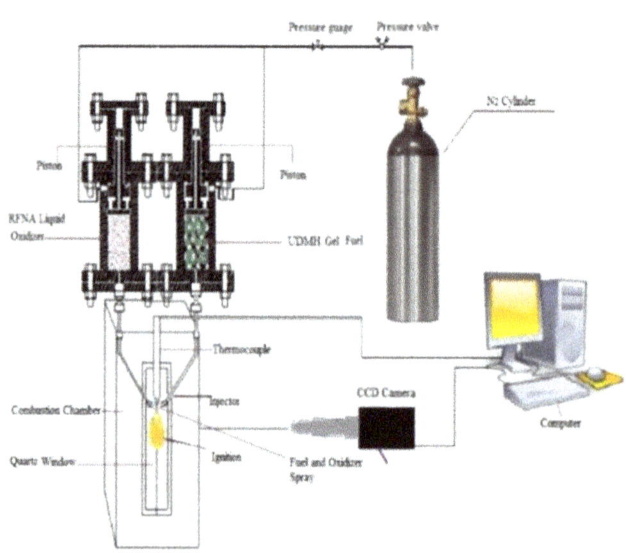

Figure 16.6. Experimental setup of twin-jet combustor.

Figure 16.7. Ignition sequence of metallized (UDMH-Mg)-RFNA system.

fuel gel to achieve a very short ignition delay. Further studies [144] were also conducted by using gels of ethanol and other alcohols with extended carbon chains till heptanol with a low (2 wt.%) gellant concentration to assess the influence of carbon chain on the enthalpy of combustion of metallized fuels using metal catalysts. It was observed that ethanol gel showed an improvement in enthalpy in comparison to liquid fuel, whereas metallized ethanol gel and gels and metallized systems of other alcohols showed a drop in enthalpy but could be ignited with the help of catalysts. The ignition delay of catalyzed ethanol gels with hydrogen peroxide (90%) are reported to be within the range of 10–40 ms, which is acceptable for operational propulsion systems [145].

An experimental investigation on hypergolicity and ignition delay on pure and energized gelled ethanol with hydrogen peroxide in conjunction with a metal catalyst (as mentioned above) was carried out [146]. A small quantity (1 wt.%) of catalyst in the fuel gel was found to be sufficient to provide hypergolicity, and violent ignition was observed with ignition delays in the range of 1–30 ms using a high-speed imaging technique. Catalytic ignition leads to very high flame temperatures in the range of 1000–1600 K under atmospheric conditions. Pure gel and energized systems formulated using nano-sized aluminum, boron, and carbon show low toxicity, high density, and short ignition delay. A minimum ignition delay time of 1.33 ms was recorded for boron-energized fuel.

The ignition and combustion characteristics of neat and gelled monomethyl hydrazine and RFNA were investigated using two different gelling agents [147]. The type of gellant was seen to have a major impact on ignition and combustion behavior, as their inert or reactive nature has a definite influence on it. The fuel with inert gellant was seen to exhibit acceptable ignition delay but a significant reduction in combustion efficiency was observed. The inconsistency of the ignition delay results of gelled fuel at higher pressures was also noted and attributed to the high sensitivity of the transient events in the combustion chamber due to the high viscosity of gels and their inconsistent atomization. Hypergolic ignition of hydrocarbon fuels with addition of reactive particles of sodium borohydride using hydrogen peroxide oxidant has also been studied [148, 149] employing doublet, triplet, and quintuplet injectors. Comprehensive models for estimating the ignition delay of gelled and metallized fuel systems are yet to be proposed, the major reason being the complexity of the involved processes, including breakup of the gel network, release of base liquid, size of gel droplet produced and its vaporization and mixing with oxidant droplets, involved chemical reaction mechanisms, heating intensity and heat transfer in the vicinity of the droplet, etc. A simplified theory to predict ignition times of hypergolic gelled propellants is presented in reference [150]. Assuming an Arrhenius reaction at the interface of the two adjacent spaces of fuel and oxidant, a simple formula with a universal constant is proposed to estimate ignition time.

The combustion characteristics of gel propellants are strongly dependent on the characteristics of the base fuel and type and concentration of gellant used, the nature and particle size of the metal powder incorporated, the rheological properties of the gel, the type of injector and pressurization system used, and the level of atomization achieved after its injection in the combustion chamber to facilitate the mixing process with the oxidant. The reactivity and nature of the oxidant, i.e. gaseous, liquid, or cryogenic, also influences the overall combustion process. The major challenge is to achieve high combustion efficiency and projected specific impulse by ensuring the near-complete combustion of the metal powder added to the fuel gel. Several approaches may be tried to harness propulsive gains with suitable modifications and optimization in the design of the injection system and combustor, which are well known to professionals working on LREs.

Recent advances in this domain have led to the production of gel propulsion systems that have shown desired performance, reliability, and safety. 'Gel engines' have been flight-tested by Northrop Grumman under the US Army's Missile Technology Integration Program. The potential of propulsion systems using gelled propellants has also been tested by Bayern-Chemie, Germany during flight tests to demonstrate variable thrust, extended mission range, and endurance maneuverability. Looking back, the research work on the combustion of gelled propellants was initiated at BIT, India in the mid-1980s and later carried out at DLR, Germany, Technion, Israel, KAIST, Republic of Korea, and Purdue University and Penn State University in the USA.

The heat of combustion of gel and metallized UDMH fuel is observed to be higher than base fuel and goes on increasing with the degree of metallization in a constant volume bomb calorimeter. However, the metallized gel lump cannot be combusted completely, and a sizable amount of incorporated metal remains unburnt [129]. The injection of metallized gels through an impinging-type injector facilitates metal combustion to a large extent even in a nozzle-less rocket motor. A higher injection pressure and suitable injector design may lead to a higher combustion efficiency. Theoretical computations indicate that metal combustion in a hypergolic propellant system at different O/F ratios follows a complex mechanism. Some part of the metal can be directly oxidized by the oxidizing species present in the combustion chamber to transform into its oxide, whereas the rest may get oxidized by reaction with water vapors produced by the reaction between the organic fuel and oxidant. This pathway yields low-molecular-weight hydrogen and may be more exothermic in nature. The increased presence of hydrogen in the combustion products brings down its mean molecular weight and attained higher combustion chamber temperature, resulting in an increase in I_s.

As stated earlier, the main attraction of using gel propellants lies in their ability to serve as a carrier of metal powder in the combustion chamber so that the advantage of the high combustion heat of metals may be used for propulsive gains. The efficient combustion of metal powder depends on generation of the adequate flame temperature in the rocket engine by the combustion of the ingredients of the bipropellant system. As a result, the selection of bipropellants and operating conditions primarily hold the key for efficient combustion of the incorporated metal, besides the nature of the metal itself. Boron may be more effective in certain applications as it undergoes surface combustion once the oxide film vanishes from its surface at a much lower melting temperature (783 K). Further, the particle size of the metal used also influences the combustion process [151], and its efficacy also depends on the metal concentration in the heterogenous formulation.

The efficacy of the combustion of a gelled propellant is strongly dependent on its formulation, extent of thixotropy, the injector type used, the breakup dynamics of the gel droplets, spray characteristics and atomization, droplet size, and collision intensity with oxidant droplets. Our understanding on oscillatory vaporization of gel droplets and the nature and propagation of spray flames in actual rocket engine conditions is limited,though a few basic theoretical models have been attempted [152, 153]. The approach appears to be simplistic as the conceptualization of gel droplets considers encapsulation of the droplets in gellant layers. Most of the stable gels are formed due to buildup of nearly symmetrical gellant networks or swelling of the gellant by soaking the base fluid within multiple layers.

Future models must consider the realistic features of gel formulation, shear-induced liquification, atomization and spray details, reaction exothermicity, flame morphology details, and spatial temperature distribution in the combustion chamber. Investigation of the secondary atomization of droplets of metallized gel fuels is of utmost importance in order to gain better knowledge of metal ignition and combustion, which is itself essential to achieve higher combustion efficiency. The experimental details in each constituent of the involved processes need to be available to provide the basis for a satisfactory model. Unrealistic and simplistic assumptions in modeling in the absence of experimental results limits the utility and usefulness of the theoretical predictions. Incidentally, nature does not respect assumptions. Performance loss due to the uncontrolled combustion process may be minimized by having detailed information on the high-pressure combustion of heterogeneous gel propellants. Scaling-up issues with such systems may bring in challenges that can only be sorted out with detailed understanding of various operational aspects in subscale gel rocket motors.

16.3.5 Rheology and flow characteristics of gelled systems

The semi-solid state of gel propellants leads to many advantages such as the ease of handling and transportation, negligible vaporization losses on storage, and convenient transfer and loading of propellants. As mentioned earlier, it allows a check on propellant oozing and leakage through joints in the feed line, as well as safe fuel storage for long duration, control on accidental spillage, inadvertent ignition of hypergolic propellants, and homogeneous distribution of incorporated metal powders in the bulk of the fuel gel. Operational advantages achievable with such systems include high performance (I_s) and density specific impulse (I_d) due to incorporation of highly energetic metals, improved shelf life of missiles, reduced rocket weight due to lesser oxidant requirement for combustion for a mission, engine restart under zero-gravity conditions as gels do not disperse to minor accelerations, and reduced heat transfer from warm tanks to low-temperature propellant. These systems have also proved to be advantageous in volume-limited micro-thrusters. Gel technology applied to cryogenic fluids may address the low density issue of liquid hydrogen, outage, geysering, and cavitation problems.

Such gels are non-Newtonian fluids, and they must exhibit a thixotropic (shear-thinning) character to be efficiently atomized and combusted in a rocket engine.

Several formulated gels have been discarded as they did not exhibit the required flow properties. Gels that show conversion of gel–sol–gel type, i.e. those which can be liquified on shearing and then may re-gel when left undisturbed for a certain re-gelation time, appear to be attractive from an application viewpoint. Metallized gels, however, are found to become inhomogeneous after shearing but exhibit shorter re-gelation time. In this context, it is essential to have an in-depth knowledge on the rheological characteristics of gels in the low, intermediate, and high shear rate ranges in order to construct an informative rheogram that can be used for production, quality control, and to provide base data for the design of the propellant feed system and injectors. The temperature, concentration of gellant, extent of metallization, and shear rate/stress control the apparent viscosity of the gel and influence its thixotropic nature. The gels should also possess adequate yield stress to remain stable during transportation and storage and to maintain homogeneity and quality under ambient temperature cycling when stored.

Rheological studies of a gel in a low shear rate range (10–$10^2\ s^{-1}$) are essential to characterize the formulated gel and measure the extent of thixotropy and yield stress [89]. This methodology can also be used to determine the required gelation time for a particular gel as viscosity buildup of the gellant dispersed fuel with time can be traced till the gel state is attained. Such information enables one to decide the processing conditions and set the influencing parameters such as the critical concentration of gellant to be used, mixing time, dissolver type, and revolutions per minute to get a desired product that can be repeatably produced. Flow behavior studies of a gel under a high shear rate range (10^4–$10^6\ s^{-1}$) [127] are of utmost practical importance as the maximum fall in gel viscosity occurs in this range, as it gets shredded and the gellant network aligns in the direction of the flow to release the entrapped liquid fuel within. Representative results are shown in table 16.5 for ethanol gel. Commercially available rheometers or suitably designed orifice or capillary rheometers can be used to investigate flow behavior and compute the extent of the fall in apparent viscosity of the gel. Since most of the gels considered fit for practical use show a time-dependent shear-thinning nature, the duration of shearing and its impact on apparent viscosity must be investigated to collect and utilize data for the effective design of injectors.

Several studies have been conducted on the rheological characterization of fuel and oxidizer gels, and they have been recently reviewed [117]. The investigations included propellants of Jet A-1 [120], unsymmetrical dimethyl hydrazine [125, 127], fuming nitric acid [126], nitromethane [154], hydrazine, JP-8 fuels and hydrogen peroxide oxidizer [155], monomethyl hydrazine [156], metallized and non-metallized ethanol [157–159], ethanolamine [160], and kerosene [161]. Voluminous experimental data are available on the abovementioned fuel and oxidant systems, which includes the effect of variation of gellant types, processing methods, concentrations, and particle sizes of metals like aluminum, magnesium and boron, temperature, and shear rate, etc, that can be used to develop necessary rheological models. Existing analytical and numerical models such as the power law, Herschel–Bulkley model, Herschel–Bulkley extended model, Carreau model, Carreau–Yasuda model, and Cross model have been used to illustrate the findings and data compliance and to compute the generalized Reynolds number using a specific flow model. Specific models [162–164]

Table 16.5. Dependence of apparent viscosity on shear rate for ethanol gel.

Sl. No.	Capillary dia (mm)	*L*/*D*	Corrected pressure (MPa)	Corrected shear stress (kPa)	Corrected shear rate (× 10^5 s^{-1})	Apparent viscosity (η) (Pa s)
1	1	10	0.79	13.67	2.29	0.06
			1.30	22.34	2.62	0.09
			1.86	32.00	2.84	0.11
			2.34	40.28	3.15	0.13
2	1	15	0.92	11.78	2.12	0.06
			1.40	17.92	2.56	0.07
			1.92	24.63	2.86	0.09
			2.39	30.65	3.22	0.10
3	1	20	0.95	9.67	1.98	0.05
			1.47	15.03	2.33	0.06
			2.06	21.01	2.52	0.08
			2.64	26.91	2.71	0.10
4	1	25	0.97	8.20	1.90	0.04
			1.61	13.66	2.00	0.07
			2.15	18.22	2.31	0.08
			2.76	23.42	2.45	0.10
5	1.5	10	0.85	14.59	1.48	0.10
			1.32	22.84	1.75	0.13
			1.81	31.19	1.97	0.16
			2.23	38.52	2.23	0.17
6	2	10	1.00	17.23	0.98	0.18
			1.53	26.44	1.16	0.23
			2.01	34.72	1.37	0.25
			2.54	43.88	1.51	0.29
7	2.5	10	1.05	18.17	0.79	0.23
			1.49	25.72	1.05	0.24
			1.93	33.34	1.26	0.26
			2.38	41.00	1.44	0.29

for the evaluation of the rheological properties of propellant gels have also been reported in the literature.

16.3.6 Spray characteristics and atomization of gels

Numerous studies have been conducted in this domain with a view to understand the achievable droplet size of a sheared gel passing through a typical orifice or given type of injector in an experiment. Details of the breaking of gels into ligaments, sheets, and finally into droplets and their distribution has been investigated to determine the Sauter mean diameter (equation (16.3)) and droplet size distribution at different injector pressures and orifice sizes. Generally, safe hydrogels, gel stimulants [165], and hydrocarbon gels [166] are used from a safety viewpoint. The gel and metallized versions of the high-energy fuels that are used in operational bipropellant LREs have rarely been studied. Some investigators have attempted to work out matching simulants for an actual gelled fuel for its assessment under cold flow conditions. A schematic setup used to study spray characterization is given in figure 16.8. The typical results of an ethanol gel spray pattern at a given injection pressure using an impinging type of atomizer is presented in figure 16.9.

$$\mathrm{SMD} = \left(\frac{\sum_{i=1}^{N} n_i d_i^3}{\sum_{i=1}^{N} n_i d_i^2} \right) \tag{16.3}$$

where SMD is the Sauter mean diameter,

n_i is the number of droplets in size range i, and

d_i is the diameter of droplet in size range i.

It should be noted that the physio-chemical properties of the base fuel largely influence the overall gel constitution, and so simulants or hydrogels offer a practical but poor approach to obtain details of the spray characteristics and atomization of energetic fuels. High-speed cameras, and in some studies Malvern laser particle size analyzers [167], are generally used to record the different stages of a gel's break-up into tiny droplets by injecting it through different types of injectors into a chamber with proper illumination and optically transparent windows. The spray

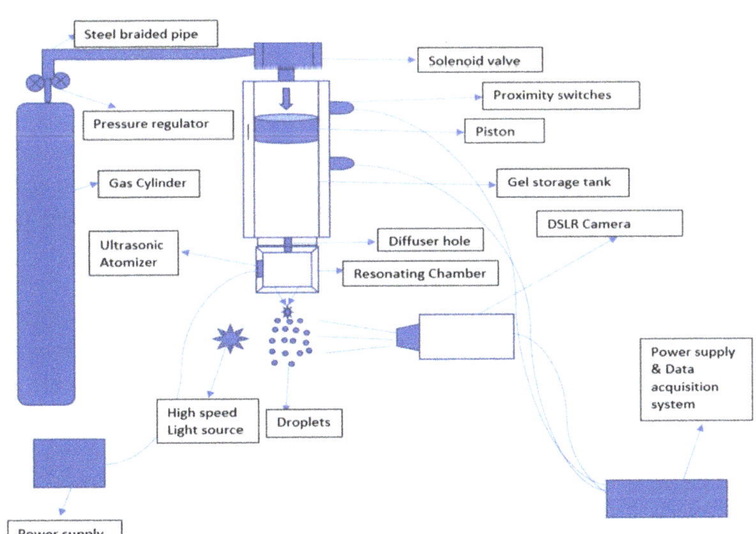

Figure 16.8. Experimental setup for gel spray characterization.

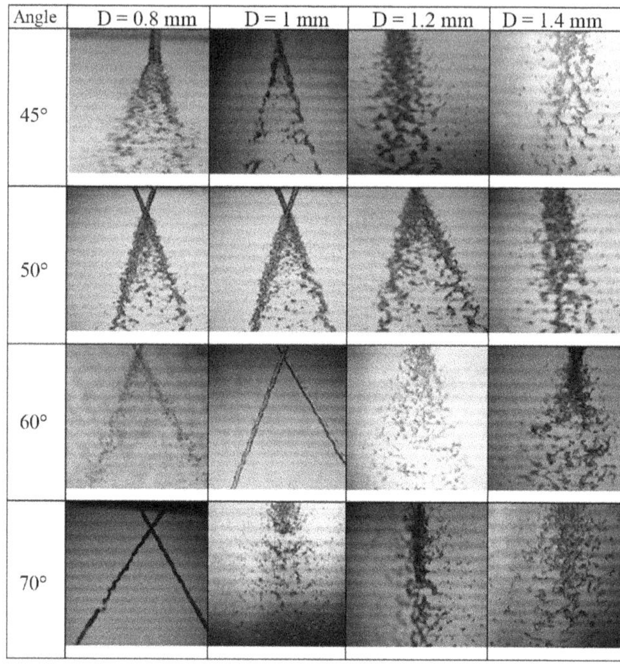

Figure 16.9. Spray pattern of ethanol gel ejecting from different orifice sizes and impingement angles (at 2.76 MPa).

characteristics are found to be strongly dependent on the nature of the gel propellant, injection pressure, configuration of injector used, viz. impinging jet [168] (self-impinging doublet or cross-impinging triplet), coaxial [165], swirl [169, 170], or pintle [114] type. Some studies have also been conducted to examine spray characteristics by assisting atomization using an air-blast atomizer [166], pulsating technique [169], or even by using a Helmholtz resonator [168, 171]. Full spray cone angles have also been determined in several studies for different types of gels and simulants as a function of momentum flux ratio, injection pressure, volumetric flow rate, and Weber number [117]. The contraction and expansion of the cone angle is primarily dependent on the rheological nature of the gel, besides the nature of the atomizer used, and influences the extent of fluid atomization.

The general constituents of a fully developed spray field downstream may have a conical liquid sheet after liquification of the gel under a high shear rate, and then conversion of some part of the liquid sheet into ligaments. Turbulent web-like ligaments, small ligaments, and droplets may coexist some distance down the injector face. The breakup length of the liquid sheet should be minimal and so the gel mass flow rate may have to be adjusted. Newtonian fluids have a parabolic velocity profile in a constant diameter injector inlet under steady state and fully developed laminar flow conditions. Thixotropic gels show different velocity profiles, with lower velocities at the center and higher gradients toward the capillary wall. The interlayer shearing of the gel during flow through the injector is affected and this may influence the subsequent breakup distance and nature of the liquid sheet formed. Detailed information on this aspect may facilitate better understanding of the operating conditions necessary to design injector hardware that can produce uniform and fine sprays for effective combustion.

The information available in the design and understanding of sprays and the design and use of atomizers for customized spray behavior is large [20]. However, the domain of non-Newtonian fluids, with an emphasis on thixotropic shear-thinning propellant gels, has only been addressed in a limited manner. Some initial attempts to assess the impact of atomizer types such as ultrasonic [172], air-blast [173], effervescent [174, 175], pentad [134], and pressure-swirl [169] atomizers on the spray characteristics of non-Newtonian fluids provide a good basis for further investigations. Other improved atomization methods, viz. aerodynamic flow perturbation [176], direct flow perturbation [173], and internal impingement [166], have also been reported in the literature. Detailed information on the effect of a typical

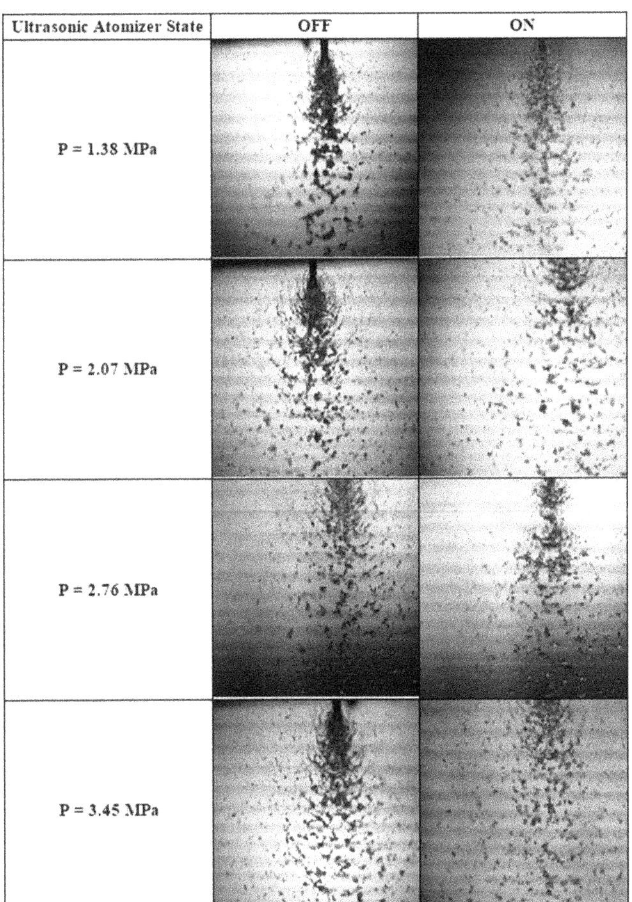

Figure 16.10. Effect of ultrasonic disturbances on spray pattern at different injection pressures.

atomizer on the breakup mechanism and atomization data is essential to optimize the injector design for a particular gel propellant. Such details on this aspect are beyond the scope of this chapter, however a good summary and review [117] may here be referred to, to extend research and development works on the subject.

Research work on the use of techniques for assisted breakup of gel droplets is limited, but this approach offers an effective way to achieve finer droplets than possible with normal injection methods. Spray patterns of ethanol gel with and without ultrasonic disturbance (figure 16.10) and for virgin and metallized gels (figure 16.11) are presented at different injection pressures. The possibility of their use in actual injection systems appears to be remote at present as it will cause complexity in system design and may pose reliability issues. More intensive work in this area is required to develop acceptable and miniaturized devices for future applications. Theoretical and numerical analysis of the atomization process of gel propellants, as well as subsequent modeling, would help to develop an effective

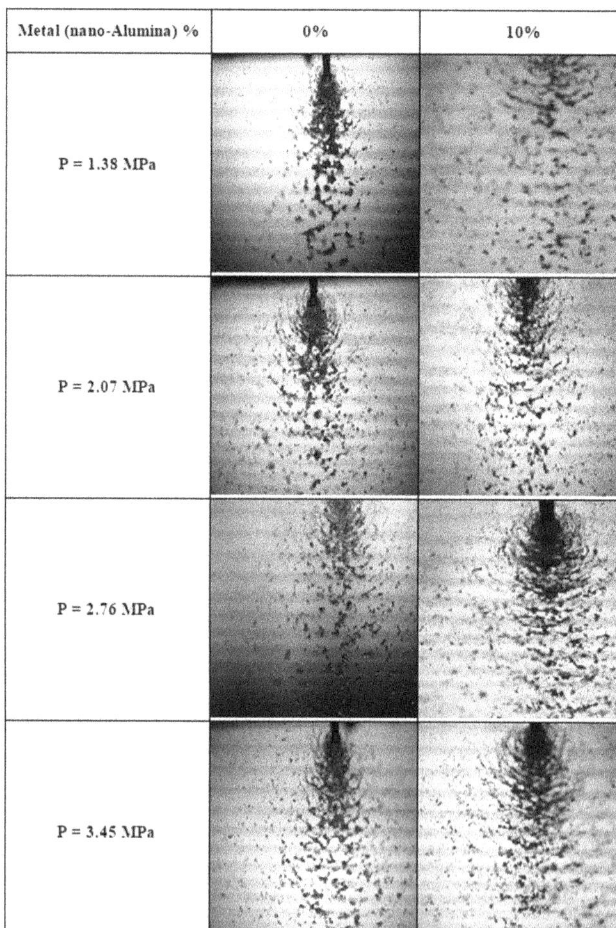

Figure 16.11. Effect of metallization of gels on spray characteristics at different injection pressures.

approach for the design of injectors. New approaches of reliable spray analysis and spray characterization will help in the development of future gel propulsion motors.

Gel propellants using particulate gelling agents, such as colloidal silica and its variants, carbon black, charcoal, types of Carbopol, synthetic hydrophilic polymers, etc, in higher concentrations may cause clogging of the injector holes, but the gels will be easily broken down due to weak inter-particulate bonding. Water-containing gels or simulants may be of academic interest, but their rheology and spray characteristics are different from fuel gels containing minor concentrations of chemical gellants. Further, the performance loss due to the presence of water in the fuel is detrimental to their use in practical propulsion systems. Chemical gellants like different types of celluloses, gums, poly (vinyl alcohol), acrylic acid, and thixatrol, etc, may react or solubilize in the fuel to entrap it in the network. Such gels are stable and their properties are suitable for propulsion application, however these gels exhibit a much higher apparent viscosity and are difficult to atomize. Further, the drop in their viscosity on shearing is time dependent, and this must be considered in the design of injection systems.

It is important to determine the spray and atomization characteristics of such gels using a proper pressure or turbopump feed system in adequate quantities and a suitable injector with acceptable orifice size, impingement angle, configuration, and injector plate thickness. The flow behavior of metallized gels with different particle sizes and concentration levels in the gel formulation also need to be investigated in detail. Progress in various aspects of the abovementioned areas will ensure realization of reliable and improved performance gel propulsion systems that can be used in tactical areas of kill vehicle technology, insensitive munitions, high-density and operationally safe propellant systems, and lightweight throttlable micro-thrusters. Properly designed gel propulsion systems may have the potential to replace semi-cryogenic engines in some missions.

References

[1] Huzel D K and Huang D H 1967 Design of liquid propellant rocket engines *NASA Special Publication* SP 125 NASA

[2] Gupta S K 2020 Analysis of liquid jet Al—gaseous oxygen based movable pintle injector *M.Tech. Thesis* Birla Institute of Technology, Mesra, Ranchi

[3] Ramamurthi K 2016 *Rocket Propulsion* 2nd edn (New Delhi: Trinity Press of Laxmi Publications)

[4] Sutton G P and Biblarz O 2001 *Rocket Propulsion Elements* 7th edn (New York: Wiley)

[5] Yang H, Fushui L, Yikai L and Han W 2019 Experimental and numerical study on the effect of dimensionless parameters on the characteristics of droplet atomization caused by periodic inertial force *Fuel* **253** 941–9

[6] Shinjo J and Umemura A 2019 Fluid dynamic and autoignition characteristics of early fuel sprays using hybrid atomization LES *Combust. Flame* **203** 313–33

[7] Alsulami R, Windell B, Nates S, Wang W, Won S H and Windom B 2020 Investigating the role of atomization on flame stability of liquid fuels in an annular spray burner *Fuel* **265** 116945

[8] Dafsari R A, Lee H J, Han J and Lee J 2019 Evaluation of the atomization characteristics of aviation fuels with different viscosities using a pressure Swirl Atomizer *Int. J. Heat Mass Transf.* **145** 118704

[9] Cui J, Lai H, Li J and Ma Y 2017 Visualization of internal flow and the effect of orifice geometry on the characteristics of spray and flow field in pressure-Swirl Atomizers *Appl. Therm. Eng.* **127** 812–22

[10] Igari N, Iso T, Nishio Y, Izawa S and Fukunishi Y 2019 Numerical simulation of droplet-formation in rotary atomizer *Theor. Appl. Mech. Lett.* **9** 202–5

[11] Kuhnhenn M, Joensen T V, Reck M and Roismn I V 2018 Study of the internal flow in a rotary atomizer and its influence on the properties of the resulting spray *Int. J. Multiph. Flow* **100** 30–40

[12] Rezayar S, Farshchi M and Ghorbanhoseini M 2019 Primary breakup dynamics and spray characteristics of a rotary atomizer with radial-axial discharge channels *Int. J. Multiph. Flow* **111** 315–38

[13] Zaremba M, Weib L, Maly M, Wensing M, Jedelsky J and Jicha M 2017 Low-pressure twin-fluid atomization: effect of mixing process on spray formation *Int. J. Multiph. Flow* **89** 277–89

[14] Markus S, Fritching U and Bauckhage K 2002 Jet break up of liquid metal in twin fluid atomisation *Mater. Sci. Eng.* A **326** 122–33

[15] Nguyen D A and Rhodes M J 1998 Producing fine drops of water by twin-fluid atomisation *Powder Technol.* **99** 285–92

[16] Yang Z, Zhang K, Hwang Y and Lian Z 2016 Performance investigation on the ultrasonic atomization liquid desiccant regeneration system *Appl. Energy* **171** 12–25

[17] Deepu P, Peng C and Moghaddam S 2018 Dynamics of ultrasonic atomization of droplets *Exp. Therm Fluid Sci.* **92** 243–7

[18] Zhang Y, Yuan S and Wang L 2021 Investigation of capillary wave, cavitation and droplet diameter distribution during ultrasonic atomization *Exp. Therm Fluid Sci.* **120** 110219

[19] Yu T, An J, Yang X, Bian X and Zhao J 2018 The study of ultrasonic vibration assisted polishing optical glass lens with ultrasonic atomizing liquid *J. Manuf. Process* **34** 389–400

[20] Lefebvre A H and McDonell V G 2017 *Atomization and Sprays* 2nd edn (Boca Raton, FL: CRC Press) pp 17–99

[21] Gad H M, Ibrahim I A, Abdel-Baky M E, El-Samed A K A and Farag T M 2018 Experimental study of diesel fuel atomization performance of air blast atomizer *Exp. Therm Fluid Sci.* **99** 211–8

[22] Vani Pooja V 2019 Studies of liquid jet atomization through non-circular orifices *M.Tech Thesis* Birla Institute of Technology, Mesra, Ranchi

[23] Vani Pooja V and Kumar R 2019 Studies of liquid jet atomization through non-circular orifices *12th Int. High Energy Materials Conf. & Exhibits (HEMCE-2019) (Chennai, India, December 16–18)*

[24] Sharma P and Fang T 2014 Breakup of liquid jets from non-circular orifices *Exp. Fluids* **55** 1666

[25] Kashyap T V, Sivakumar D and Raghunandan B N 2008 Breakup of liquid jets emanating from elliptical orifices at low flow conditions *Atom. Sprays.* **18** 645–68

[26] Kashyap T V, Sivakumar D and Raghunandan B N 2009 Flow and breakup characteristics of elliptical liquid jets *Int. J. Multiph. Flow* **35** 8–19

[27] Hong J G, Ku K W, Lee C W and Na B C 2012 Elliptical jet breakup related with the internal nozzle flow *12th Triennial Int. Conf. on Liquid Atomization and Spray Systems (Heidelberg, Germany, September 2–6)*

[28] Hashiehbaf A and Romano G P 2014 Experimental investigation on circular and non-circular synthetic jets issuing from sharp edge orifices *17th Int. Symp. on Applications of Laser Techniques to Fluid Mechanics Lisbon (Portugal, 7–10 July)*

[29] Gutmark E J 1997 Mixing enhancement in non-circular jets *Am. Inst. Aeronaut. Astronaut.*

[30] Ibrahim I M, Murugappan S and Gutmark E J 2005 Penetration, mixing and turbulent structures of circular and non-circular jets in cross flow *AIAA Paper* AIAA-2005-0300

[31] Lichtarowicz A, Duggins R K and Markland E 1965 Discharge coefficients for incompressible non-cavitating flow through long orifices *J. Mech. Eng. Sci.* **7** 210–9

[32] Yoon C, Stephen D, Xia H G and Merkle C L 2010 Simulation of injection of shear-thinning gel propellants through plain-orifice atomizer *AIAA Paper* AIAA-2010–7138

[33] Dressler G A and Bauer M J 2012 TRW pintle engine heritage and performance characteristics *AIAA Paper* AIAA 2000-3871

[34] Rezende R N, Pimeta A and Perez V C 2015 Experiments with pintle injector design and development. Propulsion and energy forum *51st AIAA/SAE/ASEE Joint Propulsion Conf.*

[35] Casiano M J, Hulka J R and Yang V 2010 Liquid-propellant rocket engine throttling: a comprehensive review *J. Propul. Power* **26** 897–923

[36] Betts E M and Frederick R A Jr 2010 A historical systems study of liquid rocket engine throttling capabilities *46th AIAA/ASME/SAE/ASEE Joint Propulsion Conf. & Exhibit*

[37] Lee S, Kim D, Koo J and Yoon Y 2020 Spray characteristics of a pintle injector based on annular orifice area *Acta Astronaut.* **167** 201–11

[38] Son M, Radhakrishnan K and Koo J 2017 Design procedure of a movable pintle injector for liquid rocket engines *J. Propul. Power* **33** 858–69

[39] Steinhauser G and Klapotke T M 2008 'Green' pyrotechnics: a chemists' challenge *Angew. Chem. Int. Ed.* **47** 3330–47

[40] Larsson A and Wingborg N 2011 Green propellants based on ammonium dinitramide (ADN) *Advances in Spacecraft Technologies* ed Jason Hall (London: InTech)

[41] Jankovsky R S 1996 HAN-based monopropellant assessment for spacecraft *NASA Technical Memorandum* 107287

[42] Courtheoux L, Gautron E, Rossignol S and Kappenstein C 2005 Transformation of platinum supported on silicon-doped alumina during the catalytic decomposition of energetic ionic liquid *J. Catal.* **23** 10–8

[43] Atamanov M K, Amrousse R, Hori K and Mansurov Z 2019 Experimental investigations of combustion: (95wt.-%) HAN–water solution with high-SSA activated carbons *Combust. Sci. Technol.* **191** 645–58

[44] Amrousse R, Hori K, Fetimi W and Farhat K 2012 HAN and ADN as liquid ionic monopropellants: Thermal and catalytic decomposition processes *Appl. Catalysis* B **127** 121–8

[45] Kang S and Kwon S 2015 Difficulties of catalytic reactor for hydroxylammonium nitrate hybrid rocket *J. Spacecr. Rockets* **52** 1508–10

[46] Oommen C, Rajaraman S, Chandru R A and Rajeev R 2011 Catalytic decomposition of hydroxylammonium nitrate monopropellant *Int. Conf. on Chemistry and Chemical Process IPCBEE* vol 10 (Singapore: IACSIT Press)

[47] Amariei D, Courthéoux L, Rossignol S, Batonneau Y and Kappenstein C 2005 Influence of the fuel on the thermal and catalytic decompositions of ionic liquid monopropellants *41st AIAA/ASME/SAE/ASEE Joint Propulsion Conf. & Exhibit 10–13 July 2005 (Tucson, AZ)*

[48] Courthéoux L, Eloirdi R, Rossignol S, Kappenstein C and Duprez D 2002 Catalytic decomposition of HAN-water binary mixtures *38th AIAA/ASME/SAE/ASEE Joint Propulsion Conf. & Exhibit 7–10 July 2002 (Indianapolis, IN)* AIAA 2002-4027

[49] Zube D, Christofferson S, Wücherer E and Reed B 2003 Evaluation of HAN-based propellant blends *39th AIAA/ASME/SAE/ASEE Joint Propulsion Conf. (Huntsville, AL, 20–23 July)*

[50] Amrousse R, Katsumi T, Bachar A, Brahmi R, Bensitel M and Hori K 2014 Chemical engineering study for hydroxylammonium nitrate monopropellant decomposition over monolith and grain metal-based catalysts *ReacKinet Mech Cat* **111** 71–88
[51] Agnihotri R and Oommen C 2018 Cerium oxide based active catalyst for hydroxylammonium nitrate (HAN) fueled monopropellant thrusters *RSC Adv.* **8** 22293–302
[52] Katsumi T, Inoue T, Nakatsuka J, Hasegawa K, Kobayashi K, Sawai S and Hori K 2012 HAN-based green propellant, application, and its combustion mechanism *Combust Explos Shock Waves* **48** 536–43
[53] Amariei D, Courtheoux L, Rossignol S and Kappenstein C 2007 Catalytic and thermal decomposition of ionic liquid monopropellants using a dynamic reactor comparison of powder and sphere-shaped catalyst *Chem. Eng. Process.* **46** 165–74
[54] Schumb W, Satterfield C and Wentworth R 1955 *Hydrogen Peroxide* (New York: Reinhold)
[55] Guseinov S L, Fedorov S G, Kosykh V A and Storozhenko P A 2020 Hydrogen peroxide decomposition catalysts used in rocket engines *Russ. J. Appl. Chem.* **93** 467–87
[56] Ventura M C and Heister S D 1995 Hydrogen peroxide as an alternate oxidizer for a hybrid rocket booster *J. Propulsion Power,* **11** 562–3
[57] Sisco J C, Austin B L, Mok J S and Anderson W E 2005 Autoignition of kerosene by decomposed hydrogen peroxide in a dump-combustor Configuration *J. Propulsion Power* **21** 450–9
[58] Pasini A, Torre L, Romeo L, Cervone A and D'Agostino L 2011 Performance characterization of pellet catalytic beds for hydrogen peroxide monopropellant rockets *J. Propulsion Power* **27** 428–36
[59] Lee S L and Lee C W 2009 Performance characteristics of silver catalyst bed for hydrogen peroxide *Aerospace Sci. Technol.* **13** 12–7
[60] Romeo L, Pasini A, Torre L, d'Agostino L and Calderazzo F 2008 Comparative characterization of advanced catalytic beds for hydrogen peroxide thrusters *44th AIAA Joint Propulsion Conf. and Exhibition (Hartford, CT, July 21–23)* AIAA Paper 2008–5027
[61] An S and Kwon S 2008 Catalyst bed sizing of 50 newton hydrogen peroxide monopropellant thruster *44th AIAA Joint Propulsion Conf. and Exhibition (Hartford, CT, July 21–23)* AIAA Paper 2008–5109
[62] Dolci S, Belli Dell'Amico D, Pasini A, Torre L, Pace G and Valentini D 2015 *J. Propulsion Power* **31** 1204–16
[63] Woschnak A, Krejci D, Schiebl M and Scharlemann C 2013 Development of a green bipropellant hydrogen peroxide thruster for attitude control on satellites *Prog. Propulsion Phys.* **4** 689–706
[64] Huh J, Jyoti B V S, Yun Y, Shoaib M N and Kwon S 2018 Preliminary assessment of hydrogen peroxide gel as an oxidizer in a catalyst ignited hybrid thruster *Int. J. Aerospace Eng.* **2018** 5630587
[65] Miraglia J, Hetem A, Burian R and Caetano C A C 2011 Pre-mixed liquid green propellant for use in rocket engines *Proc. OBEM 2011 21st Brazilian Congress of Mechanical Engineering (ABCM) (Natal/RN, Brazil, October 24–28)*
[66] Khaji Z, Klintberg L, Palmer K and Thornel G 2020 catalytic effect of platinum and silver in a hydrogen peroxide monopropellant ceramic microthruster *Propul. Power Res.* **9** 216–24
[67] Krishnan S, Hee S K and Won L C 2010 Design and development of a hydrogen-peroxide rocket-engine facility *JurnalMekanikal* **30** 24–36

[68] Bloom R, Davis N S Jr and Levine S D 1950 Hydrogen peroxide as a propellant *J. Am. Rocket Soc.* **80** 3–17

[69] Jo S, Jang D, An S and Kwon S 2011 Chugging instability of H_2O_2 monopropellant thrusters with catalyst reactivity and support sizes *J. Propulsion Power* **27** 920–4

[70] Jo S 2017 Response characteristics of H_2O_2 monopropellant thrusters with MnO_2-mixed PbO catalyst *Aerospace Sci. Technol.* **60** 1–8

[71] Baek S, Jung W, Kang H and Kwon S 2018 Development of high-performance green-monopropellant thruster with hydrogen peroxide and ethanol *J. Propulsion Power* **34** 1256–61

[72] Bonifacio S, Festa G and Sorge A R 2013 Novel structured catalysts for hydrogen peroxide decomposition in monopropellant and hybrid rockets *J. Propulsion Power* **29** 1130–7

[73] Bonifacio S, Sorge A R, Krejci D, Woschnak A and Scharlemann C 2014 Novel manufacturing method for hydrogen peroxide catalysts: a performance verification *J. Propulsion Power* **30** 299–308

[74] Pasini A, Pace G and Torre L 2015 Pulse mode operation of a 1 N 98% hydrogen peroxide thruster *AIAA Paper* AIAA 2015–4057

[75] An S and Kwon S 2009 Scaling and evaluation of Pt/Al_2O_3 catalytic reactor for hydrogen peroxide monopropellant thruster *J. Propuls. Power* **25** 1041–5

[76] Lee J and Kwon S 2013 Evaluation of ethanol-blended hydrogen peroxide monopropellant on a 10 N class thruster *J. Propuls. Power* **29** 1164–70

[77] Kang H, Lee D, Kang S and Kwon S 2017 Effect of H_2O_2 injection patterns on catalyst bed characteristics *Acta Astronaut.* **130** 75–83

[78] Huh J, Ahn B, Kim Y, Song H, Yoon H and Kwon S 2017 Development of a university-based simplified H_2O_2/PE hybrid sounding rocket at KAIST *Int. J. Aeronaut. Space Sci.* **18** 512–21

[79] Rohrbaugh E M 2001 Hybrid rockets *Aerospace America* **39** 61

[80] Maisonneuve Y, Godon J C, LeCourt R and Lengelle G 2002 Hybrid propulsion for small satellites: design logic and tests *Combustion of Energetic Materials* ed K K Kuo and L T DeLuca (New York: Begel House Inc.) pp 90–100

[81] Fisher S C 2002 Liquid propulsion *Aerospace America* **40** 64–5

[82] Scott R B 1988 *Cryogenic Engineering* (Boulder, CO: Met-Chern Research Inc)

[83] Bell J H Jr 1963 *Cryogenic Engineering* (Englewood Cliffs, NJ: Prentice-Hall)

[84] Zabetakis M G 1967 *Safety with Cryogenic Fluids.* (New York: Plenum Press)

[85] Timmerhaus K D and Flynn T M 1989 *Cryogenic Process Engineering.* (New York: Springer)

[86] Ross C C and Datner P P 1954 Combustion instability in liquid-propellant rocket motors—a survey *Selected Combustion Problems* ed W R Hawthorne, J Fabri and D B Spalding (Oxford: Butterworths Scientific Publications) pp 352–80

[87] Rayleigh J W S 1878 The explanation of certain acoustical phenomena *Nature* **18** 319–21

[88] Rayleigh J W S 1879 The explanation of certain acoustical phenomena *Notices of the Proc. at the Meetings of the Royal Institution of Great Britain with Abstracts of the Discourses delivered at the Evening Meetings* vol 8 (London: Royal Institution of Great Britain) pp 536–42

[89] Chu B T 1956 Stability of systems containing a heat source—the Rayleigh criterion *NACA Research Memorandum* NACA RM 56D27 National Advisory Committee for Aeronautics

[90] Chu B T 1965 On the energy transfer to small disturbances in fluid flow (part I) *Acta Mech.* **1** 215–34

[91] Putnam A A 1964 General considerations of autonomous combustion oscillations *Nonsteady Flame PropagationAGARDograph* vol 75 ed G H Markstein (Amsterdam: Elsevier) pp 183–98

[92] Culick F E C and Yang V 1995 Overview of combustion instabilities in liquid-propellant rocket engines ed V Yang and W E Anderson *Liquid Rocket Engine Combustion Instability, Progress in Astronautics and Aeronautics* vol 169 (Washington DC: American Institute of Aeronautics and Astronautics) ch 1 pp 3–37

[93] Harrje D T and Reardon F H 1972 Liquid propellant rocket combustion instability *NASA Special Publication* SP-194

[94] Barrère M and Williams F A 1969 Comparison of combustion instabilities found in various types of combustion chambers *Symp. (Int.) Combust.* vol 12 *(Pittsburgh, PA)* (Pittsbugh, PA: The Combustion Institute) pp 169–81

[95] Candel S 2002 Combustion dynamics and control: progress and challenges *Proc. Combust. Inst.* **29** 1–28

[96] Crocco L and Cheng S I 1956 *Theory of Combustion Instability in Liquid Propellant Rocket MotorsAGARDograph* vol 8 (London: Butterworths Scientific Publications) pp 199–200

[97] Conrad E W, Bloomer H E, Wanhainen J P and Vincent D W 1968 Interim summary of liquid rocket acoustic-mode-instability studies at a nominal thrust of 20000 pounds *NASA Technical Note* NASA TN D-4968

[98] Crocco L 1951 Aspects of combustion stability in liquid propellant rocket motors part I: fundamentals. Low frequency instability with monopropellants *J. Am. Rocket Soc.* **21** 163–78

[99] Crocco L 1952 Aspects of combustion stability in liquid propellant rocket motors part II: low frequency instability with bipropellants. High frequency instability *J. Am. Rocket Soc.* **22** 7–16

[100] Anderson W E, Ryan H M and Santoro R J 1991 Combustion instability phenomena of importance to liquid bi-propellant rocket engine *28th JANNAF Combustion Subcommittee Meeting*

[101] Hutt J J and Rocker M 1995 High-frequency injection-coupled combustion instability ed V Yang and W E Anderson *Liquid Rocket Engine Combustion Instability, Progress in Astronautics and Aeronautics* vol 169 (Washington, DC: American Institute of Aeronautics and Astronautics) ch 12 pp 345–55

[102] Bilstein R E 1980 Stages to Saturn *NASA Special Publication* NASA SP-4206

[103] Isakowitz S J 1995 *International Guide to Space Launch Systems* 2nd edn (Washington, DC: American Institute of Aeronautics and Astronautics)

[104] Feiler C E and Heidmann M F 1967 Dynamic response of gaseous-hydrogen flow system and its application to high-frequency combustion instability *NASA Technical Note* NASA TN D-4040

[105] Micci M M and Gandilhon D 2008 Shear coaxial injector LOX droplet measurements as a function of hydrogen temperature *At. Sprays* **18** 85–96

[106] Jensen R J, Claflin S E and Dodson H C 1989a Liquid oxygen/methane combustion instability investigation *The Winter Annual Meeting of the American Society of Mechanical Engineers* ed A S Hersh, I Catton and R F Keltie (San Francisco, CA: The American Society of Mechanical Engineers) pp 77–85

[107] Jensen R J, Dodson H and Trueblood B 1988 Oxygen/methane combustion stability investigations *Advanced Earth-to-Orbit Technology Conf. (Huntsville, AL)*

[108] Jensen R J, Dodson H C and Claflin S E 1989b LOX/hydrocarbon combustion instability investigation *NASA Contractor Report* NASA CR-182249

[109] Gröning S, Oschwald M and Sattelmayer T 2012 Selbst erregte tangentiale Moden in einer Raketenbrennkammer unter repräsentativen Bedingungen *Proc. 61. Deutscher Luft- und Raumfahrtkongress (Berlin, DE, 10–12 Sept)* (in German)
[110] Gröning S, Suslov D, Oschwald M and Sattelmayer T 2013 Stability behaviour of a cylindrical rocket engine combustion chamber operated with liquid hydrogen and liquid oxygen *5th European Conf. for Aeronautics and Space Sciences (EUCASS) (Munich, DE)*
[111] Gröning S, Hardi J S, Suslov D and Oschwald M 2016 Injector- driven combustion instabilities in a hydrogen/oxygen rocket combustor *J. Propuls. Power* **32** 560–73
[112] Varma M 1986 Studies on UDMH-RFNA gelled propellant systems *PhD Thesis* Birla Institute of Technology, Ranchi, India
[113] Natan B and Rahimi S 2002 The status of gel propellants in year 2000 *Int. J. Energ. Mater. Chem. Propuls.* **5** 172–94
[114] Pein R 2005 Gel propellants and gel propulsion *Proc. 5th Int. High Energy Materials Conf. & Exhibit (Hyderabad, India)*
[115] Ciezki H K *et al* 2017 Overview on German Gel propulsion technology activities: status 2017 and outlook *7th European Conf. for Aeronautics and Space Sciences (EUCASS) (Milan, Italy)*
[116] Natan B and Hasan D 2019 Advances in gel propulsion *Int. J. Energ. Mater. Chem. Propuls.* **18** 303–23
[117] Padwal M B, Natan B and Mishra D P 2021 Gel propellants *Prog. Energ. Combust. Sci.* **83** 100885
[118] Munjal N L, Gupta B L and Varma M 1985 Preparative and mechanistic studies on unsymmetrical dimethyl hydrazine-red fuming nitric acid liquid propellant gels *Propellant, Explos. Pyrotech.* **10** 111–17
[119] Varma M, Gupta B L and Pandey M 1996 Formulation and storage studies on hydrazine based gelled propellants *Def. Sci. J.* **46** 435–42
[120] Varma M and Pein R 2009 Optimization of processing conditions for gel propellant production *Int. J. Energ. Mater. Chem. Propuls.* **8** 501–13
[121] Chen A, Guan X, Li X, Zhang B, Zhang B and Song J 2017 Preparation and characterization of metalized JP-10 gel propellants with excellent thixotropic performance *Prop., Explos. Pyrotech.* **42** 1007–13
[122] Munjal N L, Gupta B L and Varma M 1991 Performance studies on UDMH-RFNA gelled propellants *J. Inst. Engrs. (India) (AS)* **71** 25–31
[123] Gupta B L, Varma M and Munjal N L 1992 Theoretical performance of metallized unsymmetrical dimethyl hydrazine-red fuming nitric acid propellant system *Ind. J. Tech.* **30** 234–44
[124] Palaszewski B 1992 Metallized propellants for human exploration of mars *J. Propul. Power* **8** 1192–9
[125] Gupta B L, Varma M and Munjal N L 1986 Rheological studies on virgin and metallized unsymmetrical dimethyl hydrazine gelled systems *Prop. Explos. Pyrotech.* **11** 45–52
[126] Gupta B L, Varma M and Goel S B 1986 Rheological characterization of fuming nitric acid gel *Prop. Explos. Pyrotech.* **11** 85–90
[127] Varma M 2017 High shear rheometry of unsymmetrical dimethyl hydrazine gel *Chemical Rocket Propulsion* ed L T DeLuca *et al* (Cham: Springer) pp 519–42
[128] Jyoti B V S and Baek S W 2016 Rheological characterization of ethanolamine gel propellants *J. Energ. Mater.* **34** 260–78

[129] Gupta B L and Varma M 1999 Ignition and combustion studies on metallized UDMH-RFNA bipropellant system *Ind. J. Eng. Mater. Sci.* **6** 13–21

[130] Varma M and Jyoti B V S 2011 Ignition and combustion studies of heterogeneous UDMH-RFNA gel propellants *Int. J. Energ. Mater. Chem. Propuls.* **10** 259–75

[131] Dennis J D, Pourpoint T L and Son S F 2011 Ignition of gelled monomethyl hydrazine and red fuming nitric acid in an impinging jet apparatus *47th AIAA/ASME/SAE/ASEE Joint Propulsion Conf. and Exhibit (San Diego, CA)* AIAA-2011–5706

[132] Feng S, He B, He H, Su L, Hou Z, Nie W and Guo X 2013 Experimental studies on the burning process of gelled unsymmetrical dimethyl hydrazine droplets under oxidant convective conditions *Fuel* **111** 367–73

[133] Palaszewski B and Zakany J S 1995 Metallized gelled propellants—oxygen/RP −1/ aluminum rocket combustion experiments *31st AIAA/ASME/SAE/ASEE Joint Propulsion Conf. and Exhibit (San Diego, CA, July 10–12)* AIAA-95-2435

[134] Rahimi S, Hasan D and Peretz A 2004 Development of laboratory scale gel propulsion technology *J. Propuls. Power* **20** 93–100

[135] Mahadev W S 2006 Development of gel propulsion system for the performance evaluation of gelled UH-25-NTO bipropellants *M.E. Thesis* Birla Institute of Technology, Ranchi, India

[136] Kirchberger C U, Kröger P and Ciezki H K 2017 Design and run-in tests of a gel combustion chamber with optical access *53rd AIAA/SAE/ASEE Joint Propulsion Conf. (Atlanta, CA, July 10–12)* AIAA-2017–5068

[137] Stierle R, Schmid K, Ramsel J and Naumann K W 2011 Free-flight demonstration of the gelled propellant rocket motor of MBDA-Bayern-Chemie *4th European Conf. for Aeronautics and Space Sciences (St. Petersburg, Russia)*

[138] Caldas Pinto P, Hopfe N, Ramsel J, Naumann W, Thumann A and Kurth G 2017 Scalability of gelled propellant rocket motors *7th European Conf. for Aeronautics and Space Sciences (EUCASS) (Milan, Italy)*

[139] Palaszewski B A 1991 Advanced chemical propulsion at NASA Lewis: metallized and high energy density propellants *NASA Technical Memorandum* NASA TM 103771

[140] Palaszewski B A 1992 Metallized propellants for the human exploration of mars *J. Propuls. Power* **8** 1192–9

[141] Palaszewski B and Powell R 1991 Launch vehicle performance using metallized propellants *27th Joint AIAA/SAE/ASEE Joint Propulsion Conf. (Sacramento, CA, June 24–27)*

[142] Varghese T L, Prabhakaran N, Thanki K P, Subramanian S, Rao S S, Ninan K N and Krishnamurthy V N 1999 Performance evaluation and experimental studies on metallized gel propellants *Def. Sci. J.* **49** 71–8

[143] Jyoti B V S, Naseem M S, Baek S W, Lee H J and Cho S J 2017 Hypergolicity and ignition delay study of gelled ethanolamine fuel *Combust. Flame* **183** 102–12

[144] Naseem M S, Jyoti B V S, Baek S W and Huh J 2018 Effect of alcohol carbon chain on enthalpy of combustion and ignition delay time for gelled hypergolic propellant system *Prop. Explos. Pyrotech.* **43** 453–60

[145] Naseem M S, Jyoti B V S, Baek S W, Lee H J and Cho S J 2017 Hypergolic studies of ethanol based gelled bi-propellant system for propulsion application *Prop. Explos. Pyrotech.* **42** 676–82

[146] Jyoti B V S, Naseem M S and Baek S W 2017 Hypergolicity and ignition delay study of pure and energized ethanol gel fuel with hydrogen peroxide *Combust. Flame* **176** 318–25

[147] Dennis J D, Willits J D and Pourpoint T L 2018 Performance of neat and gelled monomethylhydrazine and red fuming nitric acid in an unlike-doublet combustor *Comb. Sci. Tech.* **190** 1141–57

[148] Connell T L Jr, Risha G A, Yetter R A and Natan B 2014 Effect of fuel on hypergolic ignition of hydrogen peroxide with gelled hydrocarbon fuel *50th Joint AIAA/SAE/ASEE Joint Propulsion Conf. (Cleveland, OH, July 28–30)*

[149] Connell T L Jr, Risha G A, Yetter R A and Natan B 2015 Impinging jet ignition studies of hydrogen peroxide with gelled fuels rendered hypergolic by addition of reactive particles *51st Joint AIAA/SAE/ASEE Joint Propulsion Conf. (Orlando, FL, July 27–29)*

[150] Williams F A 2009 Simplified theory of ignition times of hypergolic gelled propellants *J. Propuls. Power* **25** 1354–6

[151] Negri M and Ciezki H K 2015 Combustion of gelled propellants containing microsized and nanosized aluminum particles *J. Propuls. Power* **31** 400–7

[152] He B, Nei W and He H 2012 Unsteady combustion model of non-metallized organic gel fuel droplet *Energy Fuels* **26** 6627–63

[153] Kunin A, Greenberg J B and Natan B 2008 A simple phenomenological model of an organic gel spray diffusion flame *Comb. Sci. Technol.* **180** 27–44

[154] Teipel U and Forter-Barth U 2005 Rheological behavior of nitromethane gelled with nanoparticles *J. Propulsion Power* **21** 40–3

[155] Rahimi S, Petertz A and Natan B 2010 Rheological matching of gel propellants *Prop. Explos. Pyrotech.* **26** 376–8

[156] Dennis J D, Kubal T D, Campanella O, Son S F and Pourpoint T L 2013 Rheological characterization of monomethyl hydrazine gels *J. Propuls. Power* **29** 313–20

[157] Jyoti B V S and Baek S W 2014 Rheological characterization of metalized and non-metalized ethanol gel *Prop. Explos. Pyrotech.* **39** 866–73

[158] Carer C *et al* 2021 Effect of parameter variation on viscosity of ethanol gel propellants *J. Aerosp. Technol. Mang.* **13** e1121

[159] Nandgopalan P, John J, Baek S W, Miglani A and Ardhianto K 2018 Shear flow rheology and viscoelastic instabilities of ethanol gel fuels *Exp. Therm. Fluid Sci.* **209** 181–9

[160] Jyoti B V S and Baek S W 2016 Rheological characterization of ethanolamine gel propellants *J. Energ. Mater.* **34** 260–78

[161] Yang D, Xia Z, Huang L, Ma L, Chen B and Feng Y 2020 Synthesis of metallized kerosene gel and its characterization for propulsion applications *Fuel* **262** 116684

[162] Chernov V and Natan B 2012 A simplified model for the evaluation of the rheological properties of a suspension of solids in a power law fluid *Appl. Rheol.* **22** 15163-1–10

[163] Yoon C, Heister S D and Campanella O E 2012 Plain orifice gelled propellant flow characteristics with rheological hysteresis *48thJoint AIAA/SAE/ASEE Joint Propulsion Conf. (Atlanta, GA, July 30–1 August)* AIAA-2012-4135

[164] Yoon C, Heister S D and Campanella O 2014 Modeling gelled fluid flow with thixotropy and rheological hysteresis *Fuel* **128** 467–75

[165] Green J M, Rapp D C and Roncace J 1991 Flow visualization of a rocket injector spray using gelled propellant simulants *AIAA Paper* AIAA-91–2198

[166] Padwal M B and Mishra D P 2016 Characteristics of gelled jet A1 sprays formed by internal impingement of micro-air jets *Fuel* **185** 599–611

[167] Guan H, Li G and Zang N 2018 Experimental investigation of atomization characteristics of swirling spray by ADN gelled propellant *Acta Astonautica* **144** 119–25

[168] Yang L, Fu Q, Zang W, Du M and Tong M 2013 Spray characteristics of gelled propellants in novel impinging jet injector *J. Propulsion Power* **29** 104–13
[169] Yang L, Fu Q, Qu Y, Zang W, Du M and Xu B 2012 Spray characterization of gelled propellants in swirl injectors *Fuel* **97** 253–61
[170] Fu Q, Ge F, Wang W and Yang L 2019 Spray characteristics of gel propellants in an open-end swirl injector *Fuel* **254** 115555
[171] Behal S and Varma M 2019 Spray characteristics of pure and metallized green gel propellants *12th Int. High Energy Materials Conf. & Exhibits (Chennai, India, December 16–18)*
[172] Urbon B C 1992 Atomization and combustion of a gelled metallized slurry fuel *MS Thesis* Naval Post Graduate School, Monterey, California
[173] Mansour A and Chigier N 1995 Air-blast atomization of non-newtonian fluids *J. Non-Newtonian Fluid Mech.* **58** 161–94
[174] Sutherland J J, Sojka P E and Plesnaik M W 1997 Ligament controlled effervescent atomization *At. Sprays* **7** 383–406
[175] Geckler S C and Sojka P E 2008 Effervescent atomization of viscoelastic liquids, experiment and modeling *J. Fluid Eng.* **130** 061303
[176] Morel T 1979 Experimental study of a jet-driven Helmholtz oscillator *J. Fluid Eng.* **101** 383–90

Chapter 17

Recent progress in the development of less toxic pyrotechnic smoke compositions for military applications

Anthony P Shaw, Jared D Moretti and William S Eck

17.1 Introduction

For military applications, there are two general types of pyrotechnic smoke compositions—those for visual obscuration and those for signaling. The former typically produce white or gray aerosols. These aerosols, which are usually inorganic and hygroscopic, are useful for providing concealment. Accordingly, obscuration smokes have been used extensively for troop and platform protection. In contrast, signaling smoke compositions contain a significant amount of organic material, including various organic dyes that absorb in the visible spectrum and color the resulting aerosols. These highly colored aerosols can be seen from a great distance and thus find wide application as signals to locate and direct troops.

Although both types of smoke have mission-critical applications, the consequences of human exposure and concerns regarding the environment may affect the frequency with which hazardous compositions are used, especially in training exercises. For example, the smoke produced by the aluminum-zinc oxide-hexachloroethane ($Al/ZnO/C_2Cl_6$) pyrotechnic, known as HC smoke, is a capable obscurant but its use during training exercises is generally not permitted due to its extreme human toxicity. Since military doctrine mandates that soldiers should train as they will fight, the development of effective yet less toxic smoke formulations is a priority for support of training and combat operations.

In light of these toxicity concerns, numerous formulations have been proposed as replacements for both smoke types. Although this chapter is not a comprehensive review of these developments, it does place this area of research in a broader context through clear descriptions of the physical, chemical, and biological characteristics of

doi:10.1088/978-0-7503-3943-8ch17

pyrotechnic smokes. The first major section of this chapter covers the theory of visual obscurants and the analytical methods used to characterize them, the second covers signaling smokes, and the third covers what is currently known about the toxicology of both types.

Portions of this chapter describe research that took place at the Armament Research, Development, and Engineering Center (ARDEC) and the Edgewood Chemical Biological Center (ECBC). These two centers previously resided within the U.S. Army's Research, Development, and Engineering Command (RDECOM). After this chapter was largely completed, ARDEC and ECBC were realigned into the Combat Capabilities Development Command and are now called Armaments Center and Chemical Biological Center, respectively. Toxicology investigations were carried out primarily by the Defense Centers for Public Health—Aberdeen (formerly Army Public Health Center), currently subordinate to the Defense Health Agency.

17.2 Smoke compositions for visual obscuration

17.2.1 Common evaluation criteria and general formulation strategies

There are many ways to assess visual obscurants[1]. Meaningful evaluations take into account the practical requirements of different smoke-producing munitions. The rate at which smoke is released is quite important. Devices that release smoke rapidly are useful for immediate and localized screening. Those that release smoke over a longer period of time are generally suitable for obscuring larger areas. The dispersal of a given quantity of aerosol over a short period of time produces a thick and short cloud while greater dispersal times result in thinner and longer clouds, although thinner clouds may be ineffective for screening. The qualitative behavior of the clouds is also affected by ambient weather conditions (which cannot be controlled). In the absence of any wind, which is rare, emitted smoke accumulates and provides a thick screen locally regardless of the smoke release duration. At the opposite extreme, a strong wind will elongate, thin, and quickly disperse the aerosol emitted from even the fastest-emitting device.

Beyond qualitative comparisons of field characteristics, several quantitative assessment methods have been developed. These fall into three general categories. The first involves aerosol characterization in a closed chamber. The second involves measuring the properties of a controlled aerosol flow. The third, and most challenging, involves measurements conducted in the field. In closed-chamber and controlled-flow experiments, transmittance as a function of wavelength is commonly measured. Modern quantitative field tests require contrast measurements and a substantial amount of data processing. Flow and field tests are dynamic in the sense that the results are affected by smoke release duration. Closed-chamber tests, in contrast, are static. One advantage of closed-chamber testing is that it enables the gravimetric determination of total aerosol mass.

Many types of optical measurements may be conducted within a closed chamber. Most commonly, the total attenuation of visible light is assessed by determining

[1] Portions of sections 2.1 and 2.2 of this chapter were adapted from a conference paper [1].

transmittance (T), the ratio of transmitted light intensity (I) to that of an incident beam (I_0). Each particle or droplet within an aerosol can absorb or scatter light. In many instances, the measured attenuation of visible light by bulk aerosol samples is primarily caused by scattering. The Beer–Lambert law (equation (17.1)) relates T to the aerosol concentration (c), the path length the light travels through (L), and the extinction coefficient (α). A broadband light source such as a tungsten-halogen bulb or xenon arc lamp with a collimating lens combined with a spectrometer as the detector makes a flexible experimental setup. Measurements of T can be obtained at specific wavelengths, averaged across the visible spectrum, or weighted to the photopic response of the human eye[2]:

$$T = I/I_0 = e^{-cL\alpha}. \tag{17.1}$$

However, it is the extinction coefficient, which is independent of concentration and path length, that determines the effectiveness of an aerosol as an obscurant [3]. The units of α are such that the term $cL\alpha$ (the optical depth) is dimensionless. Therefore, when L is measured in meters and c is defined as the mass of aerosol particles divided by the volume of the chamber (g m^{-3}), the units of α are m^2 g^{-1}. This is a mass-based extinction coefficient (α_m). Rearranging the terms and substituting m_a/V for c, where m_a is the aerosol mass and V is the chamber volume, gives equation (17.2):

$$\alpha_m = \frac{-\ln(T)}{cL} = \frac{-V \cdot \ln(T)}{m_a L}. \tag{17.2}$$

Since α_m is only a characteristic of an aerosol, it does not account for other important properties of smoke-producing materials. Two practical factors, unaccounted for by α_m, are the aerosolization efficiency and whether or not the aerosolized materials absorb atmospheric constituents. Various *figures of merit* were developed to account for these factors with the goal of quantifying actual smoke composition performance. In 1968, Lane and coworkers described 'total obscuring power', which was intended to represent the area in square feet that could be obscured by a pound of smoke composition. This figure of merit incorporated a factor to account for the 'discrimination capability of the human eye' [4]. A figure of merit more closely related to the Beer–Lambert law may be obtained by multiplying α_m by the yield factor (Y). Here, Y is the ratio of aerosol mass (m_a) to the mass of initial composition (m_c). Equation (17.3) shows how the resulting mass-based figure of merit, FM_m, is related to the initial smoke composition mass. Like α_m, the units of FM_m are m^2 g^{-1}:

$$FM_m = \alpha_m Y = \left(\frac{-V \cdot \ln(T)}{m_a L}\right)\left(\frac{m_a}{m_c}\right) = \frac{-V \cdot \ln(T)}{m_c L}. \tag{17.3}$$

If all of the smoke produced by a composition is retained within the chamber, the experimental technique is simplified because only T must be measured to determine

[2] The 1931 CIE standard luminosity function is often used. Many slight modifications have since been suggested [2].

FM_m (V, L, and m_c are known). The expanded form of equation (17.3) is only applicable in this situation. In practice, the aerosol generated by just one smoke grenade is too thick, even in large smoke chambers, and some of the smoke must be vented to obtain a precise transmittance measurement within a range of about 0.1–0.5. In this case, Y and α_m must be determined independently. Aerosol sampling to determine m_a is performed before and after the venting step. The first determination is used to calculate Y and the second is used to calculate α_m at the time of optical measurement. These two values are then multiplied to give FM_m. It is assumed that the fundamental properties of the aerosol do not change in the process.

Even if it is possible to determine FM_m directly without chamber venting and aerosol sampling, α_m and Y are still informative. A yield factor greater than unity indicates with certainty that the smoke has absorbed mass from the air (usually moisture). Compositions that aerosolize hygroscopic compounds perform better as relative humidity and yield factor increase. In such cases, the extinction coefficient also varies, but usually not nearly as much as the yield factor does [5, 6]. The yield factors of nonhygroscopic smokes are governed by aerosolization efficiency.

In actual devices, the amount of smoke composition is often limited by the device *volume*. Thus, for different compositions with comparable FM_m values, the one with a greater density is preferable provided a heavier device is acceptable. A volume-based figure of merit that accounts for this is obtained by multiplying FM_m by the composition density (ρ_c, g cm^{-3}). The units of FM_v are m^2 cm^{-3} (equation (17.4)). This figure of merit, which is related to the initial volume of smoke composition (v_c), has been used since the 1980s, if not earlier [7, 8]. As with equation (17.3), the expanded form of equation (17.4) is only relevant when all of the smoke produced by a composition is retained within the chamber. That is, FM_v can be determined directly from T only when it is possible to measure T without venting any smoke. The variables used in equations (17.1)–(17.4) are listed and described in table 17.1:

$$FM_v = FM_m\rho_c = \alpha_m Y\rho_c = \frac{-V \cdot \ln(T)}{v_c L}. \tag{17.4}$$

Aerosolization efficiency can be influenced by device design and configuration, however some compositions are more sensitive to these factors than others. Figures of merit suitable for device-to-device comparison are obtained by incorporating fill fractions, the ratios of m_c or v_c to the entire mass or volume of the device [9]. Multiplying FM_m or FM_v by the appropriate fill fraction provides a figure of merit related to the total device mass or volume.

A large smoke chamber is required for testing smoke grenades and other smoke-producing munitions. In the United States, the Edgewood Chemical Biological Center's chamber is 190 m^3 and has a 6 m path length [10]. Controlled-flow tests enable the use of smaller facilities. In these dynamic tests, smoke is drawn through a tunnel or a pipe as it is produced. Optical measurements are obtained as a function of time. Two examples include the United Kingdom's Multispectral Obscurant

Table 17.1. Smoke chamber measurement variables, units, and definitions.

Variable	Units	Definition
T	Unitless	Transmittance, I/I_0
L	m	Path length
V	m^3	Chamber volume
c	$g\ m^{-3}$	Aerosol concentration
α_m	$m^2\ g^{-1}$	Mass-based extinction coefficient
FM_m	$m^2\ g^{-1}$	Mass-based composition figure of merit
FM_v	$m^2\ cm^{-3}$	Volume-based composition figure of merit
m_a	g	Mass of aerosol
m_c	g	Mass of composition
v_c	cm^3	Volume of composition
ρ_c	$g\ cm^{-3}$	Density of composition, m_c/v_c
Y	Unitless	Yield factor, m_a/m_c

Assessment Tunnel and, in the United States, Pine Bluff Arsenal's Integrated Smokes Testing Chamber [11, 12]. The latter facility includes a scrubber.

The development of digital image analysis has made quantitative field testing possible. Target boards comprising alternating black and white squares are set at varying distances from a munition. Contrast measurements are performed as the smoke is carried past the target boards by the wind. One benefit of this approach is the highly realistic testing environment. However, obtaining reproducible data is challenging due to varying weather conditions. Two metrics that have been described include effective screening area and effective cover [13]. Of course, it is also possible to measure contrast reduction in a controlled-flow facility or in a closed chamber.

There is no fundamental color requirement for visual obscurants. However, those used for military applications have often produced white or gray clouds. A practical requirement is that the color should not be misconstrued as a signal. Effective obscuration can be achieved through a variety of methods and approaches, all of which are constrained by the practicalities of munition design and use. One approach involves the aerosolization of materials with high refractive indices. These materials, such as titanium dioxide, are not volatile and require explosive dissemination. For pyrotechnic smoke compositions, a high aerosolization efficiency is generally desired regardless of the refractive indices of the dispersed materials. If atmospheric components are absorbed and incorporated into an aerosol, obscuration effectiveness is often greatly increased. (This is caused by the presence of Y in equations (17.3) and (17.4).) This approach inherently requires the scattering, spraying, or volatilization of flammable or hygroscopic substances.

17.2.2 Characteristics of existing compositions and methods

Since the First World War, several strategies have been successfully employed for producing smoke on the battlefield. Table 17.2 lists and describes some of the formulations used or tested by the U.S. military. In this table, the most effective compositions tend to produce hygroscopic aerosols. These are also, coincidentally, the most problematic ones from a safety and environmental perspective.

The unique chemical and physical properties of phosphorus make it a highly effective smoke agent. White phosphorus (WP) is a waxy, pyrophoric, and highly toxic substance composed of P_4 molecules [14]. It is often yellow due to the presence of impurities including red phosphorus (RP). RP, a red powder, consists of a poorly defined amorphous network of phosphorus atoms. Unlike WP, it is not pyrophoric and is far less toxic [15]. When burned, both types ultimately form phosphorus pentoxide (P_4O_{10}). The associated increase in mass is 129%; that is, 2.29 g of the oxide is formed for every gram of phosphorus burned. Phosphorus oxides are volatile at the temperature of phosphorus combustion and are aerosolized as a result. This aerosol is extremely hygroscopic. The absorption of atmospheric moisture and eventual formation of orthophosphoric acid (H_3PO_4) corresponds to a further 38% increase in mass. Even more moisture may be absorbed, if it is available. In practice, the yield factor (Y) of WP ranges from about 3–9 depending on atmospheric temperature and relative humidity. The actual composition of phosphorus-derived aerosols is more complex than the simple explanation above would lead one to believe. The aerosol droplets are formed through a dynamic

Table 17.2. Characteristics of some U.S. military visible obscurants.

Designator or name	Component(s)	Device configuration	Mechanism	Atmospheric reactivity
WP	P_4 (white/yellow)	Bursting, neat, or felt wedges	Atmospheric burning	Hygroscopic oxides
RP	P_x (red)[a]	Bursting or can	Atmospheric burning	Hygroscopic oxides
HC	$Al/ZnO/C_2Cl_6$[b]	Core- or end-burning can	Combustion products	hygroscopic $ZnCl_2$
CA, TA	*Trans*-cinnamic acid, terephthalic acid[c]	Core-burning can	Sublimation–condensation	Minimal
Titania	TiO_2	Bursting	Dispersal	Minimal
SGF2, diesel, PEG	Hydrocarbons, polyethylene glycols	Smoke generator[d]	Vaporization–condensation	Minimal
FM, FS	$TiCl_4$, HSO_3Cl/SO_3	Bursting or spraying	Dispersal	Hygroscopic liquids

[a] Additional components are often nitrates or other oxidizers, metal additives, and organic binders.
[b] This particular mixture is Type C but there are many metal/metal oxide/chlorocarbon variants.
[c] Typically mixed with sucrose/$KClO_3$ and various carbonates.
[d] Smoke pots and grenades that aerosolize these liquids are not currently used.

Figure 17.1. A white phosphorus bomb explodes above the decommissioned U.S.S. Alabama (BB-8) as a twin-engine bomber flies overhead. The test was conducted on the Chesapeake Bay, September 23, 1921. U.S. Navy, Naval History and Heritage Command [17].

process and various polyphosphoric acids may be present, especially when the relative humidity is low [5, 16].

WP, neat or adsorbed onto felt fibers, is typically loaded into bursting smoke munitions. Figure 17.1 depicts an early test of a munition of this type [17]. RP is often combined with additives to promote efficient combustion and to provide mechanical integrity. Powdered metals such as magnesium, nitrate salt oxidizers, and polymeric organic binders are commonly added [18]. The M819 mortar projectile contains an expelling charge that ignites and ejects multiple smoke pellets. In addition to RP, these pellets contain 13.9 wt-% $NaNO_3$ and 7.4 wt-% epoxy binder [19]. RP-based smoke compositions are inherently phosphorus-rich and atmospheric combustion of the excess phosphorus is the main pyrotechnic reaction.

Particles of WP that do not burn immediately and completely can cause collateral damage in combat and the contamination of training ranges, especially wetland areas [20, 21]. Different safety and logistical issues are associated with the use of RP. RP-based smoke compositions slowly degrade in the presence of air and moisture, producing corrosive acids and phosphine gas which is extremely toxic and flammable [22]. Microencapsulation of the RP can slow the degradation but does not always prevent it [23]. Additionally, RP has a reputation for sensitizing pyrotechnic compositions with respect to unintended ignition by impact, friction, and electrostatic discharge [24].

The oxides of the nonmetals are volatile, if they are not already gases at room temperature like CO and CO_2. The oxides of most metals, on the other hand, are not. This simple fact precludes the use of many oxide-forming reactions for the purpose of smoke production. Metal chlorides tend to be far more volatile than the corresponding oxides, as illustrated by the examples in table 17.3 [25]. The most volatile metal chlorides among these examples are molecular compounds; $TiCl_4$,

Table 17.3. Vaporization temperatures of metal oxides and chlorides[a].

Metal	Oxide decomposition (°C)	Chloride boiling point (°C)
Mg(II)	3150	1351[b]
Ti(IV)	3307	136
Zn(II)	1969	745
Al(III)	3733	180[c]
Si(IV)	2874	62
Sn(II)	1548[d]	608
Sn(IV)	1974	109

[a] FactSage 7.0, FactPS, and FToxid databases [25].
[b] Forms a mixture of $MgCl_2(g)$ and $(MgCl_2)_2(g)$.
[c] Forms $(AlCl_3)_2(g)$.
[d] Vaporization of SnO from $Sn(l)/SnO_2(s)$.

$SiCl_4$, and $SnCl_4$ are liquids at room temperature. All of the metal chlorides in table 17.3 are quite hygroscopic.

Zinc is able to reduce chlorocarbons such as hexachloroethane (HC). Upon ignition, a voluminous hygroscopic cloud of zinc chloride and soot is produced (equation (17.5)). Captain Ernest É-F Berger recognized the military potential of these types of reactions. Berger's smoke compositions were used by the French Navy during the First World War [26]. They were subsequently adopted and modified by countries including the United States and the United Kingdom, among others:

$$3Zn + C_2Cl_6 \rightarrow 3ZnCl_2 + 2C. \quad (17.5)$$

In the United States, the composition used before 1941 was based on the Zn/C_2Cl_6 system and was designated 'HC Smoke.' That particular composition (later known as Type A) was susceptible to spontaneous ignition, especially upon contact with moisture. For a short time during the Second World War, the United States used a composition containing $CaSi_2$, ZnO, and C_2Cl_6 (Type B) that was reportedly safer to store and transport. The calcium silicide was later replaced by aluminum in the Type C variant (equation (17.6)) [27]:

$$2Al + 3ZnO + C_2Cl_6 \rightarrow Al_2O_3 + 3ZnCl_2 + 2C. \quad (17.6)$$

The stoichiometric amount of aluminum indicated by equation (17.6) is 10.1 wt-%. In practice, less aluminum is used to slow the burning rate and whiten the smoke. Less soot, and more CO, is produced when the system is deficient in aluminum. Values of Y range from about 1–2 and depend on atmospheric temperature and relative humidity [5, 10]. Type C compositions have been deployed in U.S. Army artillery projectiles, hand-thrown grenades, and smoke pots, which are large smoke-producing canisters. While it was still in use, the AN-M8 grenade produced a thick gray smoke screen and burned for 90 s or more. Be that as it may, HC smoke is acutely toxic and has caused several inhalation-related injuries and deaths [28]. Hexachloroethane is also a suspected carcinogen [29].

Table 17.4. Smoke compositions containing *trans*-cinnamic acid[ab].

Component	Example 1 (wt-%)	Example 2 (wt-%)	Example 3 (wt-%)
Trans-cinnamic acid	47.5	48.0	58.0
Potassium chlorate	29.0	28.5	23.0
Sucrose	12.0	12.5	10.0
Sodium bicarbonate	6.5	6.5	5.5
Diatomaceous earth	5.0	4.5	3.5

[a] Reference [30].
[b] Each composition contains an additional 2–5 wt-% nitrocellulose binder.

Organic materials may be aerosolized by pyrotechnic reactions, provided they do not decompose or oxidize in the process. Few fuel/oxidizer systems are suitable in this respect. Mixtures containing sulfur or sugars and potassium chlorate ($KClO_3$) can ignite and undergo sustained combustion at temperatures of just several hundred degrees. Added carbonate or bicarbonate salts serve as 'coolants' by diluting the compositions and undergoing endothermic decomposition reactions. This technique has long been the basis of many colored smoke signal formulations, although it is certainly not limited to the aerosolization of colored dyes. White smoke clouds can be produced with colorless organic compounds. Compositions containing *trans*-cinnamic acid (CA) were patented in 1977 (table 17.4) [30].

A similar composition is used in the U.S. Army's M83 grenade. In this case, the smoke agent is terephthalic acid (TA). The M83 was developed in the 1990s as a safer alternative to the AN-M8. The smoke produced by the M83 is not hygroscopic, the aerosolization efficiency of the composition is poor, and the grenade does not burn for as long as the AN-M8. Qualitatively, as many as *three* M83 grenades are needed to create a screen comparable to that produced by just one AN-M8 [10]. Nonetheless, the M83 is less likely to start unintended fires and terephthalic acid is not particularly toxic. Smoke pots containing TA compositions are similarly benign but are not as effective as those containing C_2Cl_6-based mixtures.

If the *pyrotechnic* compositions in table 17.2 were ranked according to their general effectiveness, the following order would apply: WP/RP > HC ≫ TA/CA. Although, in some situations, obscuration efficacy is a secondary consideration. An example is the pyrotechnic dissemination of riot control agents. For this application, the agent molecules must be aerosolized intact and unchanged. Therefore, maintaining a moderate combustion temperature is imperative. The S/$KClO_3$ and sugar/$KClO_3$ systems may be used to aerosolize any number of organic compounds, whether they are benign or not. Compounds possessing electron-withdrawing groups are less likely to be oxidized in the process (figure 17.2).

There are several nonpyrotechnic methods and materials for producing smoke screens. Some of the most effective, yet benign, materials are fine titania powder and polyethylene glycol or other oils. The former is dispersed almost instantaneously by bursting grenades, which are useful for immediate and localized screening [8, 9]. Fog oils can be aerosolized by vehicle-mounted smoke generators. This is one of the only

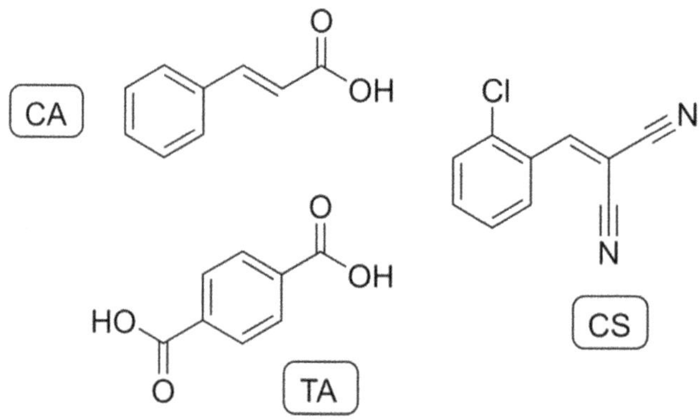

Figure 17.2. Structures of the benign smoke agents *trans*-cinnamic acid (CA), terephthalic acid (TA), and the riot control agent 2-(2-chlorobenzylidene)malononitrile (CS). All contain electron-withdrawing groups.

practical ways to create an *enormous and persistent* smoke screen. The U.S. Army's M58, a tracked vehicle, contained a 450 l tank and could aerosolize 5 l of fog oil per minute [31]. Hygroscopic liquids can be dispersed explosively or sprayed from tanks. In the United States, titanium tetrachloride and a chlorosulfonic acid/sulfur trioxide mixture were given the designators FM and FS, respectively [32]. These smoke agents can be co-aerosolized with basic chemicals, such as ammonia or carbonate salts, to reduce the acidity of the resulting clouds.

17.2.3 Recent formulation developments

17.2.3.1 Phosphorus-based alternatives

The most hazardous characteristics of phosphorus-based smoke munitions are arguably caused by the WP or RP itself, not the resulting phosphoric acid aerosol [33–35]. This has motivated the search for alternative phosphorus-containing materials. Ideally, the alternatives should not be toxic, pyrophoric, or susceptible to degradation. One strategy would be to use the other allotropes of phosphorus—violet or black phosphorus, which are much less reactive than WP or RP [36]. They would certainly need to be paired with suitable oxidizers to ensure reliable combustion. Unfortunately, there do not appear to be any major industrial uses for violet or black phosphorus and these allotropes are expensive as a result.

There are some phosphorus-containing compounds that are plausible alternatives, but few exhibit phosphorus densities comparable to elemental phosphorus (table 17.5). In burning pyrotechnic compositions, the phosphorus within these compounds could be converted into volatile phosphorus oxides. More effective compositions would produce elemental phosphorus *in situ* which would then burn upon contact with the air.

Even though phosphorus is quite oxophilic, it is possible to remove oxygen from phosphates. Indeed, WP is produced industrially by the reduction of calcium phosphate with coke and sand in a furnace (equation (17.7)). The phosphorus is produced as P_4 and P_2 vapor at the high temperatures that are involved. Magnesium

Table 17.5. Density and phosphorus content of various materials.

Material	Material density (g cm^{-3})[a]	Phosphorus content (wt-%)	Phosphorus density (g cm^{-3})
P (white)	1.82	100	1.82
P (red)	2.2–2.3	100	2.2–2.3
P (violet)	2.36	100	2.36
P (black)	2.69	100	2.69
BP	2.97	74.1	2.20
$B_{12}P_2$	2.60	32.3	0.84
TiP	4.08	39.3	1.60
Ti_5P_3	4.69	28.0	1.31
P_3N_5	2.77	57.0	1.58
P_4S_3	2.08	56.3	1.17
Li_3PO_4	2.46	26.7	0.66
$(C_6H_5)_3PO$[b]	1.21	11.1	0.13

[a] Theoretical maximum density or crystal density.
[b] Triphenylphosphine oxide (TPPO).

is able to reduce phosphate in pyrotechnic reactions. That is, certain magnesium-based pyrotechnic compositions containing phosphate salts could produce elemental phosphorus upon combustion (equation (17.8)). Lithium phosphate is not hygroscopic, making it one of the few phosphates suitable for use in pyrotechnics. The efficacy of this approach has not been established and Mg_3P_2 is another plausible product. Pyrotechnic smoke compositions containing an organic P–O compound, triphenylphosphine oxide (TPPO), have been reported [37]. However, the phosphorus content of these compositions is low and it is not clear whether elemental phosphorus is produced during combustion; the other components of the mixtures were not revealed:

$$Ca_3(PO_4)_2 + 5C + 3SiO_2 \rightarrow 3CaSiO_3 + 5CO + P_2, \quad (17.7)$$

$$5Mg + 2Li_3PO_4 \rightarrow 5MgO + 3Li_2O + P_2. \quad (17.8)$$

Phosphorus sesquisulfide (P_4S_3) has been used as a fuel in 'strike anywhere' matches since 1898 [38]. In comparison to WP, it is relatively stable and much safer to handle. The combustion of P_4S_3 and the properties of the resulting aerosol were briefly studied in the 1960s [32]. More recently, P_4S_3-coated fibers were proposed for multispectral screening [39]. To the best of our knowledge, P_4S_3 has not been used operationally for the purpose of producing smoke. If it were, sulfur dioxide formed in the process would not be desirable.

Phosphorus nitride, P_3N_5, appears to be a promising RP alternative [40]. The compound, by itself, does not evolve a detectable amount of phosphine gas in the presence of moist air. Unlike RP, the compound is not flammable. It does, however, burn rapidly in the presence of added oxidizers. P_3N_5/oxidizer mixtures are relatively insensitive with respect to ignition by impact or abrasion (table 17.6). The reported thermochemical calculations indicate that P_3N_5/KNO_3 compositions containing 40-60 wt-% P_3N_5 should produce more PN than P_2. Various phosphorus oxides are also expected. Therefore, two major processes are likely to occur at fuel-rich stoichiometries. The first involves oxidation of P_3N_5 to give phosphorus oxides and N_2. The second involves the formation of PN, P_2, and N_2 from the thermal decomposition of excess P_3N_5. The PN presumably burns upon contact with the air. The results of smoke chamber tests have not been reported.

Certain phosphide compounds could be used as phosphorus alternatives. Many, however, possess undesirable qualities. Ionic phosphides like Na_3P and Ca_3P_2 are susceptible to hydrolysis, for example. Titanium phosphides are stable in the presence of water but their phosphorus content is relatively low. Boron phosphide (BP) is remarkably resistant to chemical attack. Recently reported bulk-synthetic methods call for a purification step involving boiling concentrated hydrochloric acid [41, 42]. Yet, mixtures of BP and powdered oxidizers can be ignited (table 17.7) [43].

BP exhibits the greatest phosphorus density of any of the compounds listed in table 17.5. Crystals of BP actually contain more phosphorus per cubic centimeter than pure WP. A sub-phosphide, $B_{12}P_2$, is also known but its phosphorus content is lower. Large-scale combustion tests involving BP have not been reported. The BP/KNO_3 system has been studied computationally and the major reaction

Table 17.6. Sensitivity and thermal onset data for P_3N_5 and RP compositions[a].

Composition	Component mass ratio	Impact (J)	Friction (N)	Thermal onset (°C)
$P_3N_5/KClO_3$	34.72/65.28	7.5	>360	328
$P_3N_5/KClO_4$	38.61/61.39	>30	>360	450
P_3N_5/KNO_3	34.95/65.05	20	>360	394
RP/KNO_3	26.74/73.26	5	40	250

[a] Reference [40].

Table 17.7. Pyrotechnic properties of boron phosphide (BP)[ab].

Composition	Component mass ratio	Burning time (s)	Effects
BP/KNO_3	50/50	1.0	Green flame, white smoke
$BP/KClO_3$	50/50	0.2	Green flash, white smoke
$BP/KClO_4$	50/50	0.2[c]	Green flash, white smoke
$BP/NaIO_4$	50/50	0.3	Yellow-green flash, purple-tinged smoke
$BP/Na_2S_2O_8$	50/50	0.9	Yellow-green flame, white smoke, rotten egg odor

[a] Reference [43].
[b] Unconsolidated 0.50 g piles ignited with an electrically heated wire.
[c] Some residual material continued to burn for another 0.1 s.

predicted to occur at fuel-rich stoichiometries is represented by equation (17.9) [44]. Here, KBO_2 and P_2 should be formed as gases at the expected combustion temperature. The formation of elemental phosphorus *in situ* is probable because B–O bonds are generally stronger than P–O bonds:

$$8BP + 3KNO_3 \rightarrow 3KBO_2 + 4P_2 + B_2O_3 + 3BN. \quad (17.9)$$

17.2.3.2 Metal/metal oxide/chlorocarbon compositions

It is possible to use metal oxides other than zinc oxide in the $Al/M_xO_y/C_2Cl_6$ system. Smoke compositions containing TiO_2, for example, were tested in the 1940s [45]. More recently, compositions containing tin(II) oxide have been investigated [46]. The intended pyrotechnic reaction is illustrated by equation (17.10). It is thought that $SnCl_2$-based smoke may be less toxic than the zinc chloride aerosol produced by $Al/ZnO/C_2Cl_6$ mixtures. It remains to be seen whether the HC could be replaced with another, less objectionable, chlorine donor:

$$2Al + 3SnO + C_2Cl_6 \rightarrow Al_2O_3 + 3SnCl_2 + 2C. \quad (17.10)$$

One advantageous quality of SnO is its relatively high density of 6.45 g cm^{-3}. It should be possible to load more of the $Al/SnO/C_2Cl_6$ composition into munition cases provided some extra weight is not problematic (table 17.8). Tin(II) oxide is one of the denser binary metal oxides; others that are even more dense include NiO, CdO, PbO, and HgO with densities of 6.72, 8.15, 9.35, and 11.14 g cm^{-3}. Of course, it would not be advisable to use these other oxides considering the obvious human toxicity concerns.

Grenades containing the stoichiometric $Al/SnO/C_2Cl_6$ composition of equation (17.10) have been evaluated in smoke chamber and field tests. An averaged visual mass-based extinction coefficient of 2.29 m^2 g^{-1} and a yield factor of 0.87 were measured at 25 °C and 23% relative humidity. The α_m value is similar to that of ordinary $ZnCl_2$-based HC smoke. The yield factor, however, is less than what would be desired. Yield factors greater than unity are typical for standard AN-M8 grenades, even when the relative humidity is quite low. The discrepancy in Y may be caused by differences in smoke hygroscopicity or pyrotechnic aerosolization efficiency. It may be

Table 17.8. Density of C_2Cl_6-based compositions and materials.

Composition or material	Component mass ratio	TMD (g cm^{-3})[a]
$Al/ZnO/C_2Cl_6$	10.09/45.65/44.26[b]	3.03
$Al/SnO/C_2Cl_6$	7.77/58.16/34.07[c]	3.55
Al		2.70
ZnO		5.61
SnO		6.45
C_2Cl_6		2.09

[a] Theoretical maximum density or crystal density.
[b] Ratio corresponding to equation (17.6).
[c] Ratio corresponding to equation (17.10).

Figure 17.3. Test of an end-burning smoke grenade containing 650 g of the stoichiometric $Al/SnO/C_2Cl_6$ composition. The test was conducted at Aberdeen Proving Ground (Edgewood Area), Maryland, on November 8, 2016. Photograph courtesy of Giancarlo Diviacchi, U.S. Army RDECOM-ECBC.

possible to improve the latter by varying the ratios of Al, SnO, and C_2Cl_6. In field tests, the SnO-based prototypes produced thick clouds of white smoke and burned for about 90 s (figure 17.3). Like the analogous mixtures containing ZnO, it should be possible to adjust the burning rate by varying the aluminum content.

17.2.3.3 Other inorganic systems

A number of pyrotechnic compositions that aerosolize alkali metal chlorides have been developed since the 1970s [47–49]. Three examples are shown in table 17.9. They generally contain magnesium, an *organic* co-fuel or additive, and a strong oxidizer. Nonetheless, these are *inorganic* systems because the aerosols they produce are largely composed of salts.

The CY85A composition was formulated by researchers working for the U.S. Navy [47]. It was designed to aerosolize NaCl and KCl, the latter arising from $KClO_4$. A small amount of Li_2CO_3 or another lithium salt (in other variants) was added with the intention of increasing smoke hygroscopicity. The Nico-Pyrotechnik KM composition produces an aerosol composed mainly of potassium chloride [48]. The organic additive in that composition is a nitrogen gas generator. Compositions containing $NaClO_3$, a source of NaCl, were developed in the Netherlands at TNO, and the 11EM0745 formulation was identified as a promising candidate [49]. Natural sea salt aerosols consist mostly of sodium chloride and are thought to be benign, at least, in concentrations that occur naturally.

The aerosols produced by the smoke compositions in table 17.9 are not particularly hygroscopic. The equilibrium relative humidity above a saturated aqueous solution of a salt is an indicator of that salt's hygroscopicity. Sodium chloride and potassium chloride only absorb moisture from extremely humid air, as illustrated by table 17.10 [50, 51].

Table 17.9. Compositions for aerosolizing salts.

Component	CY85A (wt-%)[a]	KM (wt-%)[b]	11EM0745 (wt-%)[c]
$KClO_4$	65	5	–
Binder[d]	18	–	–
NaCl	10	–	–
Mg	5	8	3
Li_2CO_3	2	–	–
KCl	–	44	–
KNO_3	–	27	–
$C_2H_4N_4O_2$[e]	–	16	–
$NaClO_3$	–	–	70
$(C_6H_{10}O_5)_n$[f]	–	–	22
$CaCO_3$	–	–	5

[a] Reference [47].
[b] Reference [48].
[c] Reference [49].
[d] Hydrocarbon-based, unspecified.
[e] Azodicarbonamide.
[f] Cellulose.

Table 17.10. Relative humidity above saturated aqueous salt solutions[a].

Salt	Temperature (°C)	Relative humidity (%)
LiCl	25	11.3
NaCl	25	75.3
KCl	25	84.3
$MgCl_2$	25	32.8
$CaCl_2$	25	28.8
$ZnCl_2$	20–25	2.5–10

[a] References [50] and [51].

Another family of inorganic smoke compositions contains boron or boron carbide as a pyrotechnic fuel. The $B/LiClO_4$ and $B_4C/LiClO_4$ systems were investigated in the 1960s [4]. At the time, perchlorate salts were not yet problematic but the general need for less toxic military smokes was clearly recognized. A mixture of 60 wt-% B and 40 wt-% $LiClO_4$ produced an aerosol containing 1.3 wt-% lithium, about half of the possible amount. The lithium content of the smoke was attributed to LiCl and other compounds such as Li_2O or LiOH.

More recently, smoke compositions containing boron, potassium nitrate, and various salts were reported [52]. The B/KNO_3 igniter is known to produce a green flash and white smoke. Salts including KCl, CsCl, and NH_4Cl, were used to dilute it. Equation (17.11) is a simplified description of B/KNO_3 combustion. The $B/KNO_3/KCl$ system, with appropriate component ratios, should produce an aerosol containing potassium borates, boron oxides, and KCl:

Table 17.11. Burning times of boron carbide-based smoke grenades[abc].

$B_4C/KNO_3/KCl$ mass ratio	Calcium stearate (wt-%)	End-burning time (s)	Core-burning time (s)
13/58/25	2	70.2	21.5
13/59/25	1	41.7	13.1
13/60/25	–	13.3	3.5

[a] Reference [10].
[b] Each grenade contained 350 g of smoke composition.
[c] Compositions contained 2 wt-% polyvinyl acetate binder.

$$10B + 6KNO_3 \rightarrow 6KBO_2 + 2B_2O_3 + 3N_2. \quad (17.11)$$

Boron carbide powder is widely used as an abrasive and is much less expensive than elemental boron. Smoke compositions containing B_4C as a pyrotechnic fuel have been tested in grenade and canister configurations. In the tradition of U.S. Army two-letter smoke acronyms, the $B_4C/KNO_3/KCl$ system was designated 'BC' by its developers [10]. The burning rate of the system is slowed considerably by the addition of just 1 or 2 wt-% calcium stearate. The grenade burning time (the smoke release duration) is also affected by the geometry of the pressed pyrotechnic composition. In an end-burning configuration, a solid pyrotechnic cylinder burns from one end to the other. Core-burning grenades contain an axial core hole. They burn outward from the core and from one end to the other simultaneously. The burning time data in table 17.11 exemplify the effect of these factors. The fastest core-burning variant provides nearly immediate, localized, obscuration (figure 17.4).

A simplified description of B_4C/KNO_3 combustion is shown in equation (17.12). Here, the CO is expected to burn upon contact with the air, forming CO_2, but this has not been confirmed experimentally. The smoke produced by the compositions in table 17.11 appears to be composed of potassium borates, boron oxides, and potassium chloride. Very little residue remains after combustion regardless of the composition burning rate. Likewise, the yield factor is not affected by composition burning rate or grenade configuration. The versatility and consistency of the system are remarkable but the smoke is not especially hygroscopic. The results of smoke chamber tests are presented in table 17.12:

$$5B_4C + 14KNO_3 \rightarrow 14KBO_2 + 3B_2O_3 + 5CO + 7N_2. \quad (17.12)$$

17.2.3.4 Organic systems

Organic sublimation/condensation smoke compositions would be more effective as visual obscurants if their yield factors could be increased. For example, the yield factor of the TA composition in table 17.12 is what limits the corresponding FM_m and FM_v values, not the mass-based extinction coefficient. An increase in Y could be achieved by altering the composition such that more terephthalic acid is volatilized and dispersed. Alternatively, compositions that aerosolize other organic compounds may offer improved performance. Examples of these two strategies are shown in table 17.13

Figure 17.4. Test of a core-burning grenade containing 350 g of the third composition in table 17.11. Smoke screen formation at 0.4 s (top), 2.8 s (middle), and 8.0 s (bottom). The total grenade burning time was 3.5 s. The test was conducted at Aberdeen Proving Ground (Edgewood Area), Maryland, on June 24, 2014. Photographs by Anthony P. Shaw, U.S. Army RDECOM-ARDEC.

Table 17.12. Photopic smoke chamber performance of BC, TA, and HC grenades[a].

Composition	Y	α_m (m^2 g^{-1})	FM_m (m^2 g^{-1})	ρ_c (g cm^{-3})[b]	FM_v (m^2 cm^{-3})[b]
BC[c]	0.75 (0.03)	4.01 (0.43)	3.03 (0.42)	1.48 (0.04)	4.48 (0.62)
TA[d]	0.30	4.80	1.44	1.2	1.73
HC[e]	1.26	2.36	2.97	2.2	6.53

[a] Reference [10].
[b] Density at 72% of the theoretical maximum.
[c] Averages with standard deviations in parentheses; 23 °C–25 °C, 40%–68% relative humidity.
[d] Data for an M83 grenade at 22 °C and 61% relative humidity.
[e] Data for an AN-M8 grenade at 25 °C and 32% relative humidity.

alongside a typical TA formulation. A definitive answer as to whether these approaches are viable remains elusive. In one case, the results of laboratory-scale experiments were reported [53]. In the other, smoke grenades and canisters were built but the results of smoke chamber tests involving those prototypes have not been reported [54].

While all of the compositions in table 17.13 contain potassium chlorate as an oxidizer, it has been found that a conventional fuel such as sucrose is not required. The modified TA formulation contains 5-aminotetrazole (5-AT) as a fuel and nitrogen gas generator (figure 17.5). The results of small-scale experiments suggest that 5-AT promotes the formation of a well-dispersed aerosol that is resistant to sedimentation [53]. Both sucrose and 5-AT produce gas upon combustion, but 5-AT is expected to do so more efficiently. Sucrose is known to caramelize or char, and the resulting polymers may not be converted into CO_2 and H_2O cleanly.

The modified TA formulation in table 17.13 contains binders that are also fuels (stearic acid and polyvinyl alcohol). It remains to be seen whether these components are required or not; that is, whether 5-AT could serve as the only fuel in the composition, excluding the possible participation of terephthalic acid. Admittedly, this is an academic question because an organic binder is generally desirable for a number of practical reasons.

The MATEAB composition retains the sucrose/$KClO_3$ pair but the terephthalic acid is replaced by a mixture of melamine, acetoguanamine, and triethanolamine borate (table 17.13, figure 17.6) [54]. This composition is notable for several reasons. First, it contains no carbonate or bicarbonate coolant. Second, the triazine

Table 17.13. Compositions for aerosolizing organic compounds.

Component	Baseline TA (wt-%)[a]	Modified TA (wt-%)[a]	MATEAB (wt-%)[b]
Potassium chlorate	23.0	22.1	33.7
Sucrose	14.0	–	13.4
Terephthalic acid	57.0	51.9	–
Magnesium carbonate	3.0	–	–
Stearic acid	3.0	2.9	–
Polyvinyl alcohol	1.0[c]	1.0	–
Sodium bicarbonate	–	5.8	–
5-aminotetrazole	–	16.3	–
Melamine	–	–	21.8
Acetoguanamine	–	–	14.3
Triethanolamine borate	–	–	16.8

[a] Reference [53].
[b] Reference [54].
[c] Added in excess, making the total 101 wt-%.

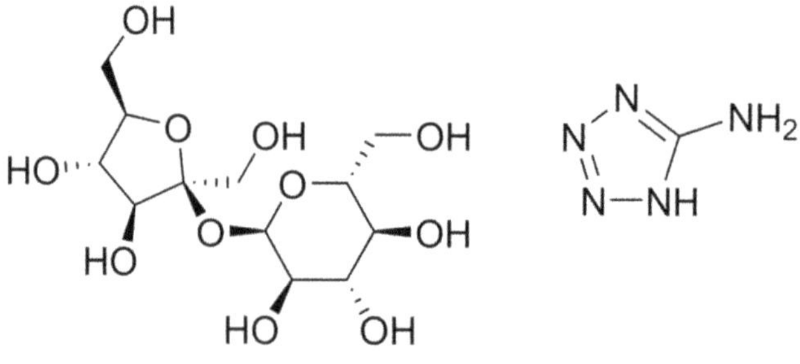

Figure 17.5. Sucrose (left) and 5-aminotetrazole (right).

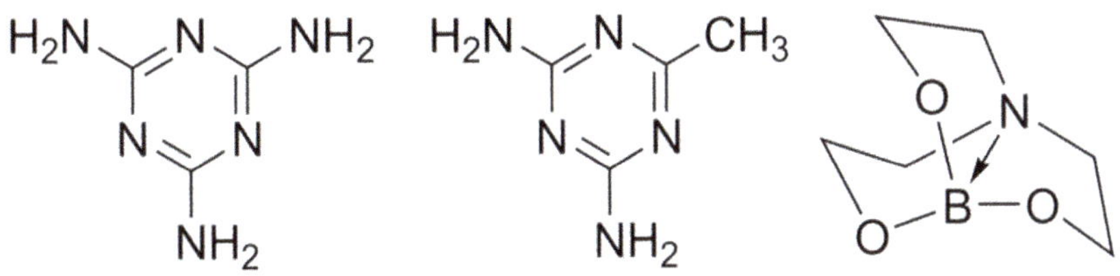

Figure 17.6. Melamine (left), acetoguanamine (center), and triethanolamine borate (TEAB, right) are used in the MATEAB smoke composition.

Figure 17.7. Test of an end-burning smoke canister containing approximately 820 g of MATEAB composition. The aerial photographs show smoke screen formation at 28 s (top left), 56 s (top right), 141 s (bottom left), and 198 s (bottom right). The canister burned for 192 s. The test was conducted at a Ventana Research Corporation facility in Tucson, Arizona, on September 30, 2016. Photographs courtesy of John L Lombardi, Ventana Research Corporation.

compounds (melamine and acetoguanamine) do not possess electron-withdrawing groups like the compounds in figure 17.2. Third, the composition may be pressed into pellets without an added binder. However, the system is compatible with nitrocellulose/acetone if additional mechanical strength is desired.

Melamine and acetoguanamine have multiple hydrogen bonding sites, which could make the smoke more effective in humid air. In theory, TEAB could be hydrolyzed by the steam produced during sucrose/$KClO_3$ combustion, forming boric acid, which could also hydrogen bond with the triazines. Melamine also has a greater refractive index than terephthalic acid. The MATEAB composition has been tested in grenade and canister configurations. Figure 17.7 depicts an outdoor canister test. It is possible to achieve long smoke release durations.

17.3 Smoke compositions for signaling

17.3.1 Characteristics of existing compositions

17.3.1.1 Mechanism and components

Colored smoke pyrotechnics operate very similarly in mechanism to organic-based obscurant smokes, such as those based on terephthalic acid. In colored smoke mixtures, the initial steps are the decomposition of the oxidizer (potassium chlorate, $KClO_3$) followed by oxidation of the fuel; this fuel is sulfur in older formulations or a sugar, such as sucrose, in newer formulations. In the case of the chlorate-sulfur pairing, the sulfur is oxidized primarily to sulfur dioxide (SO_2, equation (17.13)). For a chlorate-sugar mixture, the sugar may be oxidized by two pathways—one in which the fate of the carbon is carbon dioxide (equation (17.14)) and the other in which it is carbon monoxide (equation (17.15)) [24]:

$$3S + 2KClO_3 \rightarrow 3SO_2 + 2KCl, \tag{17.13}$$

$$C_{12}H_{22}O_{11} + 8KClO_3 \rightarrow 12CO_2 + 11H_2O + 8KCl, \tag{17.14}$$

$$C_{12}H_{22}O_{11} + 4KClO_3 \rightarrow 12CO + 11H_2O + 4KCl. \tag{17.15}$$

The heat generated by such reactions can be tuned by varying the fuel/oxidizer ratio. The stoichiometric sucrose/$KClO_3$ mixtures of equations (17.14) and (17.15) should liberate, at most, about 4.5 kJ g^{-1} and 2.9 kJ g^{-1}, respectively. In a colored smoke composition, this heat causes a volatile, colored organic dye to sublime out of the reaction mixture. Rising into cooler air, the dye immediately condenses to form a visible, often brilliantly colored aerosol. These dyes are strong absorbers of visible light and the complementary hue is reflected by the aerosol particles [24]. When deployed during daytime, the resulting smoke fountain provides a convenient means of battlefield signaling, with each color capable of encoding a preconceived message [55].

Many different organic dyes have been used for military and civilian applications, and some of their structural formulae are shown in figure 17.8. Some structural commonalities clearly exist; they tend to have rich aromatic and extended conjugated structures, with the azo and anthraquinone moieties being prevalent. Such structural features confer the broad absorption spectra necessary to produce aerosols that reflect only a narrow visible bandwidth. Also common, structurally, for such dyes is the absence of oxygen-rich functional groups that could release oxygen at pyrotechnic reaction temperatures to decompose the dyes and thus reduce the color quality of the resulting smoke fountains [24].

The last important ingredient in colored smoke compositions is the coolant, a sacrificial material which decomposes endothermically at the combustion temperature to produce an evaporative cooling effect. This is essential to ensure that the colored dye does not burn or decompose. In many modern formulations, the coolant is a basic hydrated magnesium carbonate, typically hydromagnesite, which has the

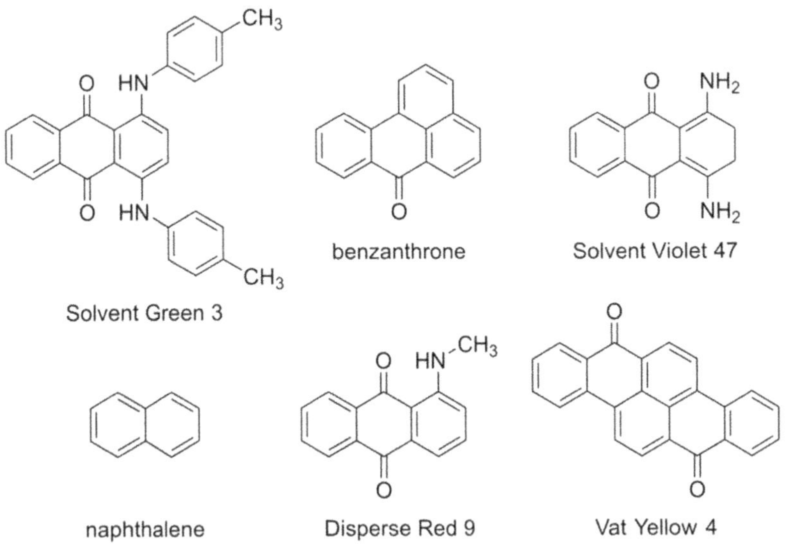

Figure 17.8. The structures of some common colored smoke dyes and naphthalene, which has been used in black smoke compositions. Solvent Violet 47 is 1,4-diamino-2,3-dihydroanthraquinone (DDA).

formula $Mg_5(CO_3)_4(OH)_2{\cdot}4H_2O$. It decomposes to form magnesium oxide, water vapor, and carbon dioxide (equation (17.16)) [56]. In addition to limiting the combustion temperature of the mixture, magnesium carbonate also functions as a retardant, with increased coolant content typically resulting in a reduced burning rate. Still, in other compositions, the coolant is sodium bicarbonate ($NaHCO_3$), which decomposes similarly into sodium carbonate (Na_2CO_3), water vapor, and carbon dioxide (equation (17.17)) [57]:

$$Mg_5(CO_3)_4(OH)_2{\cdot}4H_2O \rightarrow 5MgO + 5H_2O + 4CO_2, \tag{17.16}$$

$$2NaHCO_3 \rightarrow Na_2CO_3 + H_2O + CO_2. \tag{17.17}$$

In addition to the four most critical components of colored smoke mixtures—dye, fuel, oxidizer, and coolant—some additives may be used to promote processing in production. A binder is usually added to promote homogeneity, improve safety, and reduce dust formation. For example, the compositions within smoke grenades are typically granulated with an aqueous polyvinyl alcohol solution while some smoke signal-flare compositions contain vinyl alcohol-acetate resin (VAAR). Aside from binders, lubricants may be added to aid in pressing the compositions into their final forms and to ensure that the resulting pellets retain those forms. Stearic acid is a waxy compound that has been used for this purpose as a low percentage of the compositions (typically only 1 wt-%).

Despite the utility of colored smoke munitions, some of the components of older formulations are in need of reevaluation due to emerging hazards to human health and the environment. Most of these toxicity issues are related to specific colored dyes. However, there are also some other objectionable components of legacy colored smoke mixtures.

17.3.1.2 Problems with sulfur

As a pyrotechnic fuel in legacy formulations, sulfur undergoes combustion to form sulfur dioxide (SO_2), a toxic gas and a major air pollutant that has significant human health effects [58]. In particular, SO_2 emissions have been the subject of federal regulations [59, 60], as atmospheric SO_2 can affect the suitability of habitats for plants and animal life [61]. In addition, SO_2 emissions can cause acid rain and the formation of atmospheric particulates [62].

In addition to the health and environmental problems associated with the $S/KClO_3$ pair, this combination can be quite dangerous at large scales due to its tendency to accidentally ignite. One reason for this susceptibility is the possible presence of acidic impurities on the surface of sulfur particles which can trigger the spontaneous decomposition of potassium chlorate. Sulfur that has been melted, solidified, and crushed is generally preferred for pyrotechnic applications. Another form of sulfur known as 'flowers of sulfur' is material that has been purified by sublimation; it should not be used in pyrotechnic compositions because it may contain unacceptable amounts of acidic impurities such as sulfuric acid [24, 63].

17.3.1.3 The perchlorate issue

Potassium perchlorate ($KClO_4$) was once the oxidizer of choice in many types of pyrotechnic compositions due to its acceptable safety profile, slightly exothermic decomposition at 610 °C, and high oxygen content (46.2% oxygen by weight). However, the presence of the perchlorate anion in drinking water has been linked to thyroid disorders [24]. Perchlorate may enter drinking water sources when pyrotechnic residues or dud munitions containing $KClO_4$ are left to weather on training ranges. The perchlorate anion has been subject to state and federal advisory limits in drinking water [64, 65]. Quite a few initiatives to eliminate perchlorate oxidizers from pyrotechnic formulations have been sponsored by the U.S. Department of Defense since the year 2000 [66–70].

17.3.1.4 Colored dye toxicity concerns

Concerns have also been raised about the toxicity of the various dyes specified in legacy colored smoke formulations. A variety of risks to human health have been noted, ranging from carcinogenicity to dermal sensitization. Accordingly, several pyrotechnics technology development programs have been initiated to mitigate these risks. Now, replacement smoke dye candidates must be screened for carcinogenicity, mutagenicity, and other health effects prior to being implemented.

In older black smoke formulations, naphthalene served double duty as a pyrotechnic fuel and a smoke agent. This differs from other colored smoke compositions in which the smoke agent is a spectator that survives the fuel/oxidizer redox reaction and is only intended to undergo physical changes such as sublimation and condensation. Naphthalene forms black smoke—essentially soot—when it is partially oxidized. Naphthalene is a suspected carcinogen and is prohibited from use in consumer products in several jurisdictions. Its products of combustion and pyrolysis, including soot, are health hazards. Consequently, naphthalene-containing pyrotechnics have been targeted for replacement [24].

17.3.2 Recent formulation efforts

17.3.2.1 M18 red and violet smoke grenades

The U.S. Army's M18 colored smoke grenades use the same hand grenade form factor as the M83 white smoke grenade, which contains a terephthalic acid-based composition. On the battlefield, colored smoke fountains are useful for ground-to-air and ground-to-ground signaling. Additionally, thousands of M18 smoke grenades are used in military training exercises every year (figure 17.9). However, as M18s are ground-deployed devices, military personnel risk exposure to the smoke and gases that these devices emit. With this in mind, several reformulation efforts were initiated in recent years.

An ambitious project was completed by the U.S. Army Environmental Command in cooperation with ARDEC and the Pine Bluff Arsenal (PBA) in 2008 [71]. The pyrotechnic formulae for the red and violet M18 smoke grenades (compositions A and C in table 17.14) were reformulated to eliminate sulfur, Disperse Red 9, and 1,4-diamino-2,3-dihydroanthraquinone (DDA). In the improved formulae (compositions B and D), the chlorate-sulfur mixture was replaced with the more benign chlorate-sucrose base, Disperse Red 9 in A was replaced with a combination of Solvent Red 1, Disperse Red 11, and terephthalic acid, and, lastly, the toxic violet dye system in composition C was replaced with Disperse Red 11 and terephthalic acid.

Figure 17.9. Green smoke flows from an M18 smoke grenade at Fort Carson, Colorado, on January 17, 2012. Smoke grenades like these are often used in training exercises. U.S. Army photograph by Dustin Senger.

Table 17.14. Chemical makeup of M18 colored smoke compositions[ab].

Component	A (wt-%)	B (wt-%)	C (wt-%)	D (wt-%)
Disperse Red 9	40	–	8	–
Solvent Red 1	–	31.6	–	–
Disperse Red 11	–	13.9	–	38.03
Solvent Violet 47 (DDA)[c]	–	–	32	–
Terephthalic acid	–	6.6	–	7.67
Sulfur	9	–	9	–
Sucrose	–	14.2	–	15.5
Magnesium carbonate	–	8.7	–	10.2
Potassium chlorate	26	21.6	26	23.5
Sodium bicarbonate	25	3.4	25	5.1

[a] Reference [71].
[b] All four formulations contain an additional 2 wt-% polyvinyl alcohol and 0.50–0.63 wt-% stearic acid added in excess of the above totals.
[c] 1,4-diamino-2,3-dihydroanthraquinone (DDA).

Compositions B and D were produced on a 363 kg scale at PBA and 208 prototype grenades were fabricated from each of these batches. Each lot of red and violet smoke grenades was subsequently subjected to a battery of performance testing to assess smoke color quality, extreme-temperature functioning characteristics, and the ability to withstand an array of transportation/vibration tests (including rough handling, secure and loose cargo, and drop tests). The final results of this demonstration were mixed. While both variants passed all of the environmental and transportation tests, the red M18 grenades failed the smoke color quality test. As figure 17.10 shows, the cloud color for this lot was deemed to be outside the color range prescribed by MIL-G-12326K (EA) and MIL-STD-810F. The reformulated violet M18 lot, on the other hand, successfully passed all of the tests, including the color quality test, and was advanced into an inhalation toxicity study.

17.3.2.2 Omega 60 force-on-target battlefield effects simulator

Another significant improvement to colored smoke munitions was reported by engineers at ARDEC and ECBC for a newly designed black smoke ‘target kill’ battlefield effects simulator (BES) [72, 73]. The black smoke produced by this device signifies that a targeted vehicle has been ‘killed’ or otherwise disabled during training exercises. This simulator is required to produce a stream of black smoke for 10–30 s that is visible from a distance of 1800 meters in the daytime. The predecessor to this device was the U.S. Army’s M26 target kill training simulator, the use of which was to be discontinued due to a premature ignition problem, a major safety hazard. The original intention was to simply transfer the black smoke formulation from the M26, but this composition presents some serious environmental and health hazards. From table 17.15, it is obvious that two out of three of the components are problematic, especially naphthalene, which serves a dual role as a fuel and as a smoke agent.

Figure 17.10. M18 colored smoke fountains. In the photograph featuring two grenades (a), the grenade on the left contains the baseline red M18 composition and the one on the right contains the reformulated composition [71]. The other photograph (b) features a grenade containing the reformulated violet M18 composition.

Table 17.15. Black smoke composition of the M26 target kill simulator[a].

Component	Amount (wt-%)
Naphthalene	28
Potassium perchlorate	62
Laminac 4116/Lupersol DDM (99/1)	10

[a] Reference [72].

Figure 17.11. A sequence of images showing the black smoke fountain resulting from burning chlorate-sucrose in the presence of a green and red dye mixture [73].

There are also some concerns regarding the Laminac/Lupersol polyester binder system, which is based on phthalate plasticizer and styrene monomer.

In light of these issues, ARDEC engineers identified a range of compositions containing drug- and cosmetic-grade dyes. It is possible to use binary mixtures of green and red dyes such as Solvent Green 3, Solvent Red 23, or Disperse Red 9 to create smoke clouds that appear black. A rich black smoke forms when an appropriate dye mixture is vaporized in the presence of a burning chlorate-sucrose pyrotechnic base (figure 17.11). In the system configuration, burning times varied from about 13–30 s for the preferred compositions, which were comprised of mixtures of the red and green dyes, potassium chlorate, sucrose, magnesium carbonate, along with VAAR and stearic acid [73].

17.3.2.3 Yellow smoke hand-held signal

Shortly after the BES developmental work, the U.S. Army initiated a similar project for the M194 yellow smoke hand-held signal (HHS). The original design for this signal specified two toxic yellow dyes for the smoke flare, namely Vat Yellow 4 and benzanthrone (table 17.16). In this baseline M194 composition, the fuel is sucrose,

Table 17.16. Baseline flare composition for the M194 HHS[a].

Component	Amount (wt-%)
Potassium chlorate	35
Benzanthrone	28
Sucrose	20
Vat Yellow 4	13
Sodium bicarbonate	3
VAAR[b]	1

[a] Reference [74].
[b] Vinyl alcohol-acetate resin.

the coolant is sodium bicarbonate ($NaHCO_3$), which functions similarly to magnesium carbonate, and there is VAAR as a binder.

A report from 2013 described the development of a new M194 formulation based on Solvent Yellow 33 (SY33) instead of the two toxic yellow dyes that were formerly specified [74]. SY33 is a 2:1 cocrystalline mixture of D&C 11 and its methylated congener (figure 17.12). This commercial product consists of quinoline dyes that are thought to be safer than benzanthrone or Vat Yellow 4. There is also a pyrotechnic precedent for the use of this material, as it is now used in the yellow M18 smoke grenade.

When blended and tested in the M194 form factor, the compositions containing SY33 exhibited some interesting relationships in ingredient chemistry. Table 17.17 shows the tolerance of the new yellow smoke system to incremental changes in the amount of coolant, $Mg_5(CO_3)_4(OH)_2{\cdot}4H_2O$. Formulations A–D range in coolant content from 14.5 to just 2.5 percent by weight. Note the striking regularity that was observed in burning performance across the series from A–C (table 17.18); the burning time decreased most profoundly as the oxidizer content was increased at the expense of coolant content. This is shown most clearly by comparing the burning times of B (23 s) and C (15 s). This latter time is well within the M194 burning time requirement. However, a further trade-off of coolant for oxidizer led to degradation of the dye and erratic flame/smoke flickering in tests of composition D.

Composition C in tables 17.17 and 17.18 met the 9–18 s display time requirement and was subsequently selected for system demonstration (figure 17.13). A report

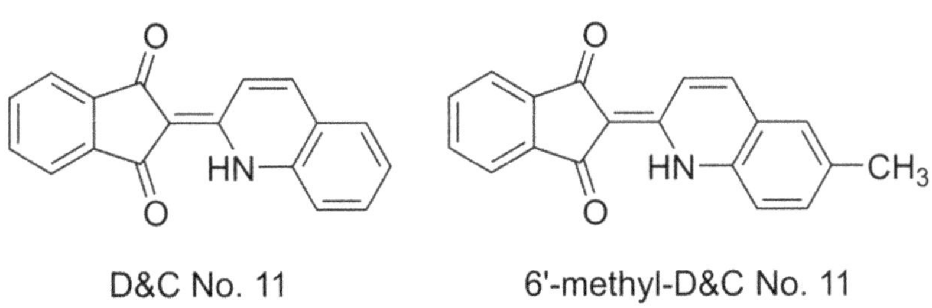

Figure 17.12. The structural congeners comprising Solvent Yellow 33.

Table 17.17. Composition tolerance study for M194 yellow smoke signal[a].

Component	A (wt-%)	B (wt-%)	C (wt-%)	D (wt-%)
Solvent Yellow 33	31	36	37	40
Potassium chlorate	29.5	29.5	34.5	34.5
Sucrose	22	22	21.5	21.5
$Mg_5(CO_3)_4(OH)_2{\cdot}4H_2O$	14.5	10.5	5.5	2.5
Stearic acid	1	1	1	1
Nitrocellulose	2	1	–	–
Fumed silica	–	–	0.5	0.5

[a] Reference [74].

Table 17.18. Performance of compositions A–D[a].

Composition	Burning time (s)
A	25
B	23
C	15
D	24[b]

[a] Reference [74].
[b] Excessive flaming noted.

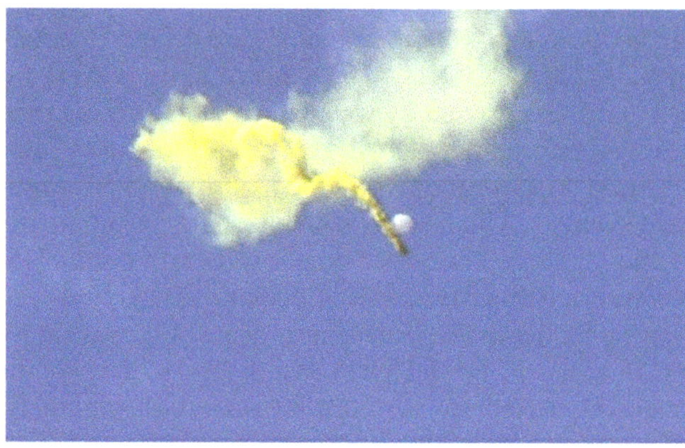

Figure 17.13. Aerial display of a parachute-suspended yellow smoke flare [75].

from 2014 indicates some moderate success in demonstrating this composition in actual system hardware, and describes some ignition train characteristics that appear to be crucial [75]. As of 2022, a fully successful demonstration of this technology involving quantities representative of typical production lots has not yet been realized.

17.3.2.4 Stoenescu compositions

Another possible solution came in the form of a U.S. patent application by Stoenescu in 2014 [76]. This patent application claims a generic colored smoke formulation containing the chlorate-sucrose pyrotechnic base, an organic dye, a nitrogen-rich compound that reportedly serves as a flame retardant, magnesium carbonate, and nitrocellulose as a binder (table 17.19). Various examples of this generic formulation were demonstrated in metal grenade cans containing 210–235 g of smoke composition. Display time, flame temperature, and can temperature were assessed.

The Stoenescu invention appears to provide a versatile solution for modern colored smoke applications. The data reported in the application suggest that smokes of high color quality can be produced using some alternative dyes, the chemical structures of which are included in figures 17.8 and 17.14. However, it is

Table 17.19. Enhanced colored smoke mixtures reported by Stoenescu[a].

Component	Amount (wt-%)
Colored dye[b]	27–40
Potassium chlorate	20–35
Sucrose	15–25
Nitrogen-rich compound[c]	5–15
Magnesium carbonate	8–18
Nitrocellulose	1–2

[a] Reference [76].
[b] Any one of the following: Solvent Red 1, Disperse Red 9, Solvent Orange 7, Solvent Orange 86, Disperse Blue 3, Solvent Violet 47, Solvent Yellow 33, or Solvent Green 3.
[c] Any one of the following: dicyandiamide, azodicarbonamide, or a 50/50 mixture of guanidine nitrate and guanidine carbonate.

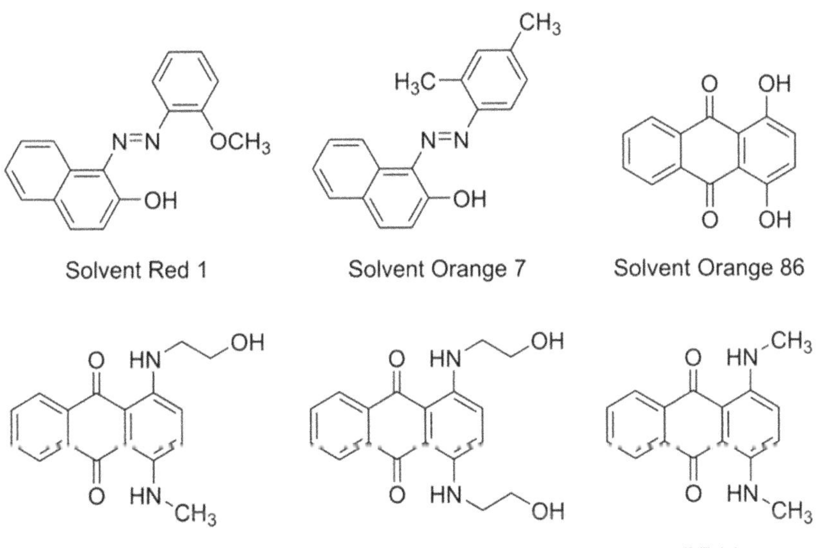

Figure 17.14. Structural formulae of some smoke agents proposed by Stoenescu. Disperse Blue 3 (DB3) is typically sold as a mixture with DB23 and DB14.

not clear what role the high-nitrogen compounds actually play in the compositions. They are supposedly flame retardants, but it is not clear why the same effect could not be achieved by simply increasing the magnesium carbonate content. Obviously, omitting these compounds would simplify the generic formulation and this could improve the manufacturability of the compositions. Additionally, there are currently no U.S. Army requirements for orange or blue smoke. Nevertheless, the Stoenescu contribution could prove useful if these colors are required in the future.

17.3.2.5 Copper complex as smoke agent

A creative way to avoid the use of Solvent Green 3 and Disperse Blue 180 was reported by the Klapötke group in 2017 [77]. The authors reported that copper(II) phthalocyanine, which is also known as CuPc or phthalocyanine blue BN, can be used as a dye in blue- and green-colored smoke compositions (figure 17.15). Small 2 g pellets were tested. Interestingly, CuPc was used as a direct substitute for Disperse Blue 180 in the chlorate-sucrose-coolant manifold to produce blue smoke, whereas green smoke was produced with a 50/50 CuPc/SY33 combination (table 17.20). Past work by Shimizu [63], Moretti [74, 75], and Chakraborty [78] inspired the development of these new CuPc-based compositions.

The performance of these CuPc-based compositions appears to benefit from the presence of an additional gas generator. In addition to hydromagnesite or sodium

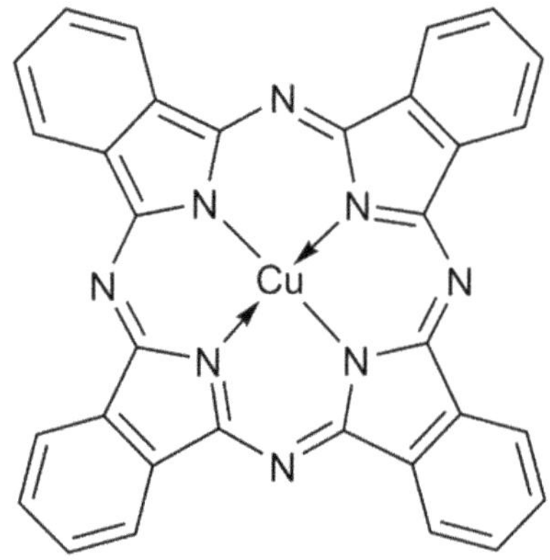

Figure 17.15. The structure of copper(II) phthalocyanine (CuPc).

Table 17.20. Chemical makeup of blue and green smoke compositions based on CuPc[a].

Component	B1 (wt-%)	G8 (wt-%)
CuPc	37.5	13.8
Solvent Yellow 33	–	13.8
Potassium chlorate	26	32
Sucrose	14.5	21.4
$Mg_5(CO_3)_4(OH)_2 \cdot 4H_2O$	14	8
5-aminotetrazole	7	10
Stearic acid	1	1

[a] Reference [77].

bicarbonate, which are gas generators, the authors also used various amounts of 5-aminotetrazole [77]. The presence of 5-aminotetrazole reportedly improves the efficiency of the compositions. Here, efficiency refers to the amount of dye dispersed as an aerosol when the mixtures burn. However, as with the Stoenescu compositions, there is little reason to believe that the presence of a high-nitrogen compound is a firm requirement. In other words, it should be possible to formulate conventional mixtures of sucrose, potassium chlorate, and hydromagnesite that effectively aerosolize CuPc.

Images of the blue and green smokes generated with the CuPc/SY33 dye system are shown in figure 17.16. Even on a small scale, the compositions clearly produce vividly colored blue and green clouds. Compositions like G8 in table 17.20 are plausible alternatives to the green smoke composition within the M18 grenade. Although any viable replacement for this application would need to meet the burning time and color quality requirements in a grenade configuration before it could be seriously considered.

Table 17.21 contains some performance data for compositions B1 and G8 [77]. Upon ignition, small pellets of these compositions released thick blue or green

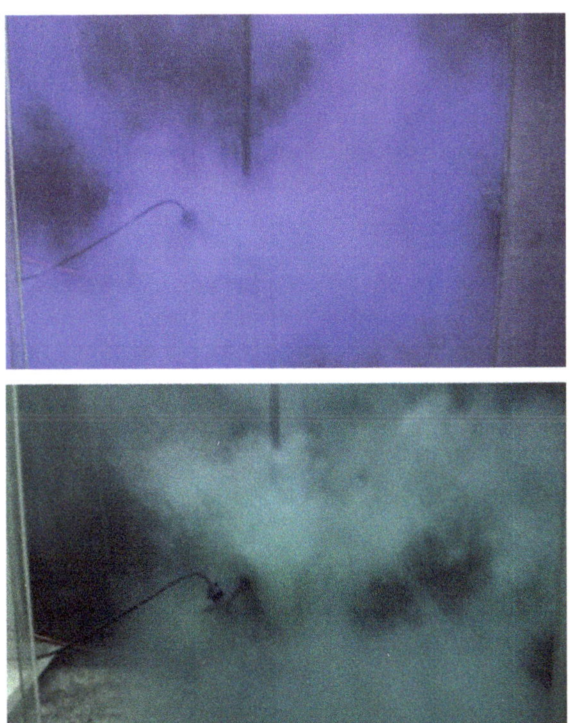

Figure 17.16. Images of the blue (a) and green (b) smokes based on CuPc [77]. Copyright Wiley-VCH Verlag GmbH & Co. KGaA, Weinheim. Reproduced with permission.

Table 17.21. Performance data for aerosols produced by Klapötke's Blue (B1) and Green (G8) smoke compositions[a].

Mixture	Burning time (s)	Impact sensitivity (J)	Yield factor	Cu content (%)
B1	19	40	0.137	not reported
G8	13	25	0.215	59.8

[a]Reference [77].

smoke for over 10 s. These types of compositions are relatively safe to prepare and handle, as they are not especially sensitive with respect to impact, friction, or electrostatic discharge. In addition, the Klapötke group determined yield factor, a metric that is typically reserved for obscurant smokes. At this time, little is known about the yield factors of signaling smokes. The copper content of the green aerosols was also determined by inductively coupled plasma atomic emission spectroscopy. The results confirmed that moderate portions of CuPc were aerosolized by the burning pellets.

17.4 Toxicology of smokes

17.4.1 Exposure and inhalation toxicity testing

The primary human route of exposure for all smokes is inhalation. Accordingly, the first toxicity issue for many of these smokes is their impact upon the respiratory system and lungs (especially respiratory tract hypersensitivity), followed by dermal or ocular exposures that could result in sensitivity and only ultimately with systemic effects such as carcinogenicity that result from incorporation of smoke particles into the body. The toxicity effects will vary depending upon how the smoke is generated.

As discussed in the previous sections, smoke particles may consist of combustion products or dyes that have been vaporized and subsequently condensed. Combustion products may include carbon oxides, nitrogen and nitrogen oxides, soot, and various inorganic compounds. Colored smoke aerosols contain the original dye materials, although the presence of products resulting from pyrolysis or partial oxidation is also possible. Toxicological effects will, of course, depend upon the details of each formulation. In general, effects are likely to be seen in the respiratory tract resulting in irritation and coughing. A significant exception to this is HC smoke, which rapidly produces a systemic response.

In order to aid understanding of the discussion that follows, some insight into the relevant inhalation toxicological testing methods is desirable. Test animal exposures can be conducted as either nose-only or whole-body. In a nose-only exposure, the animal is restrained outside of the test chamber so that only its nose projects into the chamber (figure 17.17), and the test compound may only be breathed in through the nose. In whole-body exposures, the animal is placed inside the test chamber, so that it not only breathes the test atmosphere, but may also ingest, via the process of grooming, particles that come to rest on its fur. Thus, the exposure during a whole-body test is likely to be greater than the nominal concentration of the test

Figure 17.17. Inhalation testing apparatus, nose-only exposure. Animals are inserted into the perforated stainless steel exposure cylinders so that only their heads/noses are exposed to the test atmosphere. Photograph courtesy of Arthur O'Neill, Director, Toxicity Evaluation Division, Toxicology Directorate, U.S. Army Public Health Center, Aberdeen Proving Ground, Maryland.

atmosphere. Test atmospheres are generated using different means, depending upon the physical characteristics of the test compound.

The 'limit dose' is the concentration above which no further inhalation toxicity testing is practical and often the test compound is considered nontoxic. The limit dose for inhalation testing is typically about 2 mg l^{-1} [79]. For an acute test, the standard exposure duration is 4 hr, and the animal continues under observation for a period of 14 days to check for delayed effects. To simulate longer-term exposures, repeated dose testing is also sometimes conducted. The number of repeat doses and the duration of each dose can vary depending upon the expected exposure duration in the field or use of the material.

17.4.2 Screening smokes

17.4.2.1 Phosphorus smokes

Upon combustion, both WP and RP smoke compositions produce phosphorus oxide particles that rapidly absorb moisture from the atmosphere. Consequently, a thick fog containing phosphoric and polyphosphoric acids is formed. RP smoke was reviewed by the National Research Council (NRC) in the first volume of the series *Toxicity of Military Smokes and Obscurants* [33]; WP smoke was reviewed in the second volume [34]. WP is typically soaked into felt pads that are dispersed by a munition, while RP has been deployed with polymeric binders such as butyl rubber (BR) or epoxy, which reduce the cloud pillaring effect associated with compositions that burn too rapidly. The red phosphorus in RP/BR compositions is coated with a small amount of insulating oil, and other additives such as talc or silica may be present [33]. As described previously, other RP-based compositions contain nitrate salts, metal powders, or both.

Humans exposed to WP smoke have survived exposures as high as 592 mg P_2O_5 equivalents m^{-3} for 3.5 min or 514 mg P_2O_5 equivalents m^{-3} for 15 min, but this is not the LD_{50} for the smoke as exposure levels in this experiment were kept low in order to not injure the test subjects. Coughing and throat irritation were noted at a concentration of 187 mg P_2O_5 equivalents m^{-3} for a period of 5 min [80]. Exposure levels for WP smoke have been established by the Occupational Safety and Health Administration (OSHA), the National Institute for Occupational Safety and Health (NIOSH), and the American Conference of Governmental Industrial Hygienists (ACGIH) at a time-weighted average exposure of 0.1 mg m^{-3} [34]. Human and animal data suggest that WP smoke produces respiratory irritation at lower concentrations than does RP/BR smoke. The differences in effect may be the result of uncombusted WP particles in WP smoke [34].

While it would be reasonable to assume that the products of the two phosphorus formulations are the same, it appears that they are not, possibly due to differences in combustion conditions. Table 17.22 compares the compositions of the combustion products of RP/BR and WP smokes. The slight differences in composition of the smokes may also be responsible for the differences in physiological effects, but this is unknown. There are no data indicating absorption of WP from the respiratory tract into the gastrointestinal tract, and there is no convincing evidence that WP is absorbed in sufficient quantity to cause systemic effects when humans are exposed through acute dermal contact (WP burns) or inhalation of phosphorus smokes [81].

It is likely that some uncombusted WP particles are present in the smoke clouds produced by WP munitions. Individuals that are exposed dermally to these WP particles may be subjected to burning effects and some residual WP particulates may be present in the wounds. White phosphorus burns are more severe than thermal burns, and have been demonstrated to heal more slowly. There is also the potential for systemic effects resulting from absorption of the phosphorus species [14].

Oral ingestion of WP occurs when it is either deliberately (suicide attempts) or accidentally ingested, but this is not a common route of exposure in the military. WP exhibits a higher level of toxicity than does RP due to its greater solubility and

Table 17.22. Combustion products of red and white phosphorus compositions.

Component	Formula	RP/BR (wt-%)[a]	WP (wt-%)[b]
Orthophosphate	H_3PO_4	22.8	23.8
Pyrophosphate	$H_4P_2O_7$	19.6	26.6
Tripolyphosphate	$H_5P_3O_{10}$	13.3	16.3
Tetrapolyphosphate	$H_6P_4O_{13}$	11.5	11.3
P_5–P_{13}	$H_{n+2}P_nO_{3n+1}$	32.8	22.0
P_n ($n > 13$)	$H_{n+2}P_nO_{3n+1}$	Low	Low

[a] Reference [33].
[b] Reference [34].

chemical reactivity, which were primary causes of the waterfowl toxicity observed at the Eagle River Flats (ERF) firing range in Alaska. RP-based munitions were used at ERF after the firing of WP-based munitions was halted because of extensive waterfowl mortality [82]. The toxicity of phosphorus to wildlife, especially waterfowl, has been studied extensively due to the effects of WP-based munitions; a recent analysis is available [83].

Neither WP nor RP have been found to be mutagenic in the Ames test [83]. Chronic effects, such as cancer, resulting from an acute, short-term exposure are not anticipated.

17.4.2.2 Hexachloroethane smoke

Generally, the smoke compositions collectively known as HC contain a chlorocarbon compound such as hexachloroethane, a source of zinc such as zinc powder or zinc oxide, and a pyrotechnic fuel such as calcium silicide or aluminum. The Type C variant is a ternary mixture of aluminum, zinc oxide, and hexachloroethane. The toxicity effects of HC smoke have been known for some time and were thoroughly reviewed in the first volume of the series *Toxicity of Military Smokes and Obscurants* [33]. The toxic factor in these formulations is zinc chloride ($ZnCl_2$), which is exothermically produced when the compositions burn (equations (17.5) and (17.6)).

Overexposure to HC smoke can be fatal, especially in an enclosed environment; the lethal dose has been estimated to be 50,000 mg-min m^{-3} [84]. A short-term exposure limit of 30 mg-min m^{-3} has been calculated [33]. Even in the open air, individuals who breathe HC smoke will rapidly be affected by its toxicity. The effects of zinc chloride are well known and are focused primarily on the upper respiratory and gastrointestinal tracts; it is corrosive and astringent, and causes burns to the eyes and damage to nerve endings in the nose [85]. Pulmonary effects include dyspnea (difficulty breathing), chest constriction, retrosternal and epigastric pain, hoarseness, cough, lacrimation, expectoration, and occasional hemoptysis (coughing up blood). Although an increased incidence of chromosome abnormalities has been observed in rats exposed to zinc chloride, HC smoke is not considered to be mutagenic or carcinogenic.

17.4.2.3 Terephthalic acid smoke

Terephthalic acid smoke, or TA smoke, was developed by the U.S. Army as a replacement for HC smoke in training exercises. The compositions contain terephthalic acid (figure 17.2), sucrose, potassium chlorate, a carbonate-based coolant, and other additives. Terephthalic acid is the primary smoke constituent. Terephthalic acid is used to produce polyesters and is thought to be relatively nontoxic based upon toxicity studies conducted by commercial manufacturers.

The results of toxicity studies involving pyrotechnically dispersed terephthalic acid—TA smoke—have also been reported [86, 87]. In a typical study, groups of rats were exposed to both acute and repeated inhalation doses of the smoke [86]. At low doses, the animals exhibited only a small number of toxic signs, including slight lacrimation, lethargy, and rhinorrhea. At higher dose levels, dyspnea was observed in addition to the basic symptoms. Almost all signs were reversed following 1 hr of post-exposure recovery. Necropsy revealed lesions only within the nasal cavities consisting of necrosis and acute inflammation. After a 14-day recovery period, no lesions remained. Other reaction products include formaldehyde, benzene, and carbon monoxide, all of which are expected to have a minimal effect when exposure occurs in an open environment.

17.4.3 Battlefield effects smokes

17.4.3.1 Use, exposure, and toxicity issues

BES are employed in training situations and are used to indicate whether or not a target was hit and/or the strike point of a munition. The smoke that these simulators emit is typically black, as it is intended to give the impression of burning. In some cases, other colors have been used, such as yellow to imply the presence of a chemical warfare agent. The risk of exposure depends on how the smokes are used in training exercises, with the yellow smoke being the most likely to result in human exposure. The black smoke formulation for the U.S. Army's M26 target kill simulator was targeted for replacement because it produced clouds containing naphthalene and other aromatic compounds that are known or suspected human carcinogens. Regulation of naphthalene by the State of California has affected military training conducted in the state.

17.4.3.2 Naphthalene

Exposure to naphthalene (figure 17.8) may result in hemolytic anemia caused by the destruction of red blood cells, and may result in fatigue, loss of appetite, restlessness, and pale skin. Exposure to large amounts of naphthalene may also cause nausea, vomiting, diarrhea, blood in the urine, and jaundice. In addition to blood effects, inhalation of naphthalene vapor may cause headaches, loss of appetite, and nausea. There is additional reporting of dermal irritation, optical neuritis, corneal injuries, and kidney damage. Naphthalene is considered a possible human carcinogen by the International Agency for Research on Cancer (IARC) and the U.S. Environmental Protection Agency (EPA). The ACGIH classifies naphthalene as a confirmed animal carcinogen with unknown relevance to humans [88].

17.4.3.3 COMET system black and yellow smokes

Several alternative formulations for various BES systems were evaluated in the mid-2000s. Black and yellow smoke-producing compositions were considered. One of the 'black' smoke formulations, for the COMET system, contained three dyes: Solvent Orange 86, Solvent Green 3, and Solvent Violet 47. A yellow variant contained Solvent Yellow 33. Each of these individual dyes is discussed in the following sections. The U.S. Army Public Health Center prepared a toxicity assessment of these dyes as part of the development process [89]. The existing data on the dyes were found to be limited, but the formulations were assessed favorably and were deemed to present a negligible risk to human health and the environment. There are some concerns associated with each of the dyes, however, as discussed below.

17.4.3.4 Omega 60 simulator ARDEC/ECBC black smoke

The development of this alternative, for use in the Omega 60 Force-On-Target Simulator, was described previously. This black-smoke-producing composition contains mixtures of green and red dyes such as Solvent Green 3, Solvent Red 23, or Disperse Red 9. These dyes are less of a hazard than the naphthalene-based smoke system, but still pose some hazard if inhaled. The employment scenario for this battlefield effects smoke, however, should limit human exposure.

An inhalation toxicity study on a formulation containing Solvent Green 3 and Disperse Red 9 was conducted by the Army Public Health Center on laboratory rats [90]. No mortality or signs of adverse toxicity were observed during an acute test with exposures of up to 2500 mg m^{-3}. A 4-week repeated dose follow-on study was conducted with four groups of 12 rats each, at exposures of 0 (control), 100, 700, or 2000 mg m^{-3} of the smoke. Compound-related effects were observed in rats exposed to 2100 mg m^{-3} for body weights, weight gains, food consumption, organ weights, and histopathological findings. The most sensitive effect was determined to be pulmonary alveolar histiocytosis, an accumulation of macrophages in the alveolar spaces. The no-observed-adverse-effect-level was determined to be 130 mg m^{-3} in rats.

17.4.4 Colored smoke dyes

17.4.4.1 The purity of smoke dyes

It is very common for organic dyes to be mixtures of congeners as a result of the synthetic methods used in their production. In some cases, the amount of congener 'impurity' can be significant, greatly complicating the assessment of the toxicity of mixtures. Predictive toxicity assessments are typically developed on the basis of the pure product, not on a mixture of congeners. If the composition of the mixture is fixed, then direct toxicity testing can provide for the determination of toxicity of that specific formulation. However, if the composition of the mixture is variable, the validity of a toxicity assessment will be less reliable. Generally, dyes used as additives in foods, drugs, and cosmetics are significantly purer than dyes used for other applications, such as textile dying, and such dyes are assumed to be less hazardous to humans based upon their approved use; the U.S. Food and Drug Administration (FDA) sets stringent impurity limits for these dyes.

17.4.4.2 Legacy dyes

Legacy dyes such as benzanthrone and Vat Yellow 4 have not been considered in the discussion that follows. A good discussion of the toxicology of legacy dyes may be found in the third volume of the NRC series on the toxicology of military smokes [91].

17.4.4.3 Solvent Yellow 33 [92]

Solvent Yellow 33 (SY33) (figure 17.12) is also known as 2-(2′-quinolyl)-1,3-indandione (quinophthalone), D&C Yellow No. 11, Quinazoline Yellow Spirit

Soluble, and Yellow No. 204. It has been used in the U.S. Army's M18 Yellow Smoke Grenade since the early 1980s when it replaced Vat Yellow 4 [92]. Some SY33 formulations that are not approved by the FDA for use in cosmetics contain approximately 30% of the methylated congener 6′-methyl-2-(2′-quinolyl)-1,3-indandione [93].

In studies involving laboratory rodents, inhaled SY33 was rapidly cleared from the respiratory tract with a half-life of 2–3 hr, and metabolized by the liver with 90% excretion over the course of 72 hr [94]. No tissues appear to be significant storage depots for SY33. In repeated dose testing, rats exposed for 4 weeks had reduced pulmonary gas exchange efficiency, airflow obstruction, mild pulmonary inflammation, slight Type II pulmonary epithelial cell hyperplasia, and proliferation of vacuolated macrophages. Rats exposed to the highest concentration of the dye for 13 weeks also had indications of mild pulmonary inflammation and slight Type II pulmonary epithelial cell hyperplasia. Other than effects on body weight, developmental and reproductive effects have not been noted [91].

Dermal sensitization has been noted to occur with SY33 among workers involved in the manufacture of SY33-containing smoke compositions and among the general population through nonmilitary applications of the dye, such as soaps and cosmetics. In the third volume of its review of military smokes, the NRC Committee on Toxicology [91] noted that numerous reports exist concerning the development of contact dermatitis from exposure to SY33. Many of these reports have to do with individuals exposed through the use of soaps and cosmetics such as eyeliner, rouge, and lipstick [95–105]. Worker exposure resulting in contact dermatitis from the manufacture of colored smoke compositions was reported by Noster and Hausen [95]. In controlled studies using human volunteers, Jordan found sensitization in nine of 149 volunteers exposed to SY33 at 16.4 ppm in a modified dermal Draize test [96]. Weaver found that individuals repeatedly exposed via patch tests at levels as low as 10 ppm, but not at 5 ppm, developed sensitization [97]. Rapaport induced sensitization in 15 of 56 test subjects using a petrolatum preparation containing 20% SY33 [98]. Other investigators have found that some individuals are sensitive to as little as 1 ppm in petrolatum [99] or 0.1 ppm in polyethylene glycol [100].

Overall, mixed results have been obtained in Ames *Salmonella typhimurium* mutagenicity tests of SY33 [106]. Differences in results may be due to the presence of contaminants within the test mixtures. Positive results were observed in tests employing the TA1537 R+ strain of *S. typhimurium* at concentrations of 0.195–100 μg/plate both with and without metabolic activation of the test substance. Positive tests were also observed using strain TA1538 when the substrate was activated using either rat or hamster liver S9 fractions and 30 percent Aroclor 1254. In contrast, results were negative from tests employing strains TA91, TA98, and TA1535, with or without activation of the test substance. Results with strain TA100 are conflicting, with negative results being observed when up to 10,000 μg/plate were delivered in dimethyl sulfoxide (DMSO), even with activation, but positive results at 3.3–4000 μg/plate when the dye was delivered without solvent.

All tests using the TA100 strain also employed rat or hamster liver S9 fractions and 30 percent Aroclor 1254 activation [106]. D&C Yellow No. 11 was positive in the *in vitro* Chinese hamster ovary cell assays for sister chromatid exchange and aberrations both with and without microsomal activation conducted as part of the 2-year National Toxicology Program (NTP) feeding study [107]. SY33 was negative in an *in vivo* mouse bone marrow micronucleus assay (OECD Test No. 474 compliant) at all concentrations tested (2000, 666, and 222 mg kg^{-1}) [92].

In the 2-year NTP feeding study, both male and female rats receiving 5000 ppm of SY33 daily showed an increased occurrence of hepatic adenomas and carcinomas as compared to controls [107].

17.4.4.4 Solvent Green 3

Solvent Green 3 (SG3) (figure 17.8) is used in several different pyrotechnic smoke formulations. A systematic chemical name for this compound is 1,4-bis((4-methylphenyl)amino)-9,10-anthracenedione; another chemical name is 1,4-bis(*p*-tolylamino)anthraquinone. Common names include D&C Green No. 6, Cyanine Green G Base, and Quinizarin Green SS. An extensive review of this compound, its manufacture, and its environmental fate was prepared by Davidson and Hovatter [108]. According to Davidson and Hovatter, 90%–95% of the dye in a smoke grenade formulation sublimes during the functioning of the grenade; the remaining 5%–10% decomposes into polynuclear aromatic hydrocarbons, but no ring-opening reactions occur. Some of the decomposition products were reported to be 1-*p*-toluidinoanthraquinone, 1-*p*-toluidino-4-aminoanthraquinone, *p*-toluidine, 1-aminoanthraquinone, and 1-*p*-toluidinoanthrone [108, 109].

SG3 is notable in that it tends to accumulate in the lungs more than other dye compounds. Short-term inhalation studies involving SG3 were reported by Henderson *et al* [110] and Medinsky *et al* [94]. In a typical study, Fischer 344 rats were exposed to an aerosol of a SY33/SG3 mixture for 60 min The concentration of SG3 was calculated to be 154 mg m^{-3}. Animals were sacrificed at times up to 72 hr post-exposure to determine the extent of SG3 deposition in the lungs. A total of 360 μg was found to have been deposited in the entire respiratory tract, with 310 μg in the upper respiratory tract and 50 μg in the lungs and bronchi. SG3 was not cleared from the lungs during the sampling period. The animals in these short-term studies also exhibited reduced pulmonary gas exchange efficiency, airflow obstruction, mild pulmonary inflammation, slight pulmonary epithelial cell hyperplasia, and proliferation of vacuolated macrophages.

Technical-grade formulations of SG3 have been found to be dermal sensitizers in guinea pigs. This is likely due to impurities in these formulations, as a similar response was not obtained with the purified dye [111]. SG3 is not teratogenic in rats or rabbits, so it is not likely to be a developmental or reproductive toxicant [108]. SG3 is negative in the Ames *Salmonella* mutagenicity test [108]. SG3 is not carcinogenic in the Mouse Lung Tumor bioassay [108]. Additionally, SG3 is virtually insoluble in aqueous systems in the absence of a cosolvent, and is therefore not toxic to aquatic species.

17.4.4.5 1,4-diamino-2,3-dihydroanthraquinone, Solvent Violet 47

1,4-diamino-2,3-dihydroanthraquinone (DDA) is used in combination with Disperse Red 9 to impart a violet color to smoke clouds (figure 17.8). Solvent Violet 47 (SV47) is a common synonym. Although DDA was addressed by the NRC in its review of colored smokes [91], very few experimental data are available. Toxicity predictions have been made using the TOPKAT modeling system [112].

The U.S. Army conducted an acute inhalation toxicity test of DDA/SV47, exposing rats at concentrations of 0, 0.29, 1.0, or 2.6 mg l^{-1} for 4 hr [113]. Exposures at 2.6 mg l^{-1} (the limit dose) caused 70% mortality among the test animals. In order to determine the median acute inhalation toxicity (LC_{50}), two doses below the limit dose were evaluated. Exposure to 1.0 mg l^{-1} resulted in 20% mortality to the test animals, while exposure to 0.29 mg l^{-1} resulted in no mortality. Clinical signs of toxicity at the higher concentrations just prior to death included incoordination, tremors, lethargy, and lung noise. The LC_{50} in this test was determined to be 1.8 mg l^{-1}. DDA/SV47 was assigned to U.S. EPA Toxicity Category II, and is considered moderately toxic.

DDA/SV47 is, itself, a weak mutagen but its combustion or oxidation product, 1,4-diamino-anthraquinone, is a strong mutagen [114].

17.4.4.6 Solvent Orange 7

Solvent Orange 7 (figure 17.14) is variously known as Sudan II, Sudan Red, Oil Orange, and Sudan Orange; the systematic name is 1-(2,4-dimethylphenylazo)-2-naphthol. Solvent Orange 7 is not on the FDA list of approved food additives due to oral toxicity. However, several instances of food adulteration to improve appearance have been documented in foodstuffs as diverse as red wine [115] and saffron spice [116].

Solvent Orange 7 is reported to be a respiratory, dermal, and ocular irritant, however quantitative data could not be found [117]. Ames mutagenicity testing has yielded mixed results, with or without metabolic activation. Testing in the mouse lymphoma system L5178Y is negative with or without activation [118]. No human carcinogenicity data are available. Limited evidence exists for carcinogenicity in animals. According to the IARC, Solvent Orange 7 is not classifiable with respect to human carcinogenicity [117].

17.4.4.7 Solvent Orange 86

Solvent Orange 86, or Quinizarin, is 1,4-dihydroxyanthraquinone (figure 17.14). This dye was found to be a dermal sensitizer when tested on guinea pigs [119]. Acute oral toxicity testing in rats indicated that Solvent Orange 86 has an LD_{50} exceeding 5000 mg kg^{-1} (oral limit dose), which categorizes it as nontoxic to mammals. Evidence for the mutagenicity of Solvent Orange 86 is mixed, with a positive Ames test being reported in *S. typhimurium* strain TA1537, both with and without microsomal activation [120], but other testing conducted to date has given negative results [121] and there are suggestions that Solvent Orange 86 might even demonstrate anti-mutagenic activity [122].

17.4.4.8 Solvent Red 1

Solvent Red 1 (figure 17.14) is structurally similar to Solvent Orange 7, and has several synonyms including Oil Pink, Sudan R, Sudan Red G, as well as several others [123]. Its chemical name is α-methoxybenzenazo-β-naphthol (MBN). The NRC was unable to assess the potential health effects of exposure to the dye or its combustion products due to an absence of well-controlled inhalation toxicity studies [91]. No data are available with respect to the inhalation toxicity of Solvent Red 1 by itself. Studies of dermal toxicity indicate that Solvent Red 1 is a moderate irritant and possible sensitizer, but the dye is not dermally toxic. Solvent Red 1 may be a moderate ocular irritant [91]. Solvent Red 1 is considered to be possibly carcinogenic [123].

17.4.4.9 Solvent Red 23

Solvent Red 23 (SR23) (figure 17.18) is a member of the Sudan family of dyes, all of which have been banned as food additives by the U.S. FDA and other international food safety organizations. However, these dyes are occasionally illegally used as food adulterants since they improve the color of food products and therefore enhance consumer appeal. SR23 is used externally as a component of various hair dyes. Synonyms include Sudan III, Cerasin Red, D&C Red No. 17, Tony Red, and several others [124].

Most of the Ames mutagenicity tests for SR23 are negative, however the presence of impurities may cause the occasional positive response [125].

17.4.4.10 Solvent Red 169

Solvent Red 169 (SR169) has been considered as a component of both red and violet smoke formulations. The chemical name of this dye is 1-(isopropylamino)anthraquinone. SR169 is structurally similar to Disperse Red 11 and Disperse Red 9 (figures 17.8 and 17.19).

Very little toxicological information is available for SR169. In a 4 hr acute inhalation study, the LC_{50} for SR169 was found to be greater than 2.4 mg l^{-1}, exceeding the limit test [126, 127]. Both male and female rats were exposed, nose-only, for a period of 14 days. Solvent Red 169 was reported to be positive in Ames tests with *S. typhimurium* strains TA98, TA1537, and TA1538 [126, 128]. Solvent Red 169 was considered nontoxic at the limit dose in the Xenometrix MPF Ames test [126].

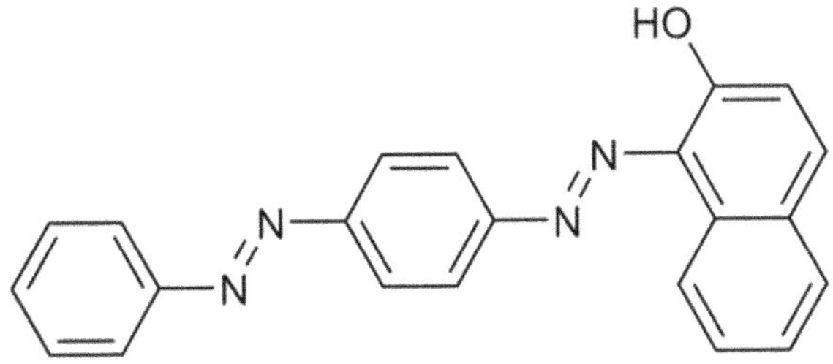

Figure 17.18. Solvent Red 23, also known as D&C Red 17.

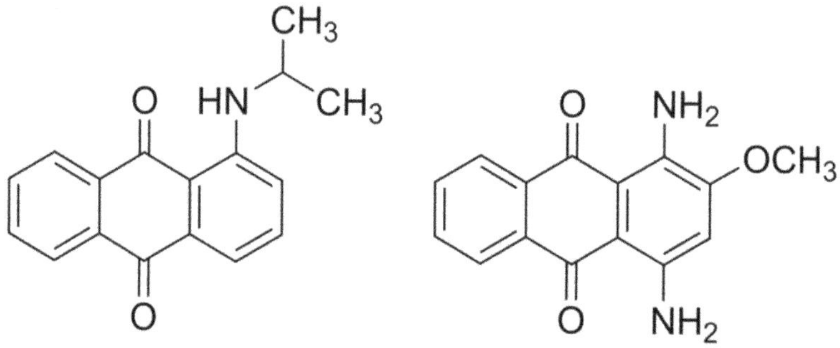

Solvent Red 169 Disperse Red 11

Figure 17.19. Solvent Red 169 and Disperse Red 11.

17.4.4.11 Disperse Red 9

Disperse Red 9 is 1-(methylamino)anthraquinone (figure 17.8). One interesting use is in bank security dye packs that explode and mark stolen bank notes [129]. Disperse Red 9 has been identified as a respiratory irritant, a dermal irritant, and a serious eye irritant [129]. Anthraquinone dyes are potentially carcinogenic [130, 131], but there is no information specifically linking Disperse Red 9 to human or animal cancers. This result is confounded because Disperse Red 9 has only been tested in combination with other dyes.

17.4.4.12 Disperse Red 11

Disperse Red 11 is 1,4-diamino-2-methoxyanthraquinone (figure 17.19). An inhalation toxicity study of a new red smoke formulation that included DR11 and MBN (or Solvent Red 1), referred to as the 'red dye mixture,' was conducted by D L Costa at the EPA Health Effects Research Laboratory, Research Triangle Park, in Fischer 344 rats [132]. Inhalation studies in rats after either acute or repeated exposures to the red dye mixture showed nasal and lung lesions, but these were rapidly resolved after exposure stopped.

Acute oral toxicity studies in rats demonstrated low toxicity after either oral or dermal exposure to the dye mixture. Acute studies of either eye or dermal irritation showed the red dye mixture to be nonirritating when applied to the skin but irritating in ocular studies [133]. Hypersensitivity studies with the red dye mixture showed no pulmonary or contact hypersensitivity in mice [134].

The results of Ames tests have been in conflict, with one study [135] concluding that observed mutagenicity might have been due to contamination of a dye lot. An increased number of mutants and micronuclei were observed in a mouse lymphoma cell system [136], and increased frequencies of sister chromatid exchange (SCE) and increases in hypoxanthine guanine phosphoribosyl transferase (HGPRTase) mutation frequencies were observed in Chinese hamster ovary cells [91].

17.4.4.13 Disperse Blue 3

Disperse Blue 3 (DB3) (figure 17.14) is an anthraquinone dye with the chemical name 1-(2-hydroxyethylamino)-4-(methylamino)anthraquinone. The commercial product is a mixture of this compound and the related dyes DB14 and DB23 in approximately a 2:1:1 ratio [137]. DB3 has been used in hair color formulations and for coloring fabrics and plastics. It has also been evaluated as a component of violet smoke formulations [126].

In a test conducted by the Army [126], no mortality was observed during a 14-day observation period for nose-only exposure of rats to a concentration of 2.4 mg l^{-1} for 4 hr. The 2.4 mg l^{-1} dose corresponds to the limit test for acute inhalation, and therefore the DB3 mixture is considered nontoxic.

The experimental acute oral LD_{50} for DB3 is 1345 mg kg^{-1} in rats and 1600 mg kg^{-1} in mice [126]. The LD_{50} for DB23 is greater than 5000 mg kg^{-1}, making it nontoxic; an experimental LD_{50} for DB14 could not be found.

The dermal LD_{50} in rabbits is reported to be greater than 2000 mg kg^{-1} [126, 133], making DB3 nontoxic by standard criteria. Dermal Draize testing of DB3 in rabbits indicated mild irritation after a 24 hr exposure to 500 mg kg^{-1} [126, 134]. DB3 was not a sensitizer as determined by the Localized Lymph Node Assay and the Mouse Earlobe Swelling Test [126, 134].

DB3, as a component of hair dye formulations (0.3% of total mixture) was tested for teratogenic effects in Charles River CD rats [138]. No biologically significant soft tissue or skeletal changes were observed and the average number of corpora lutea, implantation sites, live fetuses, resorptions, and litters with resorptions were not significantly affected by the dye treatment.

Mutagenicity testing in bacteria has generally been negative, with the exception of *S. typhimurium* strain TA1537 with or without activation, and in strain TA98 after activation but otherwise negative [126]. The TOXNET database has DB3 listed as non-mutagenic in the Ames test [139]. DB3 was found to be positive without microsomal activation in the mouse lymphoma assay, producing mutations and micronuclei [136].

DB3 is not considered to be carcinogenic in humans [126].

17.5 Conclusion

The formulation of replacement screening smoke compositions remains a challenging problem. Currently, the obscuration performance of WP, RP, and HC-based compositions is unsurpassed. Generally, the best performance is offered by hygroscopic aerosols, and these tend to be acidic and irritating even when they are not especially toxic. The search continues for benign screening smokes that are comparable to WP, RP, and HC in effectiveness but have minimal toxicity. Better progress has been made toward the development of replacement signaling smokes. In several instances, the hazardous legacy dyes in these formulations have been replaced with less hazardous (although not necessarily nontoxic) alternatives. The toxicological evaluation of smokes is a complex and challenging endeavor due to the nature of smoke formulations, the variability of exposure pathways, and the complexities of human biology. Nonetheless, such evaluations have played an important role in the recent development of several replacement formulations, and will certainly play an important role in future development efforts.

The chemicals within pyrotechnic aerosols exhibit a spectrum of toxicological properties. While much is known about the toxicity of the phosphoric acid and zinc chloride aerosols commonly used for screening, much less is known about the toxicity of the clouds produced by some of the more recent replacement candidates, including the B_4C/KNO_3, BP/KNO_3, $Al/SnO/C_2Cl_6$, and P_3N_5/KNO_3 systems, to name a few. In contrast, several signaling smoke formulations containing alternative dyes have been evaluated. However, for signaling smokes, an assessment of toxicity is complicated by the fact that the dyes are usually technical-grade substances and should be expected to contain various impurities. Importantly, the toxicological properties of these smokes may differ from those of the nominal dye substances. In most cases, casual and infrequent exposure to colored signaling smokes is not anticipated to result in serious toxicity. For individuals that are exposed on a regular and recurring basis, such as military trainers, the hazard is likely to be somewhat greater.

References

[1] Shaw A P, Diviacchi G and Black E L 2014 Advanced boron carbide-based visual obscurants for military smoke grenades *Proc. 40th Int. Pyrotechnics Society Seminar (Colorado Springs, CO, July 13–18)* pp 170–91

[2] Sharpe L T, Stockman A, Jagla W and Jägle H 2005 A luminous efficiency function, $V^*(\lambda)$, for daylight adaptation *J. Vision* **5** 948–68

[3] Bohren C F and Huffman D R 2004 *Absorption and Scattering of Light by Small Particles* (Weinheim: Wiley-VCH) pp 69–81 section 3.4

[4] Lane G A, Smith W A and Jankowiak E M 1968 Novel pyrotechnic compositions for screening smokes *Proc. 1st Int. Pyrotechnics Society Seminar (Estes Park, CO, August 12–16)* pp 263–91

[5] Milham M E and Anderson D H 1983 *Obscuration Sciences Smoke Data Compendium—Standard Smokes; accession number CBRNIAC-CB-153496* (Fort Belvoir, VA: Defense Technical Information Center (DTIC)) pp 1–54

[6] Sordoni N, Heard W and Rouse W 1993 *Pyrotechnic Smoke Analysis Volume I; accession number CBRNIAC-CB-027023* (Fort Belvoir, VA: Defense Technical Information Center (DTIC)) pp 1–49

[7] Turetsky A L and Young G P 1988 Advances in pyrotechnically based visual smoke systems *Proc. 13th Int. Pyrotechnics Society Seminar (Grand Junction, CO, July 11–15)* pp 805–10

[8] Embury J F, Walker D and Zimmermann C J 1994 *Screening Smoke Performance of Commercially Available Powders. II. Visible Screening by Titanium Dioxide; accession number ADA284072* (Fort Belvoir, VA: Defense Technical Information Center (DTIC)) pp 1–30

[9] DeLacy B, Redding D and Matthews J 2011 Mass extinction coefficients of surface modified titanium dioxide *Proc. 2011 Obscurants Symp. (Baltimore, MD, May 10–12)*

[10] Shaw A P, Diviacchi G, Black E L, Moretti J D, Sadangi R K, Grau H A Jr and Gilbert R A Jr 2015 Versatile boron carbide-based visual obscurant compositions for smoke munitions. *ACS Sustain. Chem. Eng.* **3** 1248–54

[11] Poulson G W 2011 UK obscurant developments *Proc. 2011 Obscurants Symp. (Baltimore, MD, May 10–12)*

[12] Jafar G 2011 Smoke and obscurant development programs at Pine Bluff Arsenal *Proc. 2011 Obscurants Symp. (Baltimore, MD, May 10–12)*

[13] Stewart J, Carestia R, Hale J, D'Agostino J and Redding D 2016 Metrics for evaluating obscurant grenades *Proc. 2016 Obscurants Symp. (Charlottesville, VA, November 29–December 1)*

[14] Toxicological Profile for White Phosphorus; Agency for Toxic Substances and Disease Registry, September, 1997 (http://atsdr.cdc.gov/ToxProfiles/tp103.pdf).

[15] Yon R L, Wentsel R S and Bane J M 1983 *Programmatic Life Cycle Environmental Assessment for Smoke/Obscurants Volume 2—Red, White, and Plasticized White Phosphorus;* accession number ADA135910; (Fort Belvoir, VA: Defense Technical Information Center (DTIC)) pp 1–73

[16] Rubel G O 1978 *Predicting the Droplet Size and Yield Factors of a Phosphorus Smoke as a Function of Droplet Composition and Ambient Relative Humidity Under Tactical Conditions;* accession number ADA064076; (Fort Belvoir, VA: Defense Technical Information Center (DTIC)) pp 1–43

[17] U.S. Navy, Naval History and Heritage Command (https://history.navy.mil/our-collections/photography/us-navy-ships/battleships/alabama-bb-8/NH-57483.html).

[18] Koch E-C 2008 Special materials in pyrotechnics: V. military applications of phosphorus and its compounds *Propellants Explos. Pyrotech.* **33** 165–76

[19] Wilharm C K and Montgomery F E 2004 M819 red phosphorus smoke reformulation study *Proc. 31st Int. Pyrotechnics Society Seminar (Fort Collins, CO, July 11–16)* pp 497–502

[20] Barillo D J, Cancio L C and Goodwin C W 2004 Treatment of white phosphorus and other chemical burn injuries at one burn center over a 51-year period *Burns* **30** 448–52

[21] Rivera Y B, Olin T and Bricka R M 1996 *Summary and Evaluation for White Phosphorus Remediation. A Literature Review;* accession number ADA317393; (Fort Belvoir, VA: Defense Technical Information Center (DTIC)) pp 1–72

[22] Manton G, Endsor R M and Hammond M 2014 An effective mitigation for phosphine present in ammunition container assemblies and in munitions containing red phosphorus. *Propellants Explos. Pyrotech.* **39** 299–308

[23] Collins P J D, Smit K J and Hubble B R 2004 *The Use of Red Phosphorus in Pyrotechnics—Results of an Int. Investigation;* accession number CPIAC-2004-0319AR; (Fort Belvoir, VA: Defense Technical Information Center (DTIC)) pp 1–6

[24] Conkling J A and Mocella C J 2011 *Chemistry of Pyrotechnics: Basic Principles and Theory* 2nd edn (Boca Raton, FL: CRC Press)

[25] Bale C W, Pelton A D, Thompson W T, Eriksson G, Hack K, Chartrand P, Decterov S, Jung I-H, Melançon J and Petersen S 2015 *FactSage*, version 7.0; CRCT ThermFact Inc. and GTT-Technologies (http://factsage.com)

[26] Koch E-C 2016 1916–2016: the Berger smoke mixture turns 100 *Propellants Explos. Pyrotech.* **41** 779–81

[27] McLain J H 1980 *Pyrotechnics—From the Viewpoint of Solid State Chemistry* (Philadelphia, PA: The Franklin Institute Press) pp 63–8

[28] Eaton J C, LoPinto R J and Palmer W G 1994 *Health Effects of Hexachloroethane (HC) Smoke;* accession number ADA277838; (Fort Belvoir, VA: Defense Technical Information Center (DTIC)) pp 1–60
[29] Toxicological Profile for Hexachloroethane; Agency for Toxic Substances and Disease Registry, September, 1997 (http://atsdr.cdc.gov/ToxProfiles/tp97.pdf)
[30] Douda B E and Tanner J E Jr 1977 Cinnamic Acid Containing Pyrotechnic Smoke Composition *US Patent 4,032,374*
[31] Cramer A 1997 *M56/M58 Smoke Systems—Visual and Infrared Obscuration;* accession number CBRNIAC-CB-102916; (Fort Belvoir, VA: Defense Technical Information Center (DTIC)) pp 1–4
[32] McLain W H and Evans R W 1965 *A New Smoke Screening Chemical for Use in Aerial Smoke Tanks, Report Number 6;* accession number AD0479680 (Fort Belvoir, VA: Defense Technical Information Center (DTIC)) pp 1–222
[33] Medinsky M A and Henderson R F (ed) *Toxicity of Military Smokes and Obscurants* vol 1 (Washington, DC: National Academy Press) 1997
[34] Medinsky M A and Walker B Jr (ed) *Toxicity of Military Smokes and Obscurants* vol 2 (Washington, DC: National Academy Press) 1999
[35] Shinn J H, Martins S A, Cederwall P L and Gratt L B 1985 *Smokes and Obscurants—A Health and Environmental Effects Data Base Assessment;* accession number ADA185377 (Fort Belvoir, VA: Defense Technical Information Center (DTIC)) pp 1–121
[36] Koch E-C 2005 Pyrotechnic Smoke Composition *US Patent Application US 20050252590 A1*
[37] Domanico J A 2000 TPPO—a replacement candidate for red phosphorus *Proc. 27th Int. Pyrotechnics Society Seminar (Grand Junction, CO, July 16–21)* pp 257–61
[38] Johansson S R 1986 A short pyrotechnic history of matches *Proc. 11th Int. Pyrotechnics Society Seminar (Vail, CO, July 7–11)* pp 373–93
[39] Koch E-C and Schneider J 2003 Pyrotechnic Smoke Screen Units for Producing an Aerosol Impenetrable in the Visible, Infrared and Millimetric Wave Range *US Patent 6,578,492*
[40] Koch E-C and Cudziło S 2016 Safer pyrotechnic obscurants based on phosphorus(V) nitride *Angew. Chem. Int. Ed.* **55** 15439–42
[41] Mukhanov V A, Sokolov P S, Le Godec Y and Solozhenko V L 2013 Self-propagating high-temperature synthesis of boron phosphide *J. Superhard Mater.* **35** 415–7
[42] Luan Z, Morris L A, Shaw A P, Haines C D and Poret J C 2017 High-yield Synthesis of Nanostructured Boron Phosphide by a Pyrotechnic Method *US Patent 9,790,093*
[43] Shaw A P, Brusnahan J, Poret J C, Morris L A, Haines C D and Luan Z 2018 Pyrotechnic Compositions Comprising Nanostructured Crystalline Boron Phosphide and Oxidizer *US Patent 9,938,203*
[44] Shaw A P, Brusnahan J S, Poret J C and Morris L A 2016 Thermodynamic modeling of pyrotechnic smoke compositions *ACS Sustain. Chem. Eng.* **4** 2309–15
[45] Elkins H B 1943 *HC Smoke Mixtures Containing Titanium Dioxide;* accession number CBRNIAC-CB-003466 (Fort Belvoir, VA: Defense Technical Information Center (DTIC)) pp 1–22
[46] Diviacchi G 2016 Development and testing of stannous chloride smoke *Proc. 2016 Obscurants Symp. (Charlottesville, VA, November 29–December 1)*
[47] Hanley J T, Wattle B J and Mack E J 1981 *Extinction Characteristics of Pyrotechnically-Generated Alkali-Halide Smokes*, accession number ADA112806 (Fort Belvoir, VA: Defense Technical Information Center (DTIC)) pp 1–42

[48] Krone U 1990 A non-toxic pyrotechnic screening smoke for training purposes *Proc. 15th Int. Pyrotechnics Society Seminar (Boulder, CO, July 9–13)* pp 581–86

[49] Webb R, Ramlal D R, Langenberg J P and Alblas M J 2012 *Technical Feasibility Study of an Effective Low-Toxicity Obscurant Material* Strategic Environmental Research and Development Program (SERDP) Project WP-2148 Final Report, August, 2012 (https://serdp-estcp.org/content/download/15756/180673/file/WP-2148-FR.pdf)

[50] Greenspan L 1977 Humidity fixed points of binary saturated aqueous solutions *J. Res. Natl. Bur. Stand.*, Sect. A **81A** 89–96

[51] Richardson G M and Malthus R S 1955 Salts for static control of humidity at relatively low levels *J. Appl. Chem.* **5** 557–67

[52] Blau R and Seidner N 2012 *Hexachloroethane Obscurant Replacement* Strategic Environmental Research and Development Program (SERDP) Project WP-2149 Final Report, April, 2012 (https://serdp-estcp.org/content/download/14857/171388/file/WP-2149-FR.pdf)

[53] Glück J, Klapötke T M and Shaw A P 2017 Effect of adding 5-aminotetrazole to a modified U.S. army terephthalic acid white smoke composition *Cent. Eur. J. Energ. Mater.* **14** 489–500

[54] Lombardi J, Okonya A, Stahl C, Hinote S, Canfield E, Lockwood M, Bateman J, Bonavito N, Polk A and Diviacchi G 2016 Development of novel, sublimable triazine based formulations for use in white smoke munitions *Proc. 2016 Obscurants Symp. (Charlottesville, VA, November 29–December 1)*

[55] Shidlovskii A A 1974 *Principles of Pyrotechnics* ed. and trans. Osnovy Pirotekhniki, 3rd edn, Moscow 1964; accession number ADA001859 (Fort Belvoir, VA: Defense Technical Information Center (DTIC)) pp 1–536

[56] Hollingbery L A and Hull T R 2010 The thermal decomposition of huntite and hydromagnesite—a review *Thermochim. Acta* **509** 1–11

[57] Bakirtzis D, Delichatsios M A, Liodakis S and Ahmed W 2009 Fire retardancy impact of sodium bicarbonate on ligno-cellulosic materials *Thermochim. Acta* **486** 11–9

[58] Shah P S and Balkhair T 2011 Air pollution and birth outcomes—a systematic review *Environ. Int.* **37** 498–516

[59] U.S. Environmental Protection Agency 2022 *Sulfur Dioxide (SO_2) Pollution* (https://epa.gov/so2-pollution)

[60] The National Institute for Occupational Safety and Health 2022 *NIOSH Pocket Guide to Chemical Hazards* (https://cdc.gov/niosh/npg/default.html)

[61] Hogan C M Abiotic factor *The Encyclopedia of Earth*, ed E Monosson, updated 2011 (https://editors.eol.org/eoearth/wiki/Abiotic_factor)

[62] Berresheim H, Wine P H and Davis D D 1995 Sulfur in the atmosphere *Composition, Chemistry, and Climate of the Atmosphere;* ed H B Singh (New York: Van Nostrand Reinhold)

[63] Shimizu T 1981 *Fireworks—The Art, Science, and Technique* (Austin, TX: Pyrotechnica Publications)

[64] U.S. Environmental Protection Agency 2022 *Perchlorate in Drinking Water* (https://epa.gov/dwstandardsregulations/perchlorate-drinking-water)

[65] Massachusetts Department of Environmental Protection 2022 *Drinking Water Standards and Guidelines* (https://mass.gov/guides/drinking-water-standards-and-guidelines)

[66] Chen G, Motyka M and Wejsa J 2006 Perchlorate free pyrotechnic composition and its application in M115A2 ground burst simulator and M116A1 hand grenade simulator *Proc. 33rd Int. Pyrotechnics Society Seminar (Fort Collins, CO, July 16–21)* pp 269–79

[67] Chen G 2009 Perchlorate elimination in M274 2.75″ practice rocket warhead smoke charge *Proc. 36th Int. Pyrotechnics Society Seminar (Rotterdam, NL, August 23–28)* pp 427–37

[68] Sabatini J J, Nagori A V, Chen G, Chu P, Damavarapu R and Klapötke T M 2012 High-nitrogen-based pyrotechnics—longer- and brighter-burning, perchlorate-free, red-light illuminants for military and civilian applications *Chem.—A Eur. J.* **18** 628–31

[69] Sabatini J J, Raab J M, Hann R K Jr, Damavarapu R and Klapötke T M 2012 High-nitrogen-based pyrotechnics—development of perchlorate-free green-light illuminants for military and civilian applications *Chem.—Asian J.* **7** 1657–63

[70] Sabatini J J and Moretti J D 2013 High-nitrogen-based pyrotechnics—perchlorate-free red- and green-light illuminants based on 5-aminotetrazole *Chem.—A Eur. J.* **19** 12839–45

[71] Springer M L, Rush T, Beardsley H M, Watts K and Bergmann J 2008 *Demonstration of the replacement of the dyes and sulfur in the m18 red and violet smoke grenades* Final report for ESTCP Project WP-0122; accession number ADA608283 (Fort Belvoir, VA: Defense Technical Information Center (DTIC)) pp 1–261

[72] Chen G, Showalter S, Raibeck G and Wejsa J 2006 *Environmentally Benign Battlefield Effects Black Smoke Simulator* accession number ADA481520 (Fort Belvoir, VA: Defense Technical Information Center (DTIC)) pp 1–9

[73] Raibeck G, Kislowski C and Chen G 2008 Demonstration of an environmentally benign pyrotechnic black smoke in a battlefield effects simulator *Proc. 35th Int. Pyrotechnics Society Seminar (Fort Collins, CO, July 13–18)* pp 95–102

[74] Moretti J D, Sabatini J J, Shaw A P, Chen G, Gilbert R A Jr and Oyler K D 2013 Prototype scale development of an environmentally benign yellow smoke hand-held signal formulation based on solvent yellow 33 *ACS Sustain. Chem. Eng.* **1** 673–78

[75] Moretti J D, Sabatini J J, Shaw A P and Gilbert R Jr 2014 Promising properties and system demonstration of an environmentally benign yellow smoke formulation for hand-held signals *ACS Sustain. Chem. Eng.* **2** 1325–30

[76] Stoenescu L 2014 Colored Pyrotechnic Smoke-producing Composition *US Patent Application US 20140238258 A1*

[77] Glück J, Klapötke T M, Rusan M and Shaw A P 2017 Improved efficiency by adding 5-aminotetrazole to anthraquinone-free new blue and green colored pyrotechnical smoke formulations *Propellants Explos. Pyrotech.* **42** 131–41

[78] Chakraborty S R 1991 A clean and hazard-free process for manufacture of colour smoke ammunition *Proc. 16th Int. Pyrotechnics Society Seminar (Jönköping, Sweden, June 24–28)* 665–74

[79] Chemical Hazard Classification and Labeling—Comparison of OPP requirements and the GHS (U.S. Environmental Protection Agency) (https://epa.gov/sites/production/files/2015-09/documents/ghscriteria-summary.pdf)

[80] White S A and Armstrong G C White phosphorus smoke—its irritating concentration for man and its toxicity for small animals for one-hour exposures. E.A.T.P. 190, Project A 5.2-1; 1935. Cited in reference 14

[81] Davidson K A, Hovatter P S and Sigmon C F 1987 *Water Quality Criteria for White Phosphorus;* accession number ADA186613; (Fort Belvoir, VA: Defense Technical Information Center (DTIC)) pp 1–140

[82] Eck W S and Johnson M S 2009 *Comparative Environmental Toxicology Assessment of White and Red Phosphorous Munitions, Eagle River Flats Firing Range, Ft. Richardson, Alaska;* Toxicology Report No. 87-XE-MA0B-09; U.S. Army Center for Health Promotion and Preventative Medicine (USACHPPM): Aberdeen Proving Ground, MD, 2009; pp 1–16
[83] Williams M A and Rice S W 2015 Wildlife toxicity assessment of white phosphorus *Wildlife Toxicity Assessments for Chemicals of Military Concern* ed M A Williams, G Reddy, M J Quinn Jr. and M S Johnson (Amsterdam: Elsevier)
[84] Donohue J M, Gordon L, Kirman C and Roberts W C 1992 *Zinc Chloride Health Advisory;* Interagency Agreement (IAG) 85PP5869; Office of Water, U.S. Environmental Protection Agency (Washington, DC) and the U.S. Army Medical Research and Development Command (Fort Detrick, Frederick, MD), 1992. Cited in reference 33
[85] Toxicological Profile for Zinc; Agency for Toxic Substances and Disease Registry, August, 2005 (https://atsdr.cdc.gov/toxprofiles/tp60.pdf)
[86] Muse W T Jr, Anthony J S, Bergmann J D, Burnett D C, Crouse C L, Gaviola B P and Thomson S A 1995 *Acute and Repeated Dose Inhalation Toxicity Effects of Pyrotechnically Disseminated Terephthalic Acid Smoke (XM83 Grenade);* accession number ADA300470 (Fort Belvoir, VA: Defense Technical Information Center (DTIC)) pp 1–28
[87] Thomson S A, Burnett D C, Crouse C L, Hilaski R J and Muse W T Jr 1988 *Acute Inhalation Toxicity of Pyrotechnically Disseminated Terephthalic Acid;* accession number CBRNIAC-CB-001809 (Fort Belvoir, VA: Defense Technical Information Center (DTIC)) pp 1–27
[88] *Naphthalene* Hazardous Substances Data Bank (HSDB), TOXNET, National Library of Medicine, National Institutes of Health, U.S. Department of Health and Human Services) (https://toxnet.nlm.nih.gov)
[89] Eck W S and Johnson M S 2009 *COMET System—Black and Yellow Smoke* Toxicology Report No. 87-XE-0A05–09 (Aberdeen Proving Ground, MD: U.S. Army Center for Health Promotion and Preventative Medicine (USACHPPM)) pp 1–56
[90] O'Neill A J and Crouse L C B 2014 *Acute and Four-Week Inhalation Toxicity Study in Rats Exposed to Pyrotechnically Disseminated Black Smoke (PVA);* Toxicology Study No. 87-XC-0CKC-11 (Aberdeen Proving Ground, MD: U.S. Army Public Health Command (USAPHC)) pp 1–245
[91] Medinsky M A and Walker B Jr. (ed) 1999 *Toxicity of Military Smokes and Obscurants* vol 3 (Washington, DC: National Academy Press)
[92] Adams V H and Eck W S 2011 *Environmental Health Assessment for Work Unit PYRO 07-3—Environmentally Benign Yellow Smoke Formulations in Pyrotechnics;* Toxicology Report No. 87-XE-0BMEc-10 (Aberdeen Proving Ground, MD: U.S. Army Public Health Command (Provisional)) pp 1–69
[93] Eastin W 1991 *Toxicity Studies of D&C Yellow No. 11 in F344/N Rats and B6C3F$_1$ Mice (Feed Studies)* NTP TOX 8, NIH Publication No. 91–3127; National Toxicology Program, National Institutes of Health, U.S. Department of Health and Human Services pp 1–35
[94] Medinsky M A, Cheng Y S, Kampcik S J, Henderson R F and Dutcher J S 1986 Disposition and metabolism of ^{14}C-solvent yellow and solvent green aerosols after inhalation *Fundam. Appl. Toxicol.* **7** 170–8
[95] Noster U and Hausen B M 1978 Occupational dermatitis due to a yellow quinophthalone dye (Solvent Yellow 33, C.I. 47,000) *Der Hautarzt* **29** 153–7 (article in German)

[96] Jordan W P 1981 Contact dermatitis from D&C yellow 11 dye in a toilet bar soap *J. American Academy of Dermatology* **4** 613–14
[97] Weaver J E 1983 Sensitization potentials of D&C yellow No. 10 and D&C yellow no. 11 dyes *J. American Academy of Dermatology* **9** 605–6
[98] Rapaport M J 1984 Allergy to yellow dyes *Archives of Dermatology* **120** 535–6
[99] Kita S, Kobayashi T, Kutsuna H and Kligman A M 1984 Human maximization testing of D&C yellow No. 10 and yellow No. 11 *Contact Dermatitis* **11** 210–3
[100] Björkner B and Magnusson B 1981 Patch test sensitization to D&C yellow No. 11 and simultaneous reaction to quinoline yellow *Contact Dermatitis* **7** 1–4
[101] Calnan C D 1976 Quinazoline yellow SS in cosmetics *Contact Dermatitis* **2** 160–6
[102] Calnan C D 1981 Quinazoline yellow dermatitis (D&C Yellow 11) in an eye cream *Contact Dermatitis* **7** 271
[103] Björkner B and Niklasson B 1983 Contact allergic reaction to D&C yellow No. 11 and quinoline yellow *Contact Dermatitis* **9** 263–8
[104] Monk B 1987 Allergic contact dermatitis to D&C yellow 11 in a hair cream *Contact Dermatitis* **17** 57–8
[105] Feinman S E and Doyle E A 1988 Sensitization to dyes in textiles and other consumer products. *J. Toxicol.—Cutaneous and Ocular Toxicol.* **7** 195–222
[106] *Solvent Yellow 33* Chemical Carcinogenesis Research Information System (CCRIS), TOXNET, National Library of Medicine, National Institutes of Health, U.S. Department of Health and Human Services (https://toxnet.nlm.nih.gov)
[107] Eastin W C *NTP Technical Report on the Toxicology and Carcinogenesis Studies of D&C Yellow No. 11 in F344/N Rats (Feed Studies);* NTP TR 463, NIH Publication No. 97–3379; National Toxicology Program, National Institutes of Health, U.S. Department of Health and Human Services, 1997; pp 1–190
[108] Davidson K A and Hovatter P S 1987 *Water Quality Criteria for Colored Smokes—Solvent Green 3;* accession number ADA190691 (Fort Belvoir, VA: Defense Technical Information Center (DTIC)) pp 1–98
[109] Chin A and Borer L 1983 Identification of combustion products from colored smokes containing organic dyes *Propellants Explos. Pyrotech.* **8** 112–8
[110] Henderson R F *et al* 1985 *Studies on the Inhalation Toxicity of Dyes Present In Colored Smoke Munitions. Phase II Studies—Range Finding and Toxicokinetics Studies of Inhaled Dye Aerosols accession number ADA163267* (Fort Belvoir, VA: Defense Technical Information Center (DTIC)) pp 1–84
[111] Fujii S 2003 Evaluation of hypersensitivity to anthraquinone-related substances *Toxicology* **193** 261–7
[112] *TOPKAT (TOxicity Prediction by Komputer Assisted Technology)*, BIOVIA (San Diego, CA: Dassault Systèmes)
[113] O'Neill A J and Crouse L C B 2009 *Acute Inhalation Toxicity Study (LC50) With Solvent Violet 47 in Rats; Toxicology Study No. 87-XE-0A99-08* (Aberdeen Proving Ground, MD: U.S. Army Center for Health Promotion and Preventative Medicine (USACHPPM)) pp 1–62
[114] Davidson K A, Hovatter P S and Ross R H 1988 *Water Quality Criteria for Colored Smokes—1,4-Diamino-2,3-dihydroanthraquinone; accession number ADA190690* (Fort Belvoir, VA: Defense Technical Information Center (DTIC)) pp 1–69

[115] Baggiani C, Anfossi L, Baravalle P, Giovannoli C, Giraudi G, Barolo C and Viscardi G 2009 Determination of banned sudan dyes in food samples by molecularly imprinted solid phase extraction-high performance liquid chromatography *J. Sep. Sci.* **32** 3292–300

[116] Petrakis E A, Cagliani L R, Tarantilis P A, Polissiou M G and Consonni R 2017 Sudan dyes in adulterated saffron (*Crocus sativus* L.)—identification and quantification by ^{1}H NMR *Food Chem.* **217** 418–24

[117] *Sudan II* PubChem CID 18386; PubChem, National Center for Biotechnology Information, National Library of Medicine, National Institutes of Health, U.S. Department of Health and Human Services (https://pubchem.ncbi.nlm.nih.gov)

[118] *Solvent Orange 7* Chemical Carcinogenesis Research Information System (CCRIS), TOXNET, National Library of Medicine, National Institutes of Health, U.S. Department of Health and Human Services (https://toxnet.nlm.nih.gov)

[119] Gaworski C L, Kinkead E R and Doyle R L 1979 *Acute Toxicity of a Number of Chemicals of Interest to the Air Force; accession number ADA067313* (Fort Belvoir, VA: Defense Technical Information Center (DTIC)) pp 1–32

[120] Liberman D F, Fink R C, Schaefer F L, Mulcahy R J and Stark A A 1982 Mutagenicity of anthraquinone and hydroxylated anthraquinones in the ames/*Salmonella* microsome system *Appl. Environ. Microbiol.* **43** 1354–9

[121] Kawai K, Mori H, Sugie S, Yoshimi N, Inoue T, Nakamaru T, Nozawa Y and Matsushima T 1986 Genotoxicity in the hepatocyte/DNA repair test and toxicity to liver mitochondria of 1-hydroxyanthraquinone and several dihydroxyanthraquinones *Cell Biol. Toxicol.* **2** 457–67

[122] Hao N J, Huang M-P and Lee H 1995 Structure-activity relationships of anthraquinones as inhibitors of 7-ethoxycoumarin *o*-deethylase and mutagenicity of 2-amino-3-methylimidazo [4,5-*f*]quinoline *Mutation Res.* **328** 183–91

[123] *Solvent Red 1* PubChem CID 14678; PubChem, National Center for Biotechnology Information, National Library of Medicine, National Institutes of Health, U.S. Department of Health and Human Services (https://pubchem.ncbi.nlm.nih.gov)

[124] *Sudan III* PubChem CID 62331; PubChem, National Center for Biotechnology Information, National Library of Medicine, National Institutes of Health, U.S. Department of Health and Human Services (https://pubchem.ncbi.nlm.nih.gov)

[125] *Solvent Red 23* Chemical Carcinogenesis Research Information System (CCRIS), TOXNET, National Library of Medicine, National Institutes of Health, U.S. Department of Health and Human Services (https://toxnet.nlm.nih.gov)

[126] Adams V H and Eck W S 2011 *Environmental Health Assessment for Work Unit PYRO 06–7A—Environmentally Benign Red and Violet Smoke Formulations in Pyrotechnics*, Toxicology Report No. 87-XE-0BMEe-10 (Aberdeen Proving Ground, MD: U.S. Army Public Health Command (Provisional)) pp 1–48

[127] O'Neill A J and Crouse L C B 2009 *Acute Inhalation Toxicity Study* (LC_{50}) *with Solvent Red 169 in Rats* Toxicology Study No. 87-XE-09FMa-09 (Aberdeen Proving Ground, MD: U.S. Army Center for Health Promotion and Preventative Medicine (USACHPPM)) pp 1–31

[128] *Solvent Red 169*, PubChem CID 33790, PubChem, National Center for Biotechnology Information, National Library of Medicine, National Institutes of Health, U.S. Department of Health and Human Services (https://pubchem.ncbi.nlm.nih.gov)

[129] *Disperse Red 9*, PubChem CID 6706, PubChem, National Center for Biotechnology Information, National Library of Medicine, National Institutes of Health, U.S. Department of Health and Human Services (https://pubchem.ncbi.nlm.nih.gov)

[130] Brown J P and Brown R J 1976 Mutagenesis by 9,10-anthraquinone derivatives and related compounds in *Salmonella typhimurium Mutat. Res.* **40** 203–24

[131] Doi A M, Irwin R D and Bucher J R 2005 Influence of functional group substitutions on the carcinogenicity of anthraquinone in rats and mice—analysis of long-term bioassays by the National Cancer Institute and the National Toxicology Program *J. Toxicol. Environ. Health, Part B—Crit. Rev.* **8** 109–26

[132] Lundy D and Eaton J 1994 *Occupational Health Hazards Posed by Inventory U.S. Army Smoke/Obscurant Munitions (Review Update)* accession number ADA276774 (Fort Belvoir, VA: Defense Technical Information Center (DTIC)) pp 1–84

[133] Smith S H, Doyle G L, Kreuger J C, Mellon K A and Mayhew D A 1986 *Dermal, Eye and Oral Toxicological Evaluations, Phase IV Report with Disperse Red 11, Disperse Blue 3, Solvent Red 1, and Red and Violet Mixtures; accession number ADA172758* (Fort Belvoir, VA: Defense Technical Information Center (DTIC)) pp 1–172

[134] Sailstad D M, Tepper J S, Doerfler D L, Qasim M and Selgrade M J K 1994 Evaluation of an Azo and two Anthraquinone dyes for allergic potential *Fundam. Appl. Toxicol.* **23** 569–77

[135] Brooks A L, Seiler F A, Hanson R L and Henderson R F 1989 In vitro Genotoxicity of dyes present in colored smoke munitions *Environ. Mol. Mutagen.* **13** 304–13

[136] Harrington-Brock K, Parker L, Doerr C, Cimino M C and Moore M M 1991 Analysis of the genotoxicity of anthraquinone dyes in the mouse lymphoma assay *Mutagenesis* **6** 35–46

[137] Freeman H S and Mock G N 2007 Dye application, manufacture of dye intermediates and dyes *Kent and Riegel's Handbook of Industrial Chemistry and Biotechnology* ed J A Kent 11th edn vol 1 (New York: Springer) ch 13

[138] Burnett C, Goldenthal E I, Harris S B, Wazeter F X, Strausburg J, Kapp R and Voelker R 1976 Teratology and percutaneous toxicity studies on hair dyes *J. Toxicol. Environ.* **1** 1027–40

[139] *Disperse Blue 3* Chemical Carcinogenesis Research Information System (CCRIS), TOXNET, National Library of Medicine, National Institutes of Health, U.S. Department of Health and Human Services (https://toxnet.nlm.nih.gov)

Postscript

I hope that this book has proved thought provoking and will stimulate future studies. I hope too that it has encouraged publication of research results so that the body of work can be increased for the benefit of all. Some of this may not be possible for various reasons but publication of basic research results plays an important part in learning and questioning.

No one group has a monopoly of the truth and so it is necessary to consider that you could be wrong and then check your facts. Sharing in the literature helps this.

Until recently energetic materials were considered a mature technology and one with few challenges or options for research or development. I hope that this book has shown that this is far from the truth and that there are exciting options for research. There are lessons to be learned from other disciplines but equally, as the area develops, there are lessons for others to learn. After all, working with such hazardous material systems does or should lead to greater care and self-discipline where options are carefully investigated and analysed. The results of these analyses will often have wider relevance.

After 40+ years working in the area I have seen many dead-ends and many unexpected results. More surprises are waiting and with newer tools they can be firstly found and secondly applied.

Adam Cumming
March 2022

doi:10.1088/978-0-7503-3943-8ch18

www.ingramcontent.com/pod-product-compliance
Ingram Content Group UK Ltd.
Pitfield, Milton Keynes, MK11 3LW, UK
UKHW051910100926
8556IPUK00005B/833

9 780750 33941